The Complete Resource Handbook

Increasing Renewable Energy Use in the United States

"How can the United States best reduce its fossil fuel consumption?"

Lynn Goodnight and Scott Deatherage

National Textbook Company
a division of NTC/Contemporary Publishing Company
Lincolnwood, Illinois USA

Published by National Textbook Company, a division of NTC/Contemporary Publishing Company.
©1997 by NTC/Contemporary Publishing Company, 4255 West Touhy Avenue,
Lincolnwood (Chicago), Illinois 60646-1975 U.S.A.
All rights reserved. No part of this book may be reproduced, stored
in a retrieval system, or transmitted in any form or by any means,
electronic, mechanical, photocopying, recording or otherwise, without
the prior permission of NTC/Contemporary Publishing Company.
Manufactured in the United States of America.

7890 CU 0987654321

CONTENTS

I. Determining an Energy Policy for the United States	1
II. Oil	
Outline	7
Evidence	15
III. Nuclear Energy	
Outline	125
Evidence	131
IV. Climate	
Outline	233
Evidence	239
V. Negative Approaches	
Outline	347
Evidence	357
VI. Alternative Energy Sources	
Outline	507
Evidence	511
VII. Who's Who	571
VIII. Bibliography	581

THE AUTHORS

SCOTT DEATHERAGE is Director of Debate at Northwestern University. Dr. Deatherage is also the Director of the Debate Divisions of the National High School Institute at Northwestern University. In his ten-year coaching career, he has coached four national championship teams, including back-to-back National Debate Tournament victories in 1994 and 1995, as well as four NDT Top Speakers, most recently in 1996. Dr. Deatherage has taught more than 25 groups at debate workshops around the country, including those at American University, Baylor University, Georgetown University, the Michigan National Debate Institute, and the Michigan Classic. He has been a member of the NTC writing staff for five years.

LYNN GOODNIGHT is former Administrative Director of Northwestern University's National High School Institute and a freelance communications consultant. Ms. Goodnight is author of *Getting Started in Debate* and co-author of *Strategic Debate*. While working on her master's, she was an Assistant Debate Coach at the University of Kansas. Ms. Goodnight was Director of Debate at Niles North High School from 1977-1980. She has lectured at high school debate workshops, including the University of Kansas, the University of North Carolina, the University of Houston, Baylor University, Northwestern University, and the National Federation of State High School Associations Summer Workshops. She has been a member of the NTC writing staff for twenty-one years.

Integrally involved in the preparation of this volume were Northwestern students and debaters. Matt Anderson, John Busby, Brandon Fletcher, Michael Gottlieb, Terry Johnson, Leslie Mueller, David Nemecek, Jennifer Northan, Todd Plutsky, Carrie Peril, Shorge Sato, Shuman Sohm, and Ryan Sparacino have all made substantial contributions to the completion of this handbook. Additional proofing was provided by Lisa and Pam Horstmann.

PREFACE

Once again we are about to embark on another debate season and another topic. For many it will be a new experience. For others it will be an opportunity to synthesize information gathered over the last few years. For some coaches this will be a topic which has been visited twice before—first in the mid-70s and again in the 1980s. This handbook has been designed to help both the beginning and experienced debater as he or she begins to explore the ins and outs of the United States' energy policy as well as the issues which have an impact on that policy. **Increasing Renewable Energy Use in the United States** offers analysis and evidence on such areas as climate, global warming, global cooling, oil production and oil imports, nuclear power, energy alternatives, energy use, conservation, Middle East oil, world economic issues, emission controls, recycling, energy shortages, **etc**.

This handbook has been written to serve as a resource tool. The debater should remember that no handbook can serve as a substitute for the continuous, thorough research necessary for successful debate. We have discussed issues and presented evidence on their own merit. We have presented a sampling of the ideas and evidence which might be explored on the energy topic. We have outlined issues and questions that we feel will play a role in shaping this topic. It is our hope that the issues and questions we have raised will inspire debaters in their development of affirmative cases and negative approaches.

In addition to this handbook, **Increasing Renewable Energy Use in the United States**, debaters will want to consult **United States Policy on Renewable Energy Use**, a basic overview of the topic, and **An Energy Program for the United States**, a collection of critical essays, both published by National Textbook Company for the current debate year.

It has been a pleasure working with this staff of able authors and research assistants. The coordination and editorial assistance which I have provided will hopefully enhance the structural and stylistic unity of the individually prepared materials in this book.

L.G.

CHAPTER I

DETERMINING AN ENERGY POLICY FOR THE UNITED STATES

The topic for the 1997–1998 debate season pertains to the key issues of our age, energy and the environment. There are no more important, complex, and challenging areas of argument of national and international consequence.

> **RESOLVED: THAT THE FEDERAL GOVERNMENT SHOULD ESTABLISH A POLICY TO SUBSTANTIALLY INCREASE RENEWABLE ENERGY USE IN THE UNITED STATES.**

Areas of Analysis as Defined by the Topic Committee Report

The topic committee predicated the affirmative area for debate on U.S. Department of Energy projections that "the developing world's demand for electricity—industrializations's [sic] life blood—will grow by 50 percent by 2010." The DOE's conclusion:

> Unless a massive shift toward renewable generation technologies occurs, capacity expansion will raise the toll that fossil fuel-fired plants are taking on human health, natural ecosystems, and global climate. Urban air quality, for instance,will plummet: While nitrogen oxide and sulfur dioxide emissions are expected to decline in North America and Europe over the next 25 years, they will more than double in the rest of the world, rising as much by eight fold by 2030, even if energy is used much more efficiently. (Peter Amus, "Saving Energy Becomes Company Policy," *The Amicus Journal*. Winter 1993, pp. 38–42 as cited in "Alternative Energy Sources.")

The topic committee goes on to note that the alternatives to fossil fuels—presumably oil, coal and natural gas—are renewable energy resources such as biomass, wind, solar, geothermal, and hydroelectric energy technologies that supply presently only 9 percent of US energy and by 2010 are likely only to move up to 11 percent of the market. (*Christian Science Monitor*, 31 January 1995, p. 9). It is the belief of the topic committee, and we concur, that the key issue of a post-Cold War environment is likely to be energy: its effects on new areas of industrialization, environmental impacts both global and local, consequences for economic prosperity, and links to regional and global military security issues.

In determining the intent of the framers of the topic, it is important to examine how the term "renewable energy use" is operationalized. This is the action focus of the topic from which context can be determined. On page six of the committee's document, "Energy Policy for the United States," the following is quoted from an article in *The Economist* of October 7, 1995, p. 23 ("The Future of Energy"):

> The next important point to be noted is that energy use is dominated by fossil fuel resources. "In the energy needed to move transport, oil is still the king, supplying 97% of the fuel used. Only in the production of electricity have the alternatives to fossil fuels yet made any sort of impact. Nuclear power provides 17% of the world's electricity, and hydroelectric 18%. But both are relatively old, and both are controversial. Truly modern renewables, such as solar and wind power, provide less than 1% of world electricity.

From this quotation it is clear that the world may be divided up into two types of resources: non-renewable fossil fuel energies and renewable non-fossil fuel energies. Further, the renewables fall into two categories: (1) traditional categories of nuclear, hydroelectric and (one assumes) thermal power, and (2) nontraditional categories of renewable energies developed through technological contrivance such as solar power, wind power, and other alternatives provided by science. the topic report lists these alternatives in citing a quote from Rodman Griffin ("Alternative Energy") writing in the *CQ Researcher*, 10 July 1992: "The list of potential sources of alternative energy are virtually limitless," the topic paper reads, "but the most commonly discussed are, 'those that receive federal funding from the Department of Energy.' These alternatives include solar buildings, ocean energy, wind, energy storage, solar thermal, geothermal, biofuels, photovoltaics, building conservation, shale, clean coal, magnetic fusion, and nuclear fission [p. 578]."

The topic clearly requires a shift in the United States from fossil fuel related energies to non-fossil fuel related energies. What is left highly ambiguous is (1) how much of a shift is required to meet the burden of the topic, (2) what areas are open or available to increased reliance, (3) whether the shift to increased use necessarily requires reduced use of usable energy or merely preparation for offsetting increase in energy demands with the availability of alternatives, and (4) whether the policy must result in increased use of renewable in order to be topical or whether it might simply create a technological, economic, or scientific set of supports for such policies based upon contingent triggers such as an oil embargo, trade war, or dramatic price increase.

The topic does point to two fairly well-defined areas that reflect risks incurred by fossil fuel dependency: power plant generation and transportation. Over 80% of power plant generation is fossil fueled, while most of the rest is nuclear power. Since American power plants are aging edifices, many are due to be replaced soon and the desirability of setting a precedent for renewables might encourage the alternative of more nuclear power generation. Whether this is a good idea or not is as hotly contested as the spent fuel languishing at present reactor sites. The second area of focus is transportation. While the National Energy Policy Act of 1992 does provide incentives to reduce the nation's dependence on imported oil by encouraging domestically produced fuels, an affirmative might go further and advocate for alternative sources of transportation including "electric powered cars, natural gas, propane, ethanol, and methanol" which would increase alternative fuel consumption ("Energy Policy," p. 9).

Discussion of the Burdens of the Topic and the Division of Ground

The topic is divided into several key phrases. In this section, we are concerned with discussing the overall implications of key terms for establishing the threshold burdens for an affirmative to meet its claim to support or enact a topic, and the division of ground that divides affirmative from negative positions.

FEDERAL GOVERNMENT: "The federal government" usually refers to the government in Washington, D.C. This location of the agent of change is reinforced by the scene of change which is "in the United States." Counterplans might well choose alternative agents of change which might include the private sector, state governments, or an international agency. If the international agency chosen for the counterplan requires United States participation, then one should presuppose that at least part of the counterplan is topical since it affirms a federal action—if the counterplan aims at increasing alternative energy use.

ESTABLISH A POLICY: A difficult phrase in the topic is the requirement to "establish a policy". The ambiguous term is "establish," of course. *Webster's* has two different sorts of definitions for this key term. The first includes these sorts of injunctions: "to make firm or stable," "to set on a firm basis." The second indicates "establish" means "to bring into existence," or "to institute (as a law) permanently by enactment or agreement." (*Webster's New Collegiate Dictionary*, 1973, p. 391). The difference between meanings is quite significant. In the first instance, the affirmative might argue that its ground includes the "establishment" of existing programs that are subject to budget cuts, experimental development, or the uncertainties of the market. Let us call this bolstering the status quo. In the second instance, the affirmative might argue that "establish" means the equivalent of inaugurate a new program, one that is untried. Let us call this altering the status quo. Now, consider the difference that a decision on this key term makes for the division of ground.

If "establish" is interpreted as "bolstering" the status quo, then it is the affirmative that gets to argue expansion of existing programs now funded by the DOE as a case area. Perhaps the negative then has the ground for not-establishing, that is for withholding expansion until implementation of technologies is proven, cheap or efficient. An affirmative that hedged its bets and only conducted R&D on alternatives would not thereby be meeting its burden—unless, of course, R&D was a way of securing increased USE of alternatives. This approach limits negative inherency arguments quite severely. Any expansion of existing programs that would result in the "increase of renewable energy use," including taxing fossil fuels would be essentially affirming the topic.

If establish is interpreted as institute through legislation, then the negative has broader ground. It may defend the status quo by arguing that one need not bring anything new into existence because harms and risks are overstated and because there will be a "natural" market transition to alternatives when the timing is optimal. The negative ground is linked to the float among alternatives with market forces providing price incentives for switching to alternatives as needed. Obviously, under such an interpretation the key arguments of many rounds would focus on whether or not existing energy industries are self-interested in restricting supply and driving out newer, cheaper, cleaner alternatives. Nevertheless, the negative is left with more defensible ground in the latter interpretation.

The object of establishing a policy is "to increase renewal energy use" in the United States. Notice that this is an effects wording for the topic. By this we mean to point out that the topic does not require affirmatives to directly mandate increases in energy use of a certain kind, merely to adopt a policy that results in increased renewal energy use. The difference is significant in terms of the possible expansion of affirmative ground.

There are many different ways one can substantially increase renewable energy use. We will list several here, but the topic is limited only by the imaginings of direct and indirect policies that could be used singly or in combination to increase. Here are three of the major categories:

(1) Domestic Market Incentives: One can increase alternatives by making non-renewables more expensive through taxation or price regulation. One could also subsidize uses of alternatives through bond guarantees, tax subsidy or land-use variances. Here the chief advantage of a case would hinge upon a combination of two arguments. To the extent an alternative energy source was not specified by the affirmative, taxation of nonrenewable sources would spur development into promising alternatives—wherever the market would go. The advantage of this sort of case would come in the citation of a variety of exciting new possibilities rather than a commitment to a single one. Alternatively, an affirmative might be ready to commit to a single source (like a hydrogen economy) and argue that transition barriers exist. This strategy permits development in depth.

(2) Foreign Market Restrictions: One can increase alternatives by making foreign resources more expensive through taxation. It is undoubtedly the case that if foreign oil imports were decreased, renewables would become more attractive because the price of oil would go up from its artificially low levels. This sort of case would not necessarily focus on the benefits or alternative fuels as much as it might discuss areas of world and/or United States' vulnerability due to international instability. Of course, the question of trade balance and oil interdependence has ample negative ground. Having the United States as an indebted partner may provide incentives to keep its economy strong and offset interests in oil embargoes by foreign powers.

(3) Direct Mandates for Sector Use: We have previously mentioned fossil fuel reactors and the transportation sectors as areas where one might shift to alternative fuel uses. There are several tests of whether or not such shifts are wise. These provide several major topics for consideration. We recommend that when considering renewables or fossil fuels that the advocate become familiar with these issues as related to each alternative. The key issues include: (a) is the energy resource technologically proven? (b) is the energy resource economically feasible? (c) is the energy resource expandable? (d) is the energy resource non-polluting? (e) is the energy resource politically acceptable?

The Issue of Technology: Hydrogen reactors would be a cheap, pollution free, plentiful, wonderful resource—should they be technologically feasible. They are not yet, and scientific predictions do not have a strong record in predicting availability.

The Issue of Pollution: The problem with fossil fuels is that they put carcinogens in the air through burning in smoke stacks. The pollution degrades the environment, as mid-western utility stacks belched soot high into the trade winds for decades, resulting in deforestation in Maine and Canada. Although pollution can be cleaned up, the smaller the particulates apparently the more deadly. A non-polluting source of energy is potentially wonderful—but there are trade-offs including...

The Issue of Cost: Alternatives are not cheap. Sometimes the problem is initial start up costs, as in solar power. Once the equipment is put in, there is relatively little maintenance. But inexpensive technologies tend not to produce the bulk power needed to keep industrial power plants going.

The Issue of Expendability: There are some cheap, non-polluting, proved alternatives that are great sources of energy. The problem is that these are located in a particular area or region of the country. Hydro-electric power may not do a great deal of damage (if there are not endangered species about), but the damable rivers are about used up. Alternatively, geo-thermal power is a good source of energy but there are not many geysers about.

The Political Issue: There are some resources that cannot be developed because of opposition by groups whose cultural values or community risks create opposition. One of the hot areas of this year's topics include the removal and storage of nuclear waste—an issue that is clearly going to have to be confronted if people decide that nuclear power is a benefit. On the other hand, the use of water for hydro-electric power out West has met with resistance from environmental groups.

To Substantially Increase Renewable Energy Use: The words "substantially increase" constitute guidelines for the thresholds for the affirmative. Unfortunately, this term is so vague that it is nearly completely vacuous, that is nearly empty of meaning. Is a "substantial increase" 10%, 25%, 50% or 150%? Obviously, the affirmative will say that its views are reasonable on this area and since the topic does not specify a more restrictive guideline it should be able to pump up renewable use at some level which has some sort of salutary effect on the situation.

Note that this offers a critical option for the affirmative. On the one hand, the affirmative might argue that United States use of fossil fuels ought to be reversed; that is, on the whole, at some point in time, through a combination perhaps of conservation and alternative technologies this country should switch to a predominantly renewable set of resources. This would invite a full-fledged debate on pollution and foreign dependence. On the other hand, an affirmative team might narrow its ground and argue only that alternatives should be increased as a supplement in anticipated growth. For example, an affirmative might argue that switching to a particular kind of reactor or resolving certain storage problems would remove barriers to nuclear power development—but that the effects of this would be neither to increase nor to decrease the use of nonrenewables over the long run. The importance of this strategy is that there are certain generic risks to decreasing foreign oil dependence or pollutants in the air (related to the climate debate). By not taking a position on the overall use of fossil fuels, that is, by saying that such use will continue if the plan is adopted or not, then the scope of the debate is narrowed more to the supplemental advantages of developing a particular kind of alternative fuel.

In the United States is clearly the scene of policy development. Yet it is likely that most debates will be fought out on a international and even global scale. There are several reasons for this. Most obviously, what the United States does, it does not do alone. Rather, nation-states nowadays are seen to exist in a web of interdependent relationships. These relationships furnish the basis for many topical issues and cases. Key areas of analysis are mentioned below.

1. Issues of Economic Interdependence: Non-renewable resources are chief items of world trade. Oil exports are chiefly located in the Middle East, of course. If the United States pursues a policy to reduce oil imports, then affected countries might not buy its exports. This could set off a round of trade barriers and result in economic downturns, which in turn would affect the United States. Thus a case that argues for energy independence in the name of economic security might end up hurting that security by depressing the economy. On the other hand, economic interdependence might be enhanced

if by developing alternative energy exports technology the United States would be able to strengthen its position to retire some of its international debt due to balance of trade deficits.

2. Issues of Military Interdependence: As trade goes, so does military policy. One of the chief reasons for renewable resources is that they would lessen dependence on politically volatile regions of the world. Investors like predictability, and many remember the "oil shock" economies of the early 1970s from the oil embargo and the brief price doubling that accompanied the invasion of Kuwait in the early 1990s. It might be argued that American interests are held hostage by Middle East politics, and the last time the United States was just lucky in avoiding major damage from unleashed weapons of mass destruction. On the other hand, oil diplomacy does give us leverage in an area where there are solid political commitments to Israel and other southern NATO nations. Moreover, a move toward energy alternatives might spark a renewed isolationism that removes the United States as a balancer of power in the Middle East.

3. Issues of Environmental Interdependence: Any environmental issue is perforce a global issue. Water, air, and particulates know no national boundaries, and once entered into the environment become part of a circulation system. This fact cuts both ways in terms of case development. On the one hand, since the United States is a big consumer of fossil fuels, its own use reduction should have environmental consequences that are significant. On the other, it may be the case that if the United States doesn't use fossil fuel, some other country will. There are huge oil and coal deposits in the Soviet Union, for example, and a cash-strapped post-Communist regime is likely to do as much, if not more polluting than the U.S. Sooner or later, the same amount of toxins will be afloat and large numbers will die. There is a more optimistic side to this issue, of course, as found in the next argument.

4. Issues of Political Interdependence: What the United States does (or doesn't do) provides a model of sorts for world leadership. It might be argued that it is particularly important for the U.S. to set standards for timely development of fossil fuel alternatives given the rapid industrialization of many countries no longer held in check either by the ineptitude of Communist regimes or the selfishness of their colonial masters. The so-called developing world, then, will enter advanced industrialization either with available alternative energy sources or as competitors for a limited supply. In the 1980s, Japan began to expand its international influence by selling and promoting green environmental protection equipment. Similarly, the United States might advance environmental protection and develop new exports with development of alternative energy technologies for the 1990s. Certainly, the end of the Cold War has freed up a variety of former defense researchers for development.

It is likely this year that every case will have some international ramifications. Whether debates focus on the economic, military, environmental, or political continuum will be a matter of emphasis in individual rounds. However, it should be noted that the four arenas themselves are inter-related among one another. A burgeoning world economic situation may be quite bad or quite good for the environment, depending upon how one develops the climate argument. Alternatively, a lack of economic growth may spur economic conflict which is usually thought to have great potential risk. Political signals of threat or cooperation may drive former enemies or allies further apart or closer together. This is why it is likely that experience with debating this topic will be both interesting and challenging and likely to influence how you think of world conditions for some time to come.

Scenarios of Argument Development

Energy policy can be developed by either increasing the supply of a particular energy resource or reducing the demand for a particular kind of energy resource. Why would one wish to do either?

One of the large items of discussion is whether non-renewable resources are limited in the foreseeable future. The answer to this question is not an easy one to answer. It is difficult to believe that supplies are limited when the price of oil remains at a historic low, when new reserves have been discovered in many different areas. The *Houston Chronicle* of June 23, 1996, reports that "Proven oil reserves, which total an estimated 1.017 trillion barrels, remain at or near a 43 year supply at current levels of consumption, which reached 67.93 million barrels a day last year." Even as demand has risen, over the last two decades reserves have increased by 50 percent and more are likely to be found. Optimists point to "newly opened areas in the former Soviet Union and Eastern Europe, Latin America, China, and elsewhere" as offering "a promising future" (J. Cale, *The New Global Oil Market*, p. 48.) Cale believes that "Russia has the potential to rival Middle Eastern production, although currently its reserves are more on par with the United States." Further reasons for optimism in energy supply include underestimation of Middle East reserves, advanced drilling and old well recovery techniques, the potential to exploit shale oil. And if none of these work, coal reserves are much, much larger than oil reserves. While many policy analysts harken back to times of the OPEC oil embargo of the 1970s, others claim that given the relative level of oil surpluses, combined with the Gulf War, that the threat of embargo or the shock of increased oil prices is not great.

Not everyone is salutary about the immediate prospects of energy, particularly when it comes to the short term vulnerability to price increases with the economic dislocation and threat of a deep recession that would bring. *The Atlantic Monthly* of April 1996 (p. 59) reports:

> Donald Hodel, who was a Secretary of Energy under Ronald Reagan, has said that we are 'sleepwalking into a disaster,' and predicts a major oil crisis within a few years. Irwin Stelzer, of the American Enterprise Institute, says that

the next oil shock 'will make those of the 1970s seem trivial by comparison.' Daniel Yergin says, 'People seem to have forgotten that oil prices, like those of all commodities, are cyclical and will go up again.' James Schlesinger, who was the Secretary of Energy under Jimmy Carter, has said, "By the end of this decade we are likely to see substantial price increases." In March of last year Robert Dole, the Senate majority leader, said in a speech at the Nixon Center for Peace and Freedom, 'The second inescapable reality of the post-twentieth-century world is that the security of the world's oil and gas supplies will remain a vital national interest of the United States and of the other industrial powers.

Some see that domestic oil production is going to decline in the future, that our proven reserves are not advancing or at least not advancing in such a way as to keep pace with economic growth. While it is true that relatively modest rates of growth have prevented skyrocketing oil demands, it is also the case that the United States—and its allies—are growing for the most part, more, not less, dependent upon Middle East oil. As *Industry Week* (May 6, 1996, p. 1) states: "U.S. oil demand is rising and supply is falling—a combination that inevitably has lead the nation to a greater dependence on imports."

If oil supplies are limited, then the nation might turn to coal as a substitute. It is not clear whether increased production of coal might offset either shocks due to sudden price increases or shortages due to supply restrictions. Just as it would take time to drill new oil wells and bring natural gas into production, so too it would take time to increase coal mining or convert from one fuel to the other. Coal has drawbacks, too. Coal mining has environmentally deleterious results whether one is strip mining from the surface and marring the land or deep mining with the potential of acid drainage from minerals uncovered. Additionally, coal burning is highly polluting and there are safety problems in mining that make the activity inherently undesirable.

One solution to make the most of these nonrenewable resources is alternative extraction and conversion processes. Coal gasification and liquefaction technologies exist but it is not clear whether these technologies are feasible or environmentally sound. So too, there are techniques to release oil that is locked in shale found in the American West. This is a costly process, like extraction of fuel from tar sands, and like strip mining has environmental impacts.

If development of fossil fuel reserves is limited by price, environment, and capacity issues, then an alternative that might reduce future vulnerability is efficiency measures. Some efficiency measures have already been tried out, and while many have been successful, the key issue is whether or not they can be expanded economically and reliably to larger realms of policy. We could change what we burn for fuel. In power plants, co-generation can be developed from burning garbage, using the waste heat to get steam generators going. Whether there is enough waste heat to make a difference is debatable. Others argue that one could recycle manure or grow other farm products, such as crops converted to ethanol, to produce alternative energy sources. These sources have collection problems and ethanol, already in widespread use, may catch on as the market requires. Still others argue for high tech alternatives such as creating more efficient transmission lines through superconductivity, cooling the lines to extremely low temperatures. While advocates of this process suggest that there are great energy savings to be had—as do advocates of hydrogen, solar and wind power—technological feasibility and profitability of these breakthroughs have yet to be established.

One definite area of consideration in this topic is solar power. Solar power is thought to be a potential source of unlimited power. The conversion of sunlight to electricity has been a dream of new technologies for quite a time. The dream includes everything from relatively simple passive solar collectors, to insulation, to biomass schemes, to Ocean Thermal Conversion—a technique that converts differences in water temperature into power. While some experiments have been successful, like geothermal power, solar is a resource limited by area of the country and cost. Like solar, wind power, too, has great promise as a non-polluting source but it's uneconomical in the United States.

To what extent do alternatives depend upon the oil vulnerability and shortage argument? One might argue that the two are inextricably linked from a needs perspective. Independence from fossil fuels bespeaks freedom from pollution and foreign powers. On the other hand, one could conceivably bypass the shortage issues and argue the intrinsic desirability of promoting a new source so that it becomes cheaper, better tested, and more widely accepted. There would appear little downside to this—save for the issue of climate which will be discussed below.

A key issue impacting on most every debate is nuclear power. Some believe that the U.S. nuclear power industry has collapsed, that utilities are abandoning old nuclear power plants and that new ones are not being developed out of fear of safety risks, frustration with costly legal battles, and public opposition to the nuclear power generally. Three Mile Island, the accident in Pennsylvania, damaged the credibility of the nuclear industry and it has never fully recovered the public trust. This, combined with a high risk of an accident and the inability of technology to prevent human error, natural disaster or terrorist activities, make nuclear power an enterprise that is unlikely to revive.

Conversely, some represent the nuclear industry as essential to preventing a domestic energy crisis and, with new technological breakthroughs, it may well become an answer to dwindling reserves and foreign vulnerability. The risk of significant reactor accidents, it is believed, is low and decreasing. While some indict the Nuclear Regulatory Commission as creating an unreliable system of inspection, nuclear advocates point proudly to recent reforms and claim that no American Chernobyl is possible here.

The advocate would be well-versed on the nuclear issue—on both the affirmative and the negative. It may be the case that the development of any renewable resource would discourage more risky or costly development of nuclear power which, while having undesirable features, is a best alternative—all things considered. It may be the case that despite the shortcomings of nonrenewable fuels, that these are better than the risks of nuclear power. Finally, not all nuclear technologies are the same and particular novel developments among breeders, cooling systems, regulations, and scientific research may create a mix superior to the status quo.

Climate is a very important and complex area of argument that is a potential point of discussion in every round. At the simplest level, there has been an ongoing dispute as to whether weather predictions suggest that we are on the brink of a new ice age or whether warming is on the horizon. But the consensus of experts has recently argued that we are headed toward global warming due to the increase in by-product gases from fossil fuel power plants and auto emission. Warming melts ice caps, puts cities under water, increases disease outbreaks, heat deaths, species collapse, and crop wilting. The scenarios that many authorities weave from global warming are truly apocalyptic, and theorists argue that one must act now because lead times are long and changes of reversal increasingly slim.

Not all global warming models are accepted, for the climate is a very complicated phenomenon to predict. Various sorts of data seem to contravene the theory: the temperature does not appear to be rising and the oceans are cooling. Certainly, there is no consensus yet, and many believe that it is either impossible to prevent warming due to naturalistic activities or that there are some good outcomes. There are many outcomes and nuances to this argument which we examine through evidence in our analysis of the debate.

Whether the world will end in nuclear fire or polar ice, another threat looms equally large: namely, the effect of energy policies on world trade. This is a fairly complicated issue because there are a number of prime and second order effects that could emanate from a plan, depending upon its outcomes.

First, imagine that you are arguing you will enhance America's trade position by exporting clean energy technologies. Moreover, your advantage is that you reduce the threat of warming by reducing CO_2. What impact does this have on the economies of other nations that are trying to industrialize by exporting their own energy resources, or, alternatively, are willing to buy the new energy sources and to go into debt for it. Either situation puts the trading partner at a disadvantage, for which there may be retaliation. So, too, industries within the United States that are required to use environmentally safe and renewable energy sources will either have to absorb the costs of the new source or the government will have to subsidize the costs, the result being on the first case to give foreign competition an unfair trading advantage or, in the second case, to create a situation where the foreign partner views the United States as creating an unfair trading situation. If we are talking about automobiles, for example, we are talking significant export/import markets and a touchy history of international relations.

Trade arguments generally are predicated on the portrayal of world powers as on the brink of a trade war. Suspicions are everywhere, trust nowhere. Perceptions of unfairness create internal domestic pressures to raise trade barriers and once the cycle gets going, it is self-building. The evidence is fairly strong that free trade is absolutely essential for economic prosperity and that world prosperity is a condition of avoiding more egregious human depredations in warfare. An affirmative would be well advised to take into consideration the trading advantages and impacts of new energy developments.

CONCLUSION

The three dominant scenarios that loom large over the energy debate are capacity, climate, and trade. Whereas there are a plethora of new energy alternatives that one will find represented in technological schemes and entrepreneurial dreams, all arguments on an energy topic eventually trickle back to one of these three basic positions. In development of our handbook for 1997 nd 1998, we have chosen to present the latest evidence and an in-depth analysis of these areas. In a very real way, these scenarios define the parameters of public argument for the twenty-first century.

CHAPTER II
OIL
OUTLINE

I. HIGH AMERICAN OIL DEPENDENCE CREATES FUTURE OIL SHOCKS

   A. AMERICAN DEPENDENCE HIGH AND GROWING
      1. U.S. dependence on the Persian Gulf is skyrocketing. (1)
      2. U.S. oil imports exploding. (11)
      3. IEA study proves U.S. oil demand increasing. (12)
      4. Oil dependence will increase even if U.S. production climbed. (13)
      5. Demand will inevitably increase. (14)
      6. DOE study indicates oil imports will rise. (15)
      7. Demand climbing fast. Decreasing supply risks oil shock. (16)
      8. U.S. oil imports rising. (17)
      9. Even with energy efficiency improvements, oil dependence is increasing. (18)

   B. DOMESTIC AMERICAN RESERVES LOW
      1. Despite giant expenditures and exploratory searches, domestic oil reserves have not expanded. (6)
      2. U.S. is running out of oil. No new reserves in the future. (7)
      3. Nothing can prevent future decline in domestic production. (8)
      4. A self-reliant U.S. could survive for only six years under the most optimistic scenario. (9)
      5. Little recoverable oil in the United States. (10)

   C. DEPENDENCY INDUCED OIL PRICE AND SUPPLY SHOCKS INEVITABLE
      1. Political instability in the Middle East makes oil shocks an ever present reality. (4)
      2. Oil dependency makes the U.S. vulnerable to oil shocks. (5)
      3. Foreign shut-offs guarantee oil shocks. (19)
      4. Dependence makes shocks a reality. (20)
      5. Middle East events cause oil shocks. (21)
      6. Increasing demand from developed countries will cause an oil shock. (22)
      7. Increasing Asian demand risks a shock. (23)
      8. Consensus: World is headed for another oil shock. (3)
      9. Persian Gulf control risks an oil shock in ten years. (30)
      10. Saudi stability in doubt. U.S. poised for a shock. (31)
      11. Next decade will bring an oil shock. (32)
      12. Crisis is upon us. Radical shock on the horizon. (33)
      13. Diminshed spare production will cause a shock by the turn of the century. (34)
      14. Persian Gulf political instability risks oil shocks. (43)
      15. Long-term dependency on the Gulf is unsustainable. (44)
      16. Persian Gulf is vulnerable to oil disruptions. (45)
      17. Gulf war proves danger to oil dependence. (46)
      18. Growing dependence on oil translates into price shocks and economic catastrophe. (54)
      19. Increased dependence threatens disastrous energy shocks. (55)
      20. Dependency risks oil shock as prices tighten. (56)
      21. Dependency makes U.S. vulnerable to a supply interruption. (57)

   D. OPEC POWERFUL
      1. OPEC will head a new unstable international oil regime. (2)
      2. OPEC profits increasing. (24)
      3. Decreasing oil prices pressure OPEC's reserves for future. (25)
      4. Last resort status preserves OPEC's future strength. (26)
      5. OPEC share of reserves increasing. (27)
      6. OPEC's dominance of oil trade yet to come. Current glut meaningless. (28)
      7. Increasing consumption will expand OPEC's market share. (29)

   E. BRINKS
      1. Oil key to international economy. Traders can change prices at a moment's notice. (35)
      2. Oil is traded at a high volume and rapid pace around the world. (36)
      3. All sources of energy tied to oil. Small changes in market can dramatically change prices. (37)
      4. Petroleum is the vital international commodity. (38)

F. IMPACTS
1. Indirect costs associated with shocks magnify their impact. (39)
2. Perception of crisis worsens economic effects of shock. (40)
3. Oil shocks alter production levels. (41)
4. Price instabilities cause a deflationary cycle like the 1930s. (42)
5. High oil prices transfer wealth to OPEC nations. Hurts economy. (47)
6. Oil contributed heavily to U.S. trade deficit. (48)
7. On-balance imports decrease economic security. (49)
8. Defense costs to protect oil supply hurt the economy. (50)
9. Oil dependency costs $200 billion per year. (51)
10. Oil dependency explodes the trade deficit. (52–53)
11. Vulnerability to supply disruptions decreases foreign policy flexibility. (58)
12. Protecting oil supplies uses a large part of the military. (59)
13. Oil dependency destroys foreign policy flexibility. (60)
14. Oil dependence weakens strategic position. (61)
15. Oil dependence undercuts strategic positioning options. (62)
16. Libya proves dependency destroys policy leeway. (63)
17. Every oil shock has been followed by a recession. (64)
18. The next shock will cripple the U.S. (65)
19. Oil prices vital to international economy. (66)
20. Oil shocks decrease the GNP by more than 3 percent. (67)
21. Oil is the heart of the global economy. (68)
22. Embargo causes economic shock waves. (69)

II. NO OIL CRISIS

A. NO PRICE SHOCKS
1. Status quo energy is safe and cheap. Transition to alternatives costs billions. (70)
2. Energy independence does not protect the U.S. from worldwide shocks. (71)
3. Strategic Petroleum Reserve reduces impact of a supply disruption. It's a deterrent. (72)
4. Transition to alternatives decreases the price of oil, encouraging consumption. (73)
5. Oil shocks no longer affect price. (74)
6. Conservation policies eliminate price elasticity, worsening oil shocks. (75)
7. No oil embargoes. Threat is a thing of the past. (76)
8. U.S. military ensures stable prices. (77)
9. Futures market ensures price stability. (78)
10. Economic interdependence crushes incentives to raise prices. (79)
11. Failed quotas and lack of cooperation block OPEC from setting prices. (80)
12. Exporters lost control of market. Shocks are impossible. (81)
13. No impact to dependence. Supply diversity creates oil safety. (82)

B. LOTS OF OIL
1. Oil reserves found all the time. Increasing faster than consumption. (83)
2. New oil reserves will definitely be found. (84)
3. U.S. production is an exception. The Earth has infinite reserves. (85)
4. New oil will be discovered. (86)
5. Improvements in technology means new oil will be found. (87)
6. OPEC always has excess capacity. (88)
7. Industry does not worry about oil supply. (89)
8. World can easily meet American oil demands. (90)
9. Fossil fuels are inexhaustible. No crisis. (91)
10. Middle East reserves are larger than we know. (92)
11. Oil is becoming more plentiful. (93)
12. Technology will yield new reserves. (94)
13. New technology decreases cost of oil exploration. (95)

C. AMERICAN RESERVES SUBSTANTIAL
1. DOE labs will maximize existing sources. (96)
2. Vast public lands provide oil potential. (97)
3. Gulf of Mexico will increase supply. (98)

D. OIL DEPENDENCY DOES NOT PREVENT A CRISIS
1. World energy market means energy independence does not isolate U.S. from oil shocks. (99)
2. U.S. still vulnerable even with alternative energy sources. (100)
3. Even energy isolationism does not prevent a shock from reaching the U.S. (101)

4. World energy market guarantees an oil shock reverberates in the U.S. (102)
5. Competitive U.S. oil industry does not prevent shocks. (103)
6. Absent complete isolationsim, U.S. will be vulnerable to oil shocks. (104)
7. Price stabilization efforts means U.S. is vulnerable to shock even if we do not import any oil. (105)
8. Domestic prices determined by world prices. U.S. cannot avoid an oil shock. (106)

E. STRATEGIC PETROLEUM RESERVE PREVENTS A SHOCK
1. SPR is the only solution. It adds supply to the market and reduces price. (107)
2. SPR is the perfect complement to the free market mechanism. (108)
3. Only the strategic reserve can prevent GNP loss. (109)
4. Early use of SPR can prevent GNP loss. (110)
5. SPR deters shocks from occurring. (111)
6. SPR replaces lost supply and checks runaway price speculation. (112)
7. DOE quietly upgrading SPR. (113)
8. SPR being revamped and relocated. (114)
9. Oil sharing schemes prevent shock. (115)
10. Oil sharing eliminates effectiveness of an embargo. (116)

F. LOW OIL PRICES INCREASE CONSUMPTION
1. OPEC price slash prevents shift to alternative enegy. (117)
2. Low prices stop behavioral changes necessary to shift to renewables. (118)
3. Low oil prices increase consumption and discourage search for renewable energy. (119)
4. Market always balances itself. Low prices encourage consumption. (120)
5. Low oil prices result in cancellation of alternative energy schemes. (121)
6. Cheap oil quashes conservation incentives. (122)
7. Low oil prices thwart development of sustainable energy. (123)
8. Oil price directly determined consumption. (124)
9. Global financial restraints make oil market more resilient. (125)
10. U.S. can absorb shock even under worst case scenarios. (126)

G. NO IMPACT TO FUTURE SHOCKS
1. New oil regime prevents long-lasting shocks. (127)
2. A price increase does not necessarily cause a recession. (128)
3. Production cushion quashes oil shock impact. (129)

H. NO FUTURE OIL EMBARGO
1. Diversified energy supply blocks embargoes. (130)
2. Transshipment blocks target embargoes. (131)
3. Politics don't factor into OPEC's decisions. (132)
4. Market always prevails. Political price controls fail. (133)
5. Chance of embargoes is exaggerated. (134)

I. MILITARY ENSURES PRICE STABILITY
1. Military will be used to protect oil supply. Persian Gulf proves. (135)
2. No threat to oil supply. Easy military option protects supply. (136)
3. In spite of dependence, U.S. presence stabilizes the oil supply. (137)
4. Oil stakes too high. U.S. will intervene to protect supply. (138)
5. U.S. won't abandon Saudi Arabia in Gulf. (139)
6. U.S. won't abandon military role in Gulf. Firmly committed to Israel and Saudi Arabia. (140)

J. FUTURES MARKET PREVENTS SHOCK
1. Futures trading increases market stability. Stops giant price swings. (141)
2. Market transparency increases price stability. (142)
3. Futures market eliminates price volatility. (143)
4. Futures trading protects price free-fall. (144)
5. Futures trading has exploded since last shock. Oil is number one traded commodity. (145)

K. INTERDEPENDENCE CHECKS CRISIS
1. The whole world is connected. An oil shock is not inevitable. (146)
2. Past experience and rational decision-making prevent economic harm. (147)
3. Heightened consciousness about energy means U.S. can respond to a crisis. (148)
4. Economic diversification prevents an oil shock. (149)
5. Free Market can prevent economic harm of an oil shock. (150)
6. Surge capacity, the SPR and other supply additions reduce harm of shock. (151)

L. OPEC CANNOT CONTROL PRICES
   1. OPEC no longer makes prices. Non-trading eliminates strategic advantage. (152)
   2. OPEC can support price increases. Shockproof markets weaken OPEC. (153)
   3. International markets set prices, not OPEC. (154)
   4. OPEC's effect is nothing more than supply and demand. They can't artificially set prices. (155)
   5. OPEC decisions are not the only price determinants. (156)
   6. Political developments, not OPEC, control prices. (157)
   7. OPEC must accept prices. They can't determine them. (158)
   8. OPEC can't fix prices. (159)
   9. OPEC is irrelevant as a market power. (160)

III. OIL PRICE COLLAPSE DISADVANTAGE
   A. SHELL
      1. Uniqueness: Oil prices stable now. (161)
      2. Link: Search for alternatives to oil causes OPEC to collapse prices. (163)
      3. Oil price collapse crushes world financial markets. Causes war. (162)
   B. UNIQUENESS
      1. U.S. has no viable energy policy now. Not pursuing alternative sources. (164)
      2. Government cutting funding for alternative energy. (165)
      3. Reliance on oil irreversible for near future. (166)
      4. Middle East military presence discourages conservation. (167)
      5. Energy efficiency improvements stalled since 1986. (168)
      6. Search for alternatives over. (169)
   C. LINK
      1. Energy conservation directly trades off with oil prices. (170)
      2. Search for alternative sources sparks price cutting. (171)
      3. Energy conservation drives down oil prices. (172)
      4. Energy efficiency improvements crush world oil prices. Long-run consumption increases. (173)
      5. Pursuing renewable energy decreases oil imports. (174)
      6. Pursuit of energy alternatives directly trades off with oil consumption. (175)
      7. Renewable energy will reduce oil dependence. (176)
      8. Alternative energies directly trade off with fossil fuel consumption. (177)
      9. Energy conservation policies discourage oil consumption. (178)
      10. High consumption levels magnify U.S. energy policies on world prices. (179)

IV. RUSSIAN OIL CHAOS DISADVANTAGE
   A. SHELL
      1. Brink: Russia's economy has reached critical mass. (180)
      2. Link: Narrowing of Russian oil prices encourages foreign investment. Increasing the gap destroys the Russian oil industry. (181)
      3. Impacts: A strong oil industry is key to preventing an economic meltdown. (182)
      4. Impact: Economic collapse ignites Russian-Ukrainian nuclear war. (183)
   B. U.S. ASSISTANCE KEY TO RUSSIAN ECONOMY
      1. U.S. companies have promised enough money to save Russia's oil industry. (184)
      2. U.S. oil industry critical to Russian resurgence. (185)
      3. Western technical and financial assistance can save the Russian oil industry. (186)
   C. FOREIGN INVESTMENT KEY TO ECONOMIC RESURGENCE
      1. Russian oil will collapse without foreign investment. Only investment will save the economy. (187)
      2. Lack of foreign investment causes a depression. (188)
      3. Foreign investment will stabilize the oil industry. (189)
      4. Foreign investment will lead to economic growth. (190)
      5. Without increased investment, Russian oil industry will collapse. (191)
      6. Investment will triple oil production. (192)
      7. Foreign investment will save the oil industry in the long term. (193)
      8. Oil industry is the only sector keeping the Russian economy alive. (194)
      9. Oil industry is key to all industries. (195)
   D. UNIQUENESS
      1. Russian oil prices are the same as world oil prices. (196)
      2. Gap between Russian and world prices narrowing. (197)
      3. Investment climate is significantly improving in Russia. (198)

4. Despite poor economy, investors are lining up behind the oil industry. (199)

C. IMPACTS:
1. Russian economic collapse risks global nuclear proliferation. (200)
2. Breakdown of Russian economic collapse causes war. (201)
3. Economic decline crushes Russian strength leading to multiple war scenarios. (202)
4. Economic instability leads to global war. (203)

V. PEACE PROCESS DISADVANTAGE

A. SHELL
1. Uniqueness and link: Oil is the sole reason the U.S. is engaged in the Middle East. The plan eliminates oil as an issue, causing disengagement. (204)
2. Impact: U.S. engagement crucial to peace process. Failure to lead crushes the fragile peace. (205)
3. Impact: Failure of the peace process creates a flashpoint for chemical and nuclear war. (206)

B. LINKS
1. Development of alternative energy lessens regional importance of Gulf. (207)
2. Oil is the primary reason for U.S. engagement in the Middle East. Containment is not an issue. (208)
3. Persian Gulf only strategic because of oil. (209)
4. DOD report clear: The Middle East is important solely because of oil. (210)
5. Other reasons for U.S. engagement are political creations. It's all about oil. (211)
6. Middle East military presence exists to ensure a secure supply of oil. (212)
7. U.S. forces deployed to protect oil. (213)
8. Troops in the Gulf because of oil. (214)
9. U.S. interest in the Gulf is all about oil. (215)
10. Renewable energy allows for U.S. disengagement from the Middle East. (216)
11. Even though other issues are part of engagement, the public sees oil as the only reason. (217)
12. Americans have a heightened sense of awareness towards oil dependence. (218)
13. Populist climate will bring attention to non-essential foreign policy. (219)
14. Even our commitment to the peace process is in jeopardy. (220)
15. Aid fatigue means all aid is questioned. (221)
16. Even the "sacred cow" aid to Israel is in danger. (222)
17. Israeli intransigence on "land for peace" endangers U.S. commitment. (223)

C. U.S LEADERSHIP KEY TO THE PEACE PROCESS
1. Middle East peace impossible without active U.S. engagement. (224)
2. American leadership necessary to foster peace agreements. (225)
3. Only the U.S. can check regional proliferation. (226)
4. Collective security fails without American leadership. Gulf war proves. (227)
5. Multilateralism cannot replace U.S. leadership. (228)
6. There is no substitute for American leadership. (229)
7. Security regimes are vital to solve the peace process. (230)

D. FUTURE MIDDLE EAST CRISIS WITHOUT U.S.-LED PEACE EFFORT
1. Multiple scenarios for future Middle East war. (231)
2. Despite political improvements between Middle East states, much potential for crisis exists. (232)
3. Multiple scenarios for war in the Middle East. (233)
4. Military confrontation only five years away. (234)
5. Oil resources makes crisis inevitable. (235)
6. Fundamentalism guarantees Middle East crisis. (236)
7. Nuclear proliferaton facilitates armed conflict. (237)
8. Collective security is the only solution to the peace process. (238–239)
9. Regional cooperation enhances deterrence. Rogues have no choice but to cooperate. (240)
10. Collective security discourages rogue states. (241)
11. Regional regimes increase mars transparency. This fosters verifiable arms control. (242)
12. Empirically, security regimes prevent Middle East conflict. (243)
13. Peace agreements create alliances, stoping the war coalition. (244)
14. Peace process solves war at the root cause. (245)

15. Abandoning the peace process sparks a disastrous war. (246)
16. Peace breakdown sparks a Syrian attack. (247)
17. Without Arab-Israeli Resolution CBW and nuclear war are ever-present threats. (248)
18. Breakup of peace leads to massive violence. (249)
19. Peace allows for conflict management. Deterrence failure inevitable without it. (250)
20. Territorial solution leads to security regime. (251)
21. Terrorists in the Middle East will use weapons of mass destruction if the peace process fails. (252)
22. Greatest risk of destabilization is from terrorist weapons. (253)
23. Terrorists can get weapons of mass destruction. (254)
24. Proliferation is dangerous. Poor storage causes accidents. (255)
25. Miscalculation and accidents pose the greatest risk to peace. (256)
26. Small crises escalate in the Middle East. (257)
27. Israeli insecurity of Arab war coalition causes crisis escalation. (258)
28. Lack of command and control networks forces crises to escalate.(259)
29. Middle East is unstable. Nuclear proliferation there causes war. (260)
30. Multiple scenario for war in the Middle East. (261)

VI. U.S.-ARAB RELATIONS DISDVANTAGE

A. SHELL
   1. Uniqueness: U.S. attacks on Iraq jeopardize U.S.-Arab relations. (262)
   2. Link: Oil is the cornerstone of U.S.-Saudi relations. (268)
   3. Impact: Strong U.S.-Arab relations necessary to American pre-positioning of forces. (263)
   4. U.S. pre-positioning key to prevent future Desert Storms. (264)

B. COOPERATION ESSENTIAL TO PRE-POSITIONING
   1. Forward presence vital to effective military operations. (265)
   2. Quick action is a prerequisite to military successes in the Gulf. (266)
   3. Forward presence empirically successful in the Gulf. (267)

C. OIL KEY TO RELATIONS
   1. U.S. cannot disengage from Saudi Arabia because of oil. U.S. must respect Saudi interests to maintain reltaionship. (269)
   2. Saudis demand unconditional support from the U.S. (270)

D. SAUDIS IMPORTANT TO PEACE PROCESS
   1. Saudis are playing a cooperative role in the peace process. (271)
   2. Saudi Arabia supports peace process in current form. (272)
   3. Western encouragement can entice Saudis to act on peace process. Result is Saudi leadership in the Arab world. (273)
   4. The U.S. can pressure Saudis to act on the peace process. (274)
   5. Gulf war debts and economic dependency give the U.S. leverage over Saudi Arabia. (275)
   6. Saudi Arabia is the key opinion-maker in the region. (276)
   7. Saudis are important players in the peace process. (277)
   8. Moderate political views make Saudi Arabia an effective regional leader. (278)
   9. Saudi Arabia is most influential regional power. Saudis possess all the holy sites. (279)
   10. Holding the two holy cities gives Saudis regional influence. (280)
   11. Saudi-U.S. relations are the linchpin to price stability in the new oil order. (281)
   12. Combined with the SPR, U.S.-Saudi relations keep prices stable. (282)
   13. U.S. training contains Iran and Iraq. (283)
   14. U.S. support for Netanyahu puts Saudi relations in jeopardy. (284)

E. GULF CRISES ARE INENVITABLE. MUST HAVE PREPOSITIONING AND STRONG U.S.-ARAB RELATIONS
   1. Inevitably, major conflicts will occur in the Gulf. The U.S. must be prepared. (285)
   2. Ongoing political instability makes conflicts inevitable. (286)
   3. Overpopultion in the Gulf risks tensions. (287)
   4. Territorial disputes inevitable. (288)
   5. Oil is the cornerstone of Middle Eastern economies. Collapse endangers the stability of Middle East regimes. (289)
   6. Future course of the Middle East nuclear. Changing public priorities are affecting the dynamic of the region. (290)

VII. MISCELLANEOUS
- A. INVESTMENT SHORTFALLS WILL CRUSH THE WORLD PETROLEUM INDUSTRY (291)
- B. FOREIGN INVESTMENT NECESSARY TO SUSTAIN OPEC PRODUCTION (292)
- C. NEW INVESTMENT NEEDED TO MAINTAIN AND EXPAND OIL PRODUCTION (293)
- D. IRAQ RE-ENTRY NOT LIKELY SCENARIO TO COLLAPSE PRICES (294)
- E. IRAQI OIL FLOOD WILL SNOWBALL TO REST OF OPEC (295)
- F. IRAQI RE-ENTRY COLLAPSES OIL PRICES (296)
- G. GULF ECONOMIES ALMOST 100% ON OIL (297)
- H. OIL PRODUCERS ARE BECOMING INCREASINGLY DEPENDENT ON OIL REVENUES (298)
- I. ALTERNATIVES TO OIL REVENUES NOT VIABLE FOR MIDDLE EAST NATIONS (299)
- J. GULF ECONOMIES ARE "OIL ECONOMIES" (300)

CHAPTER II

EVIDENCE

1. George Georgiou. ENERGY POLICY. August 1993, p. 835. Finally, when thinking about the world oil market, it is important to remember that the current glut tends to mask OPEC's real power. One must remember that the loss of exports from Iraq and Kuwait was offset primarily by increased production by other OPEC countries, mainly Saudi Arabia. Furthermore OPEC still dominates the international oil market: as recently as 1989, its share of net world exports of oil was 73.8%. All available evidence indicates that this share will increase over time. Thus if any independence from OPEC—specifically, from Middle East oil producers—has developed in the 1980s, it has served only to provide a false sense of security. To the extent that Western industrialized nations have failed to develop alternative energy sources and have rapidly depleted their strategically secure reserves, they have only increased their long-term dependence on OPEC oil. The lower OPEC's current share of the world market, the greater its share will be at the end of the century, when many other countries will be running out of oil. OPEC may then become a more powerful organization—smaller, more geographically concentrated in the Persian Gulf. At the same time, the world's reserves will be more limited than at any time in recent history.

In such a situation, America's position would be even weaker that it was during the 1970s, when OPEC was considered to be at the peak of its power. Since 1985, US demand for oil has grown at 2.63% per year, rising over 1 mbpd. Between 1990 and 2000, US demand for oil will grow by at least 1% per year, rising by 2 mbpd over that period. On the other hand, because of drastically reduced drilling and maturing oil fields, US crude production will fall by 2.5% per year between 1990 and 2000, resulting in a decline of over 2 mbpd. Natural gas output will also fall. Consequently, net oil imports will rise from 6.7 mbpd in 1988 to nearly 12 mbpd in 2000. Just as OPEC's control is strongest, then the USA will become more dependent on imported oil.

2. George Georgiou. ENERGY POLICY. August 1993, p. 835. What all these trends imply is that we are now witnessing the start of a transition to a more concentrated, and more powerful OPEC—smaller and more geographically concentrated in the Persian Gulf. More importantly, what is evolving is a "cartel" in the true sense of the word, and one that is not guided by ideology but market forces. That OPEC survived the Persian Gulf war intact is testament to the new pragmatic and business oriented OPEC. Certainly such a regime will have costs to the USA and other oil-importing nations. But the problem is not only limited to high oil prices or (if one regards the situation from an OPEC perspective) low oil prices; it is, rather, the oil market's overall instability. That instability has been severe under the current oil regime, and it could become even more so in the transition period ahead, causing dramatic swings in price that, in the final analysis, will benefit neither producers nor consumers.

3. Joseph Romm and Charles Curtis. ATLANTIC MONTHLY. April 1996, p. 59. Given that the most recent war America fought was in the Persian Gulf, let's start by examining the likelihood that an oil crisis will occur in the coming decade. Forecasting is always risky, especially where oil is concerned, but consider what a variety of experienced energy hands from every point on the political spectrum have said in the past year alone. Donald Hodel, who was a Secretary of Energy under Ronald Reagan, has said that we are "sleepwalking into a disaster," and predicts a major oil crisis within a few years. Irwin Stelzer, of the American Enterprise Institute, says that the next oil shock "will make those of the 1970s seem trivial by comparison." Daniel Yergin says, "People seem to have forgotten that oil prices, like those of all commodities, are cyclical and will go up again." James Schlesinger, who was the Secretary of Energy under Jimmy Carter, has said, "By the end of this decade we are likely to see substantial price increases." In March of last year Robert Dole, the Senate majority leader, said in a speech at the Nixon Center for Peace and Freedom, "The second inescapable reality of the post-twentieth-century world is that the security of the world's oil and gas supplies will remain a vital national interest of the United States and of the other industrial powers. The Persian Gulf . . . is still a region of many uncertainties. . . . In this 'new energy order' many of the most important geopolitical decisions—ones on which a nation's sovereignty can depend—will deal with the location and routes for oil and gas pipelines. In response, our strategy, our diplomacy, and our forward military presence need readjusting." The chairman of the Federal Reserve, Alan Greenspan, not known for being an alarmist, in testimony before Congress last July raised concerns that a rising trade deficit in oil "tends to create questions about the security of our oil resources."

Concerns about a coming oil crisis have surfaced in the financial markets as well. Last October, in an article titled "Your Last Big Play in Oil," *Fortune* magazine listed several billionaires and "big mutual fund managers" who were betting heavily that oil prices would rise significantly. The magazine went on to suggest an investment portfolio of "companies that are best positioned to profit from the coming boom." [ellipses in original]

4. OIL & GAS JOURNAL. 17 June 1996, p. Lexis/Nexis. Elihu Bergman, former executive director of Americans for Energy Independence who is now a Washington energy policy analyst, said, "Increasing oil

demand by itself is not likely to trigger a repeat of the crises and temporary price spikes of 1974 and 1980.

"IEA has forecasted a 50% increase in the real price of oil over the next 10 years to $28/bbl by 2005, which could be accommodated under stable economic and political conditions.

"However, an unforeseen shock in a tightening market—like the invasion of Kuwait or an insurrection in Saudi Arabia—could result in a sudden price spike with the adverse consequences experienced in the past.

"With 66% of known global oil reserves, the Persian Gulf remains the key factor in the supply/demand equation. It also remains an unstable region with considerable potential for unpleasant surprises.

"The increasing political alienation in gulf countries is real and likely to increase as political and economic grievances mount. The principal causes are unmet expectations about the distribution of economic wealth, curtailed from the heady days of the early 1980s, and the sharing of political power. Within this volatile environment, militant Islamic movements, both indigenous and foreign, provide a ready made outlet."

Bergman said Iran and Iraq continue their threatening behavior, and the Saudi Arabian government, "which functions like a family business," is experiencing increasing internal unrest while facing a change in the monarchy.

"We should realize that the principal certainty in Middle East politics is still uncertainty and our version of the rational calculations of national interest is still not employed by regimes and publics in the region."

5. Edward Fried. OIL AND AMERICA'S SECURITY. 1988, p. 237. Oil has made its mark on the world in the past fifteen years. Two oil price shocks and one price collapse exerted a powerful influence on the world economy and on international relations. Ramifications can be seen in two world recessions, in third world debt problems, as an unspoken presence in attempts to resolve Arab-Israeli differences, in the expansion of the U.S. military role in the Persian Gulf, and in the virulent boom and bust cycles in oil producing countries and in the major oil producing regions in the United States.

Throughout the period, the future course of oil prices and the economic and political consequences have been the subject of intense examination. Views have toned down remarkably as markets have eased. Nonetheless, nervousness persists, stemming principally from the unchallenged fact that the United States and the rest of the oil importing world will become increasingly dependent on oil supplied from the politically troubled Middle East. Does this mean that another oil shock is in our future? Will U.S. security policy become hostage in some respects to our growing oil import dependency? And what can we do to contain damage if oil supplies are seriously interrupted?

6. Joseph Romm and Charles Curtis. ATLANTIC MONTHLY. April 1996, pp. 61-62. What is the appropriate national response to the re-emerging energy-security threat? Abroad the Department of Energy has been working hard to expand sources of oil outside the Persian Gulf region—in the former Soviet Union, for example—and to encourage the privatization of the oil companies in Mexico and other Latin American countries.

At home the DOE is encouraging greater production by providing royalty relief in the deep waters of the Gulf of Mexico and similar incentives, so that the industry can drill wells that otherwise would not be cost-effective. The DOE is working to reduce the cost for the industry to comply with federal regulations. Finally, the department is spending tens of millions of dollars a year to develop new technologies that will lower the cost of finding and extracting oil—for example, using advanced computing to model oil fields. Still few expect to reverse the decade-long decline in U.S. oil production.

7. Howard Geller, John DeCicco, and Skip Laitner. ENERGY POLICY. June 1994, p. 482. The USA is running out of economically recoverable oil. At the end of 1992 the USA had 23.7 billion barrels (bbl) of proved oil reserves (reserves considered recoverable under current economic and technological conditions.) Proved reserves peaked in 1970 and have fallen steadily since then. Current proved reserves would only last about four years if the US relied solely on domestic sources of supply.

As well as having limited proved oil reserves, the USA also faces dim prospects of adding to them in any substantial way. Since the discovery of the Kuparuk River field in Alaska in 1969, no new major oil fields have been found in the USA. In 1992 new oil field discoveries in the USA amounted to only 8 million barrels, about one-half of *daily* US oil consumption. Of current proved oil reserves in the lower 48 states, 74% was discovered prior to 1960.

8. Howard Geller, John DeCicco, and Skip Laitner. ENERGY POLICY. June 1994, p. 45. New tax incentives, technological advances, or even a sharp increase in oil prices will not prevent US domestic oil production from declining in the future. One forecast estimates that oil production in the lower 48 states will fall too 2.5 MBD by 2000 (compared to 5.5 MBD in 1992) assuming a constant oil price of US$18 per barrel. Even if the price of oil doubles, oil production in the lower 48 states still declines to 3.2 MBD in 2000 according to this forecast. The US Energy Information Administration predicts that if world oil prices reach US$28 per barrel by the year 2000 and US$38 per barrel by 2010 (in constant dollars), US domestic oil production would still be lower than at any time during the past 20 years. As Edward Murphy, a representative of the American Petroleum Institute commented, "the trend for US crude production is pessimistic, at best."

9. Howard Geller, John DeCicco, and Skip Laitner. ENERGY POLICY. June 1994, p. 482. In 1987

the US Geological Survey (USGS) estimated that there was a 95% chance that 21 bbl, and a 5% chance that 54 bbl of economically recoverable oil remained to be discovered in the USA. Even if the USGS;s most optimistic estimate were correct, a self-reliant USA would survive for less than a decade on its yet to be discovered economically recoverable oil. Figure 6 shows the longevity of both proved and potential US oil reserves at current consumption levels.

Resource depletion is the main factor causing the decline in US oil reserves. In spite of enormous technological improvements, the return on oil drilling in terms of barrel discovered per metre drilled has fallen by more than a factor of 10 during the past 50 years. Furthermore, the downward trend in yield is expected to continue and could even accelerate if new financial incentives or other policies lead to an expansion of marginal and high-risk drilling.

10. Howard Geller, John DeCicco, and Skip Laitner. ENERGY POLICY. June 1994, p. 482. Not only is there little recoverable oil left in the USA, the oil that remains is becoming more difficult and expensive to recover. Despite increased use of advanced oil exploration and extraction techniques—such as three-dimensional seismology and horizontal drilling—US domestic oil supply continues to fall.

11. THE NASHVILLE BANNER. 10 May 1996, p. Lexis/Nexis. The United States now imports more than half the oil it uses, even more than it did at the time of the Persian Gulf war five years ago. During the same period, U.S. oil production has continued to decline.

At the present rate, by the year 2005, some 60 percent of our daily needs will be supplied by other countries, principally Persian Gulf members of the Organization of Petroleum Exporting Countries (OPEC). And our trade deficit is expected to double to nearly $100 billion a year by that time, a large drag on our economy.

If anyone doubts that rising imports pose a danger, consider this: Since the 1950s there have been six oil supply disruptions of 2 million barrels a day or more, an average of one every five to 10 years, all originating in the strife-torn Middle East.

12. Howard Geller, John DeCicco, and Skip Laitner. ENERGY POLICY. June 1994, p. 473. Some forecasts indicate that, if current policies and trends continue, both oil demand and oil prices will move significantly upwards during the next two decades. A recent reference case forecast by the International Energy Agency (IEA) projects that world oil consumption will increase from 66.3 MBD in 1990 to 92.5 MBD by 2010, an average growth rate of 1.7% per year. With oil production declining in OECD nations, oil supply from the Middle East plus Venezuela (a region used in the IEA forecast that is similar to OPEC) is expected to increase over 50% between 1990 and 2000, and more than double between 1990 and 2010 reaching 45 MBD in the latter year.

With growing oil demand and increasing reliance on Middle Eastern producers, the IEA projects that the world oil price will gradually rise to US$30 per barrel by 2005 (in constant 1993 dollars). As a lower bound price case, the IEA also considered a scenario in which the world oil price remains constant at US$20 a barrel. In this case, OPEC oil supply grows even more rapidly, potentially reaching 57 MBD and nearly 60% of total world supply by 2010.

13. Joseph Romm and Charles Curtis. ATLANTIC MONTHLY. April 1996, p. 62. Some would open the Arctic National Wildlife Refuge to drilling, a plan the Clinton Administration has opposed on environmental grounds, but not even that would change our forecasted oil dependency much. This is true even using earlier, more optimistic estimates that the refuge could provide 300,000 barrels of oil a day for thirty years. The EIA projects that within ten to fifteen years the United States will probably be importing *thirty times as much*—some 10 million barrels of crude oil a day, even if the decline in other domestic production levels off in the next few years.

14. Paul Stevens. ENERGY POLICY. May 1996, p. 394. *Demand* There is general agreement among virtually all observers of the industry that oil demand will grow for the foreseeable future. Thus the demand curve in Figure 3 will inexorably keep shifting to the right. Given the history shown in Figure 4, it is incumbent upon any who disagree to explain why the patterns of positive growth—except in the immediate aftermath of the two oil shocks—should differ in the future.

15. Howard Geller, John DeCicco, and Skip Laitner. ENERGY POLICY. June 1994, pp. 472-473. Given current trends and policies, US oil imports are expected to rise rapidly during the next 20 years. According to the most recent reference case forecast by the US Department of Energy (DOE), net oil imports are expected to increase from 6.9 MBD in 1992 to 10.8 MBD in 2000 and then to 12.2 MBD in 2010. Oil imports would rise 3.2% per year on average, compared to 0.7% per year on average during 1973–92. Net oil imports would represent 56% and 59% of total US oil consumption in 2000 and 2010 respectively.

Figure 2 includes the projected cost of US net oil imports up to 2010. Increasing oil imports by the USA and other nations would generate upward pressure on world oil prices, leading DOE to project that the world oil price will begin to rise again in the late 1990s and will reach nearly US$30 per barrel (in 1991 dollars) by 2010. The combination of rising oil imports and higher oil prices lead to a projected annual oil import cost of US$90 billion by 2000 and US$130 billion by 2010, compared to about US$45 billion in 1992.

16. INDUSTRY WEEK. 6 May 1996, p. 1. But maybe managers shouldn't be so complacent. The latest drilling statistics for oil and gas—the pivotal fuels of the industrial world–reveal a potentially troubling trend. In 1995, reports the Energy Information Administration (EIA), only 18,459 oil and gas wells were drilled in the U.S. It was the lowest number since 1945, and some 72,000 fewer than at the peak of more than 90,000 in 1981.

Statistics for exploratory wells–the measure of the industry's search for new oil and gas sources that are the seeds of a secure energy future—are equally disturbing: Only 3,543 were drilled in the U.S. in 1995, down a startling 80% from 1981's peak of 17,497. What's more, in a survey of 121 U.S. oil and gas companies by Arthur Andersen & Co. consultants late last year, only 43% indicated they expect to increase their capital spending for U.S. exploration in 1996.

The sharp decline, observes Victor A. Burk, Houston-based manager of energy-industry services at Arthur Andersen, "raises the question of whether sufficient new reserves will be discovered to meet rising demand."

And well it should raise the question, say petroleum-industry executives—especially in light of several other disconnecting statistics:

- U.S. demand for petroleum products, generally level in recent years, lately has been on the rise. A fourth-quarter burst that boosted 1995's average to more than 17.7 million barrels a day (b/d) has continued this year, indicates the American Petroleum Institute (API), Washington.
- Even as the world's proved reserves of crude oil (oil that can be economically recovered) have climbed to record levels, U.S. reserves are declining. From 34.3 billion bbl in 1975, they shrank to 23 billion bbl at the beginning of last year, reveal statistics compiled by Arthur Andersen and Cambridge Energy Research Associates, Cambridge, Mass.
- U.S. field production of crude oil, according to EIA, has tumbled from 8.97 million b/d in 1985 to 6.54 million b/d last year.

In short, U.S. oil demand is rising and supply is falling—a combination that inevitably has led the nation to a greater dependence on imports.

Although U.S. imports of crude oil and petroleum products did drop slightly last year, they've generally been on a steady ascent. In 1995, API reports, they accounted for fully 50% of the country's oil consumption. That is almost twice as high as the 29% level in 1972—just before the Arab oil embargo plunged the world into economic crisis and prompted a now-abandoned drive by the U.S. government to achieve "energy independence."

17. George Georgiou. ENERGY POLICY. August 1993, p. 835. The USA, with a GNP in 1993 approaching $6500 billion, consumes more oil than any other nation in the world—estimated in 1993 at 175 mbpd. This accounts for about one-third of total capitalist-world oil consumption and over one-fourth of total world consumption. The USA also produces a large amount of oil estimated at 9.7 mbpd in 1988—making it the second largest producer after the USSR. This implies that the USA is the world's largest importer of oil currently estimated at 7.3 mbpd. US imports have steadily risen in recent years after reaching a low of 4.5 mbpd in 1983. This trend is expected to continue with a 1987 US Department of Energy study projecting US imports in the range of 8–10 mbpd in the mid-1990s and reaching 12 mbpd by 2000. This implies a growing reliance on OPEC oil and particularly insecure Persian Gulf oil.

18. Linda Tierney. 27 March 1995, Y4:F7/2:S.HRG.104-21, p. 7. Since the 1970s, however, things have been quite different: increased energy efficiency cut demand, decontrol of energy prices allowed better economic decision making, and growth in supply diversity cut into OPEC's market power. Oil prices have trended down since their peak in 1981, and in 1985–86, oil prices crashed. OPEC, which had been 55 percent of the world market in 1973, was down to 30 percent by 1985. Oil prices and OPEC market share have since recovered somewhat.

In 1990, Iraq invaded Kuwait, and oil prices quickly doubled, only to decline just as quickly as excess production capacity was brought on line. Oil prices today, at $17.00 per barrel, are lower than they were before the invasion of Kuwait, encouraging consumption and discouraging production because domestic oil reserves are mature and therefore high cost. Although the economy continues to become more energy efficient, more consumption and less domestic production means more oil from abroad. As a result, net oil imports are again rising, closing in on 50 percent of consumption.

19. OIL & GAS JOURNAL. 17 June 1996, Lexis/Nexis. In a Senate floor speech, Helms observed that API reported the U.S. imported 8.7 million b/d during the week ending May 31, up 900,000 b/d from the same week a year earlier.

He said, "Americans relied on foreign oil for 57% of their needs last week, and there are no signs that this upward spiral will abate. Before the Persian Gulf war, the U.S. obtained about 45% of its oil supply from foreign countries. During the Arab oil embargo in the 1970s, foreign oil accounted for only 35% of America's oil supply."

Helms, Senate foreign relations committee chairman, said, "Politicians had better ponder the economic calamity sure to occur in America if and when foreign producers shut off our supply or double the already enormous cost of imported oil flowing into the U.S."

20. JOURNAL OF FINANCIAL PLANNING. April 1995, Lexis/Nexis.

A FOOL'S PARADISE

We have been living in a fool's paradise as far as energy is concerned. The first and second oil shocks of the 1970s spurred serious conservation efforts, and conversion to coal, natural gas, and nuclear power where possible. Observed *The Economist*: "So long as rich countries accounted for the lion's share of energy consumption, small changes in their economies swamped the relentless growth of demand in the developing world. This is one reason why improved energy efficiency and two recessions in the 1980s were able to keep the oil market slack, despite a rise in demand in developing Asia. But this is about to change."

"When you're 20 percent of world consumption, 6 percent growth doesn't matter; when you're 50 percent, it does," asserts Michael Grubb of the Royal Institute for International Affairs in London.

21. INDUSTRY WEEK. 6 May 1996, Lexis/Nexis.

"With imports up over 50% [industry managers] definitely should be concerned," warns Wayne Allen, chairman and CEO of Phillips Petroleum Co., Bartlesville, Okla. "The nation is increasingly vulnerable to interruption." After all, he reminds, "We [the U.S.] were willing to commit troops in Iraq a few years ago to protect oil supply."

Indeed, comments Christopher E. H. Ross, vice president of Cambridge, Mass.-based consulting firm Arthur D. Little Inc. (ADL) and managing director of its international energy directorate, "There's no argument that U.S. oil imports will increase. There's also no argument that the nation's major resource base is in historically politically unstable countries. Events in the Middle East could well occur that will cause prices to go up. We advise our clients to plan for that possibility."

22. INDUSTRY WEEK. 6 May 1996, p. Lexis/Nexis.

Yet, a repeat of the type of shortages that shook the world in the 1970s shouldn't be the main concern, contends John Rutledge, chairman of Rutledge & Co., a Greenwich, Conn., private-equity investment firm that works closely with oil and gas companies. "I worry more," he says, "about the effects of the growing energy demand and power of developing nations—particularly China and other East Asian countries. They increasingly will be buying the oil that we [the U.S.] need to grow.

"That's what will come around to bite us one way or the other—either they'll buy the resources we need, or they'll have the political power to muscle some of their neighbors around." Within two to five years, Rutledge predicts, the sharp growth of developing countries—on top of increased demand in the industrialized world—could bring a significant rise in oil prices.

If he's right, the price of oil will return as a worry for American industrial companies for the first time in years. The average acquisition cost by U.S. refiners for a barrel of crude oil—perhaps the most-meaningful oil-price measure because it influences the price of petroleum products—was $15.51 in 1994 (the last year for which figures are available), EIA data show. In inflation-adjusted dollars, that is $5 lower than five years earlier and a remarkable $40 below the peak price of $55.43 in 1981. As a result, prices industrial and other consumers paying for many petroleum products are at historic lows.

23. Joseph Romm and Charles Curtis. ATLANTIC MONTHLY. April 1996, p. 60. The final piece in the geopolitical puzzle is that during the oil crisis of the 1970s the countries competing with us for oil were our NATO allies, but during the next oil crisis a new, important complication will arise: the competition for oil will increasingly come from the rapidly growing countries of Asia. Indeed, in the early 1970s East Asia consumed well under half as much oil as the United States, but by the time of the next crisis East Asian nations will probably be consuming more oil than we do.

24. REUTER EUROPEAN BUSINESS REPORT. 3 September 1996, Lexis/Nexis. The Centre for Global Energy Studies (CGES), a London energy "think tank" headed by former Saudi oil minister Sheik Ahmed Zaki Yamani, reckons OPEC producers stand to earn $2 billion more in the fourth quarter 1996 than was previously forecast.

"OPEC has had a very good ride this quarter and things look better for the fourth quarter. We have seen a dramatic change in market sentiment," says Leo Drollas, CGES chief economist.

25. George Georgiou. ENERGY POLICY. August 1993, p. 834. Furthermore, because non-OPEC oil reserves are considerably smaller than OPEC's, the current downward spiral in oil prices is causing non-OPEC oil to be depleted much more quickly than new oil discoveries are being made. Thus, ironically, falling oil prices are actually helping to conserve OPEC oil. The depletion of the world's most abundant reserves has slowed, while depletion of some of the scarcest and most strategically important reserves has accelerated. For example, at the 1988 extraction rate, the oil reserves of the USA, the USSR, and the UK will last only 9.5 years, 12.8 years, and 5 years, respectively, while those of Saudi Arabia, Kuwait, Iraq, Iran, and the United Arab Emirates will all last over 100 years.

26. George Georgiou. ENERGY POLICY. August 1993, p. 834. From all indications, then, world oil demand will continue to grow through the 1990s, and OPEC will be the major supplier of that incremental demand, making it the world's "supplier of the last resort." In other words, oil importers will try to exhaust all other possible sources of oil before turning to OPEC, especially during periods when OPEC oil prices are on the rise. This will tend to amplify the cyclical movements in the demand

for OPEC oil. The effects will be double-edged. In times of slowing economic growth and declining demand overall, there will be a more than proportional drop in demand for OPEC oil, since consumers will exhaust other sources of energy, including non-OPEC oil, before they turn to OPEC supply. Likewise, as economic growth increases, raising demand for energy and oil, the demand for OPEC oil will grow more than proportionally, since the supply of non-OPEC oil is relatively fixed. Thus even though, as we have seen, the overall relationship between energy and GNP has been declining, the relationship of OPEC oil to oil-importing nations' GNP will actually be far stronger than in the past. At crucial points, oil-importing nations will be much more vulnerable to OPEC than they were in the 1970s.

27. Joseph Romm and Charles Curtis. ATLANTIC MONTHLY. April 1996, p. 60. Although non-OPEC nations did increase production by almost 15 percent from 1980 to 1990, they increased proven reserves of oil by only 10 percent. The net result is that the remaining years of production for non-OPEC reserves has actually fallen from eighteen years to seventeen years. On the other hand, while OPEC increased production by 20 percent in the 1980s, it increased its proven reserves by 75 percent. As a result, OPEC's reserves-to-production ratio doubled to ninety years.

28. George Georgiou. ENERGY POLICY. August 1993, p. 833. Oil will, therefore, remain the foremost source of world energy well into the twenty-first century, even as it becomes more scarce. Despite recent conservation measures, the rate of oil consumption worldwide still exceeds new discoveries—and it is likely to continue to do so. Thus, despite the current oil glut, it is a mistake to write off OPEC, for a number of reasons. OPEC nations still control three-quarters of the world's proved oil reserves, with nearly two-thirds held by Persian Gulf OPEC countries. More importantly, OPEC's dominance of trade in world oil is likely to increase. Though many countries in the world produce oil, most of them consume a great deal of it. The former USSR and the USA, for instance, are the world's two largest oil producers; but of the 12.7 mbpd the USSR produced in 1988, it consumed 8.8 mbpd itself, and although the USA produced 9.7 mbpd in 1988, it had to import 6.7 mbpd to meet its needs. What indications a nation's impact on the international economy, then, is not simply its crude oil production, but rather its trade in oil. In this connection, it is important to note that even though OPEC's net exports of oil fell from 31 mbpd in 1979 to 19.2 mbpd in 1989, OPEC still accounts for nearly 75% of world oil trade.

29. OIL & GAS JOURNAL. 17 June 1996, Lexis/Nexis. Jay Hakes, EIA administrator, told a congressional hearing, "The Organization of Petroleum Exporting Countries, with is vast store of readily accessible oil reserves, is expected to be the source of marginal supply to meet future incremental demand."

EIA said OPEC supply will be about 35 million b/d in 2000 and more than 52 million b/d by 2015, about twice its level of production in 1990.

Persian Gulf crude production will be about 40 million b/d by 2015, compared with about 20 million in 1994. With world oil consumption rising to 93 million b/d by 2015, Persian Gulf supplies will provide 43% of the world's oil consumption by 2015, compared with about 30% in 1994.

EIA said technical innovations will help push production in non-OPEC nations higher, to a little more than 41 million b/d in 2010. It will then decline slightly to 40 million b/d, near the 1994, level, in 2015.

EIA sees a substantial increase in world oil consumption during the next 20 years, rising from about 69 million b/d to more than 74 million b/d by the end of the decade and as high as 100 million b/d by 2015.

Developing nations in Asia will show demand growth greater than 5%/year, while OECD nations will have growth of less than 1%/year.

EIA predicts that U.S. oil production will drop from 6.7 million b/d in 1994 to 5.3 million b/d in 2005, then rebound during 2005-15 as a result of technology improvements and rising prices.

Therefore, it expects imports to supply 57% of U.S. demand in 2005 and remain at about that level through 2015, compared with 45% in 1994.

30. Joseph Romm and Charles Curtis. ATLANTIC MONTHLY. April 1996, p. 57. Imagine a world in which the Persian Gulf controlled two thirds of the world's oil for export, with $200 billion a year in oil revenues streaming into that unstable and politically troubled region, and America was importing nearly 60 percent of its oil, resulting in a $100-billion-a-year outflow that undermined efforts to reduce our trade deficit. That's a scenario out of the 1970s which can never happen again, right? No, that's the "reference case" projection for ten years from now from the federal Energy Information Administration.

31. Donald Hodel. 27 March 1995, Y4.F7/2:S.HRG.104-21, pp. 236–237. The world is on the brink of another oil shock. When it hits, as before, there will be an outcry for Congress to "do something" about the crisis by intervening in the oil market and fixing prices, allocating supplies, or both. It should do neither. The one thing it should do is foster the domestic oil industry, but it may be prevented from providing meaningful assistance by the adoption of GATT by a lame duck Congress.

Saudi Arabia ranks first in both oil production and proved oil reserves (nearly 260 billion barrels or 25 percent of the world's proved reserves—compared to the U.S. with less than 24 billion barrels). "The Kingdom" as it is often called, is in crisis with an economy in sharp decline. The

reason for the slide is simple. Since 1985, oil prices have plunged, falling by almost 53 percent from their 1982 peak. Therefore, while Saudi oil production is high (around 8 million barrels per day) the revenues generated by the sale of oil are almost $33 billion less annually in real terms than they were in 1982.

The proclivity of the Royal Family with its 6,000 princes (an unnumbered princesses) to spend and deposit huge sums outside the Kingdom is legendary. Despite this 42 percent drop in income, they have been unwilling to reduce their lifestyle to the reduced purchasing power of the oil revenues. Instead the Royal Family has taken such a large share of the oil income that too little is left to maintain the economy. As an even more ominous indicator of perceptions inside the Kingdom, it now appears that many other well-to-do Saudis are liquidating their assets in order to move wealth outside the country as their fear of civil turmoil grows.

The faltering economy has resulted in record unemployment, a growing, visible homeless population, and civil unrest. Rebellious talk in the street is at unprecedented levels.

A November 28, 1984, *New Yorker* magazine article *"Royal Mess,"* quotes an unnamed "British expatriate who has lived in Saudi Arabia ten years" as saying, "Economically this country is in deep, deep trouble. There are no more jobs . . ."

The same article reports, "At the end of October, N. C. M., a leading international trade risk insurer, reported to customers that it was canceling open-account credit-limit approvals for exports to Saudi Arabia, because the Saudis were not paying their bills. . . . "The situation in Saudi Arabia is very grave and continues to deteriorate."

Complicating the economic disarray is an equally volatile political landscape.

Aging King Fahd is ill, and his reputed successor, the Crown Prince Abdullah, who leans strongly towards Moslem fundamentalism, is openly and intensely anti-American. Although Abdullah's ascension to the throne would not benefit the U.S., there may be an even worse consequence following the King's death.

The level of civil unrest in Saudi Arabia is such that some close observers fear Fahd's demise could spark the onset of civil war. This risk is exacerbated by the fact that the Kingdom is made up of four nations (or tribes) who are as unfriendly to each other as the Serbs and Croats of Yugoslavia. While Fahd has been able to "play Tito" to their hostilities, when he dies civil war may erupt before his successor could assert control. Such a conflict would certainly disrupt Saudi oil production, if not halt it completely.

Of course, Abdullah could use his position as head of the Saudi National Guard to inflict sufficiently repressive measures to put down the rebellion. No matter how tightly he attempts to control press access to events inside Saudi Arabia, he will not be able to prevent the world from learning of his actions. That in turn, will build pressure to impose sanctions on Saudi Arabia. Any such moves, though, could cause Abdullah to retaliate with the one weapon he has: oil. Because of its unique position as the world's leading producer, the Kingdom can disrupt world energy markets by rapidly shifting its level of production to either flood the market with crude, causing prices to collapse, or sharply cutting back, causing them to skyrocket. Moreover, because shifts in oil prices effect all other energy sources, the economic consequences will be multiplied. But, that's not all.

Further, his antipathy towards the U.S. could generate other efforts on his part to "punish the Great Satan." No action, including cutting off oil production can be discounted. Zealots and dictators (e.g., Saddam Hussein) respond to callings higher than the mere welfare of their people.

Unfortunately, it is unlikely that the current Administration will intercede with the King to urge that the Royal Family reduce its drain on the economy and mitigate the growing unrest. Even it did, it is not clear that the King would be able in his current condition to persuade his family to accept such an edict. The die may be cast.

One serious problem in getting the U.S. to act is that in the U.S. few, if any major law firms, public relations firms, oil companies, U.S. defense contractors, airplane manufacturers, large construction firms, and their labor forces and international unions are willing to acknowledge a problem even exists lest they earn the ire of the Saudi royals, and fall off the gravy train. The impact and influence of Saudi billions on all of these constituencies is an unreported and unrecognized force on American public opinion. In truth, the reach of those billions of dollars in contracts and retainers means that few if any lobbyists on Capitol Hill who know anything about Saudi Arabia and the oil business would dare raise questions about their patrons.

The chances are, therefore, that we will go merrily on our way as if there were no potential problem to world oil supply until it is too late. Sadly, the consequences can be devastating. [ellipses in original]

32. Joseph Romm and Charles Curtis. ATLANTIC MONTHLY. April 1996, p. 60. Some argue that energy forecasts are notoriously inaccurate and that for the Department of Energy to base decisions on them is risky. We cannot, of course, say with certainty that an oil crisis will occur in the next decade, that a transition to renewable energy will occur as Shell envisions, or that industry worldwide will shift to pollution prevention. But each of these things seems very plausible.

33. Paul Tempest. ENERGY POLICY. November 1993, p. 1083. Many governments and companies are unlikely to react until the crisis is upon them. Indeed the balance of current global opinion over the next 20-30 year horizon, as indicated by the behaviour of the markets and investment intentions of the industry, would seem to

indicate that some radical adjustment is expected and that the most likely instrument of such change will be, sooner or later, a rise in the price of oil which might be both abrupt and steep.

34. Paul Tempest. ENERGY POLICY. November 1993, p. 1081. The consumer of oil and oil products is therefore likely to be faced by the turn of the century with a double pressure on price—the imbalance of supply and demand caused by much diminished spare production capacities and the needs of the industry to replenish rapidly its capital stock. Both imply heavy increases in price to the consumer. Under such conditions, most parts of the petroleum industry are likely to consider seriously the possibility of a third oil price discontinuity, again triggered by political tumult in the Middle East.

35. Robert Copaken. THE NEW GLOBAL OIL MARKET. 1995, p. 217. Oil is the world's largest cash commodity. In 1992 worldwide crude oil production totaled about 60.3 million barrels a day, which was worth about $370 billion on the free market. In the 20 years since the early 1970s, the oil market's volatility has pushed prices up by as much as 150 percent or pushed them down by 50 percent. Risk management has become a key consideration for the oil industry as well as for commercial and industrial end users. Back in 1978 the New York Mercantile Exchange (NYMEX) established an energy future trading market in heating oil. In 1981 a gasoline futures contract was added, followed by crude oil futures in 1983, propane futures in 1987, and natural gas futures in 1990. The growth in NYMEX's energy futures trading volume and open interest levels underscores the market's important role in the reduction of the significant price risk inherent in the oil industry today. NYMEX has seen the establishment of other similar commodity exchanges in other trading and financial centers around the world, such as the International Petroleum Exchange (IPE) in London, the Singapore International Monetary Exchange (SIMEX), and even an Oil Exchange in Russia. Thus, the international oil market has become globalized as an element of the world economy, with traders and speculators able to buy and sell at a moment's notice, around the clock as well as around the world.

36. Paul Stevens. ENERGY POLICY. May 1996, pp. 399–400. Today, far more oil contracts—spot or term—are being signed than ever before. Three factors explain this. There are a great many more crude oil producers than ever as a growing number of ventures come onstream. The decline of operational vertical integration mentioned earlier has also increased the number of arms length transactions. Finally, the growing role of futures and forward markets (Roeber, 1993) has also increased the number of transactions if only for paper rather than wet barrels.

There is little sign that the number of transactions will decrease in the future. Company exploration activity will continue to add to the list of new producers. The progression moving oil towards a commodity—in the sense of something which is traded using mechanisms common in other commodity markets—is likely to accelerate (Rogers and Treat, 1994). The only factor which could reduce the number of arm's length transactions is if the oil producing governments significantly expand their downstream activities and continue to operate a policy of operational vertical integration. At that point, if the reintegration was sufficiently large, markets could implode (Stevens, 1992).

37. Donald Hodel. 27 March 1995, Y.F7/2:S.HRG.104-21, p. 62.
Senator THOMPSON. Let me ask Mr. Hodel. You were talking about the problem with the domestic producers. I believe it was you who said it was a price-based problem. Does that essentially get down to a question of what policies we effect to get the prices up? Is it a matter of low prices, or is it a matter of price fluctuation that is the basic problem? And what do we do about it?

Mr. HODEL. Senator, I believe it is the latter. I believe it is the price fluctuation threat, which is a severe retardant on investment in this industry. When people have seen what happened in the 1980s with the price falling to $10 a barrel, they are very reluctant to come in and say, "Boy, it looks like a good deal to invest in an oil well" or a gas well, for that matter, if you look at the gas prices in the recent period where they have been down. So, it appears to be very risky to invest. If you talk to an investor about an oil well I am told—I do not do that sort of thing—that what you hear in response is, "Well, how do I know that next week or next month some OPEC member, like Saudi Arabia, will not open up the spigot and increase production by even a few hundred thousand barrels and drive the price down dramatically because this is a very production sensitive market? It takes a very thin change in production to affect prices sharply.

What I believe is that as long as we sit under that gun of potential price drops, we are vulnerable to not producing in the domestic level. Take Mr. Pickens' example here of natural gas vehicles. One of the big deterrents is that there is a substantial investment for the fleet operator to convert his vehicles to or to buy new vehicles that have natural gas capacity. If you can assure him that he is facing an extended period of stable world oil prices and he knows he is going to have, therefore, the cost advantages that Mr. Pickens identified, then he can afford to make the investment. But he does not know that next year or next week the oil price will not collapse and change his economics completely.

If I may respond also to the chairman's question earlier, because this leads right into it, about what the impact is on other energy sources. For all practical purposes, every energy source in the United States is pegged to oil because it is such a dominant fuel. I know this sounds crazy when

we think of coal and nuclear, but the reality is that there is a lot of fuel switching capability to generate electricity. People will use oil or natural gas to produce electricity if there is a cost advantage. If the oil price drops, they will produce with natural gas. That makes coal prices sensitive even there to oil prices. Nuclear, the same way. Every energy source is affected by it.

38. Paul Tempest. ENERGY POLICY. November 1993, p. 1080. The petroleum industry has played a very important role in these foreign trade and investment developments. Crude oil is still the largest single commodity in world trade. A relatively new international trade in nature gas has emerged in the last 10-15 years whether by pipeline or in the form of liquefied natural gas. The switch of the US industry from placing the bulk of their investment in the USA to investing elsewhere and the determined efforts of OPEC to invest in refined and marketing capacity downstream are good examples of the industry changing direction and stimulating foreign investment. The rapid development of trade in petrochemicals, in many oil and gas-based products, and in oil or gas generated electricity have also stimulated much broader international exchanges of goods and services. Indeed, petroleum can be justly regarded as remaining a key catalyst in the industrial economy of today.

39. David L. Weimer. THE NEW GLOBAL OIL MARKET. 1995, pp. 230–231. Indirect costs arise from a number of factors. The "comparative statics" for measuring the direct costs assumes that the economy can costlessly adjust from the equilibrium at $18/barrel to the one at $36/barrel. But in the real world the adjustment is not costless. One important reason is that not all prices are fully flexible. In the short run, for example, nominal wages are unlikely to be cut so that sectors that experience a reduction in demand for their products are likely to experience unemployment rather than wage reductions. Wealth transfers away from consumers create uncertainty about the new composition of aggregate demand that depresses investment. Oil price rises contribute to a higher rate of price inflation that, while speeding the adjustment to real (inflation adjusted) prices consistent wit the new equilibrium, may also distort investment. The sum of direct and indirect costs can be measured through macroeconomic models (Horwich and Weimer 1984). Macroeconomic models based on experience with the price shocks of the 1970s suggest that a doubling of the price of oil may involve indirect costs several times larger than the direct costs (Hickman, Huntington, and Sweeney 1987). These models also suggest that indirect costs grow more than proportionally with direct costs. That is, for small price rises the direct costs measure almost all of the total costs, but for large price rises they grossly underestimate total costs.

40. SOUTHERN ECONOMIC JOURNAL. January 1996, p. 121. Saunders 24; 25 addresses the different paradigms between short and long run effects of energy shocks using a theoretical model. He contends that in the short run people generally consider an effect on the economy to be dramatic, resulting in "inflation, recession, unemployment, inertia in the capital stock, and myopia of economic agents." While in the long run a relatively weak effect is observed where these same economic agents have the foresight to adjust the capital stock, and balanced growth paths exist and optimal growth theory holds.

Saunders reconciles these dichotomous views by extending the usual two factor optimal growth theory to include energy as a factor of production and notes the bridge between the two paradigms in investment and factor competition . He shows the transition from short to long term through a simulation model. In this study we have taken Saunders basic model and tested it with the use of real data. We do not distinguish between price changes and price shocks.

41. SOUTHERN ECONOMIC JOURNAL. January 1996, p. 121. On the aggregate supply side, Tatom finds that following an oil price increase the resulting effects go beyond a simple increase in the cost of output, as standard theory predicts. In addition, the shocks alter the incentives to employ energy resources and alter the optimal methods of production. These incentives are reversed during an oil price decrease leading to a symmetric analysis. The basic outcome when energy prices rise is a higher price level, lower output, lower real wages, and, over time, a reduced capital stock relative to labor.

42. George Georgiou. ENERGY POLICY. August 1993, p. 838. The consequences of oil market instability are numerous. For one thing, it creates serious international adjustment problems—disequilibria in capital and trade flows that cannot be managed without government intervention and that can disrupt both national and world economic growth. It can also cause shock effects, which may be either inflationary or deflationary. While rapidly rising oil prices contributed significantly to the cost-push inflation of the 1970s, falling oil prices had the opposite effect in the 1980s. This has caused some economists to worry about the possibility of a new deflationary era analogous to the 1930s.

Furthermore, oil market instability distorts investment, both in energy development and in industry in general. On the one hand, investments in oil production that are made during periods of skyrocketing prices have to be given up as unprofitable when prices fall. On the other hand, dramatic oil price drops lend to postpone those investment decisions needed to secure energy supplies in the future.

43. David L. Weimer. THE NEW GLOBAL OIL MARKET. 1995, p. 226. The major source of the vulnerability of the world oil market to price shocks is the concentration of production and reserves in the Persian Gulf, a politically unstable region of the world with high

risks of conflicts that can interfere with the supply of oil to the world market. In 1992 the Persian Gulf countries accounted for about 27 percent (16.1 mbd) of the total world production of crude oil. Saudi Arabia alone accounts for over one-half of this production. Events, such as the closing of the Persian Gulf to shipping or the loss of Saudi production because of invasion or insurrection, could remove sufficient quantities of crude oil from the world market to cause economically significant price shocks.

44. Paul Tempest. ENERGY POLICY. November 1993, p. 1981. In global terms, the latest International Energy Agency energy demand scenarios suggest that within 21 years we will only be able to meet those quite plausible demand requirements by asking the Gulf states—Saudi Arabia, Iraq, Iran, Kuwait and UAE—to produce two to three—perhaps four—times their current level of oil production. Is this remotely feasible?

In technical terms, the answer is certainly, yes. The Gulf has ample proven, cheaply produced reserves, a vast production infrastructure and simple technology.

In political terms, the answer is almost as certainly, no. Setting aside the millennia old struggle for ascendancy between what is now Iran, Iraq and Saudi Arabia, why should those three key states struggle in a slack market to find the huge volumes of international capital to expand production? They can see clearly that a tightening of the market is, in any case, bound to force prices up. Indeed, any expansion of Gulf production will only delay this upward adjustment of oil prices and production revenue, which is inevitable sooner or later. Equally any effective agreement to restrain OPEC production will accelerate the rise. Thus the long-term dilemma posed to the industry is that of steadily rising demand and a diminishing security of supply.

45. David L. Weimer. THE NEW GLOBAL OIL MARKET. 1995, p. 227. Oil price shocks during the last 20 years have had their origins in the Persian Gulf. The 1973–74 quadrupling of prices resulted from production reductions orchestrated by the Organization of Arab Petroleum Exporting Countries (OAPEC) in retaliation for U.S. support of Israel in the October War. The 1979–80 doubling of prices followed reductions in Iranian production during the revolution against the shah. More modest price rises caused by attacks on shipping during the war between Iran and Iraq and the recent Gulf War following the Iraqi invasion of Kuwait remind us of the region's continuing potential for conflict.

46. Pradeep Mitra. ADJUSTMENT IN OIL-IMPORTING DEVELOPING COUNTRIES. 1994, p. 2. Iraq's invasion of Kuwait in August 1990 resulted in a near doubling of the price of oil from around $16–17 to over $30 a barrel, a cessation of workers' remittances from Iraq and Kuwait for a number of developing countries, and a decline in tourism earnings for countries in the region affected by hostilities. Although oil prices returned to pre-crisis levels in the second quarter of 1991 following resolution of the crisis, the events of those critical months served as a clear reminder of the world's vulnerability to oil price shocks associated with increasing dependence on OPEC oil in general and, within OPEC, on a few countries—Saudi Arabia, Iran, the United Arab Emirates, Iraq, and Kuwait—in particular.

47. Howard Geller, John DeCicco, and Skip Laitner. ENERGY POLICY. June 1994, p. 472. When the USA pays for imported oil, it transfers wealth to oil producing countries which maintain monopolistic control over world oil prices. This transfer of wealth is based on the difference between the prevailing world oil price and the estimated price had there been a competitive market. A study by Greene and Leiby estimates that from 1972 to 1991, the USA transferred US$1.2 to US$1.9 *trillion* dollars to foreign oil producers as a result of monopolistic oil pricing. Although OPEC is much less unified and weaker today than in the past, it has averted collapse and still maintains prices well above what they would probably be in a competitive market.

High oil prices also hurt the ability of the US economy to produce goods and services. If the cost of a crucial factor of production like oil is relatively high, total economic output is reduced. Greene and Leiby estimate that periodically high oil prices during 1972–91 caused the USA to lose about US$1.2–2.1 trillion of GNP. Greene and Leiby point out that these losses are in addition to the wealth transfer noted above.

48. Howard Geller, John DeCicco, and Skip Laitner. ENERGY POLICY. June 1994, p. 472. It is also useful to compare the US oil import bill with its trade deficit with Japan, since the latter receives so much attention. During 1989–92, the total net US oil import bill was US$189 billion, while the US trade deficit with Japan was US$184 billion. Thus oil imports has contributed about as much to the overall US trade deficit as has the US trade imbalance with Japan in recent years.

A portion of US expenditures on oil imports are returned when nations such as Saudi Arabia, Venezuela, and Canada purchase US goods and services. But the USA has had a trade deficit with all its major foreign oil suppliers in recent years. Even when respending of money to purchase imported oil is taken into account, oil imports are a significant drain on the US economy.

49. William Reinsch. 27 March 1995, Y4.F7/2:S.HRG.104-21, pp. 21–22. Our analysis was not based on a strict formula. Rather we made a balanced assessment of all relevant factors and a careful judgment as to how each factor should be weighed. From that, it is clear that there are some reasons why U.S. energy security has improved since the last investigation and other reasons why it has eroded. We can also see that imports

play a role in improving overall U.S. economic security in some respects but have a deleterious impact in others. On balance, we concluded that petroleum imports threaten to impair national security because of the growing reliance of our economy on imported oil from a politically volatile region of the world. We consider economic security to be a significant part of overall U.S. national security.

50. Howard Geller, John DeCicco, and Skip Laitner. ENERGY POLICY. June 1994, p. 476. Most energy security costs are hidden within routine defence expenditures. Based on figures from four recent analyses of oil related defence costs, we estimate an average energy security cost of US$35 billion per year. This estimate is conservative because most of the cost estimates on which it is based are only for the defence of the Persian Gulf during peacetime conditions.

In addition to defence costs the USA spent over US$50 billion filling and maintaining the strategic petroleum reserve (SPR) during 1976-90. Approximately US$300 million per year is being spent to maintain and add oil to the SPR as of 1993. The Energy Policy and Conservation Act of 1975 created the SPR to reduce US vulnerability to supply disruptions. While the SPR could be used to relieve domestic supply disruptions, all past supply disruptions serious enough to warrant SPR use involved imported oil.

51. Howard Geller, John DeCicco, and Skip Laitner. ENERGY POLICY. June 1994, p. 476. Greene and Leiby estimate that total economic costs of monopolistic oil prices—including wealth loss, GNP loss, and other macroeconomic effects—to be US$4.1 trillion from 1972–91, or about US$205 billion per year (1990 dollars). The 20 year cost is equivalent to approximately one year of total US gross domestic production.

52. Howard Geller, John DeCicco, and Skip Laitner. ENERGY POLICY. June 1994, p. 472. Twenty years after the Arab oil embargo, the USA is more dependent on oil imports than it was in 1973. Net oil imports equalled 7.3 million barrels per day (MBD) in the first half of 1993. This level of imports is adversely affecting the US economy by increasing the trade deficit, reducing employment and GDP, and increasing defence costs. High oil use and oil imports also damage the environment through oil spills and air pollution. given current policies and trends, US oil imports are expected to rise by nearly 3% per year on the average during 1992–2010. Net oil imports are projected to cost over US$100 billion per year within 120 years (direct costs only), which would further harm the US economy and the environment.

53. Howard Geller, John DeCicco, and Skip Laitner. ENERGY POLICY. June 1994, p. 472. During 1974–92, net oil imports cost the USA about US$1.2 trillion (million million) (constant 1990 dollars). As of 1992 the net US oil import bill was nearly US $45 billion while its non-energy trade deficit was about US$40 billion. Thus, oil imports accounted for more than half the total merchandise trade deficit as of 1992.

54. Joseph Romm and Charles Curtis. ATLANTIC MONTHLY. April 1996, p. 60. The growing dependence on imported oil in general and Persian Gulf oil in particular has several potentially serious implications for the nation's economic and national security. First, the United States is expected to be importing nearly 60 percent of its oil by ten years from now with roughly a third of that oil coming from the Persian Gulf. Our trade deficit in oil is expected to double, to $100 billion per year, by that time— a large and continual drag on our economic health. To the extent that the Gulf's recapture of the dominant share of the global oil market will make price increases more likely, the U.S. economy is at risk. Although oil imports as a percent of gross domestic product have decreased significantly in the past decade, our economic vulnerability to rapid increases in the price of oil persists. Since the 1970 sharp increases in the price of oil have always been followed by economic recessions in the United States.

55. George Georgiou. ENERGY POLICY. August 1993, pp. 831-832. This paper will make the case that the current complacency regarding the world oil market and the weakened position of OPEC is premature as well as shortsighted. It then proceeds to argue that in the foreseeable future, a core group of OPEC countries, all located in the Persian Gulf, that hold two-thirds of the world's proved reserves of oil, may regain control of the world oil market and thereby develop into a more dominant cartel than OPEC was in the 1970s. Given the pattern of production and consumption of energy in the 1980s and those projected for the 1990s, oil will remain the most critical energy source in the year 2000 both for the USA and the world. Progress in terms of improved energy efficiency in the form of conservation, and fossil fuel substitution in the form of renewables is not expected to increase the energy security of the USA by 2000 given present circumstances and the absence of a comprehensive national energy policy. On the contrary, US energy security will be lessened and increasingly vulnerable to external energy shocks.

56. Howard Geller, John DeCicco, and Skip Laitner. ENERGY POLICY. June 1994, p. 474. A growing oil import bill is likely to have a number of negative impacts on the US economy. First, it would drive up the trade deficit and slow economic growth due to high interest rates and other factors. Second, it would mean a loss of millions of jobs and billions of dollars of income because of the dollars leaving the US economy. Third, it would increase the likelihood of another oil price shock as world oil markets tighten. Oil producing nations may find it difficult to keep up with rising demand if the IEA and similar

forecasts are correct. Fourth, it would increase US dependence on unstable regions such as the Middle East, resulting in high national security costs as well as greater risk of a supply interruption.

57. INDUSTRY WEEK. 6 May 1996, p. 6. Although U.S. imports of crude oil and petroleum products did drop slightly last year, they've generally been on a steady ascent. In 1995, API reports, they accounted for fully 50% of the country's oil consumption. That is almost twice as high as the 29% level in 1972—just before the Arab oil embargo plunged the world into economic crisis and prompted a now-abandoned drive by the U.S. government to achieve "energy independence."

"With imports up over 50%, [industry managers] definitely should be concerned," warns Wayne Allen, chairman and CEO of Phillips Petroleum Co., Bartlesville, Okla. "The nation is increasingly vulnerable to interruption." After all, he reminds, "We [the U.S.] were willing to commit troops in Iraq a few years ago to protect oil supply."

58. William Reinsch. 27 March 1995, Y4.F7/2:S.HRG.104–21, p. 21.
Vulnerability to a supply disruption.—The world's most productive oil producing region is also one of the most politically volatile. Given the growing reliance of the United States and allied countries on imported oil from the Persian Gulf, our study found that we are indeed vulnerable to a supply disruption that could cause hardships for our economy, particularly the transportation sector which cannot readily substitute alternative fuels.

This vulnerability in turn places constraints on our foreign policy flexibility. This problem is more severe for many of our allies who are even more dependent than we are on these insecure sources and who have fewer alternative sources to access in an emergency.

59. Joshua Gotbaum. 27 March 1995, Y4.F7/2:S.HRG. 104–21, pp. 24–25. While DOD's peacetime or warfighting requirements did not constitute the basis for a positive finding under Section 232, stable oil supplies and oil process nonetheless remain an important interest of our Nation and our allies. As this committee is only too well aware, the economic health of our Nation and its allies has on several occasions been severely affected by events in the Middle East and their effect on oil supplies and prices. And it is the need to defend against military threats to such national interests that gives rise to the second perspective from which DOD must address the issue of U.S. dependence on stable global oil markets.

The Department of Defense must be prepared to protect U.S. interests around the globe, wherever they may be threatened. This requires that we maintain the forces necessary to deter or defend against aggression. One of the key challenges that we face today is determining the appropriate strategy and force structure for the post-cold war era and to manage properly the drawdown of our forces without sacrificing the readiness to respond to threats in an increasingly complex world. And while that force structure is not predicated on meeting any single military threat, or protecting any single national interest, clearly, protecting against military threats to global oil supplies is an important factor for which we must be prepared.

60. James Schlesinger. OIL AND AMERICA'S SECURITY. 1988, p. 11. I believe the United States should be worried. Its role in the world is unique. Unlike Germany, Japan, or France, all of which, incidentally, worry a great deal about oil dependency, the United States is the great stabilizing power in the free world. Other nations can be oil dependent. If, however, the United States is to sustain its role in the world and to maintain the necessary freedom of action in foreign policy matters, it cannot afford to become excessively dependent on oil imports, particularly from the most volatile regions of the world.

61. Daniel Yergin. THE PRIZE. 1992, p. 14. Yet oil also has proved that it can be fool's gold. The Shah of Iran was granted his most fervent wish, oil wealth, and it destroyed him. Oil built up Mexico's economy, only to undermine it. The Soviet Union—the world's second largest exporter—squandered its enormous oil earnings in the 1970s and 1980s in a military buildup and a series of useless and, in some cases, disastrous international adventures. And the United States, once the world's largest producer and still its largest consumer, must import half of its oil supply, weakening its overall strategic position and adding greatly to an already burdensome trade deficit—a precarious position for a great power.

62. Robert Belgrave, Charles Ebinger, and Hideaki Okino. ENERGY SECURITY TO 2000. 1987, p. 283. Reliance on insecure oil supplies impinges on the military security of the United States and its allies in several other areas. Safeguarding major oil producing states against external and perhaps even internal subversion requires difficult strategic choices. The military must have the bases and operational mobility it needs to move rapidly to protect against the threat of sabotage of the energy logistics systems inside the major energy-exporting nations and on the high seas. The military must also position adequate fuel supplies for the defense of Europe, Japan and the United States, as well as to support allied interests in other conflict arenas. The West's ability to support sufficient military forces to guard oil supplies in an era of fiscal austerity is also a source of profound allied squabbling and necessitates a greater amount of burden sharing and a re-examination of whose strategic interests are most at stake in volatile areas such as the Persian Gulf. With Japan dependent for nearly 67% of its imports on oil from the Middle East and OPEC and Europe 74% dependent, it is

clear that the economic threat posed to the domestic energy industries of these regions by the collapsing prices of the 1980s, is as great as the rising prices of the last decade. Likewise, with Japan dependent on the Middle East (defined to include Arab exporters of North Africa) for 58% of its oil imports (2.6 bmd out of 4.5 mbd) and Europe for 59% (4.6 mbd out of 7.9 mbd), it can no longer be credibly argued that the US should shoulder the defense burden required to ensure access to the Persian Gulf and North African oil which is more vital to its allies than to itself.

63. James Schlesinger. OIL AND AMERICA'S SECURITY. 1988, p. 12. Let me close this historical review by citing two telling examples that suggest the influence of oil market conditions on national security policy. In April 1986 the president of the United States ordered an attack on Tripoli to punish Colonel Muammar Qadhafi for his encouragement of terrorism. It is not my purpose to examine the merits of that decision. Some may disapprove; others may be enthusiastic. My purpose here is simply to emphasize this crucial fact: the freedom that the United States enjoyed to undertake this measure was to a large degree a reflection of conditions in the oil market—slack pricing and substantial excess capacity. If instead the United States had faced a tight oil market and if it had been heavily dependent on imported oil, much of it from the Middle East, it is not clear to me that the United States would have felt sufficiently free to take that step. Certainly the policy leeway available to a president to do so would have been greatly circumscribed. And even more certainly that policy leeway will become more circumscribed in the future. Thus a tight oil market probably means the recrudescence of the oil weapon, at least in the eyes of the major players in the Middle East.

64. OIL & GAS JOURNAL. 17 June 1996, Lexis/Nexis. "The final piece in the geopolitical puzzle is that during the first oil crisis, countries that were competing with us for oil were our NATO allies. But during the next oil crisis a new, important complication will arise: Competition for oil will increasingly come from rapidly growing countries of Asia.

"Indeed, in the early 1970s, East Asia consumed well under half as much oil as the U.S., but by the time of the next crisis East Asian nations will probably be consuming more oil than we do. These nations already are establishing stronger diplomatic ties with the Persian Gulf."

He noted there have been six oil supply disruptions of 2 million b/d, an average of one every 5–10 years, all originating in the Middle East.

And in the U.S., since 1970 sharp increases in the price of oil have always been followed by economic recessions.

DOE's Oak Ridge National Laboratory in Tennessee estimates the cost to the U.S. economy during the past 25 years of overreliance on OPEC oil, including the cost of price shocks at $4 trillion.

65. Donald Hodel. 27 March 1995, Y.F7/2:S.HRG.104-21, p. 37.
THE SCOPE OF THE IMPORT PROBLEM
While estimates by authorities vary, world oil demand for 1994 was around 68 million barrels per day. Global oil production was slightly below that figure resulting in a moderate but relatively steady rise in prices. Projections are for an annual growth in demand of 1.5 to 3 percent. This would translate into an increase of roughly one to two million barrels per day (MMBD). About 75 percent of the growth will occur in developing countries to fuel their expanding economies, and the balance from the OECD nations.

The U.S. is expected to continue to consume around 25 percent of the world's oil, with imports comprising a progressively larger proportion of supplies over time. Underscoring this trend is the fact that in 1994, for the first time, more than half of U.S. petroleum supplies (50.4 percent) were obtained abroad, averaging almost 8.9 MMBD, while at the same time U.S. oil production fell to 6.62 MMBD, the lowest level in 40 years, with the balance of U.S. "oil" production being natural gas liquids. Further, U.S. demand continued to grow reaching 17.65 MMBD in 1994, a 2.4 percent increase over 1993. As a result, the potential consequences of an import disruption are even greater than they were in the past. In 1973 oil imports constituted 36. 1 percent of domestic supplies, and in 1979, 45.3 percent. Yet, even at these lower import levels both in 1973, and again in 1979, when the U.S. suffered its previous oil shocks, the effects were devastating.

According to the National Petroleum Council report "Factors Affecting the U.S. Oil and Gas Outlook," the two oil shocks of the 1970s reduced U.S. GNP by 3.5 percent, increased unemployment by 2 percent, increased interest rates by 2 to 3 percent, and added 3 percent to the general rate of inflation. Taken together, the combined impact of these effects on the U.S. economy in the decade following the 1973 Arab Oil boycott come to $1.5 Trillion!

66. Paul Stevens. ENERGY POLICY. May 1996, pp. 391–392. Between 1973 and 1987, oil prices were regarded as one of the key global economic variables. Their gyrations were crucial to consumers and producers and to the general health of the world economy. The oil price shocks of the 1970s were regarded by many as causing global recession and as triggering the debt crisis of the 1980s (Cline, 1981; Corden, 1977; Fried, 1975; Heal and Chichilnisky, 1991; Kohl, 1991; Matthiessen, 1982 Mork; 1994; Pereira et al., 1987; Ryczynski, 1976). Equally the shocks transformed many of the oil producers although for better or for worse is debatable (Auty, 1988; Mabro, 1988; Neary and Van Wijnbergen, 1986; Philip, 1994; Rowthorn and Wells, 1987; Stevens, 1986).

67. John Lichtblau. FDCH TESTIMONY. 8 May 1996, Lexis/Nexis. The Congress recognized that in the wake of the Arab oil embargo of 1973-4, the nation had suffered significantly from the higher prices and product shortages. The National Petroleum Council's 1987 report *Factors Affecting US Oil and Gas Outlook* estimated that the economy in 1973-75, after the Arab oil embargo, and in 1980-82, after the oil market disruptions that followed the Iranian revolution, suffered the worst recessions of the post-World War II period to that time. According to economic models designed to isolate the impact of the oil price shocks from other factors, GNP shrunk by about 2.5% due to the earlier upheaval and by about 3.5% due to the later one. Furthermore, during the 1979 oil disruption, private companies engaged in rational and prudent inventory behavior—husband your stocks for a rainy day.

68. Ken Matthews. THE GULF CONFLICT AND INTERNATIONAL RELATIONS. 1993, p. 193. It is not an exaggeration to say that oil is the very heart of the global economy and the foundation of the economies of the 'western' developed world. The two factors that tie the western powers inexorably to the Middle East are its geostrategic position and its vast reserves of oil. The geostrategic dimension of the west's interest in the Middle East pre-dates the oil factor. Britain's concern for the region was originally derivative of its Indian empire. During the twentieth century that concern has been inherited by the United States and reinforced by the discovery of oil and its crucial role in the development of the western industrialised economies, by the necessity to protect trade routes between Europe and the Far East, and more recently by the strategic significance of the region in the Cold War competition.

The role of oil has assumed an increasing international political dimension in the period after the Second World War due to the combination of two factors–firstly the importance of oil in the post-war success of the western economies, and secondly the fact that control of oil production has passed from the large western oil companies (known as the 'seven sisters') to the governments of the producer countries themselves. In the period up to 1970 the major oil companies were acting essentially within the strategic regimes operated firstly by the United Kingdom and then by the United Sates which ensured market stability, security of supplies, low prices (the real price of oil has declined in the post-war period) and high profits.

69. THE PLAIN DEALER. 13 October 1993, Lexis/Nexis. Now consider this. America is the world's sole remaining superpower. We won the war in the Middle East. There are even new signs of peace in the Middle East. But America isn't winning any popularity contests in the area.

Our dependence on foreign oil has reached the point where oil-producing nations of the Middle East don't have to stop oil exports to the United States. All they have to do is ratchet down in the name of energy conservation or merely to ensure the region's long-term financial stability.

If that happens, economic shock waves will ripple through the U.S. economy, and we can look back to the Arab oil embargo of 1973 for some historical perspective.

70. OIL & GAS JOURNAL. 17 June 1996, Lexis/Nexis. Glenn Schleede of Energy Market & Policy Analysis Inc., Reston Va., said world energy markets have changed dramatically and favorably since current government energy policies and spending programs were conceived.

"These changes in energy markets need to be taken into account as you consider DOE's proposals to spend another $2-2.5 billion on energy supply conservation technologies."

Schleede said EIA has substantially lowered its price forecasts. That should require the government to take a fresh look at the rationale for DOE's energy technology development programs and its claimed energy savings from conservation and renewable energy programs.

"We should not overreact to recent DOE officials' warnings about a looming energy crisis. There are many reasons to believe that another energy crisis is less likely today than previously."

He pointed out there have been many changes in energy markets since the oil price shocks of 1973-74 and 1979-81. "These points often seem to be ignored by those who want to maintain a 1970s perception of an energy crisis."

Real energy prices have declined steadily since the early 1980s. In constant dollars, Schleede said, the crude oil prices were down by 71% in 1995 from the high reached in 1981 and refinery gasoline prices, including taxes, were down 45%.

What's more, U.S. energy efficiency has improved even though energy prices have continued to decline in real terms since the early 1980s. During 1973-95, U.S. energy consumption increased 17.5% while gross domestic product increased 72.8%.

Schleede said despite the rise in U.S. oil imports, they constitute a declining share of total U.S. merchandise imports, declining from a high of 32.1% in 1980 to only 7.3% in 1995.

He also said much of the money for oil imports returns to the U.S. as payments for exported merchandise and services. The dollar outflow for oil, in constant dollars, has declined from a high of $138 billion (1994) in 1980 to $53 billion in 1994.

Schleede argued most alternatives to market pricing of oil would be more costly to the economy than continued reliance on imports. And he noted proved world oil reserves have grown significantly from 664 billion bbl in 1973 to 1 trillion bbl in 1996.

Most oil producing and exporting countries depend on oil revenues to meet domestic economic needs, and that is

a strong incentive to restart any interrupted oil production and exports. "Not all oil from OPEC is insecure."

And Schleede said projected growth in developing nations is less than certain, and they may not be able to attract required capital

71.	Gail Makinen. 31 July 1992, Y1:1/3:102–13, p. 102. The most obvious effect from an oil price shock is a run-up in the world price of oil. Would energy independence isolate the U.S. economy from these price shocks?

To answer this question, let us suppose that all of the events in the Middle East from early August 1990 through the end of January 1991 occurred exactly as they did and that, as a result, world oil prices rose exactly as they did. The only change to history that will be made is to suppose the United States to be energy independent in the sense that it imports no oil from any foreign country. Under these altered circumstances, would the U.S. economy have been immune from the rise in world oil prices?

The answer is quite simply no. To see why, consider what would happen when world oil prices rose. With world oil prices rising above the U.S. price, U.S. producers would have an incentive to sell their oil in foreign markets. As U.S. supplies were diverted from domestic to foreign markets, a situation would be created in the United States in which demand exceeded supply. This "excess demand" would then cause the U.S. price of oil to rise to restore balance or equilibrium between U.S. demand and supply. The end result of the oil supply shock whose origin was in the Middle East would be a higher domestic price of oil just as it would be if the U.S. had been an oil importer. The major difference would be that in the supposed case where the United States was self-sufficient in oil, it would no be an oil exporter.

72.	John Lichtblau. FDCH TESTIMONY. 8 May 1996, Lexis/Nexis. Designed to reduce the impact of an oil supply disruption, the SPR helps the economy during an emergency. Like an insurance policy, the benefits appear only potential, even though the very existence of a strong SPR shapes policy options available to the US, its allies and potential aggressors. When the SPR is needed, however, the benefits are incontrovertibly real and irreplaceable. Only the Strategic Petroleum Reserve can provide additional physical volumes during a disruption, and hence only the SPR can calm markets, provide time to explore solutions to international crises, and cushion the impact of hard decisions affecting the supply of oil, such as those made during the Persian Gulf conflict. Selling SPR volumes, while gaining a cash infusion, will impose a long-term cost.

73.	Paul Stevens. ENERGY POLICY. May 1996, p. 296. Alternatively, there could be a sharp increase in oil demand. If oil prices were to collapse, even if the fall was not passed on to the consumer, it could significantly affect demand. Oil demand grew dramatically in the 1960s not because the oil price was low but because it was perceived that it would remain low and probably go even lower. It was these expectations which led consumers to change the appliance stock in the static sector to oil use. The current views regarding "low" oil process could well trigger return of oil to the static sector.

74.	Akarca and Andrianacos. JOURNAL OF ECONOMICS. Fall 1995, p. 112. The statistical evidence presented in this paper indicates that it is likely the oil market structure changed after January 1986. Whereas the oil price prior to February 1986 shows no tendency to return to a particular level, it now fluctuates around a constant mean. That is, shocks to the oil market no longer have lasting effects on oil price. Also, the variance of the shocks has increased dramatically since January 1986. This brings with it the need to test the stability of models involving oil prices, estimated using data covering both sides of the structural break and the need for using ARCH models to capture the increased volatility of the series after 1986.

75.	David L. Weimer. THE NEW GLOBAL OIL MARKET. 1995, pp. 231–232. Policies to promote conservation could be evaluated in a similar way. They provide energy security benefits by reducing the preshock and postshock levels of imports. Ironically, however, they may contribute to less price elasticity of demand in the U.S. and world markets by eliminating routine energy uses that could otherwise be easily reduced during price shocks. The less elastic demand would mean larger price shocks for any given reduction in supply to the world market.

76.	OIL & GAS JOURNAL. 17 June 1996, Lexis/Nexis. Edward Murphy, American Petro-leum Institute's statistics director, said, "Oil imports were a threat in the early 1970s due to the highly structured nature of the international petroleum markets with long term fixed contracts that made it difficult to rechannel oil supplies.

"The situation has changed dramatically now as the world oil market is much more of a commodity market than it was in the early 1970s.

"And the threat of a producing country embargoing the U.S. is much less than it was then. Maybe it's no longer even worth discussing."

77.	James E. Akins. "The New Arabia," FOREIGN AFFAIRS. Summer 1991, p. 46. The reaction might not come until the next century. For as long as the gulf states remain secure the United States can reap great rewards. Saudi Arabia and the United Arab Emirates together hold almost 40 percent of the world's oil reserves, and they dominate OPEC. Both owe a debt to the United States for having saved them from the fate of Kuwait. Furthermore, the United States now occupies both countries and it will be militarily indispensable to both for the

foreseeable future. It seems unlikely that either could take action to push up the price of oil or to depress it without American concurrence. Iraq and Kuwait hold some 20 percent of the world's reserves, but neither exports oil today and neither will be a major exporter in the next few years. Their future production and exports will be contingent on American approval.

78. THE WALL STREET JOURNAL. 17 August 1995, p. A5. Perhaps the most dramatic change in international oil has been the emergence of a sophisticated crude futures market. While speculators who trade oil on commodities exchanges may contribute to short-term price volatility, experts believe the futures market's propensity for spreading risk helps reduce the possibility of a sustained surge in oil prices.

After Iraq invaded Kuwait in August 1990, nervous traders did push prices on the New York Mercantile Exchange to nearly $40 a barrel. But analysts credit those same traders with providing a speedy correction when the U.S. began bombing Iraq in January 1991; in just one day, prices fell to $21.44 from $32 a barrel.

"The market instantly adjusted its price expectation downward in a way that it couldn't do 10 years earlier," says Mr. Krapels. He notes that after the 1979 energy crisis, which was fueled by fears of shortage in the wake of the Iranian revolution, it took five years for prices to slide back down to their prerevolution levels.

79. John Lichtblau. 8 May 1996, Y4.F7/2: S.HRG. 104-21, pp.48-49. *The Role of The Middle East.*—Since the Middle East, based on its historical record, is perceived as the world's least stable oil exporting region, the question of U.S. dependence on this region has played an important role in the "oil imports and national security" debate. In 1994 19%, or 1.7 million B/D, of our gross imports came from the Persian Gulf. This was a reduction from 21% in 1993 and 23% in 1992. Those concerned with the insecurity of this supply region may view this decline as progress. Actually it is irrelevant as a national security consideration. There is a single world oil export market with a single oil price (after allowing for quality differentials). Any disruption anywhere large enough to affect world oil supplies causes the oil price to rise globally, including. domestic prices.

Given the fact that the Middle East contains two-thirds of the world's proven oil reserves and that its reserve/production ratio is nearly 100 years, compared to 21 years for the rest of the world, the Middle East's share of inter-regional oil exports (currently 47%) can be expected to grow for the foreseeable future. Will the Middle East countries provide these growing quantities of exports? There can be no doubt that they will want to. All of their economies are in a relatively poor state, they all need additional income to survive economically and politically and oil exports are practically their only source of hard currency earnings. Thus, there is true economic interdependence between the world's principal exporters and the importers of oil.

80. Khadduri. SECURITY DIALOGUE. June 1996, pp. 164-165. In fact, there is very little at the present time that producing countries can do to have stable and firmer oil prices. They have given up their role of setting prices. They have also gone ahead raising production capacity at will, and whenever possible, irrespective of the price. Moreover, some major producing countries, like Venezuela, have set their own agenda, raising actual production irrespective of the OPEC quota or the global supply-and-demand balance and the impact such a policy would eventually have on prices. They have been able to get away with it since the other producers, particularly those in the Gulf, do not want to make a big public issue out of it, as this could impact the markets negatively.

Today, OPEC countries are showing a sense of frustration because they have lost the initiative in the global oil industry, and have no influence over the production policies of the non-OPEC producers, or the international oil firms, or even other OPEC members themselves. There is very little that OPEC can do at the present time. The prime concern with quotas and production ceilings has only a passing impact on the oil market. The non-OPEC states have made it clear that they will not cooperate with OPEC and restrain production, while OPEC cannot do much to influence them. The old way of flooding the market and crashing prices, as in 1986, in order to drive away the marginal producers no longer works. The cost of production is low enough now to accommodate such temporary events and the tax system in many non-OPEC countries is drafted in such a way that the companies can still make money at low prices.

81. OIL & GAS JOURNAL. 17 June 1996, Lexis/Nexis. Philip Verleger, a Washington analyst, recently told a congressional hearing a true market for oil and energy has developed during the past 20 years.

He said, "Oil exporting countries lost control of the market, oil prices and supplies were deregulated in consuming countries, and markets for other forms of energy were deregulated.

"As a result of this, it is hard to imagine that one group of producers or one producing country will ever again be able to exert prolonged control over the price of oil.

"While dependence on oil from the Middle East is likely to increase, I believe it is extraordinarily unlikely that Middle East producers or producers from any region will ever again be able to exert control over oil prices as they did in the past."

He explained that is because the greater openness of the world economy will make users more responsive to changes in prices.

"Consumers will switch to other fuels if oil prices rise increase conservation, or change the characteristics of

82. OIL & GAS JOURNAL. 17 June 1996, Lexis/Nexis. Imports have continued to climb steadily in recent years to a level of more than 50% of the national consumption of 17.5 million b/d. That trend worries many in industry and government who are concerned about its effect on national security.

The Clinton administration has formally declared that rising imports threaten national security but has taken only modest steps to prop up domestic production. It also has sold millions of barrels of crude oil from the Strategic Petroleum Reserve, put in place as a bulwark against oil supply disruptions.

Environmental groups say rising imports are inevitable and thus are a strong argument for increasing emphasis on energy efficiency and renewable energy programs.

Many producers think the trend underscores the need to encourage domestic oil production in various ways.

By contrast, some oil analysts think the trend means nothing at all. They argue that changes in the world oil market since the 1970s—development of a futures market, diversity of supply, and increased competition, for example—have made the level of. import dependence an irrelevant matter.

83. THE HOUSTON CHRONICLE. 23 June 1996, p. 1. LONDON—New supplies of oil and natural gas continue to be found faster than they are being used up, even as world consumption reached a record last year, British Petroleum Co. said its annual review of global energy.

Economic growth sparked a 1.8 percent rise in energy consumption last year, the biggest jump since 1989, even as Russia continued a six-year collapse in consumption, the U.K. oil company said.

84. THE HOUSTON CHRONICLE. 23 June 1996, p. 1. Proven oil reserves, which total an estimated 1.017 trillion barrels, remain at or near a 43-year supply at current levels of consumption, which reached a record of 67.93 million barrels a day last year.

"This does not mean oil will run out within 43 years," said Peter Davies, BP's chief economist. "It is certain that more oil reserves will be proven before that time."

Even as demand has risen, proven oil reserves have increased by 50 percent during the past two decades, or at a rate of 1.77 barrels for every one consumed, Davies said.

85. J. Cale Case. THE NEW GLOBAL OIL MARKET, 1995, p. 48. Fortunately for the people of this world, the U.S. experience is an exception, and the earth still has a fabulous oil potential. Newly opened areas in the former Soviet Union and Eastern Europe, Latin America, China, and elsewhere all offer a promising future. Of particular significance, Russia has the potential to rival Middle Eastern production, although currently its reserves are more on par with the United States.

86. Frank Millerd. THE NEW GLOBAL OIL MARKET, 1995, p. 12. There is little doubt that further oil will be discovered, both in existing and new fields. Barnes provides a global supply profile based on current views on the total amount of oil available. The most optimistic case is an ultimately recoverable resource base of 435 billion tonnes allowing production of almost 4.5 million tonnes a year until 2050. This estimate includes proven reserves, current reserve extensions, and undiscovered reserves. Most of this oil is assumed to be available at costs below US$30 per barrel. Unconventional oil is not included.

87. Khadduri. SECURITY DIALOGUE, June 1996, p. 160. (1) The oil price collapse of 1986 resulted in fundamental changes in the outlook and operations of the industry. Oil firms have restructured their operations by cutting costs, employment, and non-essential activities. There has also been greater dependence on advancing technology to recover hitherto untapped oil from already known and producing reserves. Technology, has of course, always been the mainstay of the oil industry. However, what has happened during the past few years is that technology has moved rapidly and become available to the industry at a much lower cost than before. These technological developments have included 3-D seismic, horizontal drilling, enhanced oil recovery, and data processing.

These technological breakthroughs have led to upgrading recoverable reserves estimates in existing mature fields, as well as allowing marginal or non-commercial structures to be brought into commercial production, and resulting in higher success rates in both well established oil provinces and virgin territory. To take but one example, although offshore exploration activity in North-West Europe (excluding the UK) fell to its lowest level in 1994 since 1968, there was a sharp rise in success. In Norway, the success rate increased from 15% in 1993 to nearly 40% in 1994, while the Netherlands achieved an extraordinary rate of over 75% as opposed to 20% in 1993. Norway's oil industry has announce plans to cut overall costs by as much as 40–50% by 1998. This would, if achieve, result in an estimated saving equivalent to USD 2.6 billion and permit the exploitation of accumulations much smaller than the giant fields which account for most of the country's current production. Such industrial developments have resulted in the availability of more oil reserves, at lower costs and risks, as well as more balanced supply and demand in the world oil market.

88. Paul Stevens. ENERGY POLICY. May 1996, p. 393. Figure 3 is drawn to indicate the existence of excess capacity to produce crude oil. This is a situation which normally characterizes the international industry.

Since 1945, only on three occasions has the industry had no available excess capacity—during the Arab oil embargo of 1973; during the Iranian oil workers strike and its aftermath including the Iraqi invasion of Iran; and finally during the Iraqi invasion of Kuwait in 1990. The reasons for the existence of excess capacity are outlined in Stevens, 1984. However, the main explanation lies in the incentive to invest in capacity while the arm's length price of crude exceeds the replacement cost of developing proven reserves.

89. INDUSTRY WEEK. 6 May 1996, p. 6. Those days of shortages and crises are long gone, thanks to rigorous industrial energy conservation during the last 20 years, overproduction by oil-producing countries, and less-than-robust world economic growth. Today the supply and price of fuel ranks far down on industry's worry list.

90. DALLAS MORNING NEWS. 7 July 1996, p. 1. In 1973, American consumed 17.3 million barrels of oil a day, importing 6.2 million or 35 percent. One out of every 10 imported barrels came from Saudi Arabia. By 1980, consumption and import patterns hadn't changed.

Last year, Americans used 17.7 million barrels a day. Imports rose to 8.8 million—50 percent of consumption. Saudi Arabia accounted for 15 percent of U.S. imports and 86 percent of all. imports from the Persian Gulf.

The small rise in total consumption over 22 years shows that big strides have been made in energy conservation. Since Mr. Nixon's speech, markets have been in glut more often than in shortage. Oil companies and others use the same statistics I cite to argue that Mr. Nixon and Mr. Carter were wrong: There always has been and always will be cheap oil available for Americans, so relax.

91. Paul Ehrlich and Anne Ehrlich. HEALING THE PLANET. 1996, pp. 416–417. What are the constraints to increasing energy use? One common misapprehension can be cleared up immediately. There are no serious limitations on fossil-fuel supplies now or in the immediate future. At the 1990 rate of consumption of commercial energy, each of several fossil sources—petroleum liquids, conventional natural gas, and heavy oils—by itself could power civilization for more than 40 years (albeit with increasing costs and environmental risks). Coal alone could keep us going for over 400 years and oil shale for more than 2,400 years. Supplies of nuclear fuels are also large. Uranium in standard nuclear reactors could maintain society for perhaps 250 years, and uranium in breeder reactors for at least 250,000 years.

92. Frank Millerd. THE NEW GLOBAL OIL MARKET. 1995, p. 5. It is highly likely that Middle East reserves are much larger than reported but, with already-published plentiful supplies for most Middle Eastern countries and OPEC's wish to control prices, there is little incentive to reveal additional reserves or search for oil. Revealing additional discoveries would make OPEC's attempts to control prices more difficult (Adelman, 1986).

93. Frank Millerd. THE NEW GLOBAL OIL MARKET. 1995, p. 11. Oil price increases since 1973 have often precipitated concerns about oil scarcity. The data on reserve, however, indicate that more oil has been added to reserves in that period than has been produced from reserves. Oil is not getting scarcer, if anything it is becoming more plentiful.

94. Linda Tierney. 27 March 1995, Y4:F7/2:S.HRG.104-21, p. 32. I think that the level of reserves is very much a function of price of oil in world markets. An interesting statistic is that the level of proven reserves worldwide today is double what it was 20 years ago before price started rising. So, 20 years ago people would have thought that we would have been running out of oil by now except that we experienced some price increases and decreases during that period in which people did a lot of exploration to try to find reserves.

So, I do think that our level of proven reserves is by definition a function of price, and I would imagine that even though geology will limit it at some point, we are likely to find more if there were a different price offered.

95. Paul Tempest. ENERGY POLICY. November 1993, pp. 1081–1082. One of the most remarkable developments of the last few years has been the rapid acceleration of technology in the petroleum sector stimulated by strong competitive forces following the oil price collapse of 1986. Throughout the oil and gas business, the emphasis has been on cutting cost and on imposing rigorous financial control.

Take for example, the North Sea projects. In 1986, as the oil price fell below US$10 per barrel, the cost of inputs for drilling tumbled to very low levels and contracts were rapidly renegotiated. It has been estimated that, at this stage the savings to the companies still in business were of the order of 30%. At this point, under cost pressure, there were major changes in platform design reflecting automation (particularly in the gas fields of the Southern Basin) and the need in deeper water to move from fixed steel and concrete structures to tension leg and floating structures. The overall impact of this new technology in this second phase of cost cutting might be put at a further 20–30%. Finally, the development of subsea completions, directional and lateral drilling and further automation has prompted much field and regional rationalization: in many projects it is no longer necessary to assume the costs of a complete new production platform and pipeline system to the shore terminal: hook-ins to existing platforms, pipelines and other infrastructures probably represent another phase of saving, perhaps 30% of the original project cost. The cumulative effect of these savings has permitted the industry to push into deeper water, to exploit much smaller accumulations and more difficult geological

structures, despite the disincentives of a steadily declining real price of oil in the market and little prospect of its improvement in the short-term.

96.   Linda Tierney. 27 March 1995, Y4:F7/2:S.HRG.104-21, pp. 7–8. The Department is also retargeting our weapons laboratories to work closely with the nation's oil and gas industry in the Advanced Computational Technology Initiative (ACTRI). As part of their national security mission, DOE's national laboratories have developed extraordinary world-class computational and modeling skills. These cutting-edge computer technologies—which can aid in the search for new resources as well as optimize extraction from existing reservoirs—have not been readily available to the oil and gas industry in the past.

97.   Denise Bode. 27 March 1995, Y4.F7/2:S.Hrg, 104-21, p. 56. Before the recent price collapse, independent producers were pursuing America's substantial remaining oil and natural gas resource base, scoring impressive successes using new geological concepts, innovative drilling techniques and computerized seismic technology. In the process, these entrepreneurs are creating jobs and new economic wealth—and discovering new oil and natural gas resources—in some 40 major production areas in 20 states.

Public Lands Access.—America's vast public lands hold great potential for development of oil and natural gas resources. Independent oil and gas producers are active on federal lands. IPAA's 1994 member survey found that production on public lands accounts for nearly 12% of independents' total U.S. production and, among smaller companies, an average of 20% of their total production comes from federal lands. Reasonable access to these lands for environmentally responsible drilling and production is indispensable to maintaining a viable domestic oil and natural gas industry. Current laws governing leasing of public lands, both onshore and offshore, should be reviewed and reformed.

98.   Linda Tierney. 27 March 1995, Y4:F7/2:S.HRG.104-21, p. 8. Just last week the Administration announced its support for royalty relief to encourage the production of domestic natural gas and oil resources in deep water in the Gulf of Mexico. This step will help to unlock the estimated 15 billion barrels of oil equivalent in the deep water Gulf of Mexico, providing new energy supplies for the future, and spurring the development of new technologies and the creation of thousands of jobs in the gas and oil industry and its support industries.

These are the programs that simultaneously meet our energy needs, reduce oil imports, protect the environment, and improve our economy. That is the mission of the Department of Energy, and it is the country's energy strategy.

99.   Gail Makinen. 31 July 1992, Y1:1/3:102–13, p. 103. Disruptions to the world supply of oil and/or oil price jumps can have very serious social and economic consequences. The disruptions of 1973–74, 1979–80, and 1990-91 have produced noticeable effects on U.S. economic growth, unemployment, and inflation. The desire to isolate the United States from these disruptions is understandable as is the goal of reducing pernicious foreign influences on U.S. policies both at home and abroad. Thus, it is understandable why individuals would argue for a program stressing energy independence. However, depending on how energy independence is achieved, it may not have the intended effects. This is because many energy markets are worldwide in scope. So long as domestic suppliers of energy can participate in these markets, disruptions to the world supplies of energy will be felt even in a self-sufficient United States as domestic suppliers of the affected energy source divert their supplies to foreign markets and as suppliers of substitute energy sources do the same. Ideally, to isolate the United States from energy price shocks, U.S. energy needs should not be met from sources of supply that are poor substitutes for the supply source subject to disruption or, while potentially good substitutes, cannot be exported.

100.   Gail Makinen. 31 July 1992, Y1:1/3:102–13, p. 103. The preceding discussion demonstrates that so long as energy is sold in a *world market*, self-sufficiency in oil or not using any oil at all, would not render the U.S. economy immune from a foreign oil price shock. If the world oil price rises, the U.S. domestic price of oil will also rise. Should the United States not use oil at all, a rise in foreign markets of the prices of oil substitutes, could also increase similar prices in the United States depending on the ability of American suppliers to shift some of their output to the now higher-priced foreign markets.

101.   Gail Makinen. 31 July 1992, Y1:1/3:102–13, p. 102. The only way to isolate the U.S. economy under the circumstances posed above from supply disruptions abroad would be to forbid the exportation of oil (its diversion from domestic to foreign markets) and to prohibit domestic oil companies from raising prices.

Even if the United States were to implement such a drastic policy, a part of the oil price shock could still affect the. economy. This would occur as foreign nations adversely affected by the shock shifted to substitute energy sources, particularly coal, natural gas, wood, nuclear, and hydro. The demand for these sources of energy would rise, raising their prices, and, where feasible, U.S. suppliers would divert their output to the world market, possibly creating a situation in which domestic demand exceeded domestic supply. The resulting "excess demand" in the United States would then force up the domestic price of these substitutes for oil. Thus, ultimately, to some degree the original oil price shock would reach the United States. Thus, coal, possibly natural gas, and wood are more highly

exportable than is electricity generated by hydro and nuclear plants. the only way to prevent these price increases from occurring would be to virtually control the entire energy industry in the United States.

102. Gail Makinen. 31 July 1992, Y1:1/3:102–13, p. 103. The events that led up to the recent Middle East War have once again dramatically revealed the vulnerability of the United States to oil price shocks. Again, as in the aftermath of the 1973–74 and 1979–80 price shocks, some individuals are voicing concern that the United States has no energy policy and, particularly, lacks a strategy for achieving energy independence. They claim that energy self-sufficiency would, among other things, insulate the U.S. economy from the adverse economic effects of oil price shocks.

So long as U.S. suppliers of energy participate freely in the world energy market, this general view on the efficacy of energy independence is false. This can be seen by considering the hypothetical situation of what would happen when world oil prices rise due to a supply shock it the U.S. imported no oil. Any rise in world oil prices would bring about a simultaneous increase in American prices since American producers would be induced to divert a portion of their output to the now-more profitable foreign markets. This diversion of U.S. output would create a situation in the United States in which demand exceeded supply. This "excess demand" then would cause prices to rise to restore equilibrium in the oil market. Thus, even in the case where the United States imported no oil, a rise in world oil prices would produce a similar effect on the U.S. price. The price effect, in turn, would have macroeconomic effects that would reduce aggregate output most dramatically in the short run and possibly also over the long run. The only way to prevent this sequence of events from occurring would be to completely isolate the U.S. market from foreign markets.

103. George Georgiou. ENERGY POLICY. August 1993, p. 836. Although there is a consensus that a competitive US oil industry is vital to energy security, it is also agreed that current low oil prices have reduced US oil exploration and production, thus increasing reliance on insecure supplies particularly from the historically unstable Persian Gulf. It is important to understand that it is the global reliance on the Persian Gulf that matters, not just the level of US imports. US oil supplies will be affected during a disruption, even if the USA does not import oil from the disrupted region.

104. Gail Makinen. 31 July 1992, Y1:1/3:102–13, p. 103. Those who argue for energy independence often favor a stronger role for the Federal Government in this endeavor. Some want taxes placed on imported oil to encourage conservation; some advocate subsidies to promote the development of/or make commercially viable such alternative energy supplies as alcohol-based fuels (e.g., methanol and gasohol); some urge greater reliance on nuclear power, and some want legislation mandating such things as higher gas mileage on automobiles.

Without making a judgment on any particular public policy option, the analysis in this report shows that unless the United States were prepared to completely isolate its domestic energy supplies from the world economy, it would still fuel the effects of energy shocks elsewhere in the world even if energy independence were achieved. This would also be true even if the United States were an exporter of energy.

105. Douglas Bohi and Joel Darmstadter. "Resources for the Future," ECONOMIC TIMES. November/December 1994, p. 7. The most visible defensive measure continues to be the U.S. Strategic Petroleum Reserve (SPR), now nearly 600 million barrels. Regarding the economic consequences of major oil price shocks, it is worth underscoring two points: First, empirically, the economic damage through lost national output and inflation accompanying the second oil-price shock in 1979–80 was uncorrelated with the degree of oil-import dependence. And second, if what matters is the price of oil—the domestic price of which is determined by the world price—then reducing imports would not alone improve energy security. That recognition shifts the burden of oil-import policy to stabilizing the world price of oil during crisis situations, and, to this end, the SPR can be said to offer a sort of backstop strategy, although it begs the question of what stockpile magnitude is justified on cost-benefit terms.

106. Douglas Bohi and Joel Darmstadter. "Resources for the Future," ECONOMIC TIMES. November/December 1994, p. 7. The past twenty years have also made clear the futility of our trying to insulate the United States from the instability of the world oil market. Back in 1974, President Nixon's "Project Independence" envisaged complete self-sufficiency as a viable American objective. We have since learned that the domestic economy cannot be shielded from the world oil market, regardless of how much oil we import. Domestic oil prices are determined by world oil prices.

107. John Lichtblau. FDCH TESTIMONY. 8 May 1996, Lexis/Nexis. Some analysts have theorized that the SPR could be replaced by the diligent use of the now-mature futures market. The futures markets, vital as they are to the current oil market, would be no substitute for the SPR, because physical volumes are the only salve to an overheated market during a supply crisis. Futures transactions provide no incremental supply to the system. Thus, while futures transactions can effectively reduce a corporation's or consumer's exposure to price fluctuations (and provide ancillary benefits to the market, such as price discovery/transparency), they do not dampen the price move, i.e., change the market-clearing price.

The futures market is by nature balanced—the "long" (buy) transactions match the "short" (sell) transactions. During a sharp price change, therefore, there are winners and losers, but the system by definition comes to a zero-sum game. This contrasts with the physical market. During normal times, markets are balanced, of course. A supply disruption, however, is like a game of musical chairs—the supply loss takes a consumer's chair away. The i-market will rebalance at a higher price because a consumer drops out of the game. In a supply crisis, some companies might try to take physical delivery of volumes purchased in futures transactions, yet those volumes would effectively not be available. Only the SPR would actually put a chair back on the floor in the form of physical supply.

108. John Lichtblau. FDCH TESTIMONY. 8 May 1996, Lexis/Nexis. Establishing and maintaining the Strategic Petroleum Reserve is the perfect adjunct to our appropriate free market policy. We must all recognize that there may be times when unforeseen events will cause market disruptions of such severity that they will cause painful upheavals in the economy. These rare and temporary occurrences are exactly the target of our SPR insurance policy. Any alternative government policy carries a high multiple of the SPR's cost, and the diminution of the SPR will impose a high multiple cost on the economy in the event of a real, non-election year, disruption in the supply of oil.

109. John H. Lichtblau. 27 March 1995, Y4.F7/2:S.Hrg.104–21, p. 48. Our Strategic Petroleum Reserve (SPR) has been created for precisely this purpose. It is the only source from which we can draw "surge" supplies in case of an import disruption for whatever reason, including a technical breakdown in a country's transportation or export facilities. At its current level the SPR could supply an average of 2 million B/D for an 8-month period and a higher volume for a shorter time. If one considers that the largest disruption in recent history–Iraq's invasions of Kuwait in 1990—caused an initial global loss of 5 million B/D which was fully offset by production increases in other OPEC members within 5 months, it is clear that a single emergency supply source of the magnitude of the SPR can play a major role in mitigating the impact of a supply disruption.

As a member of the International Energy Agency, the U.S. is obliged to have some form of stand-by mechanism to deal with a supply interruption. Some IEA member countries have pledged to impose rationing and demand controls, an avenue that would lead to an unenforceable web of regulation and bureaucracy in the huge U.S. oil market. Since our domestic crude oil production is almost always at capacity and oil companies normally carry no more than their operating inventories, the SPR is the most efficient and cost effective tool for managing the impact of a supply disruption.

110. John Lichtblau. FDCH TESTIMONY. 8 May 1996, Lexis/Nexis. As noted earlier, the two oil market disruptions of the 1970s have been estimated to have penalized GNP by 2.5% and 3.5% in the years immediately following the upheaval. The negative impact of each of these shortages, and of the sharp run-up in global oil prices in the Autumn on 1990 after the UN embargoes Iraq's and Kuwait's oil, could have been countered by the early and prudent use of the Strategic Petroleum Reserve. Selling it all off for budget balancing would get a cash infusion of $8–12 billion dollars, a fraction of the cost of any one of the oil market dislocations of the last three decades. Since the 1970s, oil's role in the economy had declined, so the impact of a disruption is unlikely to be as severe. But in today's economy, each percentage point of GDP is worth nearly $70 billion of US goods and services, so even a more muted impact will impose a cost that is a high multiple of the short-term revenue gain from the sale of the SPR.

111. David L. Weimer. THE NEW GLOBAL OIL MARKET. 1995, p. 11. The determination of the optimal size of the public stockpile involves a comparison of benefits and costs. The quantifiable economic benefits result from drawdowns during oil price shocks: the reduction in the direct and indirect costs of the shock and the revenue realized from sale of the oil on the world market. Other benefits include the possible deterrence of purposeful price shocks by foreign governments, greater foreign policy flexibility during conflicts in the Middle East, and the reduction in political pressure to return to the counterproductive petroleum price controls and allocations that raised the costs of oil price shocks during the 1970s.

112. Donald Hodel. 27 March 1995, Y.F7/2:S.HRG.104-21, pp. 38-39. Several countries, which are members of the International Energy Agency, have built considerable stockpiles of crude oil or product. The U.S. has its Strategic Petroleum Reserve (SPR) which currently contains around 591 million barrels. This can be drawn down at the rate of about 2 MMBD for 90 to 120 days, and in decreasing daily amounts thereafter until it is depleted. This provides the U.S. with the ability to withstand a significant period of shortfall of imports without experiencing serious shortages.

In reality, the use of the SPR provides two benefits. It replaces crude oil supplies lost through a disruption. This is its primary and proper use. Secondarily and significantly, in replacing the lost supplies, it forestalls runaway price speculation on world oil markets. The announcement of intention to sell SPR oil has that effect even before the actual sale and deliveries can begin. Speculators have no desire to buy high priced oil in the face of at least 90 to 120 days (and perhaps longer) of price stability resulting from the sales from the SPR.

During the Gulf War, the Department of Energy did not make an early and clear statement that it would release

oil from the SPR. The result was a price spike reaching historic highs of near $40 per barrel in October of 1990. Once DOE acted and announced sales from the SPR and, the world oil market reacting positively to higher prices began to respond with additional production, the situation quickly changed. By February of 1991, prices had fallen to $16.50 per barrel—roughly 17.5 percent less than prewar levels.

113.   Linda Tierney. 27 March 1995, Y4:F7/2:S.HRG.104-21, p. 12. The Strategic Petroleum Reserve (SPR) is the Nation's stockpile of crude oil available in the event of an oil supply disruptions.

- In the event of a disruption, the Reserve is the Nation's insurance against economic damage from supply disruptions and resulting price spikes. The Reserve directly mitigates the threat to national security from either planned or naturally occurring disruptions of petroleum supplies. It contains 592 million barrels of crude oil, which could currently be made available to U.S. refiners at a rate of 3.1 million barrels per day within 125 days of notice.
- To ensure that the facilities can deliver the oil at the design rate through 2025, the Department is replacing or upgrading all of the critical components and systems of the Reserve's facilities.
- Energy emergencies can be most effectively managed if consuming nations act in a cooperative and complementary manner. The United States is a dormant country within the International Energy Program, and has agreed to a number of coordinated emergency response measures. The strategic storage of petroleum is the principal supply measure for all of the consuming countries. The International Energy Agency member countries have all agreed to maintain an inventory equivalent to a 90-day supply of imports. The combination of the Strategic Petroleum Reserve and private inventories provides the United States with an inventory that complies with that target.

114.   Linda Tierney. 27 March 1995, Y4:F7/2:S.HRG.104-21, pp. 12–13. One noteworthy requirement of the Strategic Petroleum Reserve in the near term is that of relocating the oil inventory and decommissioning of the Weeks Island storage site. Over the last three years the site has experienced water intrusion into the oil storage chambers and the development of a surface sink hole. In December, Secretary O'Leary determined that the continued deterioration of structural integrity posed unacceptable risks of oil loss and damage to the environment, and ordered the site decommissioned. The Department intends to begin removing the oil to other SPR sites in October 1995. In the long term consolidation of the oil inventory into empty storage space at the other sites will reduce manpower requirements and operating costs.

Upon completion of withdrawal, the mine will be filled with fluid, the surface restored and the site decommissioned. The entire process will be completed in FY 1998, at a cost of between $75 and $100 million. In order to fund this one time expense, the Department proposes to sell up to $100 million worth of oil and use the receipts to fund program costs.

115.   Hui-Liang Tsai. ENERGY SHOCKS AND THE WORLD ECONOMY. 1989, p. 17. Short-term emergency management included a "safety-net" oil-sharing scheme which was developed in 1974 to put IA countries in a better position to cope with major disruptions in oil supplies by the equitable sharing of available oil supplies during an emergency. In addition, an interlinked set of elements including effective programs to restrain national demand was established to be activated in an emergency to reduce oil consumption by 7 percent in the case of a supply shortfall of at least 7 percent and by 10 percent in shortfalls of greater than 12 percent.

116.   Hans Jacob Bull-Berg. AMERICAN INTERNATIONAL OIL POLICY. 1987, p. 114. From a national security perspective, on the other hand, a consumer organization like the IEA would provide an instrument which could form a buying-power wedge, countervailing OPEC cartel power in the oil market. IEA would then have to be given the necessary powers to do so, including the ability to intervene in the market. Within this line of reasoning, its insurance aspect would also be important. IEA could, with an emergency oil-sharing scheme, protect the consumers against the most grave consequences of an embargo.

117.   Stephen Zunes. MIDDLE EAST POLICY, THE U.S.-GCC RELATIONSHIP: IT'S RISE AND POTENTIAL FALL. 1993, p. 107. Despite recent setbacks, the Organization of Petroleum Exporting Countries (OPEC) is in a relatively strong position. Even the 1986 collapse of prices was not a result of splintering, but a voluntary decision based on the realization that prices were too high. The artificially high prices of previous years had resulted in a loss of market share to non-OPEC producers such as Mexico and the North Sea countries as well as from efforts at conservation and alternative energy sources. This pricing strategy worked, not only by challenging non-OPEC oil producers, but by grinding conservation programs and alternative energy research and development in the United States to a halt.

118.   George Georgiou. ENERGY POLICY. August 1993, p. 838. Between 1973 and 1985, US energy intensity declined 23%. However, there is much room for improvement. In 1986, the USA used 10% of its GNP on energy, by comparison, Japan used only 4%. Today, the USA uses virtually no more energy than it did in 1973. Higher energy prices have stimulated the use and develop-

ment of energy technologies that are more efficient. This has taken place primarily through market forces. Therefore, we can expect a decline in the pace of improvements in average energy efficiency with the recent declines in world oil prices. Fortunately, most of the improvements that have already taken place involve long-term or permanent changes to capital stock, i.e., more fuel efficient automobiles, better insulated houses, and more efficient industrial processes. Others, however, represent behavioural changes, i.e., more timely maintenance of industrial equipment and different habits in thermostat settings—changes that are more closely correlated to current world oil prices.

119. Paul Ehrlich and Anne Ehrlich. HEALING THE PLANET. 1996, p. 46. The low oil prices, and the overproduction by OPEC nations behind it, ironically were partly a result of reductions in energy consumption in the United States from 1975 to 1985. But the incentive for continuing efforts to increase the efficiency of energy use was killed by low oil prices, and momentum was lost. Meanwhile, government programs to encourage both energy efficiency and development of alternative energy sources were dropped by the Reagan administration, just as they were beginning to pay off. By mid-1990, the United States was again importing half its petroleum, domestic production was twenty years past its peak and falling, while consumption was rising rapidly. The oil import bill, already the biggest component of the trade deficit, soared even higher when oil prices rose after Iraq's invasion of Kuwait in 1990.

120. Robert Tippee. WHERE'S THE SHORTAGE? 1993, p. 1. So long as there is oil that can be extracted from the ground, the pressures of supply, demand, and price in the direction of equilibrium ensure that there can be no shortage in the sense of a measurable absence. Exceptions to this rule result from government intrusions, such as the U.S. allocation regulations mentioned earlier, that prevent oil's moving to the consumers that most need it. Otherwise, when a significant amount of oil vanishes from the market for whatever reason, prices rise as buyers compete for replacement volumes from the supplies that remain. In response to the consequently higher prices, consumers with the least intense need for foil find ways to do without it. Demand falls. The market balances at some point characterized by lower volumes and higher prices than before.

121. Ibrahim Oweiss. THE ECONOMIC DIMENSIONS OF MIDDLE EASTERN HISTORY. 1990, p. 191. Another negative effect of the drop in the price of oil is the impact on alternative sources of energy. Higher oil prices of the seventies provided the main incentive to develop alternative sources of energy, both renewable and nonrenewables, in a variety of programs funded by government and the private sector. With the drop in the price of oil such programs have either been canceled or postponed. With the drop in the price of oil, even the alcohol fuel from renewable plantations now used to supply almost one-third of all automobile fuel in Brazil is facing tough competition. Last but not least, from an economic welfare point of view, cheaper oil prices may lead to economic waste while providing no more incentive for conservation of a valuable nonrenewable source of energy.

122. Edward Krapels. INTERNATIONAL AFFAIRS, THE COMMANDING HEIGHT. 1993, p. 84. Towards the end of the 1980s, arguments that overall energy conservation—including oil—could be achieved at minimal cost gathered more political forced. US vice-presidential candidate Al Gore was prominent among them, politically if not intellectually. These assertions that painless and costless energy efficiency gains can be made usually ignored the fact that the oil conservation gains of the 1980s were attributable to a specific set of circumstances. The high oil prices of 1979–80 made most other forms of energy consumption economical, as well as purchases promoting oil and energy conservation. The oil price collapse of 1986, however, re-established oil as the world's best cheap fuel (excluding a few places where natural gas is cheaper). When the price of oil fell, demand began to grow again. Very simple and powerful forces were at play here. When oil was expensive, capital and labour were substituted for energy, and oil demand fell. When oil became cheap, these conservation incentives disappeared, and some old conservation investments were undone, proving that energy conservation was not irreversible.

123. Paul Ehrlich and Anne Ehrlich. HEALING THE PLANET. 1996, p. 46. Today, short-run economic considerations, the failure of market prices to reflect the actual relative costs of different energy sources and the unrealistically low level of energy prices (which in general are far below social costs), prevent society from fully exploring alternative energy choices—even though in the future enormous advantages would accrue from their development. In particular, the low price of oil since 1982, combined with lack of government initiatives, has substantially retarded the development and deployment of various solar technologies, despite their promise for delivering both long-term sustainability and relatively small environmental impacts.

124. Fahdil Al-Chalabi. OPEC AT THE CROSSROADS. 1989, p. 184. Furthermore, oil price fluctuations can have more important ramifications for the world energy balance and its distribution among the various energy sources. With the exception of certain usages, such as transportation, oil which is consumed can be substituted by other sources of energy. With higher oil prices, conventional energy sources, such as coal, gas and nuclear energy, could easily compete with fuel oil, radically reducing the electrical sector's dependence on oil—as, in fact,

has been the case during the 1980s. The cost of oil, therefore, has a direct bearing on its share of the total world energy mix. Not only does the substitutability of oil by domestic sources of energy have an economic significance, in the sense that it reduces dependence on what is perceived as a costly imported energy source, it is also of political and strategic significance to any energy-consuming nation, since it offers security of energy supplies.

125. Joshua Gotbaum. 27 March 1995, Y4.F7/2:S.HRG. 104–21, p. 24.

Oil Dependence and International Oil Markets

In considering the issue of U.S. oil needs, it is important to begin with certain realities and to maintain a global perspective. At the simplest level, we cannot rewrite geology. U.S. reserves of oil are finite and declining and the ratio of net imports to total consumption is growing. Yet, at the same time, the U.S. is less dependent on oil imports than are our friends and allies in Western Europe and Japan.

We have taken a number of steps to protect against this vulnerability. Begun some two decades ago, our substantial investments in the strategic petroleum reserve offer the nation a sound insurance policy against major disruptions. We must also recognize that while shocks are a recurring phenomenon in the global oil market, the financial mechanisms of the modern oil market make it much more resilient to the effects of supply disruptions that it was in the 1970s.

126. Donald Hodel. 27 March 1995, Y.F7/2:S.HRG.104-21, p. 41. Under a worst-case scenario, i.e., a total loss of Saudi production for an extended period (two to three years), a number of economic effects would immediately follow:

- Prices would begin to rise but, if the SPR is used properly, relatively slowly.
- International shortage sharing agreements would be triggered.
- Existing domestic production would begin to increase in anticipation of higher prices.
- Stripper wells, which are small produce individually, but which produce about 1 MMBD currently, would be kept in operation rather than being shut in as is now occurring.
- New wells would be drilled at a faster rate.
- Rig availability would be a constraint on new wells being drilled.
- Consumption would begin dropping. (Beginning in 1979, the high prices had an effect most analysts would not have believed possible, consumption in the world began dropping at an average of nearly 2 percent per year from 1979 to 1985).
- Surprisingly rapidly, the market would establish a new balance between supply and demand at some higher price level.

127. Akarca and Andrianacos. JOURNAL OF ECONOMICS. Fall 1995, p. 141. While the above mentioned studies focus on the persistence (or the unit root) of real GNP and other macroeconomic variables (as employment, money stock, etc.), they totally neglect to study an important variable for the U.S. economy, the oil price. The U.S. economy has experienced large oil shocks in 1973–1974 and 1979–1980 with dire consequences on growth and the price level. In each of these episodes, the real price of oil about doubled and remained above its pre-shock levels for many years, consistent with the view that oil prices have been persistent. Following Iraq's invasion of Kuwait in 1990 the real price of oil again nearly doubled. However unlike the previous two cases, the price returned to its original level within months, suggesting that a new regime may be in place in the oil market. Indeed, even a casual eyeballing of the plot of the series, given in figure 1, shows that the behavior of oil prices was dramatically different prior to 1986 than after. Whereas the real price of oil exhibits a volatile but mean-reverting behavior following its collapse in 1986, during the preceding dozen years it wanders smoothly but without returning to any fixed level.

128. Imad A. Moosa. ENERGY POLICY. November 1993, p. 1153. These conclusions should not be stated without some caveats. First, causality testing does not reveal causality in the real sense. Rather it indicates the relative timing of occurrence of the variables and, therefore, that some variables are unlikely to have caused others because they occurred later. The tests indicate that OECD output and price level could not have 'caused' the price of oil because they occurred later. However, the real causal factor of inflation and economic slowdown following a rise in the price of oil may be a policy change in reaction to this occurrence.

129. Paul Stevens. ENERGY POLICY. May 1996, p. 392. In some senses, oil prices today are less important for the oil importers. In terms of their claim on merchandise exports, their level roughly matches the level in 1970. Thus for middle income countries, energy imports accounted for 12% of merchandise exports in 1992 compared to 10% in 1970. For the OECD, the figures are 10% for 1992 compared to 11% for 1970 (World Bank, 1994). However, a key difference from 1970 is that in many countries the proportionate responsibility of crude prices for final product prices is very much less reflecting the growth of consumer government taxation. For example, in Western Europe, the crude price accounts for only 12% of the pump price of gasoline (OPEC, 1994). This provides a large cushion with which to mute the macroeconomic shock from any price increase, although of course the real resource transfer effect remains.

130. Khadduri. SECURITY DIALOGUE. June 1996, p. 157. The oil market itself is more diversi-

fied and efficient today than two decades ago. On the one hand, given the financial predicaments of most of the oil-producing countries, it is unlikely that a major oil exporter would deny supplies to another country for political or other non-commercial reasons. On the other hand, the present oil market would not permit the old embargo model to function. As a matter of fact, what we have now is a reverse situation, whereby the United Nations has enacted oil sanctions against Iraq and Libya and the USA has added an embargo on Iran. Moreover, Western industrial states have diversified their energy supplies and the sources of their oil and gas imports to such an extent that they are no longer wholly dependent on imports from the Middle East.

131. David L. Weimer. THE NEW GLOBAL OIL MARKET. 1995, pp. 227–228. Much confusion about energy security policy has resulted from the failure to understand the world oil market. Several misconceptions are worth considering. The first is that oil producing countries can effectively target an embargo against specific importing countries. The case with which oil can be transshipped from one destination to another, however, renders targeted embargoes ineffective. Despite the fact that OAPEC embargo against the United States and the Netherlands in 1973 failed to reduce supply or raise prices disproportionately in these countries, the framers of the International Energy Agency (IEA) included among its powers the imposition of an emergency sharing plan to administratively counter targeted embargoes (Smith 1988). Analysis suggests that if it were to be implemented, an extremely difficult task because of limitations of information, the sharing agreement would substantially raise the aggregate costs of the price shock to IEA members (Horwich, Jenkins-Smith, and Weimer 1988; Horwich and Weimer 1988 a&b). To the extent that any policy is needed to resist targeted embargoes, it is simple for IEA members to facilitate easy transshipment by agreeing not to restrict the export of crude oil and petroleum products.

132. Robert Tippee. WHERE'S THE SHORTAGE? 1993, p. 160. In the early 1990s, however, there was much less clamoring within OPEC for ever-higher oil prices than there had been in the 1970s and 1980s—although there was some. OPEC meetings focused more on supply and demand balances than on the political agendas that dominated earlier meetings, though politics still played a role. There were still occasional calls for production cuts designed to raise prices, but they came more from members responding to genuine needs for revenues than from firebrands seeking to punish political enemies in the consuming world.

133. Robert Tippee. WHERE'S THE SHORTAGE? 1993, p. 160. Yet the market always prevails. It certainly did so after the embargo. Just as market forces ultimately crushed U.S. efforts to limit prices and consumption, they vanquished OPEC fantasies about controlling global politics by manipulating oil prices and, therefore, economies. It just took a while for the elephant's new posture to become apparent.

134. Robert Lieber. INTERNATIONAL SECURITY, OIL AND POWER AFTER THE GULF WAR. Summer 1992, p. 159. Conversely, analyses of oil which are essentially—or even exclusively—political also provide only a one-dimensional perspective. These approaches tend to minimize the role of market phenomena. They thus have overstated the threat of oil embargoes, exaggerated the relationship between world oil supply and the Arab-Israeli conflict, and inflated the fundamental power of individual oil-producing countries. After each of the 1970s oil shocks, the political approach tended to assume that oil prices would continue to rise indefinitely.

135. Robert Copaken. THE NEW GLOBAL OIL MARKET. 1995, p. 222. In times of emergency or disruption, U.S. policymakers' attentions are focused primarily on dealing with the crisis of the moment. Yet, once that crisis is past or somehow resolved, the general public's perception tends to return to its somewhat parochial and myopic view of energy—namely, what is the price of a gallon of gasoline likely to be when we fill up the car's tank at the pump. Surely this here-and-now attitude misses the essential point about oil as a strategic commodity—namely, that it is a scarce, difficult- and expensive-to-find, depletable resource, which we in the West unfortunately must continue to import in larger and larger quantities from a politically unstable part of the world, the Middle East. As we have seen, access to this commodity has been identified as among our vital national interests, for which we are willing to employ the use of military force, if necessary.

136. Khadduri. SECURITY DIALOGUE. June 1996, p. 158. There is great doubt and much scepticism in petroleum industry circles today as to whether the security of oil supply or the fear of a disruptive rise in prices are still viable options that can be carried out intermittently or with ease. This is due both to the balance of power in the region following the Iraq-Iran war and Desert Storm, as well as the new oil market forces currently at work.

Since the start of the Cold War the impact of oil imports on national security has been a paramount issue on the agenda of the Western industrial states, particularly the USA. This was highlighted in 1973 with the shortfall of oil supplies and the instability in the Middle East. But since then, the end of the Cold War has assured that Gulf oil does not become a pawn in great power conflicts. Furthermore, the experiences of both the Iraq-Iran war and Desert Storm have shown the determination of the West to defend its oil interests in the region, at whatever cost necessary, and the resilience of the Gulf oil industry to meet crisis and

secure supplies at the most difficult of times. Finally, the US hegemony in the Gulf has meant that oil can be secured by military means at a relatively little financial cost to the Western powers, since a good part of the expenses are being paid now by the Gulf Cooperation Council (GCC) states themselves, one way or the other.

137.   Paul Aarts. "The New Oil Order: Built on Sand," ARAB STUDIES QUARTERLY. Spring 1994, p. 7. As a growing oil *importer,* the United States is quite happy with that, even though there is an inherent risk of even greater dependence on foreign supplies as oil consumption continues to grow. A permanent military presence in the Gulf region almost becomes a necessity: "US military predominance makes it more likely that the US government will favour a 'fix-it' approach over one designed to limit US vulnerability. . . . In the wake of Desert Storm, the dangers of import dependence appear to have been forgotten—if something goes wrong in the Middle East, US military might can fix it." The converse is that a low oil price does not suit the purpose of the United States as a *producer*. In sum, the "optimum" oil price is neither too high, which would hurt the American economy, nor too low, which would bring about the demise of the domestic oil industry. It would be reasonable assumption to put these upper and lower ceilings at $25 and $12 a barrel respectively. The Saudis should bear this in mind. James Akins, the oil guru and former U.S. ambassador to Saudi Arabia, even takes matters one step further, hinting at the (extorted) opposition of allowing the United States "special treatment." Even if the Saudis should decide on a price rise—because of (persisting) cash constraints—the Americans could get away scot-free and evade higher import bills. As Akins soberly observes, "Here is one of history's opportunities to have it both ways. The American balance of payments problem could be handled by a special deal whereby the United States would either pay much less than the OPEC price or the Saudis could make off-setting payments to the United States. It would be difficult for the Saudis to refuse." [ellipses in original]

138.   Edward Krapels. INTERNATIONAL AFFAIRS, THE COMMANDING HEIGHT. 1993, pp. 74–75. Internal political change is inevitable in Middle Eastern countries as it is elsewhere. And such change involves violence, occasionally in some countries, chronically in others. Relatively few countries are fortunate enough to escape interference by global powers during their domestic evolution. The European powers 'allowed' the United States to engage in its civil war with only limited intervention, but the civil wars in Vietnam and Lebanon seemed incapable of escaping world attention.

The oil reserves in the region make it inevitable that political change around the Persian Gulf will attract world attention. This has been a factor in the region's internal politics since the breakup of the Ottoman Empire after the First World War, a story that has been told often before.

Given this legacy, it is unlikely that Middle Eastern countries will be allowed to sort out their own rivalries (or, as was sometimes heard after the Iraqi invasion, that the Arabs will be allowed to 'have their own civil war') without substantial world attention and intervention. The oil stakes are simply too high.

139.   Walter Cutler. MIDDLE EAST INSIGHT. Special Edition, pp. 47–48. Anything that threatens the security of the Gulf has to be of continuing high priority to the United States. Even though the Soviet threat has dissipated, and even though we hope the Arab-Israeli dispute is on its way to some sort of a solution, serious dangers persist: Saddam Hussein is still there; Iran ever since its revolution has been a source of trouble in the region. Its arms buildup today has to be a concern to its neighbors. And it has to be a concern of ours, if for no other reason than the Gulf has two-thirds of the world's known oil reserves. Saudi Arabia has more than a fourth of those reserves, and we as a country are still heavily reliant, and increasingly so, on imported petroleum. Saudi Arabia is sitting on this tremendously rich resource, and anyone that has that magnitude of natural wealth is going to attract predators: be they Iran, Iraq, or some other country seeking fortune by force. So far for Saudi Arabia, its oil wealth is something of a mixed blessing.

140.   J. E. Hartshorh. OIL TRADE: POLITICS AND PROSPECTS. 1993, p. 140. Even before it asserted its military protection over the southern Gulf in 1990–1, US political influence in the region was widespread and pervasive. But until then, this influence had always been potentially self-frustrating. It may remain so, even as the world's sole superpower, no longer challenged. Politically, the US is now more firmly than ever locked into supporting two client states in the region with deeply conflicting interests. Its support of Israel, its strongest military ally there, and protection of Saudi Arabia, as the world's long-run marginal oil supplier, have remained hard to knit into any convincing 'peace process' between Israel and all Arab states. It is often prepared to assert a powerful military presence in support of one Middle East faction or another (though seldom for very long). But it can never be sure of backing 'the right ones.' Indeed, for Western interests in the Middle East there may never be any *permanently* right ones to back.

141.   Robert Biolsi. THE NEW GLOBAL OIL MARKET. 1995, pp. 53–54. In particular, futures contracts play an important role in providing a hedging mechanism for the oil industry. As mentioned previously, the oil industry underwent a transition from a long-term deal market to a day-to-day spot market in the 1980s. No longer could producers rely on long-term price agreements to help remove the uncertainty involved in drilling and exploration. Independent refiners faced increasing uncertainty over their costs for crude oil. In a market faced by abrupt

supply shocks and sharp swings in demand, an enormous increase in risk faced participants in the oil markets. Prices fell from $40/barrel in the early 1980s to less than $10/barrel in 1986 for crude oil. Heating oil would experience a drop from $1.15/gallon in December 1979 to less than $.30 in 1986. During the Gulf crisis, crude oil would swing from $40/barrel in October 1990 to $19/barrel in April 1991 on NYMEX.

By hedging, producers and consumers of oil could help shield themselves from the volatile nature of oil prices through much of the period after 1979. Take, for example, the position of a producer who is "long" the cash commodity. Suppose the producer anticipates selling 1,000 of crude oil a day at $20/barrel. Assuming the futures price can be locked in also at $20/barrel, a short hedge entails selling futures contracts to provide the producer with price insurance. If prices fall to $19/barrel, the producer stands to lose $1/barrel on his production output, but will gain $1/barrel on the futures position.

Of course, rising prices will entail losses on the futures position, which would then be offset by gains to the cash position. The essential point in this analysis is that the short hedge for the producer locks in the selling price for the production stream. In return for this downside price protection, however, any potential upside for rising process is foregone.

For consumers of crude oil, such as gasoline or heating oil refiners, a long hedge can be utilized to provide price insurance. Such oil consumers tend to be "short" oil; that is, they need to buy it for refining purposes. For example, a refiner could be anticipating buying 1,000 barrels/day to refine into gasoline at $20/barrel. To offset the risk of rising prices, the producer can go long the futures contracts. In this way if prices should rise to $21/barrel, and the refiner has difficulty passing along the cost increase to the consumer, the profit margin of the refiner will not be adversely impacted. The addition $1/barrel that will be paid by the refiner for the crude oil will be offset by the $/barrel that the refiner will make on the associated futures contract.

The above examples assume that the cash price of oil and the futures price move perfectly in tandem with one another. In setting the standardized terms and conditions for their contracts, futures exchanges take great care to insure "convergence" between the spot and futures markets. That is, they establish arbitrage mechanisms to help insure that the futures price at the expiration of the contract converges to the underlying cash market.

142. William Reinsch. 27 March 1995, Y4.F7/2:S.HRG.104–21, p. 21. The new factors we evaluated, Mr. Chairman, all served to improve U.S. energy security.

Disarray within OPEC.—Low world oil prices are partly due to the inability of OPEC members to coordinate production levels. The absence of cartel pricing helps ensure a competitive marketplace and clearly benefits U.S. energy security.

Transparency of oil markets.—The futures market in crude oil has become a full-fledged commodity market with computerized trading, options, and forward contracts. As a result, the market is less subject to short-term price manipulations.

143. Robert Biolsi. THE NEW GLOBAL OIL MARKET. 1995, pp. 82–83. Paradoxically, derivative contracts on petroleum became an integral part of the oil market on the heels of an attempt to fix prices by major producing countries. As will be discussed, futures and other types of derivative products are generally designed to offset the risk of fluctuating prices. In 1979 in the aftermath of the Iranian revolution, OPEC agreed to lift oil prices to $32 per barrel. Having such a target price would typically reduce the desirability of trading futures and options, since there would seem to be little profit incentive for traders to take positions on which direction the market was heading. However, attempting to establish a price floor at an unsustainably high level set off such huge turmoil and price volatility in the oil markets, that the value of futures, options, and relative derivative contracts became important for the oil industry.

144. Michael E. Humphries. ENERGY POLICY. November 1995, p. 141. The most rapidly growing derivative product is the oil price swap. The paper volume of oil traded on the major commodity exchanges exceeds 200 million barrels per day, a monetary value of approximately US$1–1.5 trillion a year, or three times the value of physically traded oil. The attraction of swaps are that they can be either short or long-term (three to seven years are the most common) and that they provide a high degree of flexibility to the user. To oil producers they provide an assurance that the price will not fall below a certain floor and end users of oil products an assurance that the price will not go over a certain ceiling at a specified time. An advantage of swaps in energy finance is that they can be customized to meet a particular borrowers' requirements. For example, an oil or gas field development financing may involve an interest rate swap, a currency swap and an oil swap. The financing will be off balance sheet and will monetize the forward flow of oil or gas. Tax efficiency is also an important factor in swap based financings. In the USA the development of the natural gas futures market has provided gas producers with an array of risk management tools at a time of intense price volatility. The ability to hedge against future price risk has also led to new sources of capital being made available to the oil and gas industry in the USA (see 'Non-traditional financings' below).

145. Robert Biolsi. THE NEW GLOBAL OIL MARKET. 1995, p. 53. Futures contracts got off to a

rather hesitant start in the oil market. Several attempts by the Cotton Exchange (1971) and the Chicago Board of Trade (1975) were unsuccessful. This was probably due to the underlying nature of the cash market, where the norm of long-term deals limited the uncertainty of price behavior for the industry. The first successful petroleum-based contract was on #2 heating oil on NYMEX in 1978. This contract called for delivery of 42,000 gallons of #2 heating oil in New York harbor. While it was not greatly utilized at first, the industry gradually became accustomed to trading futures. By the early 1980s, it became one of the most actively traded commodity futures contracts in the world. This was followed by crude oil futures in 1983 and unleaded gasoline futures in 1984, also on NYMEX. Today these contracts represent the most actively traded contracts in the world, averaging more than 200,000 contracts traded per day in 1992.

For the oil markets, the futures markets play three critical economic functions in providing a price hedging mechanism, a price discovery function, and market liquidity. The success of energy futures in the last ten years is one measure of how valuable these contracts are in today's energy industry.

146. THE WALL STREET JOURNAL. 17 August 1995, p. A12. Simply put, the world is a lot different than it was two decades ago. In the interdependent, free-trade 1990s, the big oil exporters need big oil consumers just as much as consumers need them. The old foes are now "mutual hostages," says Michael Lynch, a researcher at the Massachusetts Institute of Technology, and few analysts see that equation shifting anytime soon.

What has changed? Lots of things, illustrating how a major sector of the world economy can turn itself upside down in less than 20 years. Today, the Organization of Petroleum Exporting Countries isn't the power house it once was. There is a proliferation of new producers all over the globe, a 592-million barrel stockpile in the Strategic Petroleum Reserve in Texas and Louisiana, and crude futures markets in New York and London. And there are refineries and gas stations in more than 40 states that are owned by the very foreign producers Americans once loved to hate.

147. Douglas Bohi and Joel Darmstadter. "Resources for the Future," ECONOMIC TIMES. November/December 1994, p. 7. Forecasting failures aside, the general experience in the last twenty years reminds us that society does respond rationally to economic incentives. People alter the way the consume energy; firms invest in new technology; and new institutions arise that inject greater efficiency into world energy into world energy transactions. In other words, what is taught in Economics 101 tends largely to be true. While these lessons are no guarantee against future energy shocks, the more reliable analytical insights and stronger empirical base that we now have should help us, at the very least, to avoid doing harm and, at best, to define more judicious policy choices than those we embraced in the past.

148. Douglas Bohi and Joel Darmstadter. "Resources for the Future," ECONOMIC TIMES. November/December 1994, p. 7. In a couple of respects, however, the 1970s did represent a significant benchmark: a sobering lesson on the misplaced confidence in the effectiveness of government intervention and, conversely, an appreciation (or maybe rediscovery) that markets work and that energy is not wholly different from other economic necessities bought and sold in the marketplace. At the same time, notwithstanding the sometimes diversionary and hyperbolic preoccupation with doomsday scenarios, we have developed a heightened consciousness about the prevailing and long-term social impacts of energy that we must continue facing up to. And that is welcome.

149. John H. Lichtblau. 27 March 1995, Y4.F7/2:S.Hrg.104–21, p. 48. Since the U.S. will remain vulnerable to supply interruptions, augmenting and diversifying our supply sources should be part of our energy policy.

Many countries are now encouraging exploration and production activity through more attractive contract terms, enhanced access to potential production regions, and new development projects for existing production areas. U.S. oil companies, contract drillers and service companies are the core of this global effort. The U.S., however, has recently followed a different course in the development of its own resources. We should take advantage of U.S. company expertise and technology for new upstream initiatives domestically:

- *Allow access* to promising exploration and production areas;
- *Remove impediments* to exploration and production in existing production provinces; and
- *Implement targeted incentives* for certain high cost production.

150. Donald Hodel. 27 March 1995, Y.F7/2:S.HRG.104-21, p. 40. Without government intervention the market would have solved the worst problems in a week or two, i.e., enterprising individuals with tank trucks would have gone to where there was too much gasoline and hauled it to where there was not enough! Simple, but illegal. No one was allowed to do that without a permit from DOE, and the permit process became hopelessly overloaded in a matter of days so that it became essentially impossible to get a permit. And, the shortage worsened and persisted. The same thing or its cousins will happen again if we make the enormous mistake of trying to control that market in time of crisis.

151. David L. Weimer. THE NEW GLOBAL OIL MARKET. 1995, pp. 233–234. The most direct

response to an oil price shock is the addition of new supplies to the world market. An addition to supply causes a parallel shift to the right of the postshock supply schedule in Figure 21.1. This shift reduces the postshock price. The direct benefits to the United States from the addition to supply would be measured as the reduction in the size of the shaded trapezoid. If indirect costs do indeed grow more than proportionally with direct costs, then even small reductions in price could produce large benefits in terms of avoided indirect costs.

"Surge capacity" from existing oil wells is one possible source of additions to supply during price shocks. The upward slope of the supply schedule represents the response of existing oil producers to higher prices. Surge capacity would require wells to be shut-in during normal times so that they could be activated during price shocks. Such in situ storage of oil turns out to be extremely expensive relative to storage of extracted oil in stockpiles (Weimer 1982).

The United States and other members of the IEA have agreed to build stockpiles of oil for release during oil price shocks. By the end of 1992 the U.S. stockpiling program, the Strategic Petroleum Reserve, had 575 million barrels of crude oil stored in salt formations in Louisiana and Texas that could be drawn down at an initial rate of 3.5 million barrels/day for the first three months, followed by a gradually declining rate as reservoirs empty.

152. Khadduri. SECURITY DIALOGUE. June 1996, p. 161. OPEC has many serious problems here which remain to be tackled. As a matter of fact, OPEC for some time now has stopped making markets and has become a price-taker. OPEC crudes are priced in relation to Brent, WTI, or Dubai, rather than setting a marker for its own crudes. The lack of an OPEC pricing policy makes it dangerous to enter this volatile market.

Non-trading on the part of OPEC deprives it of the information, experience, and opportunities that these new markets are offering. The national oil companies are simply missing the operational and strategic advantages, and hence the profits, that private oil firms are now making.

153. J. Cale Case. THE NEW GLOBAL OIL MARKET. 1995, p. 47. What about the impact of OPEC? Clearly, the OPEC cartel has the potential to enormously impact prices—at least in the short term. Witness the $3 to $32 increase in the price of oil during the eight-year period from 1972 until 1980. However, prices soon renewed their decline, and today stand near pre-1972 levels (in real terms). OPEC has not been able to support substantial increases, and oil markets have become somewhat more shockproof (e.g., the surprisingly small amount of disruption brought about by the recent Gulf War). The OPEC-driven price increases stimulated exploration and brought more oil to world markets, as well as encouraged Germany, Japan, and the United States to establish strategic reserves. OPEC has, except for a brief period following the Gulf War, had to survive with chronic current excess production capacity among the larger members.

154. Khadduri. SECURITY DIALOGUE. June 1996, p. 161. The evolution of new market tools, arrangements, and institutions has wrested pricing decisions away from OPEC. The rise of the paper market, spot market, International Petroleum Exchange (IPE), New York Mercantile Exchange (NYMEX), computerization, deregulation, and 24 hour trading of oil sales, reinforced by the massive entry to the paper market of well-known hedge funds, have all had an overwhelming impact on oil prices and minimized the role of OPEC. In November 1993, an all-time record of 987,193 contracts of Brent crude oil futures was traded on the International Petroleum Exchange in London. This volume was equivalent to over 987 million barrels or around 33mn b/d, in turn roughly equivalent to nearly 50% of the total world oil demand. Average daily trade on the NYMEX in the crude oil contract for 1994 as a whole was around 60 million barrels, according to Kleinwort Benson.

155. Massood Samil. THE NEW GLOBAL OIL MARKET. 1995, p. 82. OPEC's role has changed over time from setting crude oil prices to regulating the international oil market through production programming and management. Structural changes in the international oil market have created a situation that makes it impossible for OPEC to return to fixed crude oil pricing. OPEC, at best, could only attempt to regulate the oil market through supply policy.

Even the organization's ability to regulate the market depends upon demand for OPEC oil in relation to its production capacity and the financial needs of its individual member countries. If the demand for OPEC oil reaches OPEC production capacity, oil prices will increase, irrespective of OPEC policy. When demand for OPEC oil falls far below its production capacity, it also becomes difficult for the organization to regulate the market, since financial requirements of individual member countries would entice them to exceed their allocated quotas, hence weakening crude oil price and the price structure. An analysis of the relation between the demand for OPEC oil and OPEC production capacity indicates that sharp price fluctuations have occurred when the demand for OPEC oil has reached either of the two limits.

156. Massood Samil. THE NEW GLOBAL OIL MARKET. 1995, p. 87. A number of empirical studies have attempted to test the relation between OPEC's price decision and spot crude prices or product prices. These studies—Verleger (1982), Fitzgerald and Pollio (1984), Lowinger and Ram (1984), and Loderer (1985)—have attempted to show empirically that there is a causality between official OPEC oil prices and the market price of oil. They have attempted to show that the causality runs from market prices to official prices. However, Samii,

Weimer, and Wirl (1989), have shown that while there is causality between OPEC's official prices and market prices, the causality is not unidirectional. The logical conclusion therefore, is that while OPEC decisions have been important in shaping the course of the oil market and prices, there were other factors that were as instrumental in effecting the oil market in the last two decades.

157. Massood Samil. THE NEW GLOBAL OIL MARKET. 1995, pp. 86–87. There are models that attribute the evolution of oil prices to factors other than OPEC decisions. They argue that the role of OPEC has not been crucial in shaping the course of international oil prices. In the *Property Right Theory*, Ali Johany (1979) explained the oil price adjustment of 1973–74 in light of the shift of ownership of oil reserves from international oil companies with short planning horizons to OPEC member countries with longer planning horizons. Such differences in planning horizons and its consequence on the social discount rate resulted in a slower rate of oil exploitation by oil producing countries leading to higher oil prices.

In the *Globalization Theory*, Cyrus Bina (1985) argued that internationalization of the oil market resulted in the equalization of crude oil prices internationally at the highest international price level. This represents the marginal social cost of oil production of the highest cost producing countries. The Globalization Theory eliminates price differential on the basis of variance in the average production cost of different regions.

The *Political Theory* of crude oil prices evolution argues that international oil prices and international political developments, particularly in the Middle East, are closely related. It maintains that it is the latter that has been the main factor in setting the course of crude oil prices. Regional political considerations such as the Arab-Israeli War or the Iran-Iraq War and internal upheavals and revolutions have impacted the supply of oil and consequently crude oil prices.

Finally, there is the *theory that attributes changes* in crude oil prices to supply and demand factors, rather than OPEC action. Imbalances between demand and supply of oil are, in turn, the result of excess changes in crude oil prices (MacAvoy 1982; Mead 1980). Demand for oil between 1970 and 1973 increased by about 10 mbd, while non-OPEC production only increased by 3 mbd (more than 2 mbd from centrally planned economies). This resulted in an additional 7 mbd demand for OPEC oil in three years. The demand for OPEC oil had reached the organization's maximum sustainable capacity. Therefore, any further increase in demand would have led to a shortage of oil and an increase in crude oil prices. The crisis may have been expedited by the 1972 Arab-Israeli War and Arab oil embargo.

158. Khadduri. SECURITY DIALOGUE. June 1996, p. 165. The alternative is clear. OPEC member-states in general and Middle East producers in particular must make do with the present price range of USD 14–18/B for the OPEC basket. This in itself is not a bad price considering the supply-and-demand balance, and the oil rent achieved is USD 10–12/B over the cost of production. In order to be able to live with such revenue, and until they are ready to transform their economies into more viable and diversified ones, the oil-producing countries will have to restructure and open up their societies not only to economic but also to political reforms.

159. Massood Samil. THE NEW GLOBAL OIL MARKET. 1995, p. 94. However, given the structural changes in the oil market, OPEC no longer will be able or willing to fix crude oil prices. What the organization can strive to achieve is more stable crude oil prices over the long run.

In the long run, OPEC's ability to stabilize the oil market by adjusting its output to satisfy demand for its oil will depend on the collective production capacity of the organization. To increase production capacity, massive investment in exploration and development must take place. With the current financial condition of individual member countries, the possibility of such investment to enhance production and export capacity seems to be low.

The most appropriate candidates to increase production and export capacity are the Persian Gulf oil producers with their substantial oil reserves. If demand for oil continues to grow at the rate of the late 1980s and early 1990s, OPEC will be hard-pressed to maintain a balance in the international oil market, even with increases in output from Persian Gulf oil producing countries. In such a situation, prices will increase not because of OPEC, but despite it. Such price increases should not be viewed as oligopolistic prices or as the success or failure of OPEC, but rather a reflection of the dynamic adjustment of the oil market.

160. Salah El Serafy. THE NEW GLOBAL OIL MARKET. 1995, pp. 129-130. Before attempting to touch on the economic policies of the exporting countries, it would be useful to indicate a number of developments in the oil market during the past 20 years or so, with special reference to the position of OPEC. Many analysts now believe that OPEC has become irrelevant as a market power, though a few do not rule out a possible future resurgence. OPEC's meetings no longer create much interest, and members' attempts at sharing a regulated export market are seen to be publicly flouted and ineffective. Prices seem to have come back in real terms to their long-term levels.

Writing recently on the course of oil prices during the past twenty years, *Petroleum Intelligence Weekly (PIW)* remarked that the nominal price of Mideast crude stood in December 1993 roughly at the same level as in December 1973. Correcting for inflation, however, current prices about half what they were 20 years previously. *PIW* observed that, in constant 1993 prices, and despite periods of volatility, crude sold for about $10 a barrel for most of

this century. "Only at the end of 1973, when nominal prices were raised to $11.65 a barrel for Mideast grades, did inflation-adjusted prices shoot up to around $30. Since 1986, real prices for Mideast light have hovered at around $15."

161. LOS ANGELES TIMES. 4 September 1996, p. A1. U.S. missile strikes in Iraq and the postponement of a United Nations agreement letting the rogue nation sell oil on world markets caused crude oil and gasoline prices to jump on commodity markets Tuesday. Nervous investors also bid up oil stocks.

But analysts were divided on how the events would affect gasoline prices at the pump, with some saying the loss of Iraqi oil in itself is not significant enough to make a difference.

Experts noted that Iraq's oil had yet to reach the market and would have added only 1% to the world's average daily production had shipments begun in October for the agreed-to six-month period.

"Compared with what happened to California earlier this year, when you had refinery shutdowns that cost 10% of gasoline refining capacity, this is very, very small," said energy economist Philip K. Verleger of Charles River Associates in Washington.

Instead, the effect might be to keep prices near their current level "instead of . . . falling later this year by $1 to $1.50 a barrel as we had expected," said noted oil historian Daniel Yergin, president of Cambridge Energy Research Associates in Cambridge, Mass. [ellipses in original]

162. David Barker. THE K WAVE. 1995, p. 32. An example of this is when oil began a rapid decline in the early 1980s. It collapsed in 1986 and threatened the stability of the global banking system. Major drops in the price of oil may be the force that ushers in the next international banking crisis in this long-wave decline. Governments will try and prop up the price of oil to avoid this from happening—even if it means war.

The price of oil affects the price of almost everything else. Lower prices sound good for consumers, but eventually it is lower prices that bring a leveraged global economy to its knees. Global assets are financed based on inflated values and anticipated future rising prices during the good times of the long-wave advance. When the long-wave decline sets in and asset values fall, the international banking system that financed these assets faces potential collapse.

We are all aware that a price collapse in every sector of the world economy will bring the stock markets crashing and almost all global economic expansion to a halt. The incredible amount of debt throughout every area of the economy only magnifies the collapse.

163. THE NEW YORK TIMES. 3 April 1994, p. A1. OPEC's history is textbook proof that the free market works. OPEC teaches that when you raise the price of a commodity far above its production costs—as OPEC did in the 1970s, when it yanked prices up above $30 a barrel—you trigger a counterreaction. In OPEC's case it triggered heightened oil exploration in non-OPEC countries, long-term movements toward conservation, improvements in oil-drilling technology to suck crude from the remotest spots on earth and a search for alternative energy sources—all of which combined in the 1980s to shrink demand for OPEC oil.

When this first became apparent in 1985, OPEC changed strategy. It shifted its focus from hiking prices to maintaining its market share in the face of all sorts of new competitors. To this end, OPEC, which accounts for 37 percent of world oil production, agreed on a strategy of effectively lowering prices, or holding the line on them, on the calculation that this would kill off some of the alternative energies and non-OPEC competition, while gradually rebuilding demand for OPEC oil so that prices could again be ratcheted upward.

164. THE NASHVILLE BANNER. 10 May 1996, p. 1. The trouble is that the country has no viable ENERGY POLICY, no strategy that encourages investment in natural gas research and development, advanced transportation technologies and alternative energy sources that can help make us less vulnerable to future oil shocks.

In fact, some members of Congress want to eliminate all government funding for a wide spectrum of energy technologies—even a modestly-funded design program, cost-shared with industry, to develop advanced nuclear power plants.

Others want to scale back funds for photovoltaic energy conversions, advanced wind turbines and other renewable energy technologies.

165. Joseph Romm and Charles Curtis. ATLANTIC MONTHLY. April 1996, p. 63. That the nations' and the world's dependence on Persian Gulf oil will grow over the next decade seems inevitable. This is particularly true since most projections assume continuing significant technological progress in bringing down the cost of domestic production, in developing alternatives, and in using energy and oil more effectively. But those projections have not factored in the federal government's plans to withdraw from its role in fostering the development and deployment of those technologies.

166. John H. Lichtblau. 27 March 1995, Y4.F7/2:S.Hrg.104–21, p. 47. The Commerce Department report accepts the Energy Information Administration (EIA) projection that the share of imports will rise from 45% in 1994 to 51% by 2000. Other forecasts show somewhat different growth rates but virtually all forecasts project a growth in U.S. oil imports to 2000 and well beyond. Thus, our growing reliance on foreign oil is an irreversible reality for the foreseeable future.

167. Edward Krapels. INTERNATIONAL AFFAIRS, THE COMMANDING HEIGHT. 1993, p. 74. Such a policy has implications both for US energy policy and for the countries of the Middle East. US military predominance makes it more likely that the US government will favour a 'fix-it' approach over one designed to limit US vulnerability. For decades, Washington has espoused programmes to reduce dependence on oil imports, largely because of the uncertain political stability of Middle East oil suppliers. But in the wake of Desert Storm, the dangers of import dependence appear to have been forgotten—if something goes wrong in the Middle East, US military might can fix it.

168. Edward Dowling and Francis Hilton. THE NEW GLOBAL OIL MARKET. 1995, p. 28. Although strategies differed from country to country, nearly every nation emphasized conservation, enhanced energy efficiency, domestic energy production and fuel diversification as the most effective tactics for improving energy self-sufficiency. Each of these efforts, like the overall quest for energy security, made significant progress before 1986, but then faltered. Both in their rise and in their decline, these strategies shaped the demand for fossil fuels and nuclear power.

169. Edward Dowling and Francis Hilton. THE NEW GLOBAL OIL MARKET. 1995, p. 27. Twenty years have passed since the 1973 oil crisis pummeled the global economy. That oil crisis quadrupled oil prices, eliminated economic growth, doubled inflation rates, and launched many nations into urgent pursuit of energy self-sufficiently. The 1979–80 oil crisis—a result of Iran's revolution, the Iraq/Iran War, and the West's panicked response to both events—tripled oil prices, dragged economic growth to yet another halt, reignited inflation rates, and underscored again the need for energy self-sufficiency. The 1986 crash in oil prices exerted an opposite effect. The very forces that collapsed oil prices—the fall in worldwide oil demand and the increase in non-OPEC production—suggested that there was no reason to pursue greater self-sufficiency. The 1990–91 Gulf conflict subsequently demonstrated greater cooperation between oil producers and consumers and further persuaded nations to slow the pursuit of energy independence. the fall of the self-sufficiency imperative, like its rise, reshaped the world's energy consumption patterns.

170. OIL & GAS JOURNAL. 29 August 1994, Lexis/Nexis. GRI's baseline survey departed from its previous studies in that it did not predict a significant price increase for crude oil during the long term.

"Over the last 50 years, real crude oil prices have generally either been stable or declining with the exception of four discrete years between 1973 and 1981," GRI said. "Further, with the decline in prices since 1981, the real dollar impact of the 1970s price spike has virtually disappeared.

"At this time, there is little prospect for a sustained significant increase in real crude oil prices. Price increases are likely to be offset by continued energy conservation, increased Organization of Petroleum Exporting Countries and non-OPEC production, and further technology improvement."

171. THE RECORD. 26 January 1986, Lexis/Nexis. Those factors encouraged most OPEC members to break the production quotas that the cartel had established to keep prices up. As the cheating grew, the market was glutted with oil, triggering unsanctioned price-cutting by some cartel members.

Conservation and growing use of alternative energy sources, inspired by soaring OPEC prices in the 1970s, cut into demand for oil, and OPEC faced new competition with the development of such major oil areas as Alaska's Prudhoe Bay, the North Sea, and Mexico.

172. THE RECORD. November 1985, Lexis/Nexis. Yet the world is awash in an oil glut that's unparalleled in the dozen years since the first OPEC price shock of 1973, and the glut will last at least through 1990. Energy conservation, alternative energy use, and the rise in non-OPEC oil production have conspired to drive the price of oil down.

Once as high as $40 a barrel and seemingly headed inexorably for $60, $80, $100 or more, the price of oil will probably hit $25 this winter on its way down to $22 on the Rotterdam spot market by spring.

173. Howard Geller, John DeCicco, and Skip Laitner. ENERGY POLICY. June 1994, pp. 480–481. Vigorous energy efficiency improvements, by reducing petroleum consumption, exert downward pressure on the world oil price and the price for petroleum products. This tendency was clearly seen in the mid-1980s, when the oil price collapsed following a decade of successful energy conservation efforts in the USA and other industrialized nations. And conversely when oil prices fall, conservation efforts tend to diminish, leading to rising oil use, as has been the case during the past seven years.

174. Howard Geller, John DeCicco, and Skip Laitner. ENERGY POLICY. June 1994, p. 484. On the supply side, there is little prospect for reducing oil imports through increased domestic oil production. Accelerating the introduction of fuels derived from bioenergy and other renewable energy sources, on the other hand, could make a significant contribution to reducing oil use in the post-2010 time period. Adopting a comprehensive set of energy efficiency and renewable energy initiatives could conceivably reduce US oil imports in 2010 by 5–9 MBD (50–85%) from projected levels. Even by 2000, it may be possible to lower oil imports by 2–4.5 MBD (20–25%) if vigorous and

prompt actions are taken. Reducing oil imports to this extent could greatly benefit the USA as well as other oil importing nations.

175. Edward Dowling and Francis Hilton. THE NEW GLOBAL OIL MARKET. 1995, p. 32. A second factor—the comparatively rapid rise in demand for other energy sources—also explains the marked drop in worldwide oil dependence between 1980 and 1985. As demand for other fuels increased more quickly than the demand for oil, oil's share of total primary energy consumption became significantly smaller. During this period, when worldwide oil use declined at an average annual rate of 1.3 percent, use of other fossil fuels increased impressively. Consumption of natural gas grew at an annual average rate of 2.5 percent and coal consumption increased by nearly 2.3 percent annually. Concurrently, global use of nuclear power rose 15 percent per year and the demand for hydropower increased at an average annual rate of 2.8. Thus, while oil lost nearly 6 percent of the world's energy market, gas gained 1.2 percent of it, coal picked up 1.5 percent, nuclear acquired an additional 2.3 percent, and hydro captured 0.5 percent more of the market (BP 1992: 32, and earlier editions). The gap between oil-demand growth and demand growth for other fuels narrowed significantly after 1985. Between 1985 and 1990 oil use actually grew more quickly than either coal use or hydropower.

176. Linda Tierney. 27 March 1995, Y4:F7/2:S.HRG.104-21, p. 10. The United States has an abundance of renewable energy resources that can—over time—help reduce our dependence on imported oil. During the last 15 years, intensive work by industry and the Department of Energy's national laboratories has steadily increased the reliability of renewable energy systems while dramatically lowering their costs. These systems are gradually becoming commercially competitive with conventional power sources. Within the next two decades, renewable energy resources can begin reducing our oil imports when used as a substitute for refined products, or for powering a growing segment of the U.S. transportation sector, and for heating and cooling our homes and offices.

Today, wind-generated electricity is nearly competitive with conventional electric power; the U.S. has regained its lead in world photovoltaic shipments; the geothermal heat pump industry is striving for 12 percent of the U.S. space conditioning market; the bioenergy industry is working to become a major new source of electric power and a major new source of income for America's farmers; and hydrogen energy is being developed as a major energy carrier for the future.

The overall budget for renewable energy technologies was $309 million in FY 1994 and $389 in FY 1995. The Administration request for FY 1996 is $420 million.

Renewable energy technology programs are expected to produce non-electric energy savings equivalent to 60 million barrels of imported oil per year in the year 2000 and twice that amount in the year 2010. While renewable energy technologies do not generally compete directly with petroleum in the energy sector, renewable energy is expected to provide 1.1 quadrillion BTUs of energy saving by the year 2010. In addition, pollution prevention benefits are expected to be 12 million metric tons in the year 2000 and 40 million metric tons in the year 2010. Economic benefits are $4.9 billion in 2000 and $11 billion in 2010.

177. Joseph Romm and Charles Curtis. ATLANTIC MONTHLY. April 1996, p. 64. Predicting our energy future beyond 2010 is chancy, but here we have an opportunity to rely on perhaps the most successful predictor in the energy business: Royal Dutch/Shell Group. According to *The Economist*, "The only oil company to anticipate both 1973's oil-price boom and 1986's bust was Royal Dutch/Shell." Anticipating the oil shocks of the 1970s helped Shell to move from being the weakest of the seven largest oil companies in 1970 to being one of the two strongest only ten years later. Anticipating the oil bust was apparently even more lucrative. According to *Fortune*'s ranking of the 500 largest corporations, Royal Dutch/Shell is now not only the most profitable oil company in the world but the most profitable corporation of any kind.

When such a company envisions a fundamental transition in power generation from fossil fuels to renewable energy beginning in two decades, a transition that will have a significant impact on every aspect of our lives, the prediction is worth examining in some detail. Chris Fay, the chairman and CEO of Shell UK Ltd., said in a speech in Scotland last year, "There is clearly a limit to fossil fuel . . . Shell analysis suggests that resources and supplies are likely to peak around 2030 before declining slowly. . . . But what about the growing gap between demand and fossil fuel supplies? Some will obviously be filled by hydroelectric and nuclear power. Far more important will be the contribution of alternative renewable energy supplies." [ellipses in original]

178. Linda Tierney. 27 March 1995, Y4:F7/2:S.HRG.104-21, p. 8. Energy conservation is a cornerstone of the Administration's program to reduce our energy vulnerability. Americans are now using one-third less energy than they did about 20 years ago to produce each dollar of gross domestic product—a direct result of energy efficiency measures.

New and emerging energy efficiency technologies increase the productivity and competitiveness of our businesses; prevent pollution and reduce other environmental damages; create new industries and jobs; and help reduce oil imports.

The Department of Energy operates a balanced portfolio of energy efficiency programs designed to create and commercialize new energy efficiency technologies in the buildings, utilities and industrial sectors. A sustained com-

mitment to the Department's energy efficiency programs will:

- reduce our oil imports by 241,000 barrels a day over the next five years and by 1.26 million on barrels per day by 2010;
- save an accumulated $10 billion dollars in residential energy costs by the year 2000;
- cut commercial energy costs by $6.6 billion;
- reduce the country's industrial and pollution control bill by $3 billion dollars by the turn of the century;
- produce almost 310,000 high wage jobs by the year 2000; and
- reduce pollution by 38 million metric tons.

Clearly, tax dollars-spent on energy conservation are tax dollars wisely spent. The modest 15 percent increase in the Department's energy efficiency budget request for FY 1996 will mean $14.8 billion in benefits over the next five years and will enable the Administration to carry through on its commitment to a stronger economy, a cleaner environment and a more secure future.

179. David L. Weimer. THE NEW GLOBAL OIL MARKET. 1995, p. 225. This chapter focuses on the energy security of the United States. While most of the discussion applies to other net importers of petroleum, two factors distinguish the circumstances of U.S. energy security policy from that of most other countries. First, because the United States accounts for a large fraction of the world demand for crude oil (26 percent in 1992; 15.9 million barrels/day of petroleum consumption versus world crude oil production of 60.4 million barrels/day), its energy security policies have potential for affecting the world price of oil. Second, because the United States has substantial domestic production (8.98 million barrels/day of crude oil and natural gas plant production in 1992), its energy security policies involve substantial distributional effects of political significance. These points have relevance to both the effectiveness and political feasibility of alternative U.S. energy security policies.

180. Sergei Glazyev. OFFICIAL KREMLIN INTERNATIONAL NEWS BROADCAST. 20 August 1996, Lexis/Nexis. The budget is also in critical condition. The amount of outstanding tax payments has reached more than 40 percent. Decisions are being prepared to sequestrate public spending. The infrastructure is also in its worst shape. You could have seen that from the energy crisis in the Maritime Territory. So, basically, the system of economic security is if not destroyed, then is about to fall apart.

The survivability of the country is the question of what economic measures will be taken in the immediate future. I hope that my knowledge and experience will be useful for the country in my new capacity. That's why I have accepted Alexander Ivanovich Lebed's offer to head the Economic Security Department of the Security Council.

181. James Watson. EUROPE-ASIA STUDIES. May 1996, p. 451. Certain aspects of Russia's economic transformation will also be important. If the gap between domestic and world oil prices continues to narrow, and if Russian consumers become more prompt and reliable payers, then the preference for exports over domestic sales may weaken accordingly, and competition for slots in the export schedule ease. However, if the narrowing of the gap between world and domestic prices is, as it was in 1995, due primarily to a combination of domestic inflation and a stable exchange rate, rather than to a rise in the domestic oil price relative to the general price level, then the effect on foreign investment will be negative. One of Russia's main attractions for foreign companies—low (dollar) production costs—will be slowly eroded away, and the profitability of exports and hence of foreign investments generally, will gradually decrease.

182. Vladamir Razuvayev. MOSCOW TIMES. 2 August 1996, Lexis/Nexis. It is obvious that much of the oil industry's success has to do with the government's involvement with strengthening the oil companies rights to exploitation of oil deposits in post-Soviet countries. The economic, technological and financial backwardness of Russia necessitated finding a "trump card," which would allow the country to buy time to carry out reform. It is oil and gas on which the Kremlin is mainly counting in the fight for political stability and rebuilding the economy. These resources are helping Russia to get through the transitional economic period and move toward democracy and a market economy.

183. Vladislav Zubok. WORLD POLICY JOURNAL. Spring 1992, p. 121. With the centripetal forces of economic integration stymied, historic insecurities and the divergent interests of new political elites are destined to shape the foreign policies of the Commonwealth states. Some ex-Soviet republics will never agree to Russian hegemony and will instead drift toward their more natural partners—geographically adjacent countries with similar cultures and more highly developed economies. In turn, burdened with imperial fatigue and exasperated by its newly arrogant neighbors, Russia may be tempted to turn its back on them in revenge. Territorial ambiguities—the legacy of a constantly shifting Russian and then Soviet empire—will only complicate matters. Little cold wars between Russia and these neighbors are a constant possibility; a hot war similar to the current Croatia-Serbia conflict is also possible between Russia and Ukraine.

184. James Watson. EUROPE-ASIA STUDIES. May 1996, p. 432. The Ministry of Fuel and Energy's plan envisaged that most of this investment would be undertaken by domestic enterprises, aided initially by loans

from the Russian government and, perhaps, international financial institutions. However, Russia's new vertically integrated oil companies—created subsequent to a presidential decree of November 1992, and partially privatised in 1994—have proved unable to make the necessary expenditures. Their finances have been wrecked by government price controls, which until the spring of 1995 kept the domestic oil price at around one-third of the world level, and persistent non-payment by domestic consumers. Government loans to the industry have been minimal, and, although significant sums have been forthcoming from international financial institutions, these loans will not by themselves be sufficient to turn the industry around. For this to happen, it will almost certainly be necessary for Russian entities to enter into partnerships with Western companies. The latter are prepared to make major investments in Russia. In 1994 one 'senior US administration official' suggested that half of the $40–50 billion which American companies were ready to invest in Russia by the end of the decade would be in oil and gas, while other estimates put total potential foreign oil investment at around $60–70 billion. In other words, the amount of money which Western oil companies are ready to commit to Russia roughly equates to that which the Russian oil industry needs to arrest its decline. The industry would benefit not just from the additional finance, but also from access to Western technology and managerial expertise.

185. Elgie Holstein. US NEWSWIRE. 14 August 1996, p. 1. Yes, I think the President's view was that we are now recognizing that we're facing a situation in the world where there is great sensitivity to political/economic repercussions as the affect our energy future, because it's truly a world business. Part of what the President talked about in terms of getting to your question about the future was he communicated his very strong enthusiasm for what the. oil industry has been able to do as part of the overall development—first stages, but the first stages of developing a solid economy in the former Soviet Union, which has tremendous oil resources but which has had terribly antiquated, horrendous management practices.

186. Frank Millerd. THE NEW GLOBAL OIL MARKET. 1995, pp. 8–9. Changes in reserves from 1970 to 1993, by area of the world, are presented in Table 1.3. Published proven reserves declined in the United States and Canada while large increases occurred in Latin America and, especially, the Middle East. The Middle East proportion of world reserves increased from 62.0 percent in 1970 to 65.7 percent in 1993. Despite the recent drop in the reserves of the former USSR and central Europe, there is the hope in Russia that assistance from the West in the form of technology, management, and finance can reverse the decline (World Energy Council 1992).

187. James Watson. EUROPE-ASIA STUDIES. May 1996, p. 430. The Russian oil industry is in desperate need of investment. Output has fallen by 45% since the peak year of 1987, and could fall still further if major financial commitments are not forthcoming. Domestic firms, starved of cash, and often lacking the appropriate technology and expertise, are unlikely, however, to be able to turn the industry around by themselves. Instead they will need to enter into partnerships with Western oil companies, which, attracted by Russia's skilled labour force, its relatively low production costs, and, above all, its huge reserves of oil, are ready to make major investments. It is widely accepted that such co-operation could benefit both sides. Projects could generate enough revenue to compensate both the foreign oil companies and their local partners, in the form of profits, and the Russian government, through tax revenues. Increased exports would improve Russia's trade balance and make it more able to service its foreign debt; the additional tax receipts would help to balance the state budget, and facilitate macroeconomic stabilisation. Furthermore, Russian equipment suppliers would receive additional orders, employment would be created for Russian workers, and the local economies of the oil-producing regions would benefit generally from large inflows of foreign money. Western energy consumers and governments would gain too, from added insurance against any future tightening of world oil markets and higher world prices.

188. Yevgeny Yasin. OFFICIAL KREMLIN INTERNATIONAL NEWS BROADCAST. 6 August 1996, Lexis/Nexis. Well, I don't want to go as far ahead because we are working on a new medium-term program, which will be presented by the next government. This is my assessment of the situation, an objective assessment. We are at rock bottom. But things may shape up in a way that there will be further decline. Everything depends, a lot depends on what we, the government, will do. If we do nothing, the economy will live on. There are many robust sprouts in the economy. If you tell me that the oil industry is in disarray, the picture is very mixed. For example, Surgutneftegaz has no problems. It is not indebted to the budget. Mr. Bogdanov came to me and said, "I want to build a port and I am prepared to invest 7 billion dollars in it." I said, "Go ahead." Wonderful. I often visit enterprises and I see that fertile soil has been prepared. As soon as investments start to flow in, sprouts will appear. Investment must come sooner or later because capital must circulate. So, in reality, what is the main danger in store for us? Prolonged depression. That may happen if we don't do anything and allow things to develop spontaneously especially as regards the expansion of the shadow economy, the "gray economy," the "criminal economy." That is the chief danger in this country.

189. James Watson. EUROPE-ASIA STUDIES. May 1996, pp. 431–432. The report also recognised, however, that even the achievement of these apparently modest targets would require massive capital spending. In

order to slow the decline in output in the short term it would be necessary to reduce the stock of idle wells, which had grown from 7000 in 1988 to 23,500 by early 1992, and jumped further to 31,900 in 1993. Separate studies carried out in 1993 by the Ministry of Fuel and Energy and a Texan consulting group for the US Department of Energy suggested that about one-third of these wells were potentially productive, and that their rehabilitation could perhaps add 10% to Russian oil output. In order to stabilise production in the long term, however, it would be necessary in addition to enhance recovery from existing wells, drill new wells in fields which were already being worked, and, most importantly, bring new fields into production. Diennes, writing in 1993, estimated that to stabilise production from then until the end of the century would require the drilling of 100,000 new wells. Similarly, Deputy Minister for Fuel and Energy Anatolii Shatalov predicted that meeting his ministry's targets would necessitate the drilling of between 40,000 and 76,000 wells, and the bringing on stream of 160 new fields by the year 2000. Another MFE official suggested that it would be necessary to develop 456 new deposits by the end of the decade, while the government programme itself envisaged bringing online 551 new fields.

This would cost enormous sums of money. The MFE programme calculated the total investment needs of the industry between 1992 and 2000 at about 5500 billion rubles (measured in early 1992 prices), equivalent to around $35 billion. A World Bank estimate, however, put the cost of stabilising output until the year 2000 at $50 billion, and Shatalov suggested that perhaps $75–100 billion of new capital spending would be needed in the same period. Such large investment would, though, have the potential to earn a relatively swift return. Suppose large investment would, though, have the potential to earn a relatively swift return. Suppose it could make a difference of about 100 million tons to Russia's annual oil production (a fairly conservative estimate). Even at the current, depressed world price of about $115 per ton, this could generate a revenue stream in excess of $10 billion per year which would last well into the next century, after the major spending had taken place.

190. Roland Goetz. AUSSEN POLITIK. Second Quarter 1996, p. 144. In the light of past experience and in accordance with theoretical considerations, economic growth could above all be set in motion through "autonomous front running" on the part of corporate investments or external demand. Private consumption would then follow as a result of an increase in incomes. Greater state demand is only an appropriate means of stimulating growth in exceptional cases. The corresponding preconditions in Russia are hardly suitable for this approach. Investment activity status restrained there because of the low level of capacity utilisation (at roughly 30 per cent of the extent prior to economic transition). Investments are only carried out in a few branches of economic activity with a favourable demand-side status: raw materials production, telecommunications, housing construction, and those branches of mechanical engineering activities which produce equipment for these sectors.

191. Roland Goetz. AUSSEN POLITIK. Second Quarter 1996, p. 138. The high level of crude oil production in the Eighties (500–600 million tonnes) would appear once and for all to be a thing of the past. In the long term, the output figure will probably stabilise at the level recorded in the year 1970 (300 million tonnes), which is enough to satisfy domestic demands (at 1 tonne per inhabitant, roughly 150 million tonnes) as well as provide an export potential of 150 million tonnes. However, the investments for test drilling, production equipment and pipelines would have to be extended. Otherwise, a further decline in production is inevitable.

192. RusData DiaLine. BIZEKON NEWS. 19 July 1996. Recent opinion polls suggest that Russia's oil sector performance has been erroneously perceived by various groups of the Russians as plunging with no improvement in sight for inadequate exploratory drilling.

What actually happens is that Russia has sufficient explored oil reserves with prospecting for new fields going on, although the scope of the latter has reduced over the last few years. At present, the country' exploratory drilling effort is reported at nearly 1.3 million meters a year. In the meantime, the figure can be easily tripled or even quadrupled in the years to come, according to geologists. This will, in turn, result in expanding the country's explored oil reserves to double or triple Russia's annual oil product. The matter depends on investment which has been lacking so far.

By early 1996, a total of 1,946 oil fields, oil and gas deposits and oil and gas condensate fields have been discovered with 1,093 out of the total already under development.

193. Roland Goetz. AUSSEN POLITIK. Second Quarter 1996, pp. 139–140. The extent of required investments and of the necessary capital imports, including their foreign policy implications, calls for closer examination. The analysis must begin with the question of which investments are needed to renew productive capital in Russia. There are highly divergent estimations on this aspect. If the investment requirements in the new *Laender* in eastern Germany are taken as a point of orientation and the corresponding data applied to Russian conditions the figures are astronomical. An investment volume totalling DM 5.6 million million is regarded as essential for the former German Democratic Republic (GDR) in order to align the production level there to that of the *Laender* in western Germany during the course of the next 25 years. The corresponding investment requirements for Russia would be DM 50 million million.

With realistic assumptions, i.e., if the estimate is based on the goal of only partially closing the economic gap, the figures obtained are not completely beyond the realms of possibility. Settling for just a doubling of per capita incomes within 15 years, which, assuming a constant population figure, would amount to an average growth rate for GDP of 5 per cent, and with an optimistic estimate of the (*de facto* unknown) macroeconomic capital coefficient, i.e., the ratio of capital input to production output, a capital requirement figure is calculated which could be financed by domestic savings. Taking a less favourable coefficient, annual economic growth averaging 5 per cent can only be achieved via capital imports (unless, of course, domestic savings were extremely high, something which is not to be expected in the case of Russia, with its extensive state sector and chronic budget deficit).

194. Roland Goetz. AUSSEN POLITIK. Second Quarter 1996, p. 139. Apart from natural gas, none of Russia's natural resource can offer a suitable basis for an economic upswing owing to the general decline in production. All the available natural resources, however, can keep the Russian economy "above water"—due in part to their function as a guarantor for international loans—and are already one of the main reasons why the economic situation in Russia is much better than that of the other successor states of the former USSR, which primarily suffer from a lack of sources of energy.

195. James Watson. EUROPE-ASIA STUDIES. May 1996, p. 427. Despite the barriers to investment, the Russian oil industry has still attracted more Western capital than any other sector of the economy, and remains the focus of most foreign interest. Hence an understanding of the industry is crucial for our appreciation of Russia's overall economic relations with the West. At the same time, many of the problems faced by foreign oil companies (in particular legal uncertainty and high taxes) have plagued investors in other sectors of the economy as well, and so a study of the oil industry can serve to illuminate certain more general problems with the investment climate which foreign firms face in Russia.

196. Roland Goetz. AUSSEN POLITIK. Second Quarter 1996, p. 143. If a corrective economic policy is not applied the raw materials sector can thus become a cause of de-industrialisation—a problem from which the transition economies already suffer anyway for a host of other reasons. An OECD study maintains that it is nevertheless better to export raw materials for reasonable revenue than to use them in resource-intensive Russian production to manufacture subsidised final products as exports. This only makes sense with respect to the past. As the domestic prices for raw materials have in the meantime—as opposed to the period prior to the reforms—moved close to the world market prices and as exports are no longer subsidised there is hardly any reason to fear such wastefulness today.

197. Sergei Glazyev. OFFICIAL KREMLIN INTERNATIONAL NEWS BROADCAST. 20 August 1996, Lexis/Nexis. As regards the export of oil, we should take a closer look at the property relations in the sector. After the loans for shares auctions it became unclear which oil enterprises belong to whom, and what is the function of the state.

If the oil industry continues to be managed as it has been managed, we can be quite sure that the export and extraction of oil can become low profitable and even unprofitable at all.

The prices of oil and oil products in Russia are already higher than oil prices on the world market. While inside the country oil prices are getting very close to the world level, there is not any point in introducing state monopoly on oil exports.

198. James Watson. EUROPE-ASIA STUDIES. May 1996, p. 450. As suggested above, during the first half of 1995 several events took place which promised to significantly improve the investment climate for foreign oil companies in Russia. Perhaps most importantly, the Duma passed the long-awaited laws on Oil and Gas and on Production Sharing Agreements (the approval of the latter by the lower chamber actually prompted Exxon and Sodeco to sign the Sakhalin I agreement). Several joint ventures were granted exemptions from the export tax, while the tax itself was significantly reduced, and there were promises that in 1996 it would be abolished altogether. The system of export regulation became more transparent, and the increase in the domestic price of crude oil eased, at least slightly, competition for slots in the export schedule. Also significant was a presidential decree issued in April 1995 which finally allayed foreign investors' fears that a national monopolistic oil company might be created. Instead, the decree confirmed that the government holding company, Rosneft, would be corporatised and privatised and would compete alongside the other vertically integrated companies.

199. RusData DiaLine. BIZEKON NEWS. 19 July 1996. In spite of the difficult economic situation in the country, capital investment in the oil industry has been on the increase lately as reported last year by at least by some Russia's major oil companies, including LUKoil (up 26.3 percent on 1994), Yukos (up 34 percent), Surguneftgaz (up 6.2 percent), and Tatneft (up 13.6 percent). The general trend is true for other oil producers as well. Investment in the industry is projected to grow by 10 percent in 1996.

200. Vladislav Zubok. WORLD POLICY JOURNAL. Spring 1992, p. 203. The nightmare scenarios are manifold. If the present government crumbles into anarchy,

nuclear weapons might fall into the hands of civil war combatants, terrorists, or anti-Western regimes elsewhere in the world. Several thousand individuals have top security clearance in the former Soviet nuclear complex and could, if unemployed and disgruntled, sell their services to the highest bidder. Out of economic desperation, the major nuclear and missile installations in Russia and Kazakhstan may agree to supply fissionable material and delivery systems, illegally or clandestinely, to an Iraq or Libya. If German firms could avoid all the controls of a stable, democratic society to help Iraq with the modernization of its medium-range missiles. Russian firms can with greater ease take advantage of chaotic conditions to spread technology more indiscriminately. So too could Russia, and in turn the other ex-Soviet republics, revive its interest in such weapons as a cheap substitute for large conventional armies. This potential reliance on "massive retaliation" and its impact on modern weapons production alarms Western security analysts.

201. Georgi Arbatov. FOREIGN POLICY. Summer 1994, pp. 96–97. A complete breakdown of the Russian economy, as a result of ill-designed reform, could lead to the return of a form of cold war, Western declarations that there are no alternatives do not sound persuasive to people who remember too well that only a few years ago, though they were never very affluent, they did live much better than they do now and they were not so frightened of the future. And if the West denies Russia alternatives to reform, we will look for them in our past, though that past was far from perfect. So the West should not be surprised by the results of the December elections. Today they most probably would be much worse. And if a disastrous economic policy is perceived as something imposed from abroad, in particular by America, distrust and suspicion of the West will grow.

202. Igor Khripunov. COMPARATIVE STRATEGY. Vol. 14, 1995, p. 457. The way some members of the Russian leadership currently understand the main threats to national security interests is illustrated by an interview with Andrey Kokoshin, the first deputy defense minister, who has a wide range of responsibilities including policy formation. In his view, the main threat lies "in the economic and technological sphere." According to Kokoshin, "if the vast potential of Russia . . . has not been realized in the next few years, the threat of Russia turning into the global backwoods scientifically and technologically-with all the negative consequences for the country's defense capability arising therefrom—will become real. No direct threat of major war against Russia and its allies is seen in the short term, but this is again linked in many respects to our possession of considerable military might." [ellipses in original]

203. Jennifer Mathers. CONTEMPORARY SECURITY POLICY. Vol. 16, No. 3, December 1995, p. 1288. Debates about the types of conflict and enemies which the Russian Federation's armed forces are most likely to face in the future are being conducted in the Russian military press, and in particular in the theoretical journal *Voennaya Mistl*. It is possible to identify two distinct schools of thought among the participants in these debates. The first is composed of those who emphasize threats which are emerging in the aftermath of the collapse of a bipolar international order dominated by two nuclear superpowers, such as political and economic instability within the former Soviet Union and the Russian Federation itself. These analysts therefore stress the importance of being prepared to conduct low-intensity conflicts and peacekeeping operations. The views expressed by this first group of military analysts were dominant in *Voennaya Mistl* during 1992. These analysts are challenged by others who take a more traditional position on the issue of future war and future enemies, warning of continued Western— and particularly American—hostility towards Moscow and the powerful military capabilities which the NATO alliance still possess which could be used to wage a large-scale, even global, war against Russia. This traditionalist view has been expressed much more frequently in more recent issues of *Voennaya Mistl*.

204. AGENCE FRANCE PRESSE. 17 May 1995, p. 1. The United States needs to maintain and upgrade its military presence in the Middle East for strategic and economic interests, the Pentagon said in a report published Wednesday.

"The world will be even more dependent on Persian Gulf oil in the early 21st century than it is today," said the report. US Security Strategy for the Middle East.

"Our paramount international security interest in the Middle East is maintaining unhindered flow of oil from the Persian Gulf to world markets at stable prices."

These interests justify the presence of some 20,000 US forces in the region said the report. Pre-positioned forces and agreements with six Gulf states with whom the US conducts joint exercises would facilitate the rapid deployment of additional troops if needed.

205. Farer. COLLECTIVE SECURITY IN A CHANGING WORLD. 1993, p. 173. Every plausible scheme for reducing the risk of war within the region requires the deep involvement of leading actors outside it, particularly the United States. Mirror-image perceptions among many indigenous states of relentless enmity and irreconcilable conflicts of interest discourage efforts to imagine, much less pursue, regional collective security strategies. Given the importance of the area's one great natural resource, a peace consisting of armies on hair-trigger alert glaring across contested frontiers insufficiently guards the interests both of the rich northern states of the globe and of the developing and impoverished southern ones. Furthermore, such a peace guarantees a huge continuing diversion of global capital into weapons systems, the

further proliferation of weapons of mass destruction and their deployment under problematic safeguards against unauthorized and inadvertent use. At existing levels of hostility, insecurity, and paranoia, concern over proliferation and, after it occurs, a disarming first strike will function as the hair triggers of renewed conflict. So will the acts of local antigovernment groups with transnational links and increasingly powerful weapons.

206. Robert Lieber. REGIONAL SECURITY REGIMES. 1995. Even with the Israel-PLO agreement, significant perils remain in the peace process. One is that the September 1993 agreement may prove difficult to implement successfully. Another is that the Americans (implicitly) and the Saudis (explicitly) may place exclusive emphasis on the Arab-Israeli conflict as the preponderant source of regional instability. To be sure, the Arab-Israeli conflict remains a profound source of regional instability and a possible flashpoint for a war in which missiles, chemical weapons, and even nuclear devices could be used. This logic alone is more than enough to justify a major political commitment aimed at ending the conflict. For their part, the Saudis tend to see the continuation of the conflict as the principal source of extremism in the region, and they therefore find strong reason to work toward its resolution.

207. Zalmay Khalilzad. WASHINGTON QUARTERLY. Spring 1995, Lexis/Nexis. The Persian Gulf is critically important for a different reason—its oil resources are vital for the world economy.

In the long term, the relative importance of various regions can change. A region that is critical to U.S. interests now might become less important, while some other region might gain in importance. For example, Southeast Asia appears to be a region whose relative importance is likely to increase if the regional economies continue to grow as impressively as they have done in the past several years. The Gulf might decline if the resources of the region became less important for world prosperity because technological developments provided economically feasible alternative sources of energy.

208. DALLAS MORNING NEWS. 7 July 1996, p. 1.
The initial American response to the bombing that killed 19 U.S. airmen in Saudi Arabia has been to mourn and hang tough.

President Clinton hit the right note by immediately vowing that the nation wouldn't weaken in its military presence in the Persian Gulf out of fear.

But during this past week's celebration of America's 220th Independence Day, the United States needs to make another, longer-term vow as well to remove the national dependence on imported oil that helped put those airmen in harm's way in the first place.

The slain airmen weren't there on some abstract charitable mission to contain Saddam Hussein's predatory Iraqi army and protect the Saudi monarchy. They also were protecting America's druglike reliance on cheap energy that pours out of the oil taps of the Middle East.

In choosing to station combat aircraft and about 5,000 troops in the Persian Gulf to protect the world's most important oil fields, the United States also has chosen to involve itself in the region's murderous and tangled politics. Because of the way we live now, we have given ourselves little choice.

209. JERUSALEM POST. 21 May 1995, Lexis/Nexis.
The US military must continue to be engaged in the Middle East due to America's dependence on Persian Gulf oil and freedom of navigation in the region, the Pentagon concluded in the most recent of a series of regional security strategy reports.

The report noted that threats to Israel are at "an historic low" due to Iraq's defeat and recent developments in the peace process.

210. Khadduri. SECURITY DIALOGUE. June 1996, p. 157. So as not to leave any doubt about the continuing vital importance of the region, the US-Defence Department issued a report in May 1995 outlining US enduring strategic interests in the Middle East. The study points out that the world will become even more dependent on Gulf oil in the early 21st century than it is today, and that as long as the USA is a maritime commercial nation with global interests, it will 'have a stake in protecting freedom of navigation and access to regional markets.' It goes on to say that the USA 'must therefore remain engaged in the security of the Middle East diplomatically, economically and militarily.' The report pays particular attention to Iran and Iraq, which it describes as 'the most serious' dangers to the secure flow of oil from the Gulf to world markets. It asserts that, although Iraq lost more than half of its conventional military capability in the Gulf War, it still poses a significant threat to the region; further, that Iran harbors ambitions of establishing hegemony over the Gulf and expanding its influence over radical Islamic forces.

211. James L. Larocca. THE TIMES UNION. 16 July 1996, Lexis/Nexis. Americans have again died unnecessarily in the Middle East.

If you bought President Bush's public rationale back in 1990 and 1991, the United States went to war with Iraq to protect the sovereignty of nations and to demonstrate to the world that naked aggression would not be tolerated by the world's only remaining superpower.

That explanation was, of course, baloney.

The real issue then could be summed up in a single word: oil.

Saudi Arabia sits on 25 percent of the world's known oil reserves. Since the Persian Gulf war of 1991, the United States has maintained a huge military presence in Saudi Arabia and elsewhere in the region.

The mission now, as then, is to protect the flow of oil. The cost now, as then, is high, and can again be measured in American lives. This is both tragic and unnecessary.

212. THE SAN DIEGO UNION TRIBUNE. 6 July 1996, Lexis/Nexis. After the oil embargoes of the '70s, the United States supposedly learned to lessen its dependence on oil imported from the unstable Middle East. Instead, the United States is more dependent than ever on imported oil.

The cost for this oil does not stop at the per-barrel charges levied by Saudi Arabia and other countries. To ensure a secure energy supply, the United States must maintain a large military presence in the Middle East. Terrorist attacks, including the recent bombings in Saudi Arabia, have resulted in the tragic loss of American lives. Oil was the primary interest that the United States was protecting during the Persian Gulf War. Increasing the use of oil and other fossil fuels for energy also carries potentially disastrous environmental costs, including global warming.

213. THE HOUSTON CHRONICLE. 2 July 1996, p. 1. The Saudis deserve a measure of blame for not allowing U.S. troops to widen and strengthen their base perimeter. Perhaps the Saudis should have let the FBI interrogate four convicted terrorists before they were executed recently (although Saudi agents do not have routine access to interrogate terrorist suspects in this country).

However, U.S. forces are deployed in the Middle East to protect the industrial world's oil supply. The United States cannot abandon that supply to enemy threat just because a timid Saudi government is less than completely cooperative.

214. Jesse Helms. 27 March 1995, Y4.F7/2:S.HRG. 104–21, p. 30.
THE CHAIRMAN. All right, Mr. Gotbaum. I have just got a few seconds.

It costs a bundle—meaning it costs the taxpayers of the United States a bundle—to protect foreign suppliers. We all agree on that, do we not? That is the reason we have troops over there. That is the reason we went into the Persian Gulf, et cetera, et cetera. That cost a lot of money.

215. DETROIT NEWS. 27 June 1996, p. 1. Saudi Arabia sits on one-fourth of the world's known oil reserves and remains the world's leading producer and exporter of petroleum. The United States alone annually consumes 25 percent of total world oil output. America's industrial and commercial interests in the gulf region are clear.

216. James L. Larocca. THE TIMES UNION. 16 July 1996, Lexis/Nexis. It took an oil embargo and an interruption in imports in the winter of 1973–1974 to instruct the American people in the hazards of dependence on foreign suppliers and the importance of conservation. Project Independence grew out of that experience.

But no sooner had we begun to reverse our dependence on imports than we forgot those lessons and foolishly allowed our imports to rise.

Now, for the second time, American blood has been spilled in the service of the despotic regimes of the Mideast upon whom we have allowed ourselves to become mortally dependent.

With great sensitivity, President Clinton went to a memorial service for the Americans killed by the recent terrorist attack. This was a touching expression of the nation's grief over this latest tragedy.

A better and more useful response would be to reduce our dependence on Mideast oil to the point where we do not need to risk a single American life to protect it.

To do this, we must re-examine U.S. energy policy from top to bottom, dust off the national energy plan, restore those programs that assist and increase domestic production, including conservation and renewables, and remind ourselves that oil and oil sheiks are just not worth dying for.

217. DALLAS MORNING NEWS. 7 July 1996, p. 1. Oil isn't the only reason we are in the Persian Gulf for the foreseeable future. Had George Bush listened to those who said he shouldn't take on Iraq frontally and immediately, Saddam Hussein would have a sophisticated nuclear, chemical and biological warfare capability at his fingertips today.

But that wasn't obvious at the time to the American public (or the chairman of the Joint Chiefs of Staff). Saddam Hussein's threat to Saudi oil fields was. It triggered the significant escalation of stationed American troops in the gulf that apparently has enraged Saddam Hussein, Saudi domestic extremists or whoever set off that truck bomb.

218. THE NASHVILLE BANNER. 10 May 1996, p. 1. If there is a benefit to emerge from the run-up in gasoline prices, it is reawakened awareness of America's dangerous over-dependence on Mideast oil.

Against all notions of prudence, that dependence is at an all-time high as an oil glut in the world market has lured the United States away from cost effective, energy efficient technologies and the development of alternative energy sources. Just consider that the last time America ignored the warning signs of steadily growing dependence on imported oil, the Japanese were able to seize a significant share of the U.S. auto market with fuel-efficient cars.

219. A. J. Bacevich. ORBIS. Winter 1996, p. 40. To the extent that this wooing of the neopopulists tips the balance in the politics of 1996, what will be the likely consequences for American foreign policy? To begin with, populist choler will sweep away the interminable (and thus far inconclusive) deliberations of Washington elites about the much-anticipated post-cold war reorientation of

American foreign policy. The specific content of that reorientation will derive less from any closely held principles with regard to foreign policy than from their absence. In a political climate heavily influenced by neopopulism, attention will focus on those problems—many of them laced with prickly moral and cultural connotations—deemed responsible for the unraveling of American society and prosperity. In such a climate, high officials will have neither the latitude nor the inclination to expend political capital devising comprehensive solutions for external problems.

220. Stobe Talbott. U.S. DEPARTMENT OF STATE DISPATCH. 18 September 1995, p. 1. Let me put it as simply and bluntly as I can: Every single foreign policy initiative and program we have underway in the world today—from our support of new democracies and market economies in Central and Eastern Europe, to our support for the Middle East peace process, to our fight against international crime and narcotics trafficking, to our commitment to assure the safe dismantlement of nuclear weapons that have been aimed at our cities, to our current effort to bring peace to the former Yugoslavia—every single one of those efforts and countless more are in dire jeopardy.

221. Anne O. Krueger. ECONOMIC POLICIES AT CROSS PURPOSES. 1993, p. 36. These changes have affected the operation of U.S. foreign assistance programs. Even for supporters of aid, its mission is no longer as clear as it was; as political support for foreign aid has weakened, these supporters have increasingly accepted special provisos governing the allocation and administration of aid to win the support of doubters. These provisos have arguably reduced aid's effectiveness. Expectations as to what could be accomplished with given amounts of aid have been excessively optimistic. For these and other reasons, "aid fatigue" has emerged in the United States. Confronted with the low per capita incomes and widespread poverty that remain in numerous countries, many aid supporters have come to question its effectiveness.

222. WASHINGTON POST. 23 October 1995, p. 1. But the closest Congress came to reducing aid to Egypt and Israel—which together are set to get more than 40 percent of the total foreign aid allocation—was the move in two House subcommittees to cut overall worldwide economic aid funding without specifying which programs should be trimmed. Rep. Sonny Callahan (R-Ala.), chairman of the Appropriations foreign operations subcommittee, said the "sacred cow" of aid to Egypt and Israel "is going to have to be looked at very closely next year." But he added that he "wasn't surprised" when the full Appropriations Committee and its Senate counterpart restored the Egypt-Israel allocation.

223. George McGovern. MIDDLE EAST POLICY. Vol. 3, 1992, pp. 5–6. Complicating these strategic and ideological debates, U.S. economic interests have led to some rethinking about U.S. foreign aid to Israel. This aid, which has been approximately $3 billion per year over the past decade but which climbed to $5.6 billion during the past year, is more difficult to justify in a domestic climate of yawning budget deficits, persistent recession, alarming unemployment, and cities that are crying out for help. In fact, these economic considerations have fueled a mood of growing isolationism and opposition to foreign aid in general, even aid to the republics of the former Soviet Union. Thus, the United States may not necessarily be able to sustain foreign aid to Israel at these levels forever. The United States may also be unwilling to do so if Israel is unwilling to accept the U.N. land-for-peace formula, particularly when there are so many other worthy recipients of this aid in the Middle East, around the world, and in the United States.

224. Robert Lieber. REGIONAL SECURITY REGIMES. 1995. A regional security regime in the Middle East cannot be sustained without the active engagement of the United States. None of the regional actors, individually or collectively, nor outside powers, nor collective bodies such as the European Community or United Nations has the ability to serve in this capacity. This does not mean that the United States must always act unilaterally or commit prodigious amounts of its resources or armed forces to the region. It does require, however, that the United States remain closely engaged, both to lead collective action (as in the case of the Gulf War as well as in galvanizing the Arab-Israeli peace process) and to provide reassurance and deterrence. This is vital, not only for Israel and Saudi Arabia, but also for the purpose of affecting the behavior of regimes such as that of Assad in Syria.

225. Robert Lieber. REGIONAL SECURITY REGIMES. 1995, p. ??. Indeed, in the Arab-Israeli arena the American role has been essential for nearly every significant understanding reached among the parties. The September 1975 Sinai II agreement, the 1976 red line agreement between Israel and Syria, the 1979 Egyptian-Israeli Peace Treaty, and the sponsorship of peace talks beginning at Madrid in October 1991 are cases in point. Moreover, the stature of the United States in the post-Cold War, post-Soviet world remains unique. It is evident in the evolution of positions taken by Middle East actors that the United States has been able to shape their expectations and behavior.

226. Geoffrey Kemp. GLOBAL ENGAGEMENT. 1994, p. 213. Some have suggested that military trends in the region may force Israel to turn increasingly to the United States for assistance in stopping regional proliferation. In his commentary on the Syrian military and Arab nuclear programs, columnist Ze'ev Shiff noted in *Ha'aretz*

that "the American presence in the region constitutes a clear stabilizing factor." Furthermore, *Ha'aretz* cited Israeli defense sources and reported that Israel, the United States, and other countries are already attempting to halt nuclear assistance to Iran and Libya by Western companies.

227. Robert Lieber. REGIONAL SECURITY REGIMES. 1995. Implementation of these goals rests on a combination of American capability for unilateral action and multilateral cooperation. There is a preference for "collective response" to preclude or deal with regional threats, and the revised Defense Planning Guidance puts less emphasis on unilateral action than did an original draft of the report. The document reserves the "sovereign right" of the United States for unilateral defense of its vital interests, but it describes United States military preeminence as a "catalyst" rather than an alternative to collective action. The actual U.S. performance in the Gulf crisis thus appears to embody a conception of the American role that combines leadership with collective action. "Only a nation that is strong enough to act decisively can provide the leadership that is needed to encourage others to resist aggression. Collective security failed in the 1930s because no strong power was willing to provide the leadership behind which less powerful countries could rally against fascism. It worked in the Gulf because United States was willing and able to provide that leadership."

228. Ehteshami. MIDDLE EAST IN THE NEW WORLD ORDER. 1994, p. 70. Ultimately, much in the Middle East depends on the role of those outside powers and the part international organizations and agencies are willing to play. Of the outside powers, the role and attitude of the United States is paramount in any security scheme. The United States' power is illustrated by the fact that barely a year after leading a major conflict in the Persian Gulf region, it was leading a process which brought the parties to the most intractable conflict in the Middle East (the Arab-Israeli conflict) around the same table.

Whether one can rely on Washington to act consistently is another question. Paradoxically, while the United Nations served as an efficient machinery to counter Iraq's aggression against Kuwait, and was the main vehicle of US policy during the crisis, the Arab-Israeli peace process was initiated outside of the UN machinery, even though the Madrid process is, explicitly and implicitly, based on Security Council resolutions 242 and 338.

229. Robert W. Tucker. IMPERIAL TEMPTATION. 1992, p. 141. The new world order joined a distinctive conception of the nature of international order with the idea of America's leadership. Neither element was novel. The analogy drawn between the problems of domestic and international order has been a recurrent theme of American diplomacy in this century. Equally persistent has been the conviction that international order implies this nation's leadership. The two elements are not casually related; they are not considered the accidental product of transient circumstance. They are seen, for all practical purposes, as inseparable. The essential condition for America's participation in the task of constructing a satisfactory international order has been our leadership in that task. This leadership imposes special responsibilities others do not have, but it also confers a degree of freedom others do not enjoy. In this manner, our acceptance of multilateralism has been conditioned by our ability, bordering on a right, to act unilaterally. Bush's vision of a new world order expressed what is by now a traditional outlook.

230. Bar-Siman-Tov. REGIONAL SECURITY REGIMES. 1995, p. 167. Security regimes proved to be necessary but not sufficient conditions for shifting from war to peace. Actually, their achievement prevented further advancement toward resolution in the Syrian-Israeli and Jordanian-Israeli conflicts, when both sides preferred the benefits of security regimes to the costs of resolution (the territorial and political costs even prevent the formalization of the Jordanian Israeli security regime). Nevertheless, both sides to these conflicts acknowledged that institutionalization was not the last stage, but the first. The Israeli-Palestinian conflict remains the only conflict that fails to shift to institutionalization, mainly because the two sides differ as to the possible relationship between institutionalization and resolution.

231. COMPASS. Newswire, 5 February 1996, Lexis/Nexis. Rabin's assassination underlined the fissures within Israeli society over the peace issue, differences some fear could extend to within the military as well unless the benefits of peace quickly materialize and Rabin's successor, Shimon Peres, can contain and heal the rifts within a nation on the threshold of a new and uncertain era.

In the strategic arena, Israeli planners perceive long-range threats from Iran, which they allege is seeking to acquire nuclear weapons, ballistic missiles such as North Korea's No-Dong, and a long-range strike capability, as well as an intensification of Muslim fundamentalist extremism.

In this context, they note that the long-range threat and fundamentalist terrorism could well coalesce if the secular regimes in Algeria, Egypt and other states plagued or threatened by fundamentalist insurrection collapse and fuse into a hostile Islamist bloc.

232. Yair Evon. VIEWPOINTS: THE FUTURE OF ISRAEL'S NUCLEAR POLICY. 1995, (gopher://irpsserv26.ucsd.edu:70) The changes which have now taken place in Arab-Israeli relations; the peace with Egypt, the Madrid process, the Jordanian-Israeli peace, and the agreements with the Palestinians; have diminished security threats and the likelihood of another war. In addi-

tion, the end of the Cold War, the disintegration of the Warsaw Pact and the Soviet Union, and changes in Russian foreign policy further inhibit destabilizing processes in the Arab-Israeli zone. Moreover, the global trend toward reducing the saliency of nuclear weapons could also affect Israeli considerations.

On the other hand, the growth of powerful, militant Islamic movements hostile to current regimes, the peace process, and the West; the possibility that Iran and Iraq might—though in some more remote future—acquire a nuclear weapons capability; and the continued proliferation of weapons of mass destruction throughout the Middle East, all contribute to the continued concern in Israel that the processes of instability might again be set in motion.

233. COMPASS. Newswire, 15 January 1996, Lexis/Nexis. But the danger is that if these do not appear, if they do not bring the long delayed economic, political and social reforms that will improve the quality of life of the region's 250 million people, internal conflict in many countries of the region must be expected, with the prospect of the violent collapse of entrenched regimes as populations swell while food production remains stagnant and precious water resources are drained.

What the Middle East now faces is a crisis of leadership. From Libya to Iran, republican regimes and hereditary monarchies are having to grapple with the new realities of demands for democratic reform and stagnant oil prices that are unravelling the political fabric.

This is certainly true of the Gulf region, now inescapably dependent on the West, and the Americans in particular, for its protection against Iraq and Iran, perceived as the main potential threat.

234. COMPASS. Newswire, 15 January 1996, Lexis/Nexis. The more pessimistic Arab analysts predict new military confrontation in the Gulf within five years. U.S. military strategists are a little more optimistic, but are planning for renewed conflict in the region somewhere down the line. Last May, Bill Clinton unveiled his new blueprint for protecting the United States and its interests, "A National Security Strategy of Engagement and Enlargement."

235. Martin van Leeuwen. THE FUTURE OF THE INTERNATIONAL NUCLEAR NON-PROLIFERATION REGIME. 1995, p. 125. Since time immemorial, the Middle East has been ridden with armed conflicts, and during this century, arms races have been raging in the region with special intensity. This can partially be explained by the area's recently discovered source of affluence: oil. Oil-producing countries, and regional powers subsidized by them, have spent fortunes on buying and in some cases producing advanced conventional weapons, but increasingly also on obtaining the capability to produce non-conventional systems. On several occasions during recent decades, the Middle East has witnessed the use of non-conventional—chemical-weapons in battle and against civilians.

236. JERUSALEM POST. 4 August 1995, Lexis/Nexis. The painful truth is that the pre-Islamic, "old" Middle East was at least nominally close to Israel, in the sense that the present regimes in Egypt, Jordan, Syria and even Arafat's PLO are secular in nature. But the chances for the emergence (or continued existence) of "progressive" regimes in the Middle East is rapidly fading. The violent new societal forces which have been unleashed in the Middle East bodes a brutal, uncompromising order.

237. Shahram Chubin. NUCLEAR PROLIFERATION. 1994, p. 35. The Middle East is considered one of the most sensitive regions from the point of view of proliferation, with states relatively low in capability but high in risk. Proliferation here could threaten nearby Europe and the security of allies (Israel and Turkey), as well as access to the region's oil resources. Islamic fundamentalism, which has been growing in the region, is a source of concern, and it, together with the absence of democracy, is seen as a potentially destabilizing force. The region is seen as alien, not sharing and even hostile to Western values and opposed to the Western order (political, economic, and cultural) in which the inhabitants perceive themselves victims and dispossessed.

238. Bar-Siman-Tov. REGIONAL SECURITY REGIMES. 1995, p. 141. This study argues that the establishment of formal and informal security regimes between Israel and its Arab neighbors (Egypt, Syria, and Jordan) was a necessary although not sufficient condition for the shift from war to peace. The failure of establishing such a regime in the Israeli-Palestinian conflict makes this conflict less tractable. The rationale for this argument is that a protracted conflict lingering over time with violent hostilities cannot be resolved without a prerequisite security arrangement that stabilizes the security and the strategic relationship between the parties of the conflict. Security regimes are institutions that control and reduce the conflict to the level where war ceases to be an attractive means to resolve the conflict and the sides are not ripe for conflict resolution. The effectiveness of the regimes may influence the sides to look for resolution, but the sides may also prefer conflict reduction to conflict resolution.

239. Ehteshami. MIDDLE EAST IN THE NEW WORLD ORDER. 1994, p. 70. In my view, the recognition of the need to establish viable security structures in the Middle East in the post-Cold War international system stems not from some utopian dream, which in itself is not an unworthy endeavour, but rather that since 1991 new and golden opportunities have presented themselves for making the Middle East a less unstable sub-system.

In other words, judging by what went before, the absence of superpower rivalries in the Middle East should

make the exercise a realizable one. In addition, the disappearance of the established 'norms' and the evaporation of unquestioned external patrons makes the enterprise a necessity if the post-Cold War dislocation is not to lead to another war in the Middle East in the not too distant future.

240. Kupchan. COLLECTIVE SECURITY BEYOND THE COLD WAR. 1994, p. 49. Collective security rests on the single notion of all against one. While states retain considerable autonomy over the conduct of their foreign policy, participation in a collective security organization entails a commitment by each member to join a coalition to confront any aggressor with opposing preponderant strength. The underlying logic of collective security is twofold. First, the balancing mechanisms that operate under collective security should prevent war and stop aggression far more effectively than the balancing mechanisms that operate in an anarchic setting. At least in theory, collective security makes for more robust deterrence by ensuring that aggressors will be met with an opposing coalition that has preponderant rather than merely equivalent power. Second, a collective security organization, by institutionalizing the notion of all against one, contributes to the creation of an international setting in which stability emerges through cooperation rather than through competition. Because states believe that they will be met with overwhelming force if the aggress, and because they believe that other states will cooperate with them in resisting aggression, collective security mitigates the rivalry and hostility of a self-help world.

241. George W. Downs. COLLECTIVE SECURITY BEYOND THE COLD WAR. 1994, p. 21. In clubs, the problem of free riding can be significantly attenuated if not completely eliminated. For instance, if a club owner provides club goods by collecting fees from club members, it is possible to attain Pareto optimality under some technical conditions. Of course, the analogy is nearly perfect only if the collective security system has a "manager" who provides collective forces by collecting financial and other resources from members. Whether this is true in any particular case depends on the type of system. A hegemonic collective security system has a built-in manager, a loose concert of states may have nothing that comes close.

242. Lipson. REGIONAL SECURITY REGIMES. 1995, p. 9. *Regimes can make agreements more transparent* by creating mechanisms and institutions to provide information otherwise unavailable or too costly to acquire. Greater transparency does not make betrayal impossible, but it lowers the gains from cheating and deception. In arms control, where cheating is a very real concern, agreements have aimed at raising transparency. There have been prohibitions against encrypting missile test data, for instance, and agreements to permit on-site inspections and surprise visits to military facilities. The real trick is to increase transparency enough to build confidence in the reliability of agreements but not so much that other military secrets are disclosed or sovereignty challenged.

243. Bar-Siman-Tov. REGIONAL SECURITY REGIMES. 1995, p. 1. The establishment of formal security regimes became possible in the Egyptian-Israeli and Syrian-Israeli conflicts only after the 1973 War, when the sides realized that the war ceased to be an effective means of accomplishing unilateral objectives. The informal institutionalization of the Jordanian-Israeli conflict was made even before the 1973 War, due to the strategic cooperation that saved the Jordanian regime. Formalization was not a necessary condition for reaching and maintaining security regimes. The Israeli-Jordanian regime proved to be effective. Even without formal confidence-building measures, such as buffer zones controlled by external forces, both sides succeeded in maintaining a high degree of institutionalization.

Institutionalization stabilized each dyadic conflict, and no war erupted since. The only exception was the Israeli-Syrian war in Lebanon, but it was limited to Lebanon, while the Golan Heights remained quiet. Probably the effective regime in the Golan Heights was responsible for that outcome.

244. Yair Evron. ISRAEL'S NUCLEAR DILEMMA. 1994, p. 241. It seems not unrealistic to picture an even more optimistic scenario, in which Syria would be forced to consider the possibility of actual military cooperation between Jordan and Israel, following the signing of a peace agreement. This does not lie beyond the realms of possibility. After all, a few years before it would have been difficult to imagine that the conflict-ridden relations between Israel and Egypt would improve to the extent that even prior to the signing of the Camp David agreement, Egypt would be able to react with understanding to a major Israeli military initiative. Indeed during this operation—the 1978 Litani Operation—Israel provided Egypt with a constant flow of information regarding the campaign. Furthermore, at the time of the Lebanese War in 1982, an event that caused Egypt extreme political discomfort, Egypt was still able to "accept" the war, without attempting to cancel the peace agreement. As for Jordan, in the past, when confronted by threats to her independence and security, she has acted in defense of her interests, even if these actions required the attainment of tacit understandings with Israel. Given conditions of peace between Israel and both Egypt and Jordan, it is therefore quite reasonable to assume, that in cases in which Jordanian interests would be served, such as during a war with Syria, or with a Syrian-Iraqi coalition, Jordan would be willing to accept Israeli aid either directly or indirectly.

For example, the transfer of Israeli forces through Jordanian territory in order to carry out a deep flanking attack against Syria, illustrates the kind of Israeli-Jordanian

cooperation that would improve Israel's strategic ability, and would create serious difficulties for Syria. Such a scheme would not require the activation of any Jordanian forces against another Arab army, and would be of major assistance to Israel. In the absence of a Jordanian-Israeli peace agreement, a deep flanking operation of this nature would raise the specter of war between Israel and Jordan. This strategic cost would clearly outweigh operational military benefits that the operation might furnish.

245. Yair Evron. ISRAEL'S NUCLEAR DILEMMA. 1994, p. 229. In Chapter 2 we noted the relative success, over an extended period, of Israeli conventional deterrence. Note was made of the complexity of the deterrence process, and of the existence of three axes of equilibria which comprise this process: the balance of military power, the balance of interests, and the balance of resolve, with emphasis on the first two axes. An understanding of these three balances and the interaction between them is critical to the analysis of Israeli conventional deterrence. Reducing the motivation for war as the result of political arrangements will contribute to stabilizing politico-strategic relations, and this in turn will intensify the effect of conventional deterrence.

The analysis presented in this book concerning the processes of escalation and crisis, that have preceded and eventually led to the outbreak of the various Israeli-Arab wars, is indicative of the direct correlation that exists between the level of readiness to go to war, and the degree of political grievance felt by different Arab states in their interaction with Israel. Indeed, prior to the 1967 War, while the Arab states were adamant in their refusal to accept the territorial status quo, they nevertheless refrained from initiating war with Israel. War was not at the top of their respective priorities. Moreover, war with Israel would have involved the Arab states in very high costs, and threatened to expose them to the danger of severe punishment. Since 1976, the political grievance arising from the damage to the particular interests of Egypt and Syria—the loss of Sinai and the Golan–has caused the War of Attrition of 1969–1970, and the 1973 War. These two wars resulted therefore, not from the traditional causes of the Arab-Israeli conflict, but rather from specific grievances relating to the particular interests of the respective states. It is, therefore, highly probable that the relief of this grievance will improve the balance of interests, which is fundamental to the effectiveness of the deterrence process.

246. Misha Louvish. JEWISH FRONTIER. March/April 1995, p. 8. One is the Likud's stone-walling policy, personified until 1992 by Prime Minister Ytzhak Shamir and still advocated by right-wing leaders like Binyamin Netanyahu, Ariel Sharon and "Raful" Eitan. It is founded on the principle that Israel must hold on for all eternity to the entire territory between the sea and the Jordan River. The long-term aim is the annexation of the area and an apartheid regime, in which Jews will enjoy all democratic rights, while Arabs will have no say in the future of the country in which they live.

It is difficult to understand how otherwise rational men can defend such a policy. Surely it is obvious that sooner or later there will be another uprising with the support of all the Arab states. In other words, this policy can only lead to another—and probably a disastrous—war.

247. Yair Evron. ISRAEL'S NUCLEAR DILEMMA. 1994, p. 241. The second scenario is that of war initiated by Syria alone with the objective of regaining the Golan Heights. The Syrian operation could take the form of either a surprise attack or a static war of attrition. It may be assumed that the probability of such a war will increase as political conditions become more favorable to the Syrian leadership–for example, should Arab frustration rise regarding the failure to find a solution to the Palestinian problem or should Israeli-Egyptian relations deteriorate. In such a case, Israel could concentrate greater forces against Syria than in the first scenario, thus weakening Syria's ability to block an Israeli attack. The probability for such a scenario is low, but it will rise as Syria's perception of political conditions changes—that is, when the Syrian leadership believes that additional Arab states will actively join the war.

248. Alasdair Drysdale and Raymond Hinnebusch. SYRIA AND THE MIDDLE EAST PEACE PROCESS. 1991, p. 200. So long as the Arab-Israeli conflict remains unresolved, the Middle East will not enjoy real stability, and the dispute will continue to divert the region's energies from other pressing problems. Moreover, existential fears and deeply felt grievances will continue to fuel a profligate arms race that has already made the Middle East the world's most heavily militarized, and perhaps most dangerous, region. It would be imprudent to assume that the status quo, in which another war seems unlikely, can be maintained indefinitely. Without an equitable settlement in which the legitimate interests of all parties are addressed, the possibility of military confrontation will grow. The next war—which would be the sixth since Israel's birth—would almost certainly be more destructive than any of the previous ones because of the proliferation of atomic, biological, and chemical weapons in the region. The United States, therefore, cannot afford to let the peace process die, despite formidable obstacles in the way of a diplomatic solution to the conflict.

249. Tom Idinopulos. COMMONWEAL. 16 June 1995, p. 7. Conversations with Israelis and Palestinians convince me that if the triangle for peace breaks apart, killing will resume at a greater level than what we have seen from terrorism. A failed peace will trigger more *intifada*, more military occupation, more Jewish housing construction, and a shattering of the Palestinian dream of self-determination. A failed peace will mean a victory for terrorism, for the peace rejectionists in Hamas

and Likud, and would be a severe blow to American peace initiatives.

250. David Inbar. REGIONAL SECURITY REGIMES. 1995, p. 189. The Arab-Israeli conflict is in a state of flux. It is moving toward reconciliation, though what we see is not necessarily an irreversible process. Although the parties to the conflict still consider resorting to force to realize their interests, the costs of war are climbing and its benefits seem to be in decline. Therefore, although the rationale of deterrence still has some validity, it is less satisfactory for meeting the security needs of Israel and its rivals. Indeed, several international and regional developments seem to lead to a plausible inference that Israel's ability to deter limited wars, in particular, has been hampered. The perceived reduced value of Israel as an American ally; the Arab perceptions concerning Israel's war-waging capability and its political utility; the renewed regional arms race; the increased Arab capacity to inflict damage on Israeli population centers; the growing doubts about Israeli resolve to use force; the constraints on Israel's freedom of action as a result of the peace process and its desire to maintain good relations with the United States—all indicate that Israel's traditional emphasis on deterrence has become less dependable. Effective unilateral military action has become more difficult because of political, economic, operational and intelligence constraints. The costly nature of the future battlefield has become a disincentive for military action, in particular for Israel. Concomitantly, the peace process entails promises for enhanced conflict management and the establishment of security regimes between Israel and its neighbors.

251. David Inbar. REGIONAL SECURITY REGIMES. 1995, p. 189. Fourth, when taking into consideration Israel's limited leeway in meeting the territorial demands of the Arabs, Israel might prove more flexible in negotiating the various elements required for a security regime. As a matter of fact, interim arrangements on Israel's eastern border are the most realistic goal for the present process. The peace process is important not because it offers peace—which it does not—but because it entails possibilities for the establishment of security regimes to stabilize or manage the conflict.

252. Anthony Cordesman. WEAPONS OF MASS DESTRUCTION IN THE MIDDLE EAST. 1991, p. 173. The continued paralysis of the Arab-Israeli peace process could lead to a rebirth of the most extreme forms of Palestinian covert action or terrorism. Such terrorist movements might well feel that past forms of more conventional terrorism or covert action have failed to force the West into taking dramatic action. They might conclude that chemical or biological attacks would be far more dramatic and force the West to intervene in ways that would advance their cause. At the same time, some movements might act to seek revenge or out of sheer nihilism.

Libya, Iran, Iraq and Syria have all backed such terrorist movements in the past, or given them funds and arms. The production of small amounts of chemical and biological weapons is also well within the reach of an organised and well-funded terrorist movement. Finally, one cannot dismiss the risk that a terrorist movement might see chemical and biological weapons as a means of deliberately triggering a war in an effort to force states into far more radical action than they would otherwise take. For example, they might see a successful attack on Israel in a crisis or during a conflict as a means of catalysing a broader Arab-Israeli conflict or final struggle.

253. Aaron Karp. WASHINGTON QUARTERLY. Autumn 1995, p. 45. A greater danger of destabilization comes from the threat of terrorism and fundamentalist violence. Their rise poses a direct threat to safety, but this is not, except in the most extreme circumstances, a threat that can be met with military means. The exceptions, however, are of such importance as to require serious and searching consideration. Although terrorist violence itself does not justify procurement of advanced major weapons, the danger that extremists might acquire weapons of mass destruction cannot be dismissed, nor can the potential for destabilization. Should fundamentalists take power in the Arab world as they have in Iran, the possibility of renewed escalation would be very difficult to control.

254. Aaron Karp. WASHINGTON QUARTERLY. Autumn 1995, p. 46. The latter scenario is the source of widespread fear. One need look no further than best-selling novels, in which the theme of nuclear-armed terrorists long ago became common enough to bore anyone not addicted to the genre. Since the collapse of the Soviet Union, the international community and the ex-Soviet states have taken several steps to prevent nuclear weapons from falling into the hands of substate actors. After the Russian plutonium smuggling scares of July and August 1994, action to control fissile materials also became more coordinated. Yet the possibility that nuclear weapons may reach a subnational group remains serious.

255. Lewis Dunn. AC AND THE NEW MIDDLE EAST SECURITY ENVIRONMENT. 1994, p. 235. Technical factors also contributed to the stability of the Cold War nuclear balance. For example, the United States spent a lot of money to ensure the safety of its nuclear warheads. As a result, when a technician dropped a wrench down a Titan-2 missile silo back in 1980, leading to explosion of the missile fuel and the throwing of a ten-megaton warhead 400 yards down the road, an accidental nuclear detonation was involved.

By contrast, other new nuclear powers may pay less attention to safety, putting top priority on building the bomb. Consider how the Iraqis stored chemical weapons:

they simply kept chemical agents in 55 gallon drums that were thrown around. At the very least, Iraq clearly had a very different safety culture.

256. Aaron Karp. WASHINGTON QUARTERLY. Autumn 1995, p. 32. What allowed high levels of military spending to come down was the reduction of threat perceptions. More than anything else, it was gradual acceptance of the status quo that enabled progressively more countries of the Middle East to slow their arms procurement. By the mid-1980s the process was virtually complete among Israel and its neighbors. In the mid-1990s the logic of the status quo has reached most countries throughout the Middle East. As in Europe after World War II, general and unrestrained war is increasingly unacceptable. The danger of general war in the Middle East comes not from deliberate choice, but from mistakes and miscalculation.

257. Lewis Dunn. AC AND THE NEW MIDDLE EAST SECURITY ENVIRONMENT. 1994, p. 34. Concern about dangers of proliferation has little to do necessarily with whether the Arabs or the Iranians are irrational, although one experienced "Saddam-watcher" suggested that Saddam Hussein would be prepared to use nuclear weapons even if this were to lead to his country's destruction. Setting aside the argument about the possibility of "crazy leaders" or "crazy states," two reasons for concern stand out: First, the conditions of stable deterrence may not be met in the Middle East; and secondly, "rational" national nuclear-forcebuilding choices may lead to unstable regional strategic outcomes.

258. Anthony Cordesman. WEAPONS OF MASS DESTRUCTION IN THE MIDDLE EAST. 1991, p. 110. *Conflicts involving Syria or Israel*. The increasing range and lethality of Iraqi weapons, coupled with the current crisis over Kuwait, is creating a situation where Israel must consider the risk of Iraqi attacks in support of Iraq's objectives in Kuwait. In the long run, Israel also faces the risk of Iraqi involvement in a new Arab-Israeli conflict. While Iraq is no friend of Syria, and its actions are rarely as reckless as its words, it is impossible to rule out a series of threats and counter threats, preparatory actions, or limited use of chemical weapons that could as such trigger preemption and/or escalation. If Iraq's strike capabilities remain intact after the current crisis.

259. Anthony Cordesman. WEAPONS OF MASS DESTRUCTION IN THE MIDDLE EAST. 1991, p. 112. This list of contingencies and impacts on the balance is scarcely comprehensive. It does, however, illustrate the complex interactions that weapons of mass destruction can have on the regional military balance, and how difficult it is to predict the end result. It also seems worthwhile pointing out that while neither Iran or Iraq seem to have 'irrational' leadership elites, their values are different to those of Western leaders. The history of the Iran-Iraq war, and Iraq's invasion of Kuwait, show they are willing to take significantly greater risks. Further, the experience of these leaders in crisis management is often limited, and they lack the kind of C^3I systems necessary to give them accurate information. Given the long history of failures in crisis management in the West, and the resulting military mistakes and miscalculations, there is at least equal potential in the Middle East for wars that get badly out of control.

260. Lewis Dunn. AC AND THE NEW MIDDLE EAST SECURITY ENVIRONMENT. 1994, p. 229. The nuclear agenda, both at the global level and in the Middle East, has figured prominently in public debate since the collapse of the Soviet Union and the discovery of Saddam Hussein's 'mini-Manhattan Project." In reviewing that nuclear agenda, this chapter makes six points:

First, the global nuclear order, which provides the basic context for the nuclear agenda in the Middle East, is in the midst of fundamental change; this change has mostly encouraging but some discouraging characteristics;

Secondly, the familiar Middle East nuclear status quo of the past two decades—an Israeli nuclear monopoly—is increasingly unstable;

Third, a multi-nuclear Middle East quite probably would be very dangerous, unstable and with less security for all countries in the region;

Fourth, unilateral or coercive nuclear non-proliferation measures, whether export controls or military intervention, will not suffice by themselves to prevent a multi-nuclear Middle East;

Fifth, arms control, and in particular unilateral actions by the government of Israel to indicate nuclear restraint, can buttress coercive measures in useful ways; and

Sixth, a long-term solution to head off a multi-nuclear Middle East calls for dealing with the underlying political disputes and confrontations of the region; but there may not be enough time to wait for the long-term.

261. Kam. AC AND THE NEW MIDDLE EAST SECURITY ENVIRONMENT. 1994, p. 89. This set of assumptions leads the Arabs to expect an Israeli decision to initiate a war against one or more Arab states under several possible scenarios: should the military balance change considerably in favor of the Arabs, for example due to a significant strengthening of the Syrian armed forces, which might enable them to take a military initiative;or if there emerges meaningful military cooperation between at least two central Arab states, such as Syria and Iraq, Egypt or Jordan; or if an Arab state began to produce nuclear weapons, or achieved a breakthrough of consequence toward this objective.

262. THE CHRISTIAN SCIENCE MONITOR. 4 September 1996, p. 1. Much of the Arab world reacted angrily to American missile attacks Sept. 3 against

Iraqi strongman Saddam Hussein, with declarations that the intended US message was "overkill" that jeopardized Iraq's sovereignty and the stability of the Mideast.

Even Arab leaders closest to the United States have been lukewarm about the strikes against a fellow, if distrusted, Arab nation. The strikes were a response to the weekend advance by Iraqi troops into a US-protected Kurdish zone in northern Iraq.

Jordan—which signed a US-sponsored peace deal with Israel in 1994 and hosts Iraqi opposition leaders—refused to serve as a base for this US operation against Iraq. The move was reminiscent of Jordan's pro-Iraqi tilt during the 1991 Gulf War.

Strong support came from the tiny Arab emirate of Kuwait, which was occupied by Iraqi troops in 1990 and 1991. Kuwait expressed "full understanding" for the attack.

In Saudi Arabia, a staunch US ally that provided the launching pad for Operation Desert Storm during the Gulf War, reaction was mixed. But Saudi Arabia, along with Kuwait, agreed to continue logistics support for ongoing US operations in Iraq.

Arab analysts say the negative reaction on the street is strong, and that any moral high ground that the US capitalized on during the Gulf war, when it led a coalition of Western and Arab forces to expel Iraqi occupation troops from Kuwait, is not a part of this campaign.

And the strike coincides with increased Arab anxiety about Washington's commitment to being an impartial arbiter of the Mideast peace process, especially on the Israeli-Palestinian question.

263. Feuerwerger. POST WAR POLICY ISSUES IN THE PERSIAN GULF. 1991, Y4.F76/1:P84/3, p. 308. At this time, it is clear that progress on each of these objectives will not be easy. With respect to regional security structures, the initial effort to establish an alliance between Gulf Cooperation Council states and Egypt and Syria has not borne fruit. The successful effort to impose arms reductions on Iraq by means of UN Security Council Resolution 687 will require extensive implementation efforts, and the President's prospective arms control initiative will have tough going among Mideast states and weapons suppliers. The initial robust concept of a two-track Arab Israeli peace process has been scaled back to a more modest approach, while the Lebanese problem has been placed on the back burner. Economic development and overcoming the gap between rich and poor in the Middle East is again being handled in a routine manner. And the promotion of democracy in Kuwait and Iraq can hardly be considered a triumph for American values at this juncture.

Nonetheless, American post-war policy has demonstrated some—if limited—success. While the Middle East will never enjoy a security structure akin to that of NATO, the Gulf Cooperation Council appears to be approaching regional security with a far greater degree of seriousness than in the past. In cooperation with GCC states, the United States is improving its capability to deploy forces to the Middle East in future crisis.

264. James Blackwell. PERSIAN GULF CRISIS. 1993, p. 134. It will not be easy to keep people, ideas, and systems in balance for the future. The threats we face are not diminishing, they are only shifting. As underlying tensions around the globe come to the surface in the absence of an overriding superpower rivalry, conflict is bound to increase. Where those conflicts threaten U.S. interests, the United States must be prepared to employ its armed forces to protect those interests.

Not every situation will be as demanding as the Persian Gulf deployment and war, but some will. It is likely that a future war will require even more of us than did this war. And surely we will not be able to assume that our next opponent will be as stupid as Saddam Hussein turned out to be. We must not become complacent about our military forces simply because of Saddam Hussein's miscalculations. We must strive to build a force that will cause future would-be Saddams to understand what can be brought to bear against them and not threaten our interest in the first place.

265. Michael Mazarr. DESERT STORM. 1993, p. 165. In terms of logistics, too, forward presence makes a difference. Not only does it provide a worldwide network of bases from which military operations could be launched but it also encourages the creation of logistical infrastructures in possible contingency areas. Drawing a last example from the Gulf War, the huge (and largely secret) military infrastructure built up in Saudi Arabia, partly through U.S. security assistance programs, proved essential in handling the massive influx of troops and equipment during Operation Desert Shield.

266. Anthony Cordesman. IRAN AND IRAQ. 1994, p. 290. Finally, it is important that the West and the southern Gulf be as prepared to limit escalation and terminate any future conflicts with Iran and Iraq as they are to win a major war. Delay or graduated escalation are not the answer. In most cases, it will be far better to bring superior force to bear quickly and decisively while any conflict is still limited and to encourage conflict termination on conditions Iran or Iraq can accept. Above all, military responses must be flexible and designed to achieve clear political objectives. Using force simply for demonstrative purposes simply leads to future conflict. Winning another Desert Storm is infinitely preferable to losing another Desert Storm, but it is not preferable to the kind of more limited military action that will both achieve political objectives and avoid large scale conflict. As is the case in most regions, it is necessary to be strong and decisive, but the tactics of Sun Tsu are preferable to those of Clausewitz or Douhet.

267. Michael Mazarr. DESERT STORM. 1993, p. 165.
Forward presence forces support. contingency operations in a number of important ways. In some cases they will actually provide the troops for the operation—in the Gulf War, for example, an entire U.S. corps was deployed from Europe, and such redeployments are an explicit rationale for a continued U.S. presence on the Continent. Forward presence is central to effectiveness, providing a forum for training and planning in partnership with potential allies. Again in the Gulf War, U.S., British, and French forces used procedures, tactics, and interoperable equipment drawn from forty years of NATO experience.

268. St. John Armitage. MIDDLE EAST INSIGHT.
Special Edition, 1995, p. 59. But perhaps the greatest threat to the present relationship is the fact that oil is of greater importance than all other considerations. Whatever the nature of Saudi Arabia's government, its oil will have a market with which America will deal. That certainty raises questions about the strategic importance of an American military presence in the area and the nature of the US-Saudi Arabian relationship in the long term. Both might best be answered by America addressing the problems of the situation as perceived by the Saudis and lending support to their search for solutions.

269. St. John Armitage. MIDDLE EAST INSIGHT.
Special Edition, 1995, p. 59. The strength of US-Saudi relations at the government level depends on the degree of recognition and understanding by those two countries of their respective interests which, unlike shared interests, may conflict. Saudis do not see much reciprocity in the relationship. Access to Saudi oil will remain America's vital interest second only to, if not as part of, national self-defense and one from which America cannot disengage. For Saudi Arabia, as for all countries of the region, economic well-being, inevitably oil-based, is essential to political and social stability. Whilst for many years Saudis have perceived a serious imbalance in the give and take of the relationship, they have also been very much aware that distancing themselves from America would be a destructive process.

270. Hermann Frederick Eilts. MIDDLE EAST
INSIGHT. Special Edition, 1995, p. 24. In general, the Saudi leadership tends to believe that US officialdom too often tries to be neutral in matters affecting the Kingdom and neighboring states. In its view, by virtue of the close relationship between the two countries, it is entitled to US support irrespective of the issue. The United States, it contends, should trust Saudi Arabia to be fair in such matters. The Yemeni civil war of August 1994, in which the United States supported Yemeni reunification efforts and Saudi Arabia lent unofficial support to the south Yemeni secessionists, is the latest such incident.

271. Walter Cutler. MIDDLE EAST INSIGHT.
Special Edition, p. 50. Another reason for the decline in opposition to arms sales is progress in the peace process. We have seen the Arab states, including Saudi Arabia, playing a positive, cooperative role in that process. Let us not forget that there has been a Saudi presence ever since the Madrid Conference. Even though there have been strains—such as those between the Gulf states and the Palestinians stemming from the Gulf War, as well as immense difficulties in the process itself—Saudi Arabia recognizes that confrontation does not work, has not worked, and that the only effective way to resolve these problems is through negotiations.

272. BBC SUMMARY OF WORLD BROADCASTS.
9 June 1995, Lexis/Nexis. The two **leaders** discussed current Arab, Islamic and international issues. They underlined their support for the **peace process** on all Arab-Israeli tracks. They hoped it would reach a just and comprehensive and lasting solution to the Palestinian issue and to the Arab-Israel conflict on the basis of the UN Security Council Resolutions 242, 338 and 425, and the principle of land for peace with the aim of achieving a complete Israeli withdrawal from occupied Arab areas, recovering Palestinian legitimate rights and establishing a state for them with Holy Jerusalem as its capital.

273. Omar al-Hassan. "Saudi Arabia and the New
World Order," INTERNATIONAL RELATIONS. April 1994, p. 70. The argument advanced in this article does not call for any profound policy or practical changes. Rather, what would be needed to bring about the new situation which it is suggested would be advantageous would be a new perception of the Kingdom of Saudi Arabia fostered and encouraged through a different attitude from western governments, and America in particular. The contention is that Saudi Arabia, as a result of geography, history and social structure, is a naturally moderate country which is also anxious for reasons of its own security to stem extremism and prevent any steps towards expansionism by other regional powers. With encouragement, it would be ready to act far more forcefully than it has in the past to stem aggression and encourage moderate governments. Its Islamic pre-eminence also gives it great influence with Muslim countries, where it could also be expected to throw its weight behind peaceful, middle-way policies which would be of ultimate advantage to the west as well as to those practicing them.

274. Joshua Teitelbaum. THE JERUSALEM
REPORT. 17 November 1994, p. 55. Nor should we expect any other dramatic steps toward Israel from Riyadh, though there may be some movement when the U.S. brings pressure from time to time. Saudi Arabia could well be the last Arab country to reach a peace accord with Israel. In spite of certain encouraging signs, like the boycott announcement, Riyadh's concern with its domestic

opposition will continue to dictate its position on the peace process.

275. Richard Leaver. "The Gulf War and the New World Order: Economic Dimensions of a Problematic Relationship," AUSTRALIAN JOURNAL OF POLITICAL SCIENCE. July 1992, p. 299. Economically, the blunt fact is that the Saudi budget, which was incurring both internal and external deficits before the war, has been loaded down with financial pledges and war costs that, according to some estimates, actually exceed total American disbursements in the Gulf by 20%. Additional revenue generated by increased oil prices and volumes halved the current account deficit during 1990, but covered only a quarter of total war costs. The balance of these international pledges will therefore be paid off through the coming years when prices are likely to be at or beneath the official OPEC benchmark, and high export volumes present the only source for additional revenue.

This dilemma has already turned the Saudis to their first-ever international borrowings. In May 1991, they concluded a loan of US$4.5 billion with a syndicate of 20 banks, headed by Morgan Guaranty, which paved the way for other borrowings. These sums are presumed to have been forwarded to Washington where Congressional pressure for prompt repayment was relentless. Since Congressional approval will be required for planned Saudi arms purchases, the American demand for priority was backed by a strong bargaining position. On the domestic scene, where the internal deficit ballooned, the Saudi regime expanded the quantity, range and yield of Treasury bills to cover its liquidity problem. In an environment where domestic investors traditionally exhibit a strong aversion to long-term financial instruments, yields had to be set relatively high.

276. MONEYCLIPS. 29 September 1994, Lexis/Nexis.
JEDDAH, Sept. 28—The memorandum of understanding between Canada and Saudi Arabia covers all aspects of military cooperation, including the training of each other's personnel, according to Canada's Minister of National Defense David Collenette.

Answering a question if the memorandum covered anything regarding training of Saudis, particularly in technology and modern military technology at a press conference at Inter Continental Hotel here today, Minister Collenette said, "of course, there have been discussions by respective armed forces as a result of the memorandum of understanding, about even closer relations, not only between the military but also equipment sales to Saudi Arabia." He said that "on the air side, we will be more than prepared to give pilot training and other technological training to members of the Royal Saudi Air Force."

Collenette said that Canada has had very good relations with the Kingdom for years, and these became especially close as a result of Canada's participation with **Saudi Arabia** in the Gulf War.

Stressing that Canada respects the Kingdom's role in Middle East peace, and security, he said, "it is the key opinion-maker in this area."

Collenette said, "we have seen in the last year the result of the leadership of **Saudi Arabia** in bringing together various points of views to help establish the **peace process**." He said that Saudi Arabia has been a valuable customer of Canadian goods in the military field.

277. ATLANTA JOURNAL. 12 August 1996, Lexis/Nexis. Saudi Arabia is an important player in the U.S.-led efforts to broker a comprehensive peace settlement in the Middle East and also is important in.-sponsored attempts to arm and equip the Bosnian army against potential threats from its better-armed Croat and Serb rivals.

278. Omar al-Hassan. "Saudi Arabia and the New World Order," INTERNATIONAL RELATIONS. April 1994, p. 66. Given that in the realm of realpolitik the behaviour of a country's rulers rarely affects its relations with other powers, the advances towards at least a limited form of democracy in Saudi Arabia must be seen as making the country more acceptable to public opinion in the west, enabling the kind of new status suggested here to be more readily understood. More important is the question of how Saudi Arabia would carry out its suggested role as the new world surrogate in the region, and here history must be the guide, demonstrating conclusively that in modern time Saudi Arabia has always been a force for moderation and compromise, and has used its strengths—financial and moral—to further the interests of those dedicated to the same principles. Thus the Kingdom has always backed the official Palestine Liberation Organization led by Yasser Arafat and has done nothing to help the more extreme groups such as those of Dr. George Habash or Nayef Hawatmeh. It was Saudi Arabia which produced the first official plan for recognition of Israel in return for Israeli acceptance of the need to carry out the terms of UN Resolution 242. In 1986 Saudi Arabia brokered what was hoped to be the end of the Lebanese civil war by sponsoring the Arab force which was supposed to take over from Syria and to police an accord between the warring sides. That initiative failed and the war went on, but it was again Saudi Arabia which at a conference in Taif in 1988 persuaded the Lebanese politicians to accept a compromise that enabled them to stop the fighting.

279. J. H. Binford Peay. MIDDLE EAST INSIGHT. Special Edition, 1995, p. 54. As we look to today's challenges, Saudi Arabia remains the largest and most influential nation in the region. In his role as the custodian of Islam's two most holy sites, King Fahd is entrusted with formidable responsibilities that transcend the borders of his own nation. Muslim nations worldwide

look to Saudi Arabia for spiritual leadership and guidance on far reaching issues. Saudi Arabia also contributes significantly to the well-being of fellow Muslim nations, underwriting the annual *Hajj* and supporting philanthropic, developmental, and educational activities.

280. Omar al-Hassan. "Saudi Arabia and the New World Order," INTERNATIONAL RELATIONS. April 1994, p. 64. The Saudi Arabian government has, over the past decade or so, invested huge amounts of money and effort in improving the facilities needed by the pilgrims, from building a new air terminal at Jeddah, enlarging to six-lane highways the roads to Mecca used for no more than a few weeks each year, to refurbishing and improving the Grand Mosque itself. Being, as he designates himself, the Guardian of the Two Holy Places, the King of Saudi Arabia has tremendous prestige and authority throughout the Islamic world and among the 835 million Muslims scattered all over the globe. But that position and prestige demands more than attention to the physical well-being of pilgrims and to the upkeep of the holy places.

It also demands a particular atmosphere in the whole country. Muslims used to the relative laxity of the North African or Levantine countries, or those from the secular states of Africa and the East, all prefer to make the great pilgrimage, which is the high point in the lives of many, in a country where secular distractions do not intrude. Given its special position in the Islamic community of nations, Saudi Arabia uses the Sharia as the basis of all its law and has sought to observe as far as possible the tenets and restrictions of the Islamic religion. For example, shops in the souks are left empty and commerce halts during prayer times; alcohol is banned and there are no cinemas in the Kingdom, there are no elections and the press is 'guided.'

281. Paul Aarts. "The New Oil Order: Built on Sand," ARAB STUDIES QUARTERLY. Spring 1994, p. 1. The second Gulf War has unmistakably contributed to the establishment of a "new oil order," with the American-Saudi connection as its linchpin. The shape of that new order has been visible for quite a while, but the outcome of the war, so successful for the West, has given a fresh impetus to the realization of radically different relations in the oil market. After "Operation Desert Storm," politicians and spokespersons for oil companies, with increasing frequency, can be caught speaking unashamedly of the "internationalization" of Arab oil wells. It looks as if we are living a rerun of the "golden" Fifties and Sixties, way back when the Western oil companies dictated the oil market. Hard-pressed for money, many of the oil-exporting countries today find themselves compelled to appeal to the very same oil companies for financial and technical aid. True to the motto "We pay, we control," they have regained access to the oil resources that had not been allowed to touch for twenty years. "This is just about as close to denationalization as you can get," one of them noted. It is highly doubtful, however, that this "recolonization" of Arab oil will last very long. There are just too many uncertain (f)actors at play.

282. Edward Krapels. INTERNATIONAL AFFAIRS, THE COMMANDING HEIGHT. 1993, p. 50. There is still a split between moderates and hawks over how high values should go, and here the Saudi-US connection may prove pivotal over the next few years. One need not subscribe to conspiracy theories to believe that Riyadh will continue to be sensitive to US interests, but it is unlikely that any US administration would spend political capital just to move oil prices up or down a few dollars per barrel. Washington's general aversion to very high and very low oil prices (say, over $25 or under $12 per barrel) has effectively been conveyed to Riyadh. Moreover, a desire for oil prices to remain within this range will probably survive a change in administration: Democrats want to avoid oil price crashes that adversely affect parts of the midwest, southwest, and far west. Thus, while the Saudi-US political connection is a part of the oil landscape of the 1990s, under most circumstances—i.e., short of a crisis—it is in the background. The US Strategic Petroleum Reserve is also part of that landscape; taken together, those factors make oil price explosions in the 1990s unlikely.

283. THE RECORD. 26 June 1996, Lexis/Nexis. In addition to basing planes in Saudi Arabia to keep watch on Iraq, the United States has played an important role helping to transform the Saudi military from a disorganized, ill-equipped force into a more impressive fighting organization able to counter moves by Iran or Iraq and to defend the regime from domestic insurgency.

284. APS DIPLOMATIC NEWS SERVICE. 29 July 1996, Lexis/Nexis. The Arab-Israeli peace process and the Israeli right-wing Likud party's position. A growing number of Saudis are outraged over Washington's bias towards Israel. One reason was the US failure to condemn the Qana massacre of April 18, 1996, in which Israeli shelling killed some 100 Lebanese civilians at a UN camp in south Lebanon. Since the Likud government took over and rejected the land-for-peace principle, Saudis have been amazed by the way Washington began adapting to the hardline positions of Premier Binyamin Netanyahu, demonstrated during his
4-day visit to the US from July 9 (see Recorder). The US position puts the royal family in a difficult situation vis-a-vis the Islamic opposition.

285. Feuerwerger. POST WAR POLICY ISSUES IN THE PERSIAN GULF. 1991, Y4.F76/1:P84/3, p. 306. At the same time, threats to America's friends remain. At the interstate level, Saddam Hussein remains in power in Iraq. He is likely to seek revenge against his neighbors, should he have the opportunity. In the longer run, the states of the lower gulf will continue to face secu-

rity problems caused by the combination of their wealth (which make them attractive targets), their limited military capability, and the tensions within their own societies. Israel will also face threats from its Arab neighbors unless there is greater progress toward a resolution of the Arab-Israeli conflict.

Over the next decade, the threats to internal instability in much of the Arab world are likely to intensify. Rising groups such as the army, intellectuals, and the middle class are likely to increase their demands for power. Among the masses social change may weaken traditional bonds of authority; many may turn to the social agenda and religious-psychological security offered by Islamic fundamentalism.

In the broader Middle East, unresolved conflicts and the persistent threats of terrorism will also create security challenges. Despite the reduction in international terrorism which has occurred in the Middle East over the past two years, the most likely threat U.S. citizens will face in the 1990s is the threat of terrorism.

It is important to remember that the United States has had a poor track record of predicting threats to American interests and allies during the past twenty years. Few anticipated the downfall of the Shah, the Soviet invasion of Afghanistan, the outbreak of the Iran-Iraq War, or the Iraqi conquest of Kuwait. With this history, the best and wisest course is to anticipate that threats will emerge—but to avoid overconfidence about whether it will be possible to determine how and where they will arise.

286. David Long. MIDDLE EAST POLICY. Vol. 2, No. 1, 1993, p. 115. The greatest physical threat to. interests in the Gulf is the prospect of armed conflict, either within a Gulf state in the form of externally supported civil strife, or involving two or more gulf states. Because of the ongoing political and social volatility of the region, policies seeking to resolve such conflicts once and for all are not realistic. For the United States and other major oil consumers to protect their oil interests, therefore, they must develop policies seeking conflict management, not conflict resolution—that is, policies that seek to keep the endemic conflicts of the region within manageable proportions.

287. David Long. MIDDLE EAST POLICY. Vol. 2, No. 1, 1993, p. 115. Even the smaller Gulf Arab states, with small populations and relatively abundant oil revenues, are not immune to the problem of population growth. Bahrain and Oman, with relatively modest oil resources, are the most likely to feel the political effects of overpopulation the soonest. The others could experience problems, however, since the concept of material well-being is not based on an absolute scale, but rather on one's expectations. The degree to which overpopulation lowers per capita incomes and causes economic expectations to be unmet, will have a major impact on political stability.

Thus, overpopulation remains a major source of potential conflict.

288. David Long. MIDDLE EAST POLICY. Vol. 2, No. 1, 1993, p. 115.
TERRITORIAL DISPUTES
Since habitable areas in the Gulf are limited by available water, population distribution was relatively static until the twentieth century opened the door to rapid urbanization. For millennia, the population centers and outlying areas were contested by both petty local rulers and imperial potentates. Over the centuries, empires waxed and waned, with corresponding ebb and flow of freedom of action by the petty rulers on their peripheries. Ethnic and national groups inevitably claimed the maximum area ever under their control as their "historic boundaries," creating a large number of territorial disputes, and over the centuries, territorial wars became an integral part of Gulf politics. Thus, boundaries meant little more than the ability of a major ruler or a petty one to maintain a degree of political control over neighboring tribes and principalities. In the twentieth century, however, with tremendous amounts of oil wealth at stake with the shifting of boundaries by only a few degrees, territorial disputes became a major threat to regional security.

289. Carlos Elias. THE NEW GLOBAL OIL MARKET. 1995, pp. 140–141. It is very important to note that economic growth has not been followed by political transformation. Six of the eight countries with the biggest oil reserves have traditional conservative governments–the only exceptions are Iraq and Iran. An important question is why traditional governments have survived in the Middle East. The main explanation is that oil reserves are big enough to placate the demand for political reform. The population of these countries is small enough for the government to provide free housing, education, and medical care. In some countries the native population is very small: in Kuwait, for example, only 40 percent of the 1.6 million inhabitants were native-born.

This policy has reduced the demand for political reform but is not without cost or risk. In the case of Saudi Arabia, the government has been able to keep the status quo by keeping taxes very low and providing its citizens with social benefits such as health care and education. All this has been subsidized by the oil revenues that Saudi Arabia obtained during the boom years. With the oil boom over and the price of oil getting weaker, the government has run a budget deficit in the last ten years to finance spending. With this type of deficit, reserves of foreign currencies have fallen from $121 billion in 1984 to $51 billion in 1992 (Gorth et al., 1993). This could postpone economic crises, but not forever. When most of the oil wealth is gone, the government will have to increase taxes to reduce the budget deficit. This could lead to more demand for political liberalization.

Several factors could cause political problems in the near future for those countries (Bill and Springborg 1990; 372). The biggest political risk for the established political order is the gap that exists between economic modernization and political openness. This is especially true because of the growth of a professional middle class interested in more political power. The growth of Islamic fundamentalism is always a threat. The shortage of native labor that forces some of these countries to import a large part of their labor force could threaten the stability of the conservative governments in Middle East oil exporting countries. Still the most important problem that these countries face is the lack of diversification in their economies.

290. Khadduri. SECURITY DIALOGUE. June 1996, p. 155. Major historical changes are taking place in the Middle East today, all happening at the same time, leaving many questions about the future unanswered. Regionally, the area now finds itself at a crossroad, with diametrically opposed forces pulling in all directions: a peace process with Israel at the helm; political Islam dominated by vociferous armed militants; and, an unfranchised majority aspiring to achieve stability and democracy. There is also a slow and gradual movement across the Arab world to change public priorities, with greater emphasis on the economic and social agenda rather than military hardware and armed adventures.

291. Paul Tempest. ENERGY POLICY. November 1993, pp. 991–992. Two further major interrelated issues preoccupy the petroleum industry today. Again, both evoke strong memories of the problems of the 1870s.

In the last 21 years, the oil and gas industries have changed from being largely self-financed to industries where the debt/equity gearings are quite high. The oilmen can no longer simply back their own hunches. They have to listen to their bankers. Yet few bankers today can find acceptable rates of return on new-energy projects without making assumptions of a rising oil price. They look at the current expansion of capacity, surpluses in the market and the absence of Iraq (with a potential of 3–6 million barrels per day) pending the lifting of UN sanctions—there is very little to comfort them in the short term.

As a result, the degree of investment shortfall can already be accurately measured sector by sector and country by country worldwide. Take the ageing stock of very large crude carriers (VLCCs) (160,000 to 320,000 tonnes), launched mainly in the mid-1970s with a 20 year design life. They are now approaching scrapping or extensive rebuilding to prolong their life. Only a much higher oil price can provide the industry and its financial backers with the rationale for investment on an adequate scale. Even then there is a long 5–10 year time lag before those investments can begin to give good value. For most bankers investing in as volatile a price environment as the oil market with a 5-10 year lag on returns, the risks are too great. Yet underinvestment on the current scale cannot persist for very long without a considerable weakening of the industry worldwide.

292. Paul Stevens. ENERGY POLICY. May 1996, p. 388. The future capacity among OPEC members and hence the extent and distribution of excess capacity is entirely a matter of financial ability and willingness to invest. The reserve base undoubtedly exists. In 1993, the reserve production ratio of the World outside of the Gulf was 20.7 years. If such a ratio were to be applied to the Gulf, assuming no investment constraints, the Gulf could support a production of 88 million bl/d. The only issue is will the investment take place? There are widely differing estimates of just how much investment is needed (Adelman and Shahi, 1989; Browne, 1991; CGES, 1993, Seymour, 1993; Stauffer, 1993). However, all of the governments are facing financial constraints to a greater or lesser degree, which may constrain ability. Against this is the apparent willingness of all the governments with the exception of Saudi Arabia to allow in foreign companies to provide both capital and technology. In some countries this policy does meet local resistance. For example, Kuwait has so far failed to get the support of the National Assembly for such a move.

293. Michael E. Humphries. ENERGY POLICY. November 1995, pp. 999–1000. The sheer magnitude of the capital required to maintain current world oil production levels, replace reserves, fund exploration in the frontier areas and meet the environmental challenges of the refining and marketing sector, call upon both the industry and the financial community to create innovative financial structures which ensure that capital is made available on terms which meet the requirements of both the borrowers and the lenders. The continued growth of the derivatives market will be crucial in meeting this challenge. The role of the multilateral agencies will also be crucial in ensuring that sovereign risk is underwritten and that capital can flow to the former USSR and newly emerging regions. Governments themselves have an important role to play in ensuring an operating and fiscal environment which is conducive to international investment and which ensures a fair rate of return in the prevailing oil price environment.

294. Paul Stevens. ENERGY POLICY. May 1996, pp. 400–401. The second consequence of the new market environment, the squeezing of rent, carries even more serious implications for producing governments. A central tenet of economic theory is that competition in a market derived from a large number of buyers and sellers drives out super normal profit, defined as anything above average cost with a normal rate of return being embodied in costs. This is reinforced rapidly if sellers are interested in maximizing volume in a world of surplus capacity. Figure 7 illustrates the experience of copper. Occasionally, events occur to push prices above average costs but such

windfalls are short lived. Competition quickly squeezes out temporary rent.

Oil is now traded in such a competitive market once the range maker has made the range. At current prices there is significant rent in the system. How much rent is clearly a matter of considerable debate but it would be difficult to argue for an average cost above US$10 per barrel. The last period when such a threat to rent was perceived was in the ten years preceding 1970 (Adelman, 1972). The consequences for price can be seen in Figure 2. The average price in this period was US$8 per barrel (US$ 1993). The actual comparable price in 1993 was US$14.90 (BP, annual).

This leads to two questions. What might persuade the price to fall and how long would the price stay low? At the moment, based upon the earlier arguments, it is only belief which is maintaining prices above average cost. It has been described using the computer term WYSIWYG—'what you see is what you get'—i.e., the longer the price is maintained at a given level—'burnt into the traders' screens'—the longer it is likely to survive. Therefore, the question arises as to what might cause a change in belief to lower the prices substantially. The most obvious trigger is likely to be the return of Iraq. The removal of Saddam could well lead to a very dramatic fall in oil price in a very short space of time, especially if it occurs in conjunction with some other event such as an already weakening market.

295. Paul Stevens. ENERGY POLICY. May 1996, p. 398. For the future, the issue is whether other sources of excess capacity will emerge thereby increasing the number who must control. The key issue concerns Iraq. When Iraq does return, it will do so rapidly and with large volumes. It is quite clear that foreign companies will enter on some form of equity crude basis and will invest heavily with the result that production capacity will expand. It is not inconceivable that Iraqi capacity could rapidly rise to 5-6 million bl/d. Assuming no constraint on export capacity, it is likely that Iraq will *de facto* join the base load suppliers even if *de jure* it remains part of the residual suppliers, i.e., OPEC. In these circumstances, if the range is to be made, OPEC must agree to reduce its production to accommodate Iraqi reeentry. The basis of such a reduction will be the cause of much dispute, positioning and negotiation.

The most likely outcome, assuming some form of joint cutbacks can be agreed, is individual OPEC members yielding to the temptation to produce a few more barrels on the grounds that a few more can be slipped in with minimal impact—the fallacy of composition which dominated the 1980s.

296. Lowell Feld. THE NEW GLOBAL OIL MARKET. 1995, p. 113. Despite the absence of Iraq from world oil markets since 1991, however, world oil prices have remained well below OPEC's official goal, largely due to weak demand caused by anemic economic growth in major oil consuming countries. At the same time, revenue-strapped OPEC countries have been reluctant to cut production. As a result, even with Iraq still excluded from exporting oil, OPEC meetings since 1991 have mainly struggled with the challenge of reining in *overproduction*. Ironically, therefore, OPEC's greatest challenge today lies not in coping with any new oil supply disruption, but with the final resolution of the last one (Crisis 4). Some have speculated, in fact, that OPEC's failure to deal effectively with Iraq's return to world oil markets could cause another collapse in oil prices, and even lead to the disintegration of OPEC. Already, one country (Ecuador) has officially left OPEC, while others (Kuwait, Iraq, Venezuela) have threatened to do so. Thus, 20 years after the first oil shock helped OPEC grab control of world oil markets, the organization now faces a potential threat to its very existence.

297. MONEYCLIPS. 26 June 1996, Lexis/Nexis. Reducing the Regions Dependence on oil: There is a general recognition in the Gulf region today that economic difficulties are unlikely to be solved by manipulating the price of oil. The perception in the past was that if only oil exporters could coordinate policy more effectively, prices could be raised and economic difficulties would disappear. This is no more the case now that OPEC's power to determine the price of oil is substantially less than in earlier years. OPEC does not control more than 35% of world oil production and its strategy to maintain a production ceiling of 24.5%/mb/d since September 1993 failed to boost oil revenues. Even if oil prices do rise substantially in the next few years it would not solve any of the fundamental problems of the region.

298. Khadduri. SECURITY DIALOGUE. June 1996, p. 189. In fact, the oil-producing countries have become increasingly dependent on oil revenue, at a time when the industrial countries have been able to diversify their sources of energy-supply and reduce their dependence on oil consumption. For environmental and/or security reasons, the industrial countries continue to introduce measures to cut their dependence on oil, while the producing countries of the Middle East are asked to continue in the role of residual supplier and to invest billions of dollars in raising their production capacity, without a commensurate rise in demand or a call on their crude.

The inability of the oil-producing countries to de-link their economies from this one natural resource plays havoc with their societies when prices fluctuate upwards or downwards. Moreover, the fact that the oil market has been oscillating sharply in recent years makes proper budgeting and planning almost impossible in these countries.

299. Salah El Serafy. THE NEW GLOBAL OIL MARKET. 1995, p. 133. Taking advantage of their new wealth, the exporting countries managed to make great advances in social development as reflected in their

social development indicators. Data are not complete and, if available, are not always reliable. Besides, the great influx of poorer-country labor in many of the sparsely populated oil exporting countries distorts the social indicators. However, significant advances were made in many of these countries in life expectancy, infant survival, and education and health—advances that would not have been possible without the oil earnings. Between 1970 and 1990 (or 1991) life expectancy at birth, measured in years, rose for Saudi Arabia from 52 to 65; for Ecuador from 65 to 70, for Nigeria from 44 to 52, and for Algeria from 53 to 66. The illiteracy rate also dropped in the same period: in Saudi Arabia from 49 to 38 percent; Ecuador from 18 to 14 percent; Nigeria from 58 to 49 percent; and Algeria from 50 to 45 percent. Remarkable improvements occurred in under-five mortality rates, calorie intake, and school enrollment. Total fertility rates (i.e., births per woman) also reflected advances in women's education and social status, falling from 6.3 to 3.6 in Ecuador; from 7.4 to 4.9 in Algeria; though in Saudi Arabia and Nigeria, despite falling, remaining quite high at 7.0 and 5.9, respectively. In several exporting countries signs of progress abound, with modern airports, highways, well-equipped hospitals, airlines, universities, systems of communications, and government structures. But except in rare instances, alternatives to oil as a source of income and a sector that would provide employment have not been successfully developed.

300. MONEYCLIPS. 26 June 1996, p.1. The focus of economic planning in the region has therefore shifted to non-oil revenues and the various calls on government expenditures. This does not mean that the importance of oil revenues to these economies has diminished. All the GCC states will remain "oil economies" well into the next century, despite their present attempts at economic diversification. While the contribution of the hydrocarbon sector to GDP has been on the decline, the sector still accounts for 30–40% of the gross domestic product of several GCC countries. Oil also influence nearly all other economic sectors especially the government sector where 70–90% of revenues are derived from oil.

CHAPTER III
NUCLEAR ENERGY
OUTLINE

I. THE STATE AND FUTURE OF NUCLEAR ENERGY

   A. NUCLEAR INDUSTRY ON THE BRINK OF DEVELOPING NEW TECHNOLOGIES (1)

   B. EMERGING DESIGNS ARE BEING EXPORTED NOW (2)

   C. NUCLEAR POWER IS COST COMPETITIVE COMPARED TO ALTERNATIVES (3)

   D. NOW IS THE CRITICAL PERIOD. INDUSTRY IS DECIDING ON THE FISCAL VIABILITY OF NUCLEAR POWER (4–6)

   E. EFFICIENCY REFORMS PROVE INDUSTRY COMMITMENT TO NUCLEAR ENERGY (7–8)

   F. PUBLIC SUPPORTS NUCLEAR POWER OPTION (9)

   G. U.S. NUCLEAR INDUSTRY HAS COLLAPSED (10)

   H. UTILITIES ARE ABANDONING NUCLEAR POWER (11,18–20)

   I. CONTRACTS FOR NEW NUCLEAR REACTORS ARE BEING CANCELED (12)

   J. FEAR OF LIABILITY BLOCKS NUCLEAR REVIVAL (13)

   K. SAFETY RISKS STOP EXPANSION OF NUCLEAR ENERGY (14)

   L. LEGAL BATTLES BLOCK REVIVAL OF NUCLEAR POWER (15)

   M. PUBLIC OPPOSITION BLOCKS NUCLEAR REVIVAL (16–17)

II. NUCLEAR POWER AND ENERGY SUPPLY

   A. OIL INDEPENDENCE IS IMPOSSIBLE WITHOUT EXPANDING NUCLEAR ENERGY (21–22)

   B. NUCLEAR POWER IS ESSENTIAL TO PREVENTING A DOMESTIC ENERGY CRISIS (23–25)

   C. NUCLEAR ENERGY IS MORE COST EFFICIENT THAN ALTERNATIVE ENERGY SOURCES (26–27)

   D. NUCLEAR POWER IS IRRELEVANT TO FUTURE ENERGY NEEDS (28)

   E. ALTERNATIVE FUELS ARE MORE COST COMPETITIVE. THEY ARE THE KEY TO FUTURE ENERGY SUPPLY (29–31)

III. ACCIDENTS AND RADIATION HARMS

   A. ACCIDENT RISK HIGH
      1. Risk of nuclear accidents has increased. (32)
      2. Technological advances will not prevent accidents. (33)
      3. Technology cannot control for human error and natural disaster. (34)
      4. Emergencies prove risk of accidents is high. (35)
      5. NRC regulations are not enforced. (36–37)
      6. Safe technologies are not cost effective. (38)

   B. ACCIDENT RISK LOW
      1. Scientific consensus disproves accident risk. (39)
      2. New reactor design solves accident risk. (40, 49)
      3. Benefits of nuclear power outweigh risk of accidents. (41)
      4. Accident risk non-existent. (42, 45)
      5. NRC reforms resolve corruption and improve safety. (43–44)
      6. NRC operational philosopphy makes an accident here impossible. (47–48)
      7. U.S. nuclear power safety record perfect despite maximum capacity utilization. (46)
      8. No American Chernobyl possible. (50–51)

   C. CALDICOTT INDICTS
      1. Caldicott's analysis empirically disproven. (261)
      2. Shouldn't base policies on Caldicott's recommendations. (262, 266)
      3. Caldicott also said that Reagan made nuclear war inevitable. (263)
      4. Caldicott demonizes industry. Blames them for all problems. (264)
      5. Caldicott's rhetoric causes numbing. Proves exaggerated rhetoric. (265)
      6. If she were right we would all be dead by now. (268)
      7. Caldicott scapegoats America and industry. Ignores reality. (269)

8. Numbing hypothesis would end education. We would never discuss crisis. (270)
9. Numbing hypothesis is a distortion. (271)

IV. WASTE

A. STATE OF THE WASTE CRISIS
1. Massive waste crisis emerging. (52–53)
2. On-site storage space is exhausted. (54–55)
3. Dry cask storage space is exhausted. (57)
4. No more re-racking capacity. (58)
5. Temporary storage is the only immediately available option. (56, 60–61)
6. By 2010 all space will be filled. (59)
7. Status quo waste management is costly. (188)
8. Nuclear Waste Fund provides sufficient monies. (189–190)

B. YUCCA MOUNTAIN STORAGE
BAD:
1. Yucca is vulnerable to earthquakes. Explosions a certainty. (62–63)
2. Groundwater heating ensures Yucca explosions. (64)
3. Yucca density assures nuclear blow-ups. (65)
4. Close population centers guarantee substantial radiation harms. (66)
5. Radiation leaks make human exposure inevitable. (67)
6. Yucca is too small. It can't solve the waste crisis. (68)

GOOD:
1. Improvements in technology make Yucca disposal safe. (69, 78)
2. Yucca disposal facility will only be constructed if it passes rigorous safety requirements. (70)
3. New studies prove no risk of Yucca earthquake. (71)
4. No Yucca transportation until 2010. (72)
5. Evacuation protections protect the population. (73)
6. Skepticism unfounded: Scientific consensus denies explosion theory. (74)
7. Long time frame to Yucca leakage. (75)
8. Yucca isolated: No health risks. (76)
9. Yucca site insulated from earthquake risk. (77)
10. Yucca is the best and most feasible storage option. (79)
11. No danger to humans or the environment. (80)
12. Comprehensive studies prove it is safe. (81)

C. INTERIM AND ON-SITE STORAGE
INTERIM STORAGE GOOD/ON-SITE STORAGE BAD:
1. Interim storage best prevents transportation accidents and terrorism by improved monitoring (82–83)
2. Transportation infrastructure exists now. Only political will is missing. (84)
3. Interim storage empirically proven as the safest alternative. (85)
4. On-site storage fails because power plants lack necessary experience. (86)
5. Dry casking at interim storage facilities best solves waste problems. (87)
6. Interim storage allows planning and decision-making for long term waste management. (88)
7. Clean-up efforts have made Hanford and Savannah interim storage sites safer. (89)
8. New technology makes DOE sites safer and cleaner. (90)
9. New studies prove no earthquake risks at interim storage facilities. (91)
10. Savannah River has substantial additional waste storage capacity. (92)
11. Plenty of space at Oak Ridge interim storage site. (93)
12. Acceptance of foreign spent fuel proves DOE has capacity to take more reactor fuel. (94)
13. On-site storage increases terrorism risk. (127)
14. On balance, interim storage is better than on-site storage. (95–96)
15. On-site storage causes major price hikes and kills the nuclear industry. (97)
16. On-site storage costs passed on to the consumer. Only interim storage avoids this. (98–102, 109, 111)
17. Increased electricity prices destroy economic growth. (103–104)
18. On-site storage destroys the nuclear power industry. (105–106, 117)
19. Centralized storage critical to saving the nuclear power industry. (107–108)
20. On-site storage causes plant shutdowns. (110, 112–115)
21. On-site storage will force shutdowns at 80% of nuclear facilities by 2010. (116)
22. On-site storage destroys the energy benefits of nuclear power. (118)
23. On-site storage increases the relative competitiveness of alternative energy industries. (119–120)

ON-SITE STORAGE GOOD/INTERIM STORAGE BAD:
1. Mobile Chernobyl: Centralized storage causes transportation accidents. (122, 128–129)
2. Interim storage will become *de facto* permanent storage. (121)
3. Interim storage too slow: No facilities feasible before 2002. (123, 130)
4. Centralized storage site decisions destroy incentives to develop safe repository techniques. (124)
5. Political infighting will prevent the development of centralized facilities. (125)
6. Centralized storage will cause future reprocessing. (126)
7. Empirically, NIMBY syndrome makes it impossible to locate sites. (131)
8. On-site storage will not produce health harms. (132–133)
9. Studies prove that dry storage prevents health harms for on-site storage. (134)
10. NRC safeguards check terrorist theft from on-site storage facilities. (135)
11. Improved technology makes on-site storage feasible. (136–137)

D. STORAGE/DISPOSAL TRADEOFF
1. On-site storage development collapses political will to invest in disposal technologies. (138)
2. Interim storage costs will trade-off with research funds for new disposal technologies. (139-141)
3. Empahasis on any storage alternatives diverts funding, attention, and political will from disposal alternatives. (142–143)

E. DRY CASKING
   GOOD:
1. Stops radioactive leaks. (144)
2. Decreases the risk of accidents. (145–146)
3. Saves nuclear energy. (147)

   BAD:
1. Will impose massive costs on the nuclear industry. Prevents competitive pricing. (148)
2. Costs billions in compliance costs. (149–150)
3. Regulatory burden from dry cask development will kill nuclear industry. (151–152)

F. NUCLEAR VERSUS CONVENTIONAL WASTE:
1. Energy waste is inevitable. Nuclear power is the best waste crisis solution. Causes sustainable development. (272)
2. Reprocessing and easy decommissioning make nuclear waste much easier to handle. (273)
3. Conventional fuel preparation generates huge amounts of waste. (274)
4. Fly ash from fossil fuels, even with recycling, causes huge waste. (275)
5. Nuclear power generates a tiny fraction of all energy waste. (276)
6. Nuclear waste disposal is super-safe compared to conventional hazard waste. (277)
7. Fossil fuel transportation risks are substantial. (278)
8. Natural gas is not a better alternative. Produces substantial production waste. (289)

V. PUBLIC OPINION AND NUCLEAR ENERGY

A. WASTE TRANSPORTATION AND SITING
1. Public opposes transportation of waste. (153, 158–159)
2. Yucca Mountain siting popular. (154)
3. Yucca Mountain siting has bipartisan congressional support. (160, 166)
4. Public opposes Yucca siting. (181–183)
5. Public fears that Yucca will become a permanent disposal site, no disposal development. (184–185)
6. No public fear of permanent storage. Legal efforts reassure opponents of the commitment to develop disposal methods. (161)
7. Storage more popular than disposal. (162)
8. Public supports disposal over storage. (167)
9. Disposal has bipartisan congressional support. (172, 187)
10. Nuclear Waste Fund unpopular. (163)
11. Nuclear Waste Fund popular with the public. (168, 186)
12. Public opposes on-site storage risks. (164, 177–178)
13. Ground-based dry cask storage is popular. (165)
14. Ground-bsed dry cask storage is unpopular. (179–180)
15. Public opposes reprocessing. (170)
16. Public fears health risks from waste crisis. (173)
17. Clinton's waste disposal policy in a shambles. (176)

B. NUCLEAR REVITALIZATION
   1. Government commitment to nuclear power unpopular. Public wants commitment to alternatives. (155)
   2. Media spin assures that any commitment to nuclear energy will be unpopular. (156)
   3. Public hysteria insures that nuclear revitalization will be unpopular. (157)
   4. Decreased cost of nuclear operation popular. (169)
   5. Public supports revitalization of nuclear industry. (171)
   6. Actually, little public notice of nuclear power issues. (174)
   7. Clinton Administration committed to nuclear power in the status quo. (175)

C. FEDERAL/STATE RELATIONS AND THE NUCLEAR WASTE CRISIS.
   1. Failure to provide interim storage causes massive federal-state conflicts. (222, 225)
   2. States are demanding DOE acceptance of spent fuel. (223)
   3. States relying on Federal government action. (224, 226)
   4. State power is preempted by the Atomic Energy Commission in the nuclear arnea. (227–228)
   5. Supreme Court decisions preempt state autonomy in nuclear power regulation. (229)
   6. Yucca siting preempted state control. (230)

VI. NUCLEAR WASTE REPROCESSING AND U.S. INTERNATIONAL LEADERSHIP

   A. DOE WILL REPROCESS SPENT FUEL ILLICITLY (191)

   B. REPROCESSING STANDARDS INEVITABLY FAIL (192)

   C. ALL EXPERTS AGREE THAT REPROCESSING POSES GRAVE SECURITY RISKS (193)

   D. U.S. IS PUSHING OTHERS TO STOP REPROCESSING (194–197)

   E. USE OF REPROCESSING ELIMINATES PRESSURE FOR OTHER DISPOSAL ALTERNATIVES (201)

   F. REPROCESSING IS THE WORST ALTERNATIVE (200)

   G. U.S. DECLARATION OF PLUTONIUM AS WASTE STOPS REPROCESSING (204)

   H. SHOULD BAN PYRO-PROCESSING AS A DISPOSAL OPTION TO PRESERVE LEADERSHIP (205)

   I. DOE IS USING MOX PROCESSING NOW (199)

   J. MOX OPTION DESTROYS U.S. CREDIBILITY (198)

   K. NEED 100% BREAK FROM REPROCESSING ALTERNATIVES TO SOLVE (202–203)

   L. SEPARATION MAKES RISK OF NUCLEAR PROLIFERATION MORE LIKELY (215–216)

   M. U.S. CAN EASILY PERSUADE RUSSIA TO ABANDON ITS COMMITMENT TO REPROCESSING (206)

   N. RUSSIA WON'T STICK WITH PLUTONIUM (207–208)

   O. GREAT BRITAIN AND FRANCE WILL ABANDON PLUTONIUM IF THE U.S. DOES (209–210)

   P. GERMAN WASTE CRISIS INEVITABLE. THEY WILL LIKE A DISPOSAL ALTERNATIVE (211)

   Q. GERMANY ABANDONING REPROCESSING NOW (212–214)

   R. VITRIFICATION SENDS PROPER INTERNATIONAL SIGNAL AGAINST PLUTONIUM USAGE (217-218)

   S. NEW TECHNOLOGIES MAKE VITRIFICATION MOST FEASIBLE SOLUTION (219–220)

   T. VITRIFICATION IS TECHNICALLY FEASIBLE AND SOLVES PROLIFERATION RISKS (221)

VIII. BENEFITS OF NUCLEAR POWER

   A. KEY TO GLOBAL SURVIVAL (231)

   B. ESSENTIAL TO REDUCING THE TRADE DEFICIT (232–233)

   C. CRITICAL TO U.S. NON-PROLIFERATION LEADERSHIP (234)

   D. COLLAPSE OF NUCLEAR POWER WILL CAUSE A MASSIVE INCREASE IN CO_2 (235, 239, 241-242)

   E. WASTE SOLUTION WILL CAUSE A PHASE OUT OF FOSSIL FUELS (240)

- F. WASTE SOLUTION KEY TO EXPANDED GLOBAL USE OF NUCLEAR ENERGY (251–252)
- G. EXPANDED NUCLEAR POWER DEVELOPMENT KEY TO THIRD WORLD SUSTAINABLE DEVELOPMENT (243–245)
- H. AMERICAN GLOBAL LEADERSHIP DEPENDS ON ROBUST NUCLEAR POWER PROGRAM (236)
- I. AMERICAN COMPETITIVENESS DEPENDS ON CONTINUED NUCLEAR DEVELOPMENT (237–238)
- J. GLOBAL DEMAND FOR NUCLEAR ENERGY INCREASING (246–247, 250)
- K. ASIAN DEMAND EXPLODING (248–249)

IX. MISCELLANEOUS NUCLEAR POWER ISSUES

- A. GLOBAL DEMAND AND MARKET SHIFT
    1. Collapse of U.S. nuclear industry causes flight to Asia. (253, 255)
    2. Nuclear industry will turn to Asian market. (256)
    3. Industry considering Asian market. (257)
    4. Asian demands for nuclear power increasing. (258–259)
    5. Shift of Asia causes South Korean reprocessing and war with North Korea (254)
    6. Asian shift causes Chinese re-export and proliferation. (260)
- B. MARKET SOLUTIONS
    1. Performance-based regulations best solve waste crisis. (279–280)
    2. Performance-based regulations solve best and save the nuclear industry. (281)
    3. Private control best for energy needs. (282)
    4. Market approach avoids politization. (283)
    5. Government role distorts market. Decreases efficiency and collapses economy. (284-285)
    6. Private industry has experience with waste storage. (286)
    7. DOE shifting to private waste control in the status quo. (287–288)
- C. NUCLEAR POWER BETTER COMPARED TO FOSSIL FUELS
    1. No accidents. (290–295)
    2. U.S. plants are the best in the world. (296–297)
    3. Nuclear waste easier to process than waste from fossil fuels. It's minimal. (298–299)
    4. Now is a critical decision-making time for fiscal viability of nuclear power. (300)

CHAPTER III

EVIDENCE

1. Phillip Bayne. FEDERAL NEWS SERVICE. 29 February 1996, Lexis/Nexis. To help maximize NRC resources, the industry recommends that the NRC revise its approach to regulation and allow utilities to meet regulatory requirements without step-by-step prescriptive regulation. In the area of advance nuclear research and development, the industry supports appropriating $64 million in FY '97. The Advance Reactor Corporation, a consortium of 12 utilities, participates in the development of new reactor designs and represents one of the largest sources of private sector cost-sharing at DOE's energy supply research and development activities.

   In recent years, the nuclear industry through ARC has front-loaded its cost share with the ALWR program. Now, with the program on the verge of producing market-ready nuclear design, the industry expects the federal government to fulfill its funding obligation with the came commitment. These reactor technologies are between 50 and 90 percent complete. Furthermore, payback provisions in the program will provide a return to the US Treasury when these advance reactor designs are sold.

   This new generation of reactor technology represents a growing high-tech industry and trade opportunities that will strengthen both the US economy and the nation's leadership.

2. NEI PRESS RELEASES. 16 January 1996, (www:nei.org/main/ pressrm/release). "Building plants based on standardized designs significantly cuts construction timetables and costs. TEPCO's experience with the Kashiwazaki plant is proof that advanced plants can be built on schedule and on budget to be competitive producers of electricity," Bayne said. "U.S. utilities that are considering the purchase of new plants are taking note of the success the U.S.-designed advanced plants are enjoying abroad."

   According to the Japan Atomic Industrial Forum, at least 125 nuclear power plants are either under construction or definitely planned worldwide—the vast majority of them in Asia. The International Energy Agency estimates that electricity demand in eastern Asia will increase by 5.3 percent per year between now and the turn of the century.

   According to the U.S. Department of Energy, current U.S. electric power supplies should be adequate through the remainder of the decade. However, it is estimated that the U.S. will need approximately 200,000 megawatts of new generating capacity by the year 2010. The advanced-design nuclear plants being developed by American companies will make nuclear—already the nation's second-largest source of electricity—an economic option for future electricity generation.

3. Phillip Bayne. FEDERAL NEWS SERVICE. 29 February 1996, Lexis/Nexis. The nuclear industry is ready to work with the administration and Congress in the development of an integrated spent nuclear fuel management system that can satisfy DOE's commitment to electricity consumers.

   In 1995, the US nuclear industry continued to push performance to record levels and to cut production costs. As the electricity utility industry becomes more competitive, we must closely examine the budgets in all areas. Yet, this fiscal year, 97 percent of the Nuclear Regulatory Commission budget is being recovered through user fees to the electric utilities. We believe the percentage of NRC cost paid by our members should be commensurate with the agency's resources dedicated to regulating the industry. We encourage the subcommittee to examine the commission's budget closely and assign the industry only those costs that are appropriate to protect public health and safety and to regulate nuclear power plants.

4. Ivan Selin. FEDERAL NEWS SERVICE. 16 May 1995, Lexis/Nexis. This improved license renewal option comes at a key time for industry, when the prospect of open competition has forced utilities to face the prospect of large financial unknowns. These financial pressures are leading our reactor licensees to face major licensing decisions that most thought would not have to be made for decades. Among their hard choices are: whether to shut down plants prematurely due to high operating costs or the high costs of needed capital improvements, or whether to consider license renewal in order to spread the cost of capital investments. The NRC must be, and is, ready to respond to this changing situation by working efficiently with industry to process license renewal applications, or to oversee plant decommissionings.

5. LOS ANGELES TIMES. 28 April 1996, p. B8. The Chernobyl accident occurred at a time when construction of new U.S. power plants was at a standstill, due in part to spiraling costs. Now a new cycle is at hand. Within the next 10 years, most U.S. plants will approach the end of their 40-year operating licenses. Utilities must either get a federal extension to continue operating or build new plants; either way they will face a public sometimes skeptical about nuclear safety.

6. FINANCIAL TIMES. 26 June 1996, p. 8. The skills required in the industry have never been greater: competition in technology, marketing, finance, service and international alliance-building, is putting enormous pressures on power equipment makers in the 1990s.

   So far, most have reacted positively, improving their performance to levels few would have thought possible 10

years ago. But some groups are better placed than others for global competition. The next few years could see big differences emerging in financial performance as the winners capitalise on their advantages at the expense of the losers.

7. ST. LOUIS POST-DISPATCH. 1 April 1996, p. 7B. Today, U.S. nuclear plants have achieved a 77 percent capacity factor, meaning that their actual performance is 77 percent of the theoretical maximum. This is a remarkable accomplishment. Ten years ago 60 percent seemed unrealistic, but utilities are now targeting 80–85 percent capacity, and in 1994 and 1995 at least 20 of the 109 U.S. nuclear plants had capacity factors above 90 percent.

Not only are electric utilities pushing plant performance higher, they are controlling production costs as well. For example, the median production cost (operation and maintenance cost plus the cost of fuel) of nuclear-generated electricity from plants with a capacity factor of at least 70 percent is currently about 1.72 cents per kilowatt-hour, with some companies as low as one cent per kilowatt-hour. These costs compare favorably with the production costs from base-load coal plants.

8. THE CHRISTIAN SCIENCE MONITOR. 4 June 1996, p. 1. This decline has contributed to an overall improvement in the efficiency of US nuclear power, according to industry figures. Last year, US nuclear plants ran at an average of 82.6 percent of capacity, up from 75.1 percent in 1994.

"Several years ago people finally realized that if you put more effort into improving safety and reliability, you would reduce unplanned outages," Colvin says.

He says increased efficiency is helping the industry become more competitive with other suppliers of electricity. NEI calculations show that the inflation-adjusted average price for a kilowatt-hour of electricity produced by nuclear power fell 12 percent in 1994, the last year for which full data are available.

9. Gregg Taylor. NUCLEAR NEWS, August 1996, p. 75. Corey Radtke, a researcher at the Idaho National Engineering Laboratory trained in toxicology, reviewed the current market for non-power uses of nuclear science and technology. Industrial uses include gauging, radiography, analytical measurements, radiation processing, heat and lighting sources, alarm sensing, and raidioisotope tracers. Medical uses include brachytherapy, external beam therapy, diagnostic imaging, in vitro studies, and unsealed source therapies.

In most of these areas, nuclear technology is the only option. In a few, nuclear technology is even seen by the public as a good thing, for instance in radiography, which is used to find cracks in bridge welds or airplane components as a public safety measure. Another field that has benefited from positive public perception is alarm sensing, although some members of the audience questioned whether there really was a positive public perception, considering that most people probably do not know that many smoke alarms have radioactive sources in them.

10. THE COURIER JOURNAL. 1 May 1996, p. 13A. Thirty years ago, nuclear power was being touted as a cheap, clean, safe and virtually inexhaustible source of electricity that soon would supplant coal and other fossil fuels.

Then came the 1979 meltdown at Three Mile Island, which put the U.S. nuclear industry on life support. In 1986, Chernobyl pulled the plug.

That's the conventional, condensed version of the history of nuclear power in this country.

Advocates of nuclear power like to blame the industry's problems on anti-nuclear hysteria, which made it impossible to find any community willing to accept a nuclear power plant anywhere in its vicinity.

Its opponents often portray nuclear power as an enterprise that collapsed under the weight of a mistaken faith in technology and an inability to recognize the human propensity to make mistakes.

There is no question that the American nuclear industry is moribund. Since Three Mile Island, there have been no new nuclear power plants ordered in the United States. Many existing orders were cancelled. More plants have gone into mothballs than have come into operation.

11. Kathleen Hart. NUCLEONIC WEEK. 18 January 1996, p. 4. A new survey confirms that fewer electric utility officials than ever believe there will be a revival of nuclear power in the U.S. "The 1996 Electric Industry Outlook" the fifth annual survey of power industry executives by the Washington International Energy Group (WIEG), found that a paltry 8% of respondents thought there will be a resurgence of nuclear power, down from 31% in 1995, and 37% in 1994. The 1996 WIEG study is based on responses of nearly 400 executives from electric utilities and independent power producers in the U.S. and Canada, a response rate of 11% of the 3,557 executives contacted. The survey was performed with the Canadian Electrical Association. There was no breakdown on how many of the respondents, if any, came from nuclear utilities.

12. SAN DIEGO UNION-TRIBUNE. 12 May 1996, p. G4. At the top of the list of countries no longer building nuclear plants is the United States, which started the whole business in the 1950s. The last U.S. plant, in Watts Bar, Tenn., was completed in February. The last Canadian plant was finished in 1993. Since the late 1970s, 120 North American nuclear plants have been canceled. Nuclear power now produces one-fifth of North America's electricity, but with no new reactors being built, it is unlikely to go any higher.

13. SAN DIEGO UNION-TRIBUNE. 12 May 1996, p. G4. Although nuclear power provides 17 percent of today's electricity worldwide, it represents less than 5 percent of all power-generating capacity now under construction. Most new capacity will be fueled by natural gas, coal and hydropower.

The world's largest companies would have been bankrupted by a Chernobyl-class accident. Even a "minor" accident like the one at Three Mile Island required a decade-long, billion-dollar cleanup effort.

Although governments have exempted nuclear-power-plant owners in the United States and many other countries from liability for most off-site contamination, the immediate damage to the plant could be devastating to its owners. A multibillion dollar asset can turn into an even larger liability in seconds.

14. SAN DIEGO UNION-TRIBUNE. 12 May 1996, p. G4. When the sad history of nuclear power is written, April 26, 1986, will be recorded as the day the dream died. The explosion at the Chernobyl plant was a terrible human tragedy. And it delivered a stark verdict on the hope that nuclear power will one day replace fossil fuel-based energy systems. Nuclear advocates may soldier on, but a decade later it is clear that nuclear power is no longer a viable energy option for the 21st century.

15. THE COURIER JOURNAL. 1 May 1996, p. 13A. The safety record of U.S. reactors has improved enormously since Three Mile Island. But as U.S. reactors age, there will be a need for increased vigilance.

Safety issues also had an impact on the cost of U.S. nuclear power. The already astronomical price tag for building a reactor has been boosted by increased attention to safety and, in many cases, legal battles begun by safety-minded local opposition.

16. SAN DIEGO UNION-TRIBUNE. 12 May 1996, p. G4. For nearly two decades, the nuclear industry has periodically proclaimed the imminent renaissance of nuclear power. Safety records will be improved, they say, costs reduced and the public educated about the benefits of nuclear energy. None of this has even begun to happen.

While many nuclear scientists remain convinced t' nuclear power's problems can be solved and it will eventually mount a comeback, the odds of a nuclear revival appear to be getting slimmer with each passing year. Maybe nuclear power will make a comeback in the 22nd century. But don't count on it.

17. Gregg Taylor. NUCLEAR NEWS. AUGUST 1996, p. 75. Radtke was optimistic about the future of food and water sterilization using radiation, but right now there is a bigger market for sterilization of medical products. It is difficult to do by other methods, and is more convenient with radiation, which can be used to sterilize medical products through packaging.

Radtke provided further confirmation that the nuclear industry has a lot to do in the area of public education. He said public perception is the primary force that will drive the markets for nuclear science and technology, through political events and regulation affecting its cost and convenience. Radtke could find no positive mention of nuclear science in the grade school textbooks he checked, only references to nuclear bombs. He likened this to a forest with no young trees in it—a dying forest.

18. Kathleen Hart. NUCLEONIC WEEK. 18 January 1996, p. 4. The sharp decline in the number of utility executives expecting a nuclear revival "leaves little hope that the nuclear generation will remain an option for utilities in a time frame that has any practical significance," a draft report released by WIEG January 10 stated.

19. STAR TRIBUNE. 19 July, 1996, p. 17A. The problem is, nobody wants them. No American utility has bought a nuclear plant since 1973, and 89 percent of utility executives recently polled said they never would.

How come is simple: Nukes don't work. The Safe Energy Communication Council reports that nuclear energy costs 5 to 10 cents per kilowatt hour; coal costs between 1.5 and 3.5, natural gas between 3 and 4, and windmills 5 cents and going down.

20. Paul Barton. GANNETT NEWS SERVICE. 21 April 1996, Lexis/Nexis. Similarly, John Ahearne of Duke University, chairman of a National Academy of Sciences committee on the future of nuclear power, said a lot will have to happen for nuclear plant orders to come in again.

Among other things, power generators of the next century will have to be convinced nuclear power is the least-cost alternative for energy needs.

"Right now, no utility believes that," he said.

21. John M. Hood, Jr. SAN DIEGO UNION-TRIBUNE. 20 April, 1996, p. G4. The alternatives to nuclear-power generation other than oil include solar energy, batteries (which need charging from the central plant), wind (another form of solar energy) and so-called renewable resources such as bio-mass, corn, etc. These latter sources can only provide a tiny fraction of our needs and in fact have a net negative impact on the environment and human society in the long run.

Properly designed and managed nuclear-power sources can bestow real independence and riches on this nation.

22. Larry Craig. ROLL CALL. 15 July 1996, Lexis/Nexis. America's 110 nuclear power plants are this nation's second largest source of electricity. Nuclear energy has supplied more than 40 percent of all

the new electricity required by the American people since 1973. Nuclear energy has allowed this nation's electric utility sector to virtually eliminate its dependence on imported oil, in an environmentally sound manner. Furthermore, our nuclear power plants will make by far the largest contribution of any technology toward meeting the Clinton Administration's goals for reducing greenhouse gas emissions by 2000.

23. Reilly, Kathleen. 70 INDIANA LAW JOURNAL 679. Spring 1995, p. 699. Many argue that our lack of non-nuclear energy sources is not great enough to concern ourselves with the societal costs associated with nondevelopment of nuclear power. In his concurring opinion, Judge Williams cheerfully states that a lack of alternatives "appears remote," assuming that the flow of fossil fuels and natural gas will continue unhindered. Many experts disagree, however.

Since nuclear power currently provides almost one-fourth of all electricity produced in the United States, the country could not easily replace any reduction in this energy supply. Although wind and solar energy, due to their lack of polluting agents and radiation threats, have staunch advocates in the environmental movement, these sources have heretofore provided only a minuscule share of national energy. Moreover, many experts believe that these energy sources are simply unworkable. Finally, scientists also point out that the nation needs to do more than simply maintain its current energy supply. Demand for energy continues to grow and is expected to increase at a significant rate over the next few years.

24. Reilly, Kathleen. 70 INDIANA LAW JOURNAL 679. Spring 1995, p. 699. A $5 million study completed by representatives from 100 countries resulted in a 1993 report entitled "Energy for Tomorrow's World." The report projects massive increases in energy demand fueled by dramatic population growth centered in the developing world. It also concludes that the "most important requirement for supplying an adequate global energy supply up to 2020 will be the efficient and responsible use of fossil fuels and nuclear energy." The report, however, qualifies its enthusiasm for fossil fuels by emphasizing their contribution to air pollution. See Massive Growth in Energy Demand Seen by 2020; Fueled by Developing Nations, ELECTRIC UTIL. WK., Oct. 25, 1993, at 14. For this study's report on nuclear energy's environmental advantages, see Kaplan, supra note 9.

25. THE VIRGINIAN-PILOT. 8 December 1995, p. A18. Unfortunately, many who oppose the use of nuclear energy have seized on the spent-fuel issue to justify their opposition. Yet nuclear energy provides about 23 percent of currently electrical generating capacity in the United States—48 percent in Virginia—and does so safely and benignly without filling the air with carbon dioxide and other greenhouse gases as do coal, oil and natural gas. This electricity production is essential to our economy, and as electrification increases, we will need more.

26. ST. PETERSBURG TIMES. 15 August 1996, p. 15A. Fortunately, a shift to nuclear energy does not require an economic sacrifice. The average cost to produce electricity from nuclear plants in the United States in 1995 was 1.91 cents per kilowatt hour, a bit cheaper than coal and far cheaper than oil or natural gas. In this rare circumstance, selecting the environmentally sound course of action directly saves consumers money.

27. ELECTRICITY JOURNAL. May 1996, p. 32. We find a similar production cost profile for nuclear plants (also shown in Figure 1), but unlike that of fossil plants, the O&M component represents 60–80 percent of the total. Fuel costs for nuclear plants are only about five mills per kWh.

On the basis of variable production costs—which are arguably the cash-flow basis of future competition—most nuclear plants appear competitive. Again, technology is playing a critical role in containing and reducing costs. Over the last decade, substantial progress has been made in improving the reliability of major components—from steam generators to control systems.

28. THE HUMANIST. July/August 1996, p. 22. Ten years after the Chernobyl nuclear disaster on April 26, 1986, the nuclear power industry itself has become a casualty, according to an article that appeared in the May issue of *World Watch* magazine. Nuclear power generating capacity rose less than 3 percent between 1990 and 1995. As a result, nuclear power's contribution to the world's electricity supply has peaked at 17 percent.

29. ELECTRICITY JOURNAL. May 1996, p. 32. In the emerging climate, only a few things seem certain. As market forces lower the price of electricity, and new technologies expand its appeal, electricity will increasingly become the energy form of choice at the point of consumption. Gas will become an alternative and highly competitive path of energy delivery, as well as the preferred source of clean primary energy for new generation in the next decade. And technology will remain at the forefront of change, both as a driving force and leading indicator of structural change.

30. THE HUMANIST. July/August 1996, p. 22. The biggest problem facing nuclear power is its high cost, according to the article: "In an increasingly competitive power market, nuclear energy is two to three times as expensive as the least-cost alternatives, including a new generation of small natural-gas-fired generators that can be installed inside a factory or even in an office building: the 'PCs' of the electricity industry."

Nuclear power could be reinvented with some completely new technology or fuel cycle, or perhaps "cold

fusion" will eventually pan out, the authors conclude. "But such ideas are little more than speculative dreams. Meanwhile, a host of renewable technologies that do not rely on carbon-based fuels are entering the market. If current trends continue, the annual growth of wind-generating capacity will overtake that of nuclear capacity by the end of this decade."

31. THE HUMANIST. July/August 1996, p. 22. "Just 34 nuclear power plants are under construction worldwide today, compared to 160 a decade ago," says Nicholas Lenssen and Christopher Flavin, the article's authors. "The nuclear power industry is in a meltdown."

Most countries are instead building natural-gas-fired power plants, which are far more efficient and economical, or turning to wind turbines and solar cells to meet their electricity needs. Germany and India, for example, are each installing more wind-generating capacity than nuclear capacity in the mid-1990s.

32. FINANCIAL TIMES. 26 June 1996, p. 8. It's likely that most Americans seldom give this federal agency, the Nuclear Regulatory Commission, a second thought. That's because it has worked relatively well over the years, if the safety record of the 110 nuclear power plants in the US is any measure. There's been only one major mishap on the NRC's watch: the 1979 Three Mile Island partial core meltdown.

But the NRC, like the facilities it oversees, today faces an uncertain future. Whistleblowers say the agency is lax in its oversight and too cozy with the industry it regulates. The result, they say, is a growing danger of another serious nuclear accident.

33. Paul Barton. GANNETT NEWS SERVICE. 21 April 1996, Lexis/Nexis. But watchdog groups such as the Safe Energy Communication Council doubt the trumpeted new designs will be a cure-all for the industry.

Because of the amount of lethal material concentrated in a nuclear plant, claims of safety always have to be taken with a grain of salt, the group contends in one of its reports.

"Claims of inherent safety are unsubstantiated in the design of nuclear power plants," it said.

34. LOS ANGELES TIMES. 4 July 1996, p. A1. "Human error is at some point unavoidable," says Jinzaburo Takagi, a nuclear chemist turned antinuclear activist. He predicted the accident at Monju more than 1-1/2 years before it happened, in an interview with The Times.

Although proponents say scientists are bound to discover safer ways to dispose of nuclear waste in coming decades, even today's most secure method—in which the material is mixed with molten glass and buried deep underground—causes worries in quake-prone areas like Japan.

35. Jason Aamodt. 16 ENERGY LAW JOURNAL 181. 1995, p. 181. Although serious accidents are uncommon, emergencies are not. For instance, when a seemingly small event takes place, the emergency plans under the Final Safety Analysis Report will activate to ensure that the event does not proceed to something more serious as it did at TMI. 1979 NRC Investigative Report No. 50–320/79, at 1 app. A (1979).

36. CHARLESTON GAZETTE. 20 March 1996, p. 4A. In the Millstone case, *Time* said that "the NRC's double-barrelled oversight system shot blanks from both barrels." Regulators were aware of the full-core offloads for years, but did not realize the practice was a violation.

NRC Inspector General Leo Norton said: "The agency completely failed. We did shoddy work. And we're concerned that similar lapses might be occurring at other plants around the country."

Pressure to cut corners on expensive safety measures will grow as utilities begin to deal with deregulation and competition. Nuclear power can be twice as expensive as power from fossil fuel plants.

37. THE CHRISTIAN SCIENCE MONITOR. 4 June 1996, p. 1. Such criticism isn't necessarily new—but it's increasing in volume at a time when fundamental change in the nuclear business has the potential to undercut the industry's own safety efforts.

A review of NRC documents and transcripts, as well as conversations inspectors and industry insiders across the country, indicates that the agency may, indeed, have problems. Specifically, the *Monitor* has found that:

- NRC inspectors may too often be willing to take a utility's word that it is addressing troubling problems, instead of doing inspections at the plant or verifying calculations to ensure the problems are fixed.
- Higher-level NRC managers sometimes downgrade the severity of safety problems identified by on-site inspectors without giving reasons for the change.
- The agency is too slow to act when confronting potentially dangerous problems that could affect plants using similar reactor designs.
- NRC inspectors who persist in pressing safety issues have been subjected to harassment and intimidation by their supervisors.

38. SAN DIEGO UNION-TRIBUNE. 12 May 1996, p. G4. What's more, nuclear energy is simply not an economically competitive means of generating electricity. In the 1960s and 1970s nuclear was touted as the low-cost power option. But the cost of nuclear power, unlike that of other technologies, actually rose as the industry matured in the 1970s and 1980s.

Nuclear power requires the installation of complex, expensive safety systems. And after Three Mile Island,

even more costly control systems were added. By the late 1980s, severe market pressures had forced most U.S. utilities out of the nuclear construction business.

When a Wall Street analysts was asked what he would do if a utility decided to build a nuclear plant, he said, "Sell its stock."

And in a 1995 poll, only 2 percent of U.S. utility executives said the would even consider ordering a nuclear plant.

39. Kathleen Reilly . 70 INDIANA LAW JOURNAL 679. Spring 1995, p. 679. Since the highly publicized episode at the Three Mile Island nuclear reactor in 1979, nuclear power has evoked fear in the hearts of Americans. The accident at Three Mile Island, however, resulted in no injury or death, and scientists claim that very little radiation actually escaped from the plant. There has been no scientific evidence disproving these statements. While many people turned against nuclear power after the Three Mile Island accident because of its perceived risks, scientists, presumably those persons most knowledgeable about nuclear power, do not take these risks very seriously. A professor of physics writes:

> The present risk to the average American from the nuclear-power industry is equivalent to that of smoking one cigarette in one's life . . . of an overweight person increasing his weight by 0.004 ounces . . . crossing a street one extra time every three years . . . or increasing the national speed limit from 55 to 55.003 miles per hour. [ellipses in original]

40. Paul Barton. GANNETT NEWS SERVICE. 21 April 1996, Lexis/Nexis. With construction on the wane in recent years, the industry has concentrated on improving performance and safety, Unglesbee said.

"The next generation (of plants) is going to incorporate a lot of lessons learned in the first generation," he added.

The industry is counting heavily on new reactor designs, including ones being developed by General Electric and Westinghouse that already are being marketed overseas, to improve the industry's safety image and provide for more cost-efficient construction.

The new designs, like the AP600 by Westinghouse, incorporate so-called "passive" designs that are supposed to simplify operations and cut down on the amount of human interaction needed to deal with an emergency.

For instance, emergency cooling of the reactor vessel is provided by gravity-fed water from tanks located above it.

Bob Vijuk, project manager for Westinghouse on the AP600, said the new design provides a level of safety that is "far better than NRC (Nuclear Regulatory Commission) requirements and the design goals of the industry."

41. Kathleen Reilly . 70 INDIANA LAW JOURNAL 679. Spring 1995, p. 698. Clearly, federal regulators must evaluate nuclear energy safety procedures in light of their cost-effectiveness. Undoubtedly, there are procedures which, though economically burdensome, are too beneficial from a safety perspective to forgo. For example, a very expensive procedure which eliminates all possibility of radiation leakage would be worthwhile for nuclear plants to implement. On the other hand, such a procedure may be so costly that the utility could not survive financially if it implemented the procedure. In this case, the procedure may be too costly from a societal standpoint to make implementation feasible.

In cases where a nuclear plant shuts down because its safety costs are too great, one must consider opportunity costs—the value of benefits forgone in favor of another benefit. Naturally, the opportunity costs of forgoing nuclear power include the elimination of energy the nuclear plant would have provided. However, this cost will vary under different circumstances. The fewer energy alternatives to nuclear power that exist, the higher the value of the forgone nuclear energy. In the unlikely circumstance that no energy sources other than nuclear power exist, the choice would be between a less-safe nuclear plant and zero electricity. Arguably, zero electricity could prove to be "more expensive" to society than an operating nuclear plant that is not as safe as it might possibly be. Another opportunity cost of forgoing clean-burning nuclear energy would be the use of coal or fossil fuels as nuclear substitutes which contribute to air pollution. The more severe the air pollution problems that already exist, the higher the costs of additional air pollution.

42. THE CHRISTIAN SCIENCE MONITOR. 4 June 1996, p. 1. All the industry's indicators to measure plant problems are trending down and, by the end of last year, bettered five-year goals set in 1990, says Joe Colvin, president of the Nuclear Energy Institute (NEI) in Washington. Take automatic "scrams"—unplanned, automatic reactor shutdowns caused by equipment problems. The number of such malfunctions has fallen markedly, Mr. Colvin notes. In 1990, the industry adopted a goal of an average of one automatic scram for every 7,000 hours (roughly one year) of plant operation. Last year, it bettered the goal by achieving 0.9 scrams per 7,000 hours.

43. THE CHRISTIAN SCIENCE MONITOR. 4 June 1996, p. 1. While conceding some problems, NRC officials insist their agency has been effective. Moreover, the NRC is tightening its processes under its new chairwoman, Shirley Jackson. Individual incidents notwithstanding, she says in an interview, "the staff is a very technically competent, very able group of people who are very focused on safety. I think one should not lose sight of that."

44. THE CHRISTIAN SCIENCE MONITOR. 4 June 1996, p. 1. "We've had no new license applications for years, and maybe we'll get none for the rest of the NRC's life," says one insider at NRC headquarters. "Whether it's said or not, an awful lot of managers in the NRC have in the back of their mind the viability of the nuclear industry."

It's just this kind of problem that may lie at the heart of some inherent conflicts at the NRC. Many critics claim, for instance, that higher-ranking NRC managers are often less strict than on-site inspectors when it comes to problems at plants. On-site overseers have a more by-the-book approach to their work. They say: "This is the regulation, and you've passed or not passed," says the NRC insider. "For resident inspectors, it's a black-and-white issue. If there's a serious accident, they know the first question will be: Why didn't the resident pick up on that?"

"[NRC] managers interact with utility vice presidents," he continues. "They're more likely to say: This is a technical violation, not a threat to public health and safety. There's a concern that if you apply regulations written in the abstract, you'll get into a situation where plants shut down prematurely for minimal safety issues."

45. NEI PRESS RELEASES. 15 April 1996, (www:nei.org/main/ pressrm/release).
WASHINGTON, D.C., April 15, 1996—U.S. nuclear power plants are exceeding ambitious performance goals for safety and reliability, according to new information released by the Institute of Nuclear Power Operations (INPO).

Based in Atlanta, INPO was established after the Three Mile Island accident in 1979 to promote the highest levels of safety and reliability—to promote excellence—in the operation of nuclear power plants. Since 1983, the industry-sponsored organization has been collecting, monitoring and reporting data on America's nuclear power plants to measure their performance against industry-established goals.

"This year's outstanding performance by U.S. nuclear plants is the result of 15 years of dedication and hard work by the people who run them," said Phillip Bayne, chief executive officer of the Nuclear Energy Institute (NEI). "These professionals have demonstrated a commitment to continually seeking excellence in plant performance, and the proof of their success is in the 1995 numbers.

The World Association of Nuclear Operators (WANO), the nuclear industry's international safety organization founded after the 1986 accident at the Chernobyl nuclear power plant, adopted these performance indicators in 1989 for use by nuclear plants worldwide.

Nuclear plants with good performance, as measured by such performance indicators, are more reliable, have higher margins of safety and are generally recognized as well-managed plants. Utilities use performance indicator data to set both annual and long-term performance goals.

46. NEI PRESS RELEASES. 15 April 1996, (www:nei.org/main/ pressrm/release). In 1995, the U.S. nuclear industry met a number of challenging five-year goals set in 1990. Highlights include:

- **Unit Capability Factor**—Expressed as a percentage of maximum electricity a plant is capable of producing limited only by factors within the control of plant management. Last year, U.S. plants again achieved the highest capability factor ever—a median value of 82.6 percent–exceeding the 1995 goal of 80 percent.

- **Unplanned Automatic Shutdowns**—Nuclear plant "scrams," or unplanned automatic shutdowns, have declined by almost 90 percent since 1980. The 1995 median of 0.9 unplanned shutdowns is better than the 1995 goal for the third consecutive year.

- **Safety System Performance**—This indicator measures the availability of three key plant safety systems. Last year, 94 percent of safety systems met challenging industry goals for availability. The percentage of safety systems meeting industry goals for availability has improved since tracking began in 1989 and has been better than the 1995 goal since 1992.

- **Low-level Solid Radioactive Waste**—Minimizing the volume of tools, rags, clothing and other materials that come in contact with radiation reduces the need for storage, transportation and disposal. Last year, pressurized water reactors produced a median value of only 30 cubic meters of solid low-level waste—well under the 1995 target of 110 cubic meters. Boiling water reactors produced a median of 107 cubic meters—also well under their 1995 target of 245 cubic meters.

47. NEI PRESS RELEASES.1996,(www:nei.org/main/safe/site.htm)
Keeping It Safe
How safe are nuclear power plants? They do, after all, contain radioactive materials—the stuff left over after uranium atoms are split to create heat to generate electricity. To prevent any harmful release of **radiation** into the environment, U.S. nuclear power plants follow a philosophy of "safety in depth." This philosophy begins with the **design** of nuclear plants, which takes advantage of the laws of nature and incorporates backup safety systems and protective barriers, including a thick steel and concrete containment building. For example, it is physically impossible for a nuclear power plant to explode because of the low concentration of fissionable uranium 235 in the fuel. The philosophy is carried forward in plant **operations** through extensive and continuous **training** of plant personnel.

An additional level of safety is provided through **plant regulation** by the Nuclear Regulatory Commission, an independent federal agency. As a result of this safety-in-depth philosophy, even the 1979 **Three Mile Island** accident—the only major accident in the history of U.S. commercial nuclear energy—caused no injuries, deaths or

discernible direct health effects to the population in the vicinity of the plant. The 1986 **Chernobyl** accident in the former Soviet Union was the product of a badly flawed plant reactor design which never would have been licensed in the United States.

The nuclear industry exercises the same degree of vigilance for safety in its handling, **transportation** and disposal of the relatively small and manageable amounts of **high-level waste** and **low-level waste** produced at nuclear power plants.

48. Ivan Selin. FEDERAL NEWS SERVICE. 16 May 1995, Lexis/Nexis. First, I am confident that the safety performance of the civilian nuclear power industry is, by and large, excellent. Our evaluations of operating reactor performance, as well as industry's own indicators, clearly show that safety is at an all-time high. The number of plants with which we have significant safety concerns has decreased steadily. Today only two plants are on our watch list, and one of them is showing a positive trend toward removal. Furthermore, the number of plants evaluated as superior performers has approximately doubled in the past few years.

Such positive performance by the nuclear power industry has allowed the agency to shift its focus in safety regulation. Ten years ago, our primary role was to promote and accelerate performance improvements at numerous plants with marginally adequate safety records. More recently, at the start of my term, we were focusing on a few outlier plants which had not achieved the excellent performance standards found in the rest of the industry. Today, as I leave the NRC, we are almost entirely focused on ensuring that the excellent performance levels—found across the country—do not deteriorate.

49. Poong Eli Juhn. IAEA BULLETIN. Vol. 38, No. 1, (http://www.iaea.or.at/worldatom/inforesource/bulletin) In a number of countries, great emphasis is being placed on the development of advanced nuclear power plant designs. These new generations of nuclear power plants have been, or are being, developed by building upon experience and applying lessons learned from existing plants. Hence, the new, advanced designs are anticipated to become even more safe, economic, and reliable than their predecessors.

The advanced designs generally incorporate improvements of the safety concepts, including, among others, features that will allow operators more time to perform safety actions, and that will provide even more protection against any possible releases of radioactivity to the environment. Great attention is also paid to making new plants simpler to operate, inspect, maintain, and repair, thus increasing their overall reliability and economy.

Advanced designs comprise two sub-categories: evolutionary and developmental designs. The first encompasses direct descendants from predecessors (existing plant designs) that feature improvements and modifications based on feedback of experience and adoption of new technological achievements. They also take into account possible introduction of some new features, e.g., by incorporating passive safety functions. Evolutionary designs are characterized by requiring, at the most, engineering and confirmatory testing prior to commercial deployment. Developmental designs consist of those that deviate more significantly from existing designs, and that consequently need substantially more testing and verification, probably including also construction of a demonstration plant and/or prototype plant before large-scale commercial deployment.

50. NUCLEAR ENERGY INSTITUTE. 1996, (main/pressrm/special/differ.htm) No Western nuclear power plant has a positive void coefficient—the most significant of the design flws that led to the Chernobyl accident. Western light water plants have both a thick reactor vessel containing the nuclear fuel, and a massive containment structure around the reactor vessel and the coolant system with its associated piping.

In addition, Western plants are designed to handle a complete rupture of the largest pipe in the coolant system. The Chernobyl plant was designed to accommodate an accident involving the rupture of no more than about four pressure tubes out of a total of approximately 1,660. In the Chernobyl accident, a substantial number of tubes ruptured, which led to the rupture of all of them. As a result, the reactor core was exposed to the environment. Such core exposure could not happen in a Western plant.

51. NEI PRESS RELEASES. (www:nei.org/main/pressrm/special/differ.htm). The Chernobyl plant's RBMK design uses graphite in place of water as a moderator to control the nuclear chain reaction. The graphite surrounds vertical pressure tubes that hold the nuclear fuel and the water that will be boiled to steam. Unlike Western-designed nuclear power plants, the RBMK's nuclear chain reaction and power output increase when cooling water is lost. This design flaw—called a "positive void coefficient"—was a major cause of the uncontrollable power surge that led to the Chernobyl accident.

52. THE CHRISTIAN SCIENCE MONITOR. 17 July 1996, p. 1. In a recent letter to President Clinton, Governor Chiles warned that three of the five nuclear power plants in Florida will run out of on-site storage space two years before the US Department of Energy, in its most optimistic prediction, says it can find a permanent waste repository.

Other nuclear power plants across the US face a similar threat. The end result could be to shut down these plants. "Nuclear power plants were not designed to be permanent waste-storage facilities. So, in many cases, fuel storage facilities at these plants are nearly full," warns the Nuclear Energy Institute in Washington, which represents utilities and companies that design and build nuclear power plants.

53.   John Cantlon. DISPOSAL AND STORAGE OF SPENT NUCLEAR FUEL—FINDING THE RIGHT BALANCE. March 1996, p. 34. By the end of 1995, approximately 32,000 metric tons of spent fuel had been generated by commercial reactors located at 70 sites nationwide (See Figure 1.) Unless a significant number of reactors shut down early, spent fuel will continue to be produced at a rate of roughly 2,000 metric tons per year through the year 2010. If there is not a significant number of reactor license extensions, the rate of spent fuel production will slowly decline thereafter until the last of the presently operating reactors reaches the end of its scheduled 40-year lifetime sometime in the 2030s. By that time, the amount of commercial spent nuclear fuel will total approximately 85,000 metric tons (DOE 1994). The practice at all commercial reactors is to store the newly discharged spent fuel in pools on site for at least five years to allow for initial cooling. However, the total pool storage capacity nationwide is only about 60,000 metric tons. This means that, if a repository does not become available, storage capacity of approximately 25,000 metric tons *in addition to pool storage* will have to be provided somewhere over the next 35 years to accommodate commercial spent fuel.

54.   OMAHA WORLD HERALD. 2 January 1996, p. 7. How much waste is enough? The warning signs are sufficiently serious to require the attention of Congress.

Every year the amount of highly radioactive spent fuel at nuclear power plants in the United States grows. By the end of the decade, the inventory will be 44,000 tons.

For now, the spent fuel is being stored safely in water pools and concrete casks at 110 nuclear plant sites. However, nuclear plants were designed to generate power—not to hold large quantities of waste. By 1998, about two dozen nuclear plants will have no more space to hold the waste. By 2010, more than 70 units–including the Cooper nuclear plant at Brownsville, Neb., and the Fort Calhoun nuclear plant—will be out of room.

55.   Larry Craig. FDCH TESTIMONY. 15 July 1996, Lexis/Nexis. Nuclear power plants provide more than 20 percent of the United States' electric generating capacity. Nuclear reactors were designed with on-site storage pools that were intended to serve as temporary storage facilities. Currently, no permanent storage exists, and so nuclear power plants must store their nuclear waste on site. At this time, about 30,000 metric tons of spent fuel are now in temporary storage at nuclear power plants in some 75 sites in 34 states.

According to the committee report, 23 reactors will run out of room in their temporary storage pools by 1998, and an additional 55 will run out of storage space by 2010.

56.   Phillip Bayne. FEDERAL NEWS SERVICE. 29 February 1996, Lexis/Nexis. NEI represents approximately 275 members, including all US electric utilities that operate nuclear power plants. My testimony today focuses on three issues: The civilian high-level nuclear waste disposal program, the Nuclear Regulatory Commission user fees, and advanced reactor research and development. As this nation confronts a new budgetary climate, Congress must not lose sight of previous commitments made by the federal government. The federal government has a responsibility to begin managing used nuclear fuel from the nation's commercial nuclear power plants beginning in 1998. To fulfill that commitment, Congress is crafting legislation to authorize a central storage facility in Nevada in the Nevada Desert until a permanent disposal facility is ready. In recent months, DOE has adjusted its milestones, indicating another five-year delay for repository operation until 2015.

By that date, 93 of the nation's 110 reactors will have exhausted existing capacity to store used fuel at their plant sites. In this year's budget the subcommittee must balance two major imperatives: Preparing for federal acceptance of used fuel beginning in 1998 and continuing tightly-focused scientific studies at Yucca Mountain. Assuming an authorizing legislation is passed, the federal government must move forward with work at a central storage facility and on other components of the integrated program.

57.   John Upton. CONGRESSIONAL PRESS RELEASES. Lexis/Nexis. Thus, the solution that was fifteen years away at the passage of the first Act's still fifteen years away. We haven't gained so much as a year on the problem since 1982. Meanwhile, high-level nuclear wastes are accumulating at more than seventy locations in thirty four states. One of these sites is in the Sixth District of Michigan, where the utility licensee has been forced to begin storing spent fuel assemblies in so-called "dry casks" less than two football field lengths from America's largest body of freshwater.

Given the rate of progress on creating such a place to go, my constituents understandably fear that this temporary necessity will outlive us all.

58.   Daniel Dreyfus. FDCH CONGRESSIONAL TESTIMONY. Federal News Service, 28 June 1995, Lexis/Nexis. By the early 1980s, utilities began designing and retrofitting nuclear power plants to provide spent fuel storage for the life of the plant. In some cases, utilities with reactors under construction modified their storage pool designs. The utilities with older reactors began adopting management options for long-term storage, including reracking of pools, transshipment to other utility sites with greater storage capacity, rod consolidation, and on-site dry storage. At most of the sites, life-of-plant storage for the full 40-year period of the operating license can be accomplished by the two most economical options, reracking and dry storage. Reracking, refitting spent full storage pools with racks that hold more fuel assemblies than the initially installed racks, has been the utilities' pre-

ferred method of increasing at-reactor spent fuel storage. Most utilities have already expanded their pools' capacities at least once using reracking. About 75 percent of the reactors can provide sufficient storage to serve through the year 2000, but by the year 2005 approximately 40 sites in 28 States. All need additional storage capacity.

59. THE CHRISTIAN SCIENCE MONITOR.
27 October 1994, p. 19. Nationally, electricity customers have paid nearly $10 billion into the Nuclear Waste Fund created by the Nuclear Waste Policy Act of 1983. California ratepayers alone have contributed $358.7 million to the fund. The money was to go for spent-fuel storage and disposal. But DOE has made little progress in establishing an underground repository for the waste; the department's date for opening the facility is not until 2010, even assuming legal and political hurdles are overcome.

Hence, the spent fuel—contained safely and securely in steel rods—continues to be stored temporarily in cooling pools at nuclear plants around the country. By 1998, about two dozen nuclear units will have no more space in their pools. By 2010, over 70 units will be out of room.

60. Gary Gates. FEDERAL NEWS SERVICE.
14 December 1995, Lexis/Nexis.
A Comprehensive Approach Is Needed
The Nuclear Waste Policy Act of 1982 required the Department of Energy (DOE) to construct and operate a permanent storage facility for high-level nuclear waste by 1998. DOE estimated in 1989 that a permanent storage facility will not be in operation until at least 2010. Many of the nation's 109 nuclear plants, including OPPD's Fort Calhoun Station, will run out of on-site storage capacity long before this target date. In light of this significant delay, Congress must act to authorize an interim storage facility and ensure this vitally important program is brought back on track. If enacted this Congress, S.1271 would allow for operation of a storage facility by the target date established in the Nuclear Waste Policy Act of 1982, and would lay the groundwork for future operation of a permanent site. Most importantly, S.1271 accomplishes these objectives by providing an integrated approach to nuclear waste management through a comprehensive transportation plan and requirements for spent fuel container technology.

61. Susan Clark. FEDERAL NEWS SERVICE.
28 June 1995, Lexis/Nexis. The purpose of my testimony is to convey the message that our nation's radioactive waste program is at a critical juncture. It desperately needs program changes and priority setting to provide a centralized interim storage facility by 1998, to facilitate waste transportation, and to continue a program to provide a permanent repository in the longer term. Prompt Congressional action must be taken to secure storage for the spent nuclear fuel generated by this nation's nuclear power plants.

62. THE CHRISTIAN SCIENCE MONITOR.
17 July 1996, p. 1. The Lawrence Livermore National Laboratory in California gave opponents more ammunition this week when it warned that some nuclear-weapons production and storage sites—including Yucca Mountain—are vulnerable to earthquakes and other natural disasters. At Yucca Mountain, the report states, hot volcanic material "could ascend directly through the repository . . . compromising the integrity of the waste isolation system." [ellipses in original]

63. WASHINGTON POST. 31 January 1996, p. A1.
The problem with Carter's approach to nuclear waste issues is that it lacks scientific basis and ignores the impact such a decision would have on the fastest-growing state in the union. Carter makes the claim that "there is no place other than the Nevada Test Site to store all these various radioactive and proliferation-sensitive nuclear materials" because of the site's remoteness and dry climate. What he conveniently omits is that the part of Nevada that contains the NTS is also a geologically active region with high earthquake activity. There is the possibility of renewed volcanic activity during the required lifetime of any long-term storage facility, and the area contains large and potentially vulnerable groundwater resources.

64. LOS ANGELES TIMES. 17 July 1996, p. A3.
Scientists are trying to determine whether Yucca Mountain, a ridge of ancient volcanic rock, is a suitable place to entomb thousands of tons of high-level radioactive waste. In the plan, shown below, canisters would be placed in the mountain with several hundred feet of rock above and below.
Danger Scenario
There are fears that ground water—pushed upward by shifts in the Earth's crust—may reach the waste. This could force water, superheated by the radioactivity, to gush to the surface, carrying the threat of contamination with it, or cause water to seep back down and contaminate the ground water supply.

65. WASHINGTON POST. 31 December 1995,
p. A1. After nearly a decade of study, "we know a whole lot about that mountain, but we don't have very many answers to very specific questions," said Dreyfus. And according to a recent DOE report, it is "not realistic to assume that all information needed to make a final judgment on the long-term safety of the repository" can be gathered before deciding to move ahead. "Some of that information can only be gathered and analyzed after a repository is built and loaded," the report concludes.

A few experts—including some federal researchers—believe that within several thousand years the facility might act like a giant nuclear reactor, achieving "criticality" and erupting in an explosion that would spew radioactive material above or below ground.

66. Mike Zapler. STATES NEWS SERVICE. 2 May 1996, Lexis/Nexis. Bill proponents, however, say the measure's radiation standard is safe. They cite groups such as the International Council on Radiation Protection, which recommends a 100 millirem per year ceiling on radiation exposure.

The Council's standard, however, covers radiation exposure from all sources. The nuclear waste bill, by contrast, sets a 100 millirem level for exposure to the Yucca repository alone. It does not include radiation intake from other sources such as the Test Site, where the United States tested nuclear weapons during the Cold War.

Further, the bill only vaguely identifies to whom the 100 millirem standard would apply, saying it should cover "an average member . . . in the vicinity of the Yucca Mountain site."

"Since the average member is undefined, it could include Las Vegas, the closest major population center. this would allow high doses (of radiation) to a large number of people," according to an Environmental Protection Agency draft document outlining the agency's concerns with the bill. [ellipses in original]

67. Mike Zapler. STATES NEWS SERVICE. 2 May 1996, Lexis/Nexis. One of the most egregious provisions, according to environmentalists, would set a radiation exposure standard of 100 millirems per year for "an average member of the general population in the vicinity of the Yucca Mountain site." That level of exposure, applied over a 70-year lifespan, would cause one of every 286 people in the area to die, according to the Nuclear Regulatory Commission.

The bill's radiation standard for Yucca is four times the Nuclear Regulatory Commission's standard for low-level nuclear waste, 25 millirems per year. Other countries have implemented—and the National Academy of Sciences has approved—radiation standards ranging from two to 20 millirems per year.

68. WASHINGTON POST. 31 December 1995, p. A1. Since 1987, DOE has paid $1.7 billion to investigate whether it is feasible to bury the nation's used reactor fuel there, 1,200 feet beneath the surface. After a couple more years and another $1.5 billion, DOE officials say, scientists may have a pretty good answer—but still not a conclusive one. Even if the site eventually is found suitable, the regulatory process is so complex that Yucca could not receive its first ounce of radioactive material for at least 15 years.

The project's director, however, gives it only a 50-50 chance of ever being approved by the Nuclear Regulatory Commission, which must license any such facility. And even if it were constructed as planned—at an estimated cost of $33 billion to build and fill—it would still be far too small to hold the estimated 93,000 tons of spent fuel that will have built up by the time all the nation's nuclear reactors are decommissioned in 2033. Not to mention another 8,000 tons of radioactive waste from defense plants that also is slated for burial at Yucca.

69. John Cantlon. DISPOSAL AND STORAGE OF SPENT NUCLEAR FUEL—FINDING THE RIGHT BALANCE. March 1996, p. 7. Since 1987, the DOE's civilian radioactive waste management program has progressed more slowly than expected. However, the Board is quite encouraged by recent progress and believes that the DOE program has reached a pivotal phase. Significant progress in excavating the underground exploratory facility is being made. The tunnel-boring machine reached the level of the proposed repository in November 1995, and important underground data about the suitability of the site are now being acquired. Progress also is being made on repository design and on the engineered barrier system. The DOE finally is developing a waste isolation strategy, which, in combination with recent advances in performance assessment, should permit a better delineation of priorities and a more efficient allocation of funds among program activities. However, fiscal year 1996 funding for the program was cut severely—from $520 million to $400 million, $85 million of which was set aside for future development of a federal spent fuel storage facility. In addition, legislation has been introduced in Congress that could change the focus of the program from disposal to storage at the Yucca Mountain site. (See Note 3.)

70. CONGRESSIONAL PRESS RELEASES. 18 December 1995, Lexis/Nexis. Permanent Storage Facility. The bill requires DOE to continue studying the intended site at Yucca Mountain to determine if the area meets scientific tests of reliability. If the study turns out positively, DOE must submit, prior to December 31, 2001, an application for construction of a permanent repository to the NRC. The bill establishes the date of January 17, 2010, for opening the permanent storage facility. The bill also places full responsibility for determining the safety of the site on the NRC, revoking the authority of the Environmental Protection Agency established as part of the 1982 law. The permanent storage facility must limit the radiation emissions to national-average levels, defined as 100 millirems, for at least 1,000 years under worst-case scenarios. The site must also be expected to meet the standard radiation requirement under more conservative expectations for the next 9,000 years.

71. NEI PRESS RELEASES. 29 May 1996, (www:nei.org/main/ pressrm) The work of DOE's scientists "is confirming our expectations about conditions" at Yucca Mountain, Curtis said. "We have learned that the quality and dryness of the rock at the repository level are even more favorable than we had anticipated. We have discovered no unexpected geological features, such as major faults, and there are indications that the flow of groundwater at the repository horizon has been very limited for at least 100,000 years. We still believe that we can

maintain momentum so that we can address the major unresolved technical issues allowing us in 1998 to make an assessment on the viability of the site."

72. Daniel Dreyfus. FDCH CONGRESSIONAL TESTIMONY. Federal News SERVICE. 28 June 1995, Lexis/Nexis. Transportation planning cannot be completed and many key planning activities cannot proceed until a site is selected for storage or disposal. Current plans are to begin shipments from reactor sites to the repository in 2010, assuming the successful licensing of the Yucca Mountain site. Shipping to an interim storage facility at an earlier date would require a greatly accelerated effort. It will take several decades to remove the projected quantities of spent fuel to be produced by existing reactors. Assuming the availability of high-rapacity transportation containers, we estimate there will be about 175 to 200 shipments per year of spent fuel, predominantly by rail, with some truck shipments. Depending on the site of the storage or disposal facilities, transportation routes could potentially go through every state except Hawaii, and Alaska. Specific transportation routes have not yet been designated.

74. Emmit George. FEDERAL NEWS SERVICE. 12 July 1995, Lexis/Nexis. The NARUC has considered alternatives for improving the program. First, the licensing process for the repository could be conducted more efficiently while addressing health, safety and environmental concerns. In this context, distinctions could be made between repository construction licensing and licensing for the emplacement of waste in the repository. With respect to the latter proceeding, perhaps we should consider allowing the waste to be retrievable during the initial 100-year period until it is determined that a Permanent repository may safely be permanently sealed. This would allow the DOE answer many of the scientific questions that have arisen during this process regarding the geologic composition of the repository and its suitability for permanent disposal of high-level nuclear waste that are difficult to determine today. H.R 1020 takes similar approaches while streamlining certain aspects in the performance of reviews under the National Environmental Policy Act (NEPA). NARUC supports these approaches in concept.

74. CONGRESSIONAL PRESS RELEASES. 18 December 1995, Lexis/Nexis. Nevertheless, fears persist about the Yucca Mountain site. In March of this year, two scientists at the Los Alamos National Laboratory released a study outlining conditions under which the plutonium from the waste could seep into the surrounding rock formations and over a lengthy period of time could create a nuclear explosion. While other scientists around the country and even at Los Alamos itself have discounted the theory, skepticism remains.

75. Mike Zapler. STATES NEWS SERVICE. 2 May 1996, Lexis/Nexis. To be sure, the Yucca repository isn't expected to be completed until 2010 at the earliest, assuming the Nuclear Regulatory Commission finds it safe enough to license. And radiation exposure from the waste may not occur for hundreds or even thousands of years, if it happens at all.

"We're talking about long periods of time, which makes it difficult to prove with certainty" how much exposure will occur, said a Senate aide who has worked on the bill. "The 100 millirem standard is the maximum. We expect there will be less" exposure.

76. Mike Zapler. STATES NEWS SERVICE. 2 May 1996, Lexis/Nexis. But the measure's proponents counter that Yucca Mountain is so isolated and dry that few people would be harmed by the relaxed environmental standards.

"Nobody lives there," said Scott Peters of the Nuclear Energy Institute, which represents the nuclear power industry. "The only folks who have any reason to be there are the people who work there."

77. WASHINGTON POST. 31 December 1995, p. A1. Others—including the State of Nevada Nuclear Waste Project Office—argue that the site is unsafe because of the risk of earthquakes, which are not uncommon in the area. Dreyfus counters that "the overwhelming scientific opinion is that it's not a problem," and DOE officials believe that a dump can be built in such a way as to minimize the danger. A National Research Council panel has emphasized that the area is expected to be relatively stable geologically on a time scale of about 1 million years.

Critics point to a 1992 magnitude 5.6 quake that rocked the area. Shock waves from miles beneath the surface hit the two-story Yucca Mountain Project Field Office, causing about $1 million in damage and breaking one-third of the windows, according to Derek Scammell of the DOE's Las Vegas public affairs office.

But the effect was far less in underground tunnels, said geologist J. Russell Dyer, DOE's deputy project manager for the Yucca site. Coffee cups weren't even knocked off tables.

78. BULLETIN OF THE ATOMIC SCIENTIST (BAS). 11 January 1997, p. 13. But repository performance studies are complex, and even the site's advantages can also present disadvantages. For instance, the fact that the proposed repository would be built in relatively dry rock, high above the water table, is in most respects all to the good. But it also means that oxygen is present and that a corrosive "oxidizing environment" may shorten the life of waste packages.

Whatever the site's disadvantages, they might be compensated for over time through various engineering strategies—improving repository design or providing even more durable waste containers. (Performance studies indicate

that the robust double-walled spent fuel casks now planned would, under most credible assumptions, last 10,000 years. These casks are to have three-fourths-inch-thick nickel alloy inner wall and a four-inch-thick carbon steel outer wall.)

79. BULLETIN OF THE ATOMIC SCIENTIST (BAS). 11 January 1997, p. 13. Technically, the reality is that even if other geologic sites were politically available, probably nothing could be gained by choosing one of them over Yucca Mountain. Experience with geologic site exploration in the United States and abroad shows that all sites present problems, complications, and uncertainties.

No site can offer permanent and absolute containment of all radioactivity. The one irrefutable advantage of geologic disposal over surface storage is that geologic repositories are less subject to the vagaries of natural forces and human intrusion, especially over long reaches of time after institutional control has been lost.

The Energy Department now appears genuinely optimistic about the ongoing investigation of Yucca Mountain, and the department expects to make a positive "viability" determination in 1998 as a step toward a formal license application in 2001. A much-needed political lift for the project may come as early as next spring, when an impressive 25-foot-diameter tunnel, now being bored through the mountain in a five-mile exploratory loop, is completed.

80. ATOMIC ENERGY INSIGHTS. Vol. 1, Issue 12, March 1996, http://www.ans.neep.wisc.edu/~ans/point_source/AEI/mar96/Opposers.htm) The arguments related to the long term behavior of Yucca Mountain are simply unanswerable questions. That is not to say that they have any validity, but no scientist will ever categorically state that he knows what will happen in the distant future. At best, legitimate scientists will use evidence of what has happened in the past to make some probabilistic predictions about the future.

When people demand answers to unanswerable questions, it is a pretty good indication that they are simply interested in delays and obstruction instead of solutions.

Even if the worst possible conditions occur, there is a very small chance that there will be any danger to humans or other living creatures. There are several basic facts that back up this statement.

- Radioactive material is not dangerous unless it is ingested or unless people get too close to unshielded material.
- It does not regenerate itself or pass from one individual to another like a contagious disease.
- Spent nuclear fuel is in a form that cannot be easily distributed over a large area.
- In order for the waste to cause even minor health impacts, society as a whole must somehow forget how to sense radioactive materials and it have to lose

records and warning signs indicating existence of the facility.

In this discussion it is also important to compare the possible health impacts of spent nuclear fuel with the health impacts from the waste products of competitive power sources. Routine emissions from burning coal, for example, cause several different kinds of serious respiratory ailments, and have been implicated as contributing factors to lung cancer rates.

81. The Nuclear Energy Institute. HIGH LEVEL NUCLEAR WASTE. August 1996, http://www/nei/org/main/pressrm/facts/hlw.htm) Site characterization at Yucca Mountain is one of the most thorough and comprehensive scientific investigations ever conducted, lasting years, costing more than $6 billion and involving thousands of scientists, engineers, and technicians.

The start of site characterization was delayed for several reasons, among them the refusal of Nevada to issue the environmental permits need for surface-disturbing work such as maintaining roads, digging trenches and drilling boreholes. By mid-1992, following several court cases involving DOE and Nevada, the state had issued the permits needed to begin a full-scale study of the site.

Since then, site characterization has proceeded without interruption. The LM 300, a specially developed rig that can drill to a depth of 3,500 feet, has drilled three deep boreholes. The rig uses no drilling fluids, so samples are clean of materials that might be introduced by drilling. Other activities include the digging of test pits for use in volcanism studies, the completion of trenching to gather data on potential geologic faulting in the area, and the drilling of more than 20 shallow boreholes to study rainwater infiltration and faulting in the area, gather data for constructing the Exploratory Studies Facility. The surface-based investigations have been largely completed.

DOE began construction of the Exploratory Studies Facility—a system of tunnels that will allow scientists to conduct seismological, geological and hydrological studies—in September 1994, using a tunnel boring machine. By the end of May 1996, the machine had excavated 18,000 feet, and the digging was two-thirds complete.

82. Emmit George. FDCH TESTIMONY. 5 February 1997, Lexis/Nexis. In the NARUC resolution, the Association also urged the designation and location of above-ground, centralized, interim storage capacity for spent nuclear fuel by a date which, incidentally, has already passed (June 1995), and that such a site should be "de-linked" from the location or licensing of a permanent repository. This principle, while outdated, still expresses the urgency of moving forward with an program for spent management and disposal as it is the only means by which DOE can meet its statutory and contractual obligation. Centralized, interim storage will add a measure of flexibil-

ity to the overall spent fuel management system, and as DOE data indicates, would provide for a more cost-effective and efficient waste program. Additionally, an interim storage capability would not only allow the nuclear waste management system to deal with situations where utilities are unable to develop additional at-reactor storage due to physical, economic or other constraints, but it would also provide DOE and other parties the opportunity to experience transportation, handling and managing spent nuclear fuel that could, ultimately, meet the needs of the final repository system. Another benefit would be the advantages in the areas of environmental protection, health and safety, and would increase cost-effectiveness as a result of standardization of procedures, operations and equipment. Finally, the development of an interim storage capability would build confidence among ratepayers, regulators and utilities in the DOE's capability to ultimately provide for nuclear waste disposal functions.

83. Ivan Selin. FEDERAL NEWS SERVICE. 16 May 1995, Lexis/Nexis. On the issue of spent fuel, I believe that both centralized interim storage and at-reactor storage would protect public health and safety. However, I also believe that a centralized facility would provide advantages relative to dispersed storage at 74 sites throughout the country. Most significantly, centralized storage lessens the likelihood of accidents occurring at non-operational plants storing spent fuel. In addition, considering the potential for decades of licensed storage, a centralized facility would allow for a more focused inspection and surveillance program by both DOE and NRC and would also offer operational and programmatic benefits on DOE's program for accepting waste from utilities. Therefore, because there are clear advantages to centralized interim off-site storage, I personally believe that legislation directed towards the establishment of such a facility could partially resolve one of the most vexing environmental issues facing this nation.

84. INSIDE ENERGY WITH FEDERAL LANDS. 20 May 1996, p. 8. Sandra was referring to the White House threat last year to veto the energy and water development appropriations bill if it contained a provision designating a particular state as the location for an interim storage facility for civilian spent fuel.

The Nuclear Energy Institute, which represents nuclear utilities, also weighed in the issue. In a May 15 letter to O'Leary, the group said it endorsed DOE's decision to accept the foreign fuel, but "strongly believes the department should demonstrate the same commitment to managing domestic spent nuclear fuel." The decision, NEI continued, shows that the United States has the technology to safely transport and store nuclear waste until a repository is operating. "The only missing element is [DOE's] support for legislation that will allow the federal government to honor its promise to electricity consumers to manage domestic spent nuclear fuel."

The 20 tons of foreign spent fuel involved in last week's announcement is dwarfed by the 30,000 tons that is currently held at civilian nuclear plants across the country.

85. THE CHRISTIAN SCIENCE MONITOR. 27 October 1994, p. 19. There is a better solution to managing nuclear wastes called monitored retrievable storage (MRS)—developing an interim storage facility where spent fuel from the nation's 110 nuclear plants could be consolidated and stored until a permanent repository begins operation. An MRS would be managed by specially trained staffs. It is a highly successful approach used by other countries.

For example, Sweden has a central storage facility located in a rock cavern. Spent fuel from Sweden's 12 nuclear plants are shipped to the cavern after about one year. The Swedish government's plan is to store the spent fuel there for about 30 or 40 years. Transport to a deep geologic repository will begin around the year 2020.

A strategy of this kind is more likely to succeed than our own. It would best serve the interests of utilities and electricity customers. Congress should amend the Nuclear Waste Policy Act to permit development of an MRS so that DOE has a place to store the waste temporarily until a permanent repository is in operation. Also, Congress needs to correct flaws in the 1983 law that are needlessly slowing the study and licensing of a permanent repository.

The disposal of high-level nuclear waste is yet another example of an environmental issue for which there is a sound technical solution that other countries are pursuing. America, too, needs to demonstrate it can keep its promises and follow through with its plans. A good beginning would be for DOE to develop an interim storage facility and accept power plant waste, as it agreed to do years ago.

86. CHICAGO SUN-TIMES. 17 September 1996. p. 26. For Illinois, this is no small matter. Illinois gets 54 percent of its electricity from 13 nuclear reactors. Not only would there be a large added expense for building alternative fuel storage facilities at every plant, but one centralized, federally operated site is certainly more prudent than having interim facilities at every reactor in the country.

Although safe, nuclear plants never were designed to store large amounts of highly radioactive used fuel. Electric utilities are in the business of running power plants and generating electricity, not storing radioactive waste.

87. William McCormick. FDCH CONGRESSIONAL TESTIMONY. Federal News Service. 28 June 1995, Lexis/Nexis. Consumers Power's experience with dry cask storage also has demonstrated that this technology is a safe, simple and secure method for storing spent nuclear fuel, characteristics which make the technology ideally suited for use at a central Federal interim storage facility. The cask system is simple, with no moving parts. After two years of detailed monitoring of the dry cask stor-

age containers at Palisades, we can say with confidence that the system meets or exceeds design expectations for removing heat and shielding radiation. Radiation from spent fuel is shielded by a dense steel container and steel and steel-reinforced cask. One would need to hug a cask for an entire hour to receive 3 millirems of radiation, the equivalent of one-third of a chest x-ray. Thirty feet away from the casks, behind the fence where the public can view them, only background radiation is detectable.

88. William Sherman. FEDERAL NEWS SERVICE. 14 December, 1995, Lexis/Nexis. However, based on past performance and present appropriations, there is no confidence in the completion schedule, presently stated as 2010, for the spent fuel repository. Spent fuel will have to be in temporary storage for a long time, even after many nuclear plants have ceased operation. A single interim storage area will allow careful resolution of repository concerns without the pressure for hasty decisions / responsibilities: Monetary Responsibility, Environmental Responsibility and Governmental Responsibility.

89. Federico Pena. FDCH TESTIMONY. 5 February 1997, Lexis/Nexis. DOE has had similar success treating and disposing of the tons of waste generated and stored from over 40 years of weapons production. The risk of explosions from the liquid wastes in the tanks at Hanford, Washington has been greatly reduced. Waste processing is ongoing in the Savannah River Site canyons, stabilizing materials that the Defense Nuclear Facilities Safety Board identified as presenting a near-term hazard. In addition, we have poured 100 canisters of vitrified waste at the Defense Waste Processing Facility.

90. Federico Pena. FDCH TESTIMONY. 5 February 1997, Lexis/Nexis. DOE has had a number of success in speeding up the cleanup. For example, at the Savannah River Site in South Carolina the Department deployed three new technologies to achieve a 75% percent increase in the removal of organic solvents, including more than 90,000 pounds removed from soil and groundwater within the "AM" area. At the Oak Ridge Reservation in Tennessee, the Department's cleanup program has capped 121 contaminated areas, remediated 38 acres of radioactively contaminated collection ponds and incinerated over 1.5 million kilograms of waste.

91. THE ENERGY DAILY. 22 August 1996, Lexis/Nexis. New studies have put to rest earthquake safety concerns at one of Savannah River Site's aging nuclear reprocessing canyons, clearing the way for increased waste stabilization operations, the Energy Department said Wednesday.

The department released a new seismic risk analysis that concluded the F Canyon building would confine radioactive material adequately in the event a major earthquake shook the South Carolina weapons site.

The new analysis was conducted after Westinghouse Savannah River Co., which operates the Savannah River for DOE, uncovered information in January 1996 suggesting that the F and H canyons might be more vulnerable to earthquake damage than previously believed.

92. CHARLESTON POST & COURIER. 5 January 1997, p. A1. A spokesman for the Department of Energy says DOE has "high confidence" that the European waste won't remain at SRS, but will eventually be disposed of elsewhere. DOE also has said it has no plans to store additional waste at SRS, despite a study that concluded that more nuclear waste could be handled by the site's contractor.

93. KNOXVILLE NEWS-SENTINEL. 28 January 1997, p. A1.
OAK RIDGE—A report by the Department of Energy's Inspector General concludes that nearly half of Oak Ridge federal reservation is "non-essential" for DOE's missions and should be sold at fair-market value or transferred to other state or federal agencies.
It continues . . .
The Inspector General report said about 16,000 acres is not needed for DOE's work, which includes scientific research and development, nuclear material storage, nuclear weapons dismantlement, environmental cleanup, and management of hazardous wastes.

94. THE ENERGY DAILY, 4 February 1997, Lexis/Nexis. And to ensure that DOE is considering all available options, the company reminded the department that it currently is accepting spent research reactor fuel for storage, and that there is no reason the department cannot also begin accepting its reactor fuel next year as stipulated in the contract.

"Based on DOE's acceptance of certain research reactor fuel at federal sites," the company said, "we have concluded that it is possible to do the same for Yankee and propose that the shipments be made in truck casks which are available to you for lease from a domestic vendor. Thus, we believe that DOE possesses the capability of accepting Yankee's spent fuel by January 1988."

95. Emmit George. FEDERAL NEWS SERVICE. 14 December 1995, Lexis/Nexis. However, this principle still expresses the urgency of moving forward with an integrated program for spent management and disposal. Centralized, interim storage will add a measure of flexibility to the overall spent fuel management system, and as DOE data indicates, would provide for a more cost-effective and efficient waste program. An interim storage capability would not only allow the nuclear waste management system to deal with situations where utilities are unable to develop additional at-reactor storage due to physical or other constraints, but it would also provide DOE with valuable experience in transportation, spent fuel con-

tainer handling and management operations that could meet the needs of the final repository system. Another benefit would be safety advantages and increased cost-effectiveness as a result of standardization of procedures, operations and equipment. Finally, the development of interim storage capacity would build confidence among ratepayers, regulators and utilities in the DOE's capability to ultimately provide for nuclear waste disposal functions.

96.   Emmit George. FEDERAL NEWS SERVICE.
        14 December 1995, Lexis/Nexis.

II. The DOE Program Lacks an Interim Strategy.
Under the DOE's repository suitability and licensing strategy, the DOE expects the final repository to begin accepting spent fuel sometime on or after the year 2010—if the site is, indeed, found to be suitable. There is no certainty that this 2010 date will be achieved. It is important to note, however, that the past experience of the program has shown how difficult it is to anticipate and keep to time schedules in this first-of-a-kind project. Additionally, and perhaps most importantly, the Yucca Mountain site may not prove suitable. That situation would extend the interim period for a decade or more. The NARUC believes that provisions need to be approved to develop a centralized interim storage facility. Also, as I mentioned earlier, the storage capacity at nuclear reactors will be largely depleted during this time period without expanding capacity with the costs supported by ratepayers. Both the NRC and the DOE have stated that centralized interim storage would provide programmatic advantages over dispersed at-reactor sites.

97.   Samuel Skinner. FEDERAL NEWS SERVICE.
        12 July 1995, Lexis/Nexis. Based on repeated promises from the federal government, electric utilities that built nuclear power plants expected spent fuel to be stored for a short period of time on site, then shipped off-site either for reprocessing or storage and disposal at federal government facilities. On-site spent fuel storage facilities were never intended to provide life-of-plant storage capacity. Consequently, by the time DOE starts accepting spent fuel in 1998, 26 reactor sites will have exhausted existing spent fuel capacity. Continued delays in the Energy Department's program to accept spent fuel will subject utility customers to significant added costs for additional interim storage capacity at plant sites. According to industry analysis of Department of Energy data, nuclear utility ratepayers would be forced to pay an additional $5 billion (1993 dollars) for on-site spent fuel storage if DOE fails to build a central interim storage facility and start accepting spent fuel in 1998. That assumes a permanent disposal facility opens in 2010 as currently scheduled. This $5 billion is, of course, in addition to the billions of dollars already being paid into the Nuclear Waste Fund.

If the permanent repository schedule slips beyond 2010 and the federal government does not have an interim storage facility, costs to consumers will soar. For example, if the permanent repository does not begin operating until 2020, and DOE does not build a central interim storage facility, electricity consumers will pay an additional $10 billion.

98.   Susan Clark. FEDERAL NEWS SERVICE.
        28 June 1995, Lexis/Nexis. In addition to providing a safe solution to the immediate disposal needs of many nuclear generators, a centralized interim storage facility would provide many benefits to utility ratepayers and the nuclear waste disposal program as a whole. Swill run out of existing on-site storage. By the year 2010, the earliest date for opening a permanent repository 80 plants will have exhausted their existing on-site storage capabilities. Interim storage is the only way that the Department of Energy can begin accepting spent nuclear fuel in 1998, and reduce the need for additional ratepayer investment in on-site storage capacity.

99.   Kris Sands. FEDERAL NEWS SERVICE.
        29 February 1996, Lexis/Nexis. Congress is supposed to appropriate this money from the fund for safe and timely storage and disposal of waste from our nuclear power plants. They produce 20 percent of our nation's electricity. We, the consumers have kept our end of the bargain but not the federal government. By failing to appropriate more than a fraction of our annual payments, some in Congress have effectively diverted, stolen, if you will, up to two-thirds of consumer's annual payments for other spending.

Collecting money for one purpose for the nuclear waste fund but using it for another and flat out DOE program failure has created the worst of all situations for American consumers. This year, Congress has severely cut funding for the DOE waste program, expressing legitimate concern over falling two years further behind every single year it operates.

So what do our friends at DOE do? First, they raise the percent of program overhead; they raise cost estimates, and then they extend the repository completion date, out yet another five years, to the Year 2015. Since then, come in Congress have threatened to cut all funding while still taking our money. If Congress slashes our funds, consumers cannot be expected to continue paying and paying and paying hundreds of millions of dollars for services that we do not yet. Any progress will slip even further, creating billions of dollars in unnecessary costs; up to 10 billion, I figure, and that's what I'm here to talk about today, avoided costs.

100.   ENERGY REPORT. 5 August 1996, p. 1. The industry has taken a high profile role in promoting the legislation that would allow interim storage, which is currently prohibited. The facility will save an additional $5 billion in on-site storage that ratepayers would have to shoulder, industry says.

101. Krista Sanda. FEDERAL NEWS SERVICES. 29 February 1996, Lexis/Nexis.

THE BOTTOM LINE: COSTS OF FAILURE
First, let's look at the costs for additional storage at nuclear power plants. A sampling of data from around the country indicates that implementing dry cask storage facilities at 73 power plant locations could easily result in costs of $6 billion to $8 billion—for the first twenty years. Unless the DOE program reverses its history of failure—and the opening date for a repository gets closer instead of farther away with each passing year—waste may remain at plant sites for many more decades. In addition, we should note that present DOE contracts with utilities assume an acceptance rate of 3000 MTU per year over a 45 year period. Additional costs of maintaining 73 spent fuel facilities for additional decades appears to be between $1.5 and $3.3 billion per decade. Even in the value of today's dollars, these dry cask storage costs approach, or exceed $10 billion. They would likely be even higher at the time of performance. Associated delay of planned plant decommissioning could also escalate costs. In a paper presented by Energy Resources International at a May, 1995, NARUC meeting, it was noted that keeping reactor pools operational for extra years could add tens of millions of dollars to each facility's total cost. Costs associated with premature retirement and replacement of power plants because of the waste issue could raise costs even further.

102. Krista Sanda. FEDERAL NEWS SERVICES. 29 February 1996, Lexis/Nexis. Chairman Myers, members of the Subcommittee, I appreciate this opportunity to testify today and to alert you to the urgent need for Congress to reform, and fund, the U.S. civilian nuclear waste program.

This coming year's appropriations will have a major impact on how this country deals with nuclear waste for decades to come. Congress' action, or inaction, will determine whether the United States will fund an effective civilian nuclear waste program, or continue to squander billions of dollars to create a relatively small, temporary infusion of trust fund money into the present budget process. Failure to reform and fund the U.S. civilian nuclear waste program will create a tremendous financial burden for electricity consumers and taxpayers. Relying solely on the Federal Courts to address this has powerful, and painful, implications for the Federal budget and appropriations process. We must take immediate steps to provide centralized, temporary storage of high-level nuclear waste. We also must reform and fund a program lending to safe, timely, and cost-effective permanent disposal of nuclear waste from our power plants. Electricity consumers will not continue to quietly pay their money into the Nuclear Waste Fund and get nothing in return. There is a growing realization by this country's consumers, regulators, elected officials and representatives of the electric industry that we must act to change the bumbling manner in which this program has been managed over the past 25 years. Now initiatives are attempting to address the management problems but it will be difficult to succeed without the commitment of this subcommittee to support the reform process with adequate funding.

103. Carole Gorney. COMPETITIVENESS AND AMERICAN SOCIETY. 1993, p. 167. Both the steel and chemical industries are subject to a double whammy when it comes to environmental legislation. Both are heavy users of electricity—electric power production being another industry reeling under environmental and safety regulatory costs. According to Douglas Biden, economist and secretary-treasurer of the Pennsylvania Electric Association, the United States used to have the second lowest electricity costs in the world. "It is now fifth or sixth," he said. "Canada has always had the lowest costs because it has a wealth of hydro (water) resources and it is run by the government and subsidized by it. France has passed us as it is 75% nuclear, and it is also subsidized. Sweden is 40 to 45% nuclear."

104. Eric Marshall Green. ECONOMIC SECURITY & HIGH TECHNOLOGY COMPETITION IN AN AGE OF TRANSITION: THE CASE OF THE SEMICONDUCTOR INDUSTRY. 1996, p. 103. The need to manufacture is not only an economic necessity, but an imperative for national security. The Pentagon relies heavily on domestic industry to source the materials necessary in times of war. It would be politically and militarily untenable to rely on foreign technology and material for national defense requirements. The question then becomes not whether the United States should maintain a manufacturing capacity but what type of capability will be retained. Given the challenge to American dominance in high technology, the erosion of our presence in traditional smokestack manufacturing industries, and the systemic problems associated with trade and budget deficits, a more structuralist view in the debate over national competitiveness has emerged. Stephen Cohen and John Zysman observe:

> A decline of manufacturing presages a decline in associated high wage services.... Links that promote ongoing market adaptation and technological innovation mean that advantage in a national economy is embodied not simply in the capacities of individual firms but in the web of interconnections that establishes possibilities for all. Further since making links within the national economy creates real advantages and speeds the development of the most advanced technologies and the applications of these new possibilities even to traditional industries.... Strategic sectors are those whose products and processes alter or transform the goods and production arrangements throughout the economy. [ellipses in original]

105. Emmit George. FDCH TESTIMONY. 5 February 1997, Lexis/Nexis. Moreover, continuing the pro-

gram as currently planned could lead to the indefinite storage of spent nuclear fuel at more than 70 nuclear reactor sites now operated nationwide. The impact of this situation would severely compromise the future viability of operating nuclear power plants and the resources they generate, and adversely affect electricity supply around the country. Furthermore, without resolving many of the program's problems, ratepayers will ultimately bear the burden of the costs associated with expanding on-site storage of spent nuclear fuel or for the decommissioning nuclear power plants that were unable to afford the cost of storing spent nuclear fuel. In a time of increased competition in the electric industry, this situation may detract from nuclear utilities' ability to compete in the new marketplace to the detriment of local customers.

106.   Emmit George, Jr. FEDERAL NEWS SERVICE. 5 February 1997, Lexis/Nexis. The problems with expanding at-reactor storage are not trivial, particularly in this time of potentially great economic change in the electric power industry. Most utilities have already paid to re-rack and expand pool storage capacity. At least ten facilities have already or are now planning to establish dry-cask storage as their spent fuel pool capacity is depleted. It is estimated that 30 percent of the nation's spent nuclear fuel pools will reach capacity by 1998 and approximately 80 percent of nation's pools will reach capacity by the year 2010—the earliest year that the DOE expects to begin receiving spent fuel for disposal in a repository. In addition, for nuclear reactors that have already shut down, continued at-reactor storage adds substantial costs to nuclear plant decom- missioning activities. These plants must maintain active operational spent fuel pool systems, and cannot complete their decommissioning activities until all spent fuel is removed. With the advent of greater competition in the electric industry, many nuclear facilities will be have to find a place in the new competitive marketplace. Any added costs or significant planning uncertainties are certain to compromise their competitive edge and increase the complexity of moving to more competitive markets. Moreover, storing fuel at more than 70 de facto interim storage sites around the country is not acceptable public policy and clearly was not intended when Congress passed the NWPA.

107.   James Rhodes. FDCH CONGRESSIONAL TESTIMONY. 5 February 1997, Lexis/Nexis. Despite a recent decision by the U.S. Court of Appeals that DOE is obligated by law to begin accepting used nuclear fuel, the department is unchanged in its position that it is "unable to begin acceptance" of the fuel next year. The Energy Department's failure to act throws into jeopardy the future of America's 109 nuclear power plants, which supply about 20 percent of our country's electricity. By 1998 27 nuclear units will have exhausted their existing on-site pool storage capacity for used fuel.

Mr. Chairman, in less than one year, the federal government must assume responsibility for managing used nuclear fuel at North Anna, Surry and nuclear power plants in 33 other states. The nuclear industry and electricity customers have held up their side of contracts with the federal government by committing almost $13 billion to fund this program. Our most recent return on that investment, Mr. Chairman, is a December 17, 1996 letter from DOE to every utility with a nuclear power plant advising them that the agency can't meet its obligation to begin accepting used fuel in 1998.

108.   BUFFALO NEWS. 25 September 1996, p. A1. More is at stake than one lightly populated Western state. No more nuclear plants can be built anywhere until a place is found for existing nuclear waste. Environmental purists are adamant against atomic power and so oppose a solution to the waste problem. When the Senate on July 31 voted 63 to 37 to break a filibuster on the storage bill, environmentally militant senators voted no.

109.   THE CHRISTIAN SCIENCE MONITOR. 27 October 1994, p. 19. The nation's nuclear-waste problem would be quickly and painlessly eased if the US Department of Energy (DOE) were required to keep a commitment it made 12 years ago to take spent nuclear fuel from power plants beginning in 1998. The burden on electricity customers would be reduced significantly if DOE developed a central facility to hold the spent fuel until an underground waste repository is ready.

110.   THE PATRIOT LEDGER. 25 January 1997, p. 1. Other utilities, however, say they're running out of space. In any case it's not a sound policy to have 110 nuclear power plants in 35 states storing about 30,000 tons of highly radioactive waste when the federal government should be taking over this responsibility.

Electricity buyers should care, because this is also a pocketbook issue. Ratepayers are being charged an extra fee for their electricity to pay for the federal government's costs of managing spent-fuel wastes from nuclear power plants. Since 1982, they've been assessed a fraction of a cent per kilowatt hour for a Nuclear Waste Fund from which they've so far received no benefit. About $13 billion has been collected nationwide for this purpose, including $94 million from Massachusetts ratepayers, most of whom are contributing to the Nuclear Waste Fund.

Even if Pilgrim should be shut down before its license runs out, ratepayers wouldn't escape paying for spent-fuel storage. These expenses are not significant now, being incidental to normal operating expenses. However, they would zoom into millions of dollars a year if Pilgrim closes without the federal government assuming responsibility for the plant's nuclear fuel.

That is what is happening at Yankee Electric's plant in Rowe. Although Yankee Rowe, the nation's oldest operat-

ing nuclear power plant, closed in 1992 and has been largely dismantled, its fuel continues to be stored there, at ratepayers' expense.

111. Emmit George. FEDERAL NEWS SERVICE.
5 February 1995, Lexis/Nexis. The nuclear waste program has, for many years, been a source of deep concern and enormous frustration to our nation's utility ratepayers and regulators for two reasons. First, ratepayers pay approximately $600 million per year into the Nuclear Waste Fund. Since 1983, this Fund, which is supported solely by the Nation's electricity consumers, has accumulated close to $13 billion. State regulators, therefore, have a compelling interest in the cost-effectiveness and success of the Federal government's civilian waste management program, which is supported by this Fund. Second, effective waste management and eventual permanent disposal are essential to minimize the life cycle costs of existing nuclear plants which comprise more than 20 percent of the electric energy produced in the United States. Cost increases for expanding on-site storage, reactor decommissioning and centralized disposal of nuclear wastes increases the costs of nuclear energy overall, which in turn, can have a significant adverse effect on energy costs to consumers.

112. Gary Gates. FEDERAL NEWS SERVICE.
14 December 1995, Lexis/Nexis. More broadly, failure to enact legislation that would allow construction and operation of an interim storage facility by 1998 threatens our nation's nuclear generation capacity and unfairly requires consumers and local governments to bear the cost burden for bureaucratic delays in this program area. APPA strongly supports S.1271, as it would enable the federal government to fulfill its commitment to ratepayers, while providing for an integrated and responsible approach to high-level nuclear waste management.

113. Emmit George. FEDERAL NEWS SERVICE.
12 July 1995, Lexis/Nexis. Electricity utility ratepayers are the primary source of revenue for the civilian nuclear waste disposal program. Therefore, an effective waste management and disposal system is needed to minimize the life-cycle costs of nuclear power plants, which comprise 2055 of the U.S. electric energy production.

- The NARUC adopted a Resolution in November 1994 setting forth legislative policy principles resulting from a constructive dialogue among stakeholders.

- The NARUC believes the Department of Energy (DOE) must act on its responsibility to take title to and begin accepting spent nuclear fuel by the Jan. 31, 1998, statutory deadline. Otherwise, capacity for storing spent nuclear fuel will continue to diminish, making costly expansions necessary.

- The DOE needs to be given authorization to develop a centralized, interim storage facility as part of an integrated nuclear waste disposal system. Such a facility should be "de-linked" from the location of a permanent repository. A facility of this kind would relieve pressures on utilities to develop at-reactor storage while providing DOE with valuable experience that could meet future needs of the permanent repository.

114. NUCLEAR WASTE NEWS. 23 March 1995,
p. 1. "Essentially what we are looking for is for the federal government to live up to its obligation to start taking spent fuel beginning in 1998," Cathy Roche, vice president of public and industry communications for the Nuclear Energy Institute, said.

Utilities see a central federal monitored retrievable storage (MRS) facility as the way to achieve this. The MRS would serve as interim storage until a permanent repository at Yucca Mountain, Nev., is completed. Environmental groups, however, see a host of problems associated with this strategy.

Utilities want the spent fuel removed from reactor sites because on-site cooling ponds are reaching maximum capacity, threatening plant shutdowns. Some say such shutdowns are exactly what anti-nuclear groups want.

115. Emmit George. FDCH CONGRESSIONAL TESTIMONY. Federal News Service. 12 July 1995, Lexis/Nexis. The NARUC believes the DOE must act on its responsibility to being accepting spent nuclear fuel on January 31, 1998. On April 28, the DOE formally concluded that because a repository is not available, it does not have responsibility to accept spent fuel according to contracts entered into pursuant to the Nuclear Waste Policy Act (NWPA) beginning on January 31, 1998. Unfortunately, this is not a satisfactory response to ratepayers who win be asked to bear the additional costs of at-reactor, out-of-pool storage and beyond the one mill per kilowatt hour fee being paid to the Federal government.

116. Emmit George. FEDERAL NEWS SERVICE.
14 December 1995, Lexis/Nexis. The problem with expanding at-reactor storage are not trivial and could prematurely force nuclear energy plants out of service or prevent them from returning to service. This is not mere speculation. At least five facilities have already established dry-cask storage and more are planning to do so as their pool reserves continue to diminish. It is estimated that 30 percent of the nation's spent nuclear fuel pools will reach capacity by 1998 and approximately 80 percent of nation's pools will reach capacity by the year 2010—the earliest year that the DOE expects to begin receiving spent fuel for disposal in a repository. In addition, for nuclear reactors that have already shut down, continued at-reactor storage adds substantial costs to nuclear plant decommissioning activities. This situation occurs because these plants can not complete their decommissioning activities until all spent fuel is removed from pool storage. Storing fuel at

more than 70 de facto interim storage sites around the country is not acceptable public policy and was not intended when Congress passed the NWPA.

117. Emmit George. FEDERAL NEWS SERVICE. 14 December 1995, Lexis/Nexis. Moreover, continuing the program as currently planned could lead to the indefinite storage of spent nuclear fuel at more than 70 nuclear reactor sites now operating nationwide. The impact of this situation would severely compromise the future viability of operating nuclear power plants and the resources they generate. Furthermore, without resolving many of the program's problems, ratepayers will ultimately bear the burden of the costs associated with expanding on-site storage of spent nuclear fuel or for decommissioning nuclear power plants that were unable to afford the cost of storing spent nuclear fuel. The NARUC believes after having already committed approximately $11 billion on developing the nuclear waste program over a 12-year period while the inventory of spent nuclear fuel at nuclear reactors continues to grow, this situation can no longer be tolerated. This series of hearings could not have come at a more timely moment.

118. Samuel Skinner. FDCH CONGRESSIONAL TESTIMONY. Federal News SERVICE. 2 March 1995, Lexis/Nexis. Nuclear power plants have supplied over 40 percent of all the new electricity required by the American people since 1973. They have allowed this nation's electric utility sector to virtually eliminate its dependence on imported oil. And our nuclear power plants will make by far the largest contribution of any technology toward meeting this Administration's year 2000 goals for reducing greenhouse gas emissions.

Without passage of new legislation—and particularly without an interim storage facility and a means of satisfying the existing federal obligations to accept spent fuel starting in 1998—more and more nuclear power plants will run out of on-site storage space for spent fuel. By 1998, 26 nuclear units will have exhausted their storage capacity. By 2010—the earliest date for opening permanent disposal facility—80 units will be out of space.

Without passage of this legislation, more and more nuclear units will need to build additional on-site storage. More and more companies will have to navigate through a minefield of state and local politics and organized, antinuclear fear-mongering. More and more companies will be forced to explain to their state regulators why consumers should pay twice for spent fuel management, and why anyone should trust the federal government to keep its word in this area when its track record is so dismal.

119. NATION'S CITIES WEEKLY. 5 August 1996, Lexis/Nexis. The outcome of this legislation could have impact on companies during the electric utility deregulation debate. If companies have to continue to pay for the on-site temporary storage of nuclear waste, their costs are higher than non-nuclear power companies. One of the main debates in the electric utility deregulation debate is stranded costs, past costs of companies that the customer currently pays. If standed costs are not included as part of the deregulation process, companies with nuclear power will not be as competitive.

120. Gary Gates. FEDERAL NEWS SERVICE. 14 December 1995, Lexis/Nexis. Since 1982, electric consumers have made significant financial contributions toward a responsible solution to the need for a high-level waste storage facility. The Nuclear Waste Fund was created by the Nuclear Waste Policy Act of 1982 to collect fees from consumers of nuclear energy to cover the costs of constructing and operating a permanent storage facility for nuclear waste. To date, over $10 billion has been collected to enable DOE to effectively address this important issue. In my home state of Nebraska, consumers have paid a staggering $136 million into the Nuclear Waste Fund. Yet, while consumers have met their obligation, the federal government has failed to hold up its end of the deal. As a result, utilities and their customers must either construct new on site storage facilities, or face possible power shortages and drastically increased electricity rates if plants are required to shut down due to inadequate storage for spent fuel. For example, at our Fort Calhoun Station, OPPD has already spent nearly $7 million to provide for additional on-site storage. In effect, the consumers of nuclear energy are being doubletaxed. A significant amount of the funding they have provided to support DOE's high-level waste storage program has been redirected toward general deficit reduction—and they are left to find alternative solutions to the problem at a very high cost. Sharing these concerns regarding management of the Nuclear Waste Fund, the Consumer Federation of America has also taken a strong position urging Congress and the Administration to ". . . ensure that the billions of dollars already collected from consumers, as well as the funds that will be collected in the future, are used for purposes specified in the Nuclear Waste Policy Act and not retained in the Treasure for deficit reduction." [ellipses in original]

121. Chuck Broscious. IDAHO STATESMAN. 8 May 1996, p. A1. The current Nuclear Waste Policy Act of 1987 prohibits opening an interim facility until the Nuclear Regulatory Commission licenses a permanent repository. This provision is important because once nuclear waste is shipped to any centralized location, the site is likely to become a de facto repository. Maintaining this prohibition is important to ensure that a permanent facility is not, in effect, a "done deal" and that the site suitability studies will be honest.

122. STAR TRIBUNE. 1 August, 1996, p. 17A. Opponents also say the cross-country transportation of the deadly waste would endanger millions of people along routes through 43 states. Bryan said each 125-ton

cannister shipped by rail "contains the radiological equivalent of 200 Hiroshima bombs." Reid, coining the shipment the "mobile Chernobyl" after the failed Soviet nuclear reactor, said there have been "train wrecks all over the place in the United States."

123. John Cantlon. DISPOSAL AND STORAGE OF SPENT NUCLEAR FUEL—FINDING THE RIGHT BALANCE. March 1996, p. 34. The process of licensing and developing a large federal centralized storage facility and the transportation infrastructure that goes with it will take time; estimates range from five to seven years. Even if passed into law now, none of the proposals before Congress would enable operation of a centralized storage facility to begin much before 2002—and then not at full scale. With the spent fuel stockpile currently at 32,000 metric tons and growing at 2,000 metric tons per year, it will take as long as 30 years to empty the inventory at all the individual reactor sites. So, developing a centralized storage facility at Yucca Mountain now would only *reduce*, but not eliminate, the need to continue adding spent fuel storage capacity at reactor sites. The Board's suggested approach differs from currently proposed strategies only by the time it will take to determine site suitability—at most five years.

124. John Cantlon. DISPOSAL AND STORAGE OF SPENT NUCLEAR FUEL—FINDING THE RIGHT BALANCE. March 1996, p. 28. Congress is now considering proposals to locate a centralized storage facility in the vicinity of Yucca Mountain. In the NWPA, Congress established a process to ensure that sound technical judgment plays the primary role in determining whether a particular site could be used to host a permanent repository. Furthermore, the act specifically prohibits the development of a federal centralized storage site in the same state where a site is being characterized for potential repository development. Deciding now to develop a storage facility at or near Yucca Mountain, *prior* to a decision about the suitability of that site for repository development, could undermine seriously the credibility of the process established by Congress. Institutional momentum to develop a repository there could increase by creating the perception that the suitability of the site has already been determined.

125. John Cantlon. DISPOSAL AND STORAGE OF SPENT NUCLEAR FUEL—FINDING THE RIGHT BALANCE. March 1996, p. 29. Increasing the level of confidence in predictions of repository performance will require additional testing, analysis, and exploration after the site-suitability determination. given the inherent difficulties associated with proving safe performance over many thousands of years, a site-suitability decision would not be an iron-clad guarantee that the site could be developed as a repository.

126. Kent Hansen. THE CHRISTIAN SCIENCE MONITOR. 30 May 1996, p. 16. Storage also allows for the future recovery of spent fuel, which is a valuable resource. The spent fuel contains plutonium and unused uranium, both of which can be used as fuel to produce electricity, as is currently done in France, Japan, and a number of other countries. In years to come, as prime uranium sources become depleted, spent fuel in the United States should be worth tens of billions of dollars.

127. WASHINGTON POST. 31 December 1996, p. A1. In a recent interview, Energy Secretary Hazel R. O'Leary called it "a gerrymandered, short-term effort to force-fit an interim storage facility" into the Yucca region. On Dec. 14, O'Leary told a Senate hearing that the administration is opposed to the legislation.

The power industry strongly disagrees. "It will cost the nation's consumers about $5 billion to build repositories at the various [nuclear plants] that were never selected to be permanent sites," said Bill Lee of Duke Power. "It's a terrible waste of our resources." And because plutonium can conceivably be used to make weapons, he added, "you have the added expense of surveillance without any end in sight."

128. LOS ANGELES TIMES. 1 August 1996, p. A3. They also say the cross-country transportation of the deadly waste would endanger millions of people along routes through 43 states—invoking the notion of a "mobile Chernobyl," the notorious Russian nuclear reactor that exploded 10 years ago in the Ukraine.

129. THE CHRISTIAN SCIENCE MONITOR. 17 July 1996, p. 1. Reid also warns about plans to transport nuclear waste through 43 states. "Nuclear waste is not just a Nevada problem. Far from it. Every town and community along the transportation rail lines would be in danger as lethal radioactive waste barrels through," he says.

130. John Cantlon. DISPOSAL AND STORAGE OF SPENT NUCLEAR FUEL—FINDING THE RIGHT BALANCE. March 1996, p. 38. The process of licensing and developing a large federal centralized storage facility and the transportation infrastructure that goes with it would take time. Developing a transportation system will require the acquisition of sufficient numbers of trucks or rail cars and casks, the establishment of transportation routes, and the development of emergency preparedness plans at the affected state and local levels. As a result, the federal government could not begin accepting spent fuel at a facility much before 2002, and then not in significant amounts. Congressional action to begin developing centralized storage now would only *reduce*, but not eliminate, the need to continue storing spent nuclear fuel at reactor sites.

131. Charles Timms, Jr. NATIONAL LAW JOURNAL. 10 June 1996, p. C12. Connecticut's efforts to site an LLRW disposal facility have lagged behind those of California. The state's current plan calls for a "volunteer approach" to siting. No viable volunteers, however, have stepped forward thus far. Other states, such as New Jersey and Pennsylvania, have adopted variations on the "volunteer" approach and have offered incentives, such as direct municipal payments and property-tax exemptions.

132. Arthur Coulter. NEWS AND RECORD. 14 July 1996, p. 1. After nuclear reactors finally shut down and the high-level wastes (irradiated fuel rods) are removed, the reactor vessels and some components become LLRW. Yet, they are so highly radioactive they cannot economically be handled safely for decades.

They are kept on site in what the industry calls "SAFSTOR." High-level waste could also be stored on power plant sites for 50 to 100 years, in pools and dry casks, to allow for dissipation of heat and radioactivity before handling.

133. John Cantlon. DISPOSAL AND STORAGE OF SPENT NUCLEAR FUEL—FINDING THE RIGHT BALANCE. March 1996, p. 43. The board sees no compelling *technical* or safety reason to move spent fuel to a centralized storage facility *for the next few years*. The methods now used to store spent fuel at reactor sites are safe and are likely to remain safe for decades to come. Despite some recent public opposition to utility efforts to develop additional storage, so far, utilities have been able to add new storage capacity at their sites when needed.

134. John Cantlon. DISPOSAL AND STORAGE OF SPENT NUCLEAR FUEL—FINDING THE RIGHT BALANCE. March 1996, p. 18.
Health and safety risks associated with spent fuel storage
During the last ten years, several studies have concluded that storing spent nuclear fuel in pools or dry storage presents low risks to workers and even lower risks to the general public. Pool storage (required for at least five years after removal of spent fuel from the reactor) involves more monitoring and management than dry storage. Projected radiation exposures to workers and the public resulting from pool or dry-storage activities would equal only a small fraction of exposures due to background radiation. Essentially the same dry-storage technologies would be used to store spent fuel whether at reactor sites or at a centralized facility. And the NRC found in its 1990 *Waste Confidence Decision Review* that spent nuclear fuel could be stored safely for as much as 100 years (NRC 1990). the same basic NRC requirements would have to be met regardless of storage facility location.

135. John Cantlon. DISPOSAL AND STORAGE OF SPENT NUCLEAR FUEL—FINDING THE RIGHT BALANCE. March 1996, p. 20.
Sabotage and security
One consideration related to safety is that of security. Requirements for minimizing the likelihood of theft or sabotage of nuclear materials, referred to as "safeguards," are set by the NRC (10 CFR 73). Based on its experience and analysis, the NRC has concluded that the physical security of either on-site or off-site spent fuel storage is not a significant problem. The NRC reached a similar conclusion about transportation safeguards risks (NRC 1979).

136. Arthur Coulter. NEWS AND RECORD. 14 July 1996, p. 1. Allowing time to seek long term solutions for safer management of these wastes. Some scientists believe a method for deactivating the radioactivity may one day be found. The rush for burial would practically eliminate this option for future generations.

Some nuclear power plants have capacity to store LLRW or could build it. The Electric Power Research Institute has published guidelines for long term on-site LLRW storage.

137. John Cantlon. DISPOSAL AND STORAGE OF SPENT NUCLEAR FUEL—FINDING THE RIGHT BALANCE. March 1996, p. 4. In 1977, the Department of Energy (DOE), one of the AEC's successor agencies, again proposed developing one or more away-from-reactor storage facilities. The DOE proposal would have provided for federal acceptance of commercial spent nuclear fuel for storage until a repository could begin operating (DOE 1978). Improved storage technologies just coming on-line, however, enabled utilities to rerack their pools and expand their pool storage capacity, making federal storage unnecessary.

138. William Sherman. FEDERAL NEWS SERVICE. 14 December, 1995, Lexis/Nexis. The Congress made an implicit contract with the towns and localities which hosted nuclear facilities that spent fuel disposal would be solved. Since then, it has become clear that spent fuel disposal is far more difficult than the 79th Congress knew, and the federal promotion of nuclear power without this clearly worked out in advance may have been ill advised. Still, the federal responsibility over the spent fuel exists. We are now at a crossroads where more and more sites are building dry cask storage. Once all the sites have dry cask storage, and then all the reactors cease operation, it will be extremely difficult to muster the will to move the waste. These sites will become what has been labelled "Nuclear Stonehenges," memorials to a mistake in governing and to a lack of political will.

139. John Cantlon. DISPOSAL AND STORAGE OF SPENT NUCLEAR FUEL—FINDING THE RIGHT BALANCE. March 1996, p. 27. Headway has been made, especially during the last couple of years, in assembling a more focused and disciplined management team. Asking the OCRWM to develop and operate a cen-

tralized storage facility and the related transportation system would create major additional responsibilities. At the very least, management structures—both within the agency and within its contractor family—would have to be modified. Diverting limited funds and other resources now would slow the process of characterizing Yucca Mountain and could even jeopardize support for the disposal program. Deferring disposal ultimately increases the risk, and creates the perception, that any storage facility could become a de facto disposal site. Deferring disposal has additional implications for the fate of the nation's government-owned spent fuel and defense high-level waste.

140.   John Cantlon. DISPOSAL AND STORAGE OF SPENT NUCLEAR FUEL—FINDING THE RIGHT BALANCE. March 1996, p. 26. The costs for disposing of commercial spent fuel are paid from the Nuclear Waste Fund. But, because the disposal program must compete for funding against other energy programs both inside the DOE and before Congress, competition has been and will continue to be intense. This already constrained financial situation could be squeezed even more severely by the possible diversion of funds from the disposal program to develop and operate a centralized storage facility. (This problem would be exacerbated if the facility is located at Yucca Mountain, which has no rail access.) The DOE has estimated that it will cost about $600 million to construct a centralized storage facility and approximately $250 million a year to operate it at full capacity, including transporting 2,000 metric tons of spent fuel to the facility annually.

141.   John Cantlon. DISPOSAL AND STORAGE OF SPENT NUCLEAR FUEL—FINDING THE RIGHT BALANCE. March 1996, p. 22. Uncertainties about the fund's adequacy could increase as a result of at least two additional circumstances. (1) If some nuclear power plants shut down earlier than now anticipated, the fund would collect less money. (2) If Congress decided to develop and operate a large centralized storage facility now and pay for it out of the Nuclear Waste Fund, the fund will be used up sooner. Should the fund prove inadequate to pay for both storage and disposal, either the utilities and their ratepayers would have to contribute more into the fund, or federal taxpayers would have to absorb the costs of final disposal of commercial spent fuel. As illustrated in Note 4, views on how storage should be paid for differ. Adequacy of a different kind relates to the potential competition for resources between developing centralized storage and the goal of disposal.

142.   John Cantlon. DISPOSAL AND STORAGE OF SPENT NUCLEAR FUEL—FINDING THE RIGHT BALANCE. March 1996, p. 44. In the past whenever there has been a choice between storage and disposal, disposal has always been made the primary focus of the federal high-level waste management program. This is because the storage of commercial spent fuel is not an acceptable substitute for disposal. Ultimately, spent fuel (commercial and defense) as well as sizable amounts of high-level radioactive defense waste will have to be disposed of. The Board believes that the nation needs *both* a repository development program and a plan to address future spent fuel storage needs. However, efforts now to refocus the program from disposal to storage, especially at a time when budgets are tight, could jeopardize site-characterization and repository development efforts in three ways: (1) by competing with the disposal program for resources, (2) by causing a real or perceived prejudicing of a future decision about the suitability of the Yucca Mountain site, and (3) by eroding the impetus and political support for repository development.

143.   John Cantlon. DISPOSAL AND STORAGE OF SPENT NUCLEAR FUEL—FINDING THE RIGHT BALANCE. March 1996, p. 4. The AEC then proposed constructing at one or more of its existing nuclear sites a retrievable surface *storage* facility for the high-level radioactive reprocessing waste while continuing to pursue disposal on a developmental basis (AEC 1974). The Environmental Protection Agency, however, strongly criticized the AEC proposal, arguing that its emphasis on storage would divert attention from the search for a permanent disposal solution (EPA 1974). As a result, the AEC proposal was withdrawn, and efforts were renewed to find a disposal site.

144.   The Nuclear Energy Institute. HIGH LEVEL NUCLEAR WASTE. August 1996, http://www.nei.org/main/pressrm/facts/hlw.htm)
The Dry Cask Option. Dry casks are large, rugged containers made of steel or steel-reinforced concrete, 18 or more inches thick. The casks use materials like steel, concrete and lead—instead of water—as a radiation shield. Depending on the design, a dry cask can hold from seven to 56 12-foot-long fuel assemblies.

The NRC has approved several designs for use by utilities. The casks have a 20-year license. After 20 years they must be inspected, and with NRC approval the license could be extended. Various dry cask storage technologies are being successfully used by utilities today.

The approved dry-cask designs use one of three storage systems. The first system stores spent fuel in steel canisters that are inserted horizontally into a steel-reinforced concrete vault. The second uses steel canisters placed vertically inside a concrete storage building. Each of these systems uses the concrete storage containment as a final radiation barrier. The third system uses vertical casks made of steel or reinforced concrete that stand outside on a three-foot-thick pad of reinforced concrete.

Loaded casks are filled with an inert gas, sealed, and stored either on reinforced concrete pads or inside steel-reinforced concrete bunkers. The casks are designed to withstand natural disasters such as tornadoes, hurricanes, and floods, and to prevent the release of radioactivity. All

the designs are passive, in that they require no mechanical devices for cooling and ventilation.

During a typical plant refueling, usually scheduled 12–24 months apart, about one-fifth to one-third of the fuel assemblies in a nuclear reactor are replaced. Utilities using dry-cask storage use two or three containers per refueling. To make room in the fuel pool for the hot and highly radioactive assemblies removed from the reactor, the oldest stored fuel—which has been cooling in the pool for at least 10 years—is transferred to dry casks.

145.   Richard Bryan. FEDERAL NEWS SERVICE. 29 February 1996, Lexis/Nexis. Congressman Riggs, if I could just add to this. Look at the progress we've made since 1982 when we started this process of trying to site geologically the nuclear waste. Dry cast storage is a perfect example of the progress we've made and we did that kind of by accident. It was done as a result of trying to figure out a safe way to transport this poison, and as a result of their figuring out a safe way they think they can transport it, people said, well, if you can transport it safely, wouldn't it be even safer to put these spent fuel rods in these dry cask storage containers and leave them where they are; and somebody said, yeah, that's what we should do.

So there's a lot of the scientific community that deals with nuclear waste says we should leave it where it is, just put them in these dry cast storage containers.

146.   Representative Ensign. FEDERAL NEWS SERVICE. 28 June 1995, Lexis/Nexis. Another option, put forth by our colleagues in the Senate, concerns the abilities of on-site dry cask storage. Since site characterization studies were initiated at Yucca Mountain, the technology for dry cask storage has been perfected. We need only look to Maryland to see the phenomenal job they did with dry cask storage at Calvert Cliffs, and to Virginia Power which recently began dry cask storage at one of their facilities.

147.   THE CHRISTIAN SCIENCE MONITOR. 27 October 1994, p. 19. As a stopgap measure, a few utilities have received approval from state regulators to store spent fuel in above-ground concrete casks. Other utilities are likely to do the same. Thus, for a growing number of nuclear plants, dry-cask storage could mean the difference between continued operations and shutting down.

148.   Susan Clark. FEDERAL NEWS SERVICE. 5 February 1997, Lexis/Nexis. Why are there costs associated with delay? Developing an on-site independent spent fuel storage installation using dry cask storage requires capital expenditures for site development (a concrete pad and support building), the casks, and a vehicle for transporting those casks from the spent fuel pool to the storage pad. Costs for such a facility depend on the amount of spent fuel to be stored. In Florida, the cost for such a facility is estimated to be $25 million. Continuing operating costs for the facility consists of radiation monitoring and security, and amount to around $1 million annually. Capital and operating costs are not the only costs incurred. There are other "indirect" costs that utilities and consumers may be required to pay if the federal government fails to deliver the services it promised. By 1998, 26 nuclear power plants will run out of existing on-site storage. By the year 2010, 80 plants from 27 different states will have exhausted their existing on-site storage capabilities. As you can see, on-site storage may become a significant political issue affecting over one-half of the states in this country by 2010.

149.   Susan Clark. FEDERAL NEWS SERVICE. 5 February 1997, Lexis/Nexis. The federal government's inaction and failure to fulfill its obligation to begin disposing of nuclear waste by January 31, 1998 will have significant monetary impacts on consumers. First, industry estimates conclude that nuclear utilities will pay an additional $6 to $8 billion for dry cask storage needed at 73 power plants across the nation as a result of continued delays in the Department's acceptance, and this is just for the first 20 years. The Department's contracts with utilities assume an acceptance rate of 3,000 metric tons of uranium (mtu) per year over a 45-year period. Additional costs of maintaining 73 spent fuel facilities for additional decades appear to be between $1.5 and $3.3 billion per decade. Even in the value of today's dollars, these dry cask storage costs cold exceed $10 billion.

150.   James Rhodes. FDCH CONGRESSIONAL TESTIMONY. 5 February 1997, Lexis/Nexis. In some cases, building a dry-storage facility has subjected companies to unacceptably high political and financial costs. Precipitating these battles on a state-by-state basis is not what federal policy intended. It is not fair to expose utility customers and state legislatures to this kind of highly charged conflict, and it is avoidable. As difficult as they are today, these issues will only get more complex.

151.   William McCormick. FDCH CONGRESSIONAL TESTIMONY. Federal News Service. 28 June 1995, Lexis/Nexis. Continued delay in the Energy Department's program to accept spent fuel will subject utility customers across America to significant added costs for additional interim storage capacity at plant sites. Because of no other alternatives, a number of utilities— including my own—already have been forced to build additional dry cask storage capacity for spent nuclear fuel. The difficulties Consumers Power and other utilities have experienced in negotiating the financial, regulatory and legal hurdles to expanding on-site storage capacity demonstrate that a plant-by-plant, state-by state approach to spent fuel policy is undesirable, inefficient and costly.

Construction of additional on-site storage imposes unacceptably high political and financial costs on compa-

nies forced to take this step. Failure to resolve this issue affects each company's relationship with its regulators, the public, the investment community and Federal and state lawmakers.

152. NUCLEAR WASTE NEWS. 23 March 1995, Lexis/Nexis. There is general agreement that dry cask storage is technically feasible and safe, and several utilities have turned to this option as storage space in their fuel pools dwindled. Nuclear utilities, however, say they do not want to deal with the political problems of storing waste on-site, and do not want to take the pressure off DOE to provide interim storage.

There also are economic reasons for moving spent fuel off-site, as on-site storage is considerably more expensive, utilities claim. Utilities must now pass on-site storage costs on to ratepayers, even as they are compelled to contribute to the Nuclear Waste Fund.

Utilities fear if they agree to dry cask storage they will have to wage a site-by-site battle for expanded storage. "The same people are there in force, opposing any kind of temporary centralized solution, and they're also opposing—everywhere the issue now comes up—expanded storage on-site. The anti-nuclear community has decided to make it a political issue in an effort to scare people into shutting down plants," Roche said.

153. John Cantlon. DISPOSAL AND STORAGE OF SPENT NUCLEAR FUEL—FINDING THE RIGHT BALANCE. March 1996, p. 34. Often of a much higher concern than calculated transportation risks, *perceived transportation risks* will play an important role during any discussion of moving spent fuel. Large segments of the public fear radiation and carry those fears over to any activity involving radioactive materials (Weart 1988, Slovic 1987). Those concerns will understandably grow once people realize that current levels of spent fuel transportation would increase several-fold once a full-scale storage facility (or repository for that matter) begins operating. A decision to build a centralized facility now would raise the public's perception of transportation risks sooner rather than later.

154. UNITED PRESS INTERNATIONAL. Wednesday BC Cycle, 31 July 1996, p. 1. Chief advocates of the measure warned that the proposal was a measured response intended to relieve the concerns of 41 states, now temporarily storing high-level nuclear wastes at the point of generation at about 80 sites, and provide for the centralized disposal of wastes from nuclear power plants. "We have a choice—a responsible solution or more delay," complained Sen. Frank Murkowski, R-Alaska and a chief advocate of the measure. "If the president vetoes this bill, he's saying in effect we'll have 80 waste dumps around the nation instead of one." Murkowski also warned the bill could carry election-year consequences for Clinton, promising that Republicans would publicize a veto in those 41 states. The bill, introduced earlier this month, would set up an interim storage facility at the Nevada Test Site near Yucca Mountain, Nev., with work to begin by December 1998, unless the president, the Energy Department and the Nuclear Regulatory Agency determine it is unacceptable. The site also would be considered as a permanent site ready to accept wastes in that capacity by 2010. If the site is found to be unsuitable, work at the Yucca Mountain site would cease with the president designating a new interim storage facility within 18 months and Congress approving that choice within the succeeding two years.

155. FINANCIAL TIMES. 26 June 1996, p. 8. While nuclear energy doesn't produce carbon dioxide, it creates long-term problems that are just as intractable, as anyone listening to the bitter debate over what to do with our high-level radioactive waste would agree. Simply trading one form of pollution for another is not much of a solution.

The solution is the greater use of more energy-efficient and renewable energy technologies. Not only does most of the scientific community recommend this approach, but as nearly every public opinion poll shows, so do the American people. A recent poll by the Sustainable Energy Budget Coalition found that 60 percent of the public would be more likely to vote for candidates who back policies that increase the use of these technologies.

156. Gregg Taylor. NUCLEAR NEWS. August 1996, p. 75. Bernard L. Cohen, professor emeritus of the University of Pittsburgh, talked about turning the tide of public opinion, an important factor in the fate of the nuclear industry. He observed that "Darwinian selection [survival of the fittest] forced environmental groups and their leaders to attack nuclear power." They found that scaring the public about nuclear power and radiation attracts media coverage, people to their rallies, dues paying members, and foundation grants, he said. "After all, financial support comes from attacking the Establishment, not supporting it," Cohen observed.

157. Margaret Kriz. THE NATIONAL JOURNAL. 6 April 1996, p. 763. Carretta argues the antinuclear groups are trying to dominate the debate over whether to build new low-level nuclear waste sites with emotional arguments. "A lot of it is scare tactics and non-sciences," he said.

In the debate over building new waste facilities, "we're not dealing with logic," he said. "We're dealing with highly emotional, political issues."

158. Mike Zapler. STATES NEWS SERVICE. 18 January 1996, Lexis/Nexis. "This underscores what we've known all along: Americans are wisely turning their backs to the nuclear power option, and do not support shipping poisonous radioactive wastes by train or truck

across the nation's transportation arteries," said Bill Walsh, who works on energy issues for Greenpeace USA.

159. John Cantlon. DISPOSAL AND STORAGE OF SPENT NUCLEAR FUEL—FINDING THE RIGHT BALANCE. March 1996, p. 19. The Board is mindful that the *public's perception* of transportation risk is of a much higher concern and, therefore, becomes a factor in public policy decisions. If a centralized storage facility were developed in the near future, transportation operations would begin much sooner than previously anticipated by repository operation schedules. The level of spent fuel transportation activity and the complexity of the total set of operations would be distributed more widely than in the past.

160. Larry Craig. ROLL CALL. 15 July 1996, Lexis/Nexis. That legislation has strong bipartisan support in the Senate, with 30 Democratic and Republican co-sponsors from around the country.

Recently, S. 1271 was substituted by another bill, S. 1936, which is expected to draw even more bipartisan support. It would create a well-organized, integrated program to manage nuclear byproducts.

The substitute legislation would accomplish two major objectives.

First, it would continue the ongoing scientific investigation of a possible site for a permanent disposal facility. Some of the top scientists in the country have been studying Yucca Mountain, in the vast, barren Nevada desert, for nine years. During all that time, they have found no reason why nuclear material cannot be isolated for thousands of years by natural geology and manmade structures.

Second, the legislation would consolidate spent commercial nuclear fuel and defense fuel at one location in the Nevada desert. This central storage center would be built at the Nevada Test Site, where hundreds of nuclear weapons tests have been conducted since the 1950s. It would be licensed by the Nuclear Regulatory Commission to ensure protection of the environment and public health and safety.

161. John Cantlon. DISPOSAL AND STORAGE OF SPENT NUCLEAR FUEL—FINDING THE RIGHT BALANCE. March 1996, p. 35. The possibility of creating a de facto disposal has been an important focus of community opposition during past attempts to site a storage facility. When in 1974 the AEC proposed building a retrievable surface storage facility while pursuing disposal on a developmental basis, the concern was raised that moving spent fuel to one central site would erode support for finding a permanent disposal solution (EPA 1974). The implication was that without a disposal option, storage could become de facto disposal. Indeed, the quantity and timing linkages contained in the 1987 NWPA amendments addressed those concerns.

162. John Cantlon. DISPOSAL AND STORAGE OF SPENT NUCLEAR FUEL—FINDING THE RIGHT BALANCE. March 1996, p. viii. The Board sees no compelling *technical* or safety reason to move spent fuel to a centralized storage facility *for the next few years*. The methods now used to store spent fuel at reactor sites are safe and likely to remain safe for decades to come. Despite some recent public opposition to utility efforts to develop additional storage, so far, utilities have been able to add new storage capacity at their sites when needed.

163. Krista Sanda. FEDERAL NEWS SERVICE. 29 February 1996, Lexis/Nexis. Electricity consumers will not continue to quietly pay their money into the Nuclear Waste Fund and get nothing in return. There is a growing realization by this country's consumers, regulators, elected officials and representatives of the electric industry that we must act to change the bumbling manner in which this program has been managed over the past 25 years. New initiatives are attempting to address the management problems but it will be difficult to succeed without the commitment of this subcommittee to support the reform process with adequate funding.

164. Larry Craig. CONGRESSIONAL PRESS RELEASES. 15 July 1996, Lexis/Nexis. Given a choice, most Americans would choose to have one safe, environmentally sound storage site for the disposal of thousands of tons of nuclear waste existing in this country, as opposed to having that waste stockpiled in more than 80 facilities around the nation—which is the current situation. A single storage site is clearly the pro-environmental option. Moreover, millions of Americans have paid for this long-needed and long-promised consolidated waste storage site.

165. Mike Zapler. STATES NEWS SERVICE. 18 January 1996, Lexis/Nexis. The industry says keeping the waste on-site will increase energy costs for consumers, since they will have to pay the companies to build additional storage casks.

"In order to keep the reactors running, utilities have to move the waste to dry casks," said Scott Peters, spokesman for the Nuclear Energy Institute. "That is double jeopardy for the ratepayer."

166. LOS ANGELES TIMES. 17 July 1996, p. A3. A plan to create a nuclear dump to store radioactive waste in Nevada within 100 miles of Las Vegas cleared a preliminary but important hurdle Tuesday in the Senate.

By a vote of 65 to 34, the Senate agreed to open debate on the proposal, which is considered one of the most important environmental tests in the months leading up to election day.

The bitterly disputed legislation has drawn the threat of a presidential veto and it may tie the Senate in knots,

167. John Cantlon. DISPOSAL AND STORAGE OF SPENT NUCLEAR FUEL—FINDING THE RIGHT BALANCE. March 1996, p. 32. Although utilities and public utility commissions strongly support the development of a federal centralized storage facility now, other stakeholders are not so enthusiastic. For example, some think that keeping spent fuel at reactors will keep the utilities and, thus, the federal government actively pursuing the repository development program. Some believe in intergenerational equity—the generation that produces the waste is responsible for its disposal. To the extent that storage might substitute for disposal, waste management burdens could be passed on to future generations.

168. Larry Craig. ROLL CALL. 15 July 1996, Lexis/Nexis. Consumers paid this money specifically to finance construction of storage and disposal facilities for used nuclear fuel. The Nuclear Waste Policy Act is an opportunity for Congress to demonstrate that the federal government can actually meet its commitments and deliver on its promises.

The common-sense principles in this legislation enjoy widespread support from Senators, governors, state utility regulators, and labor unions. More than 200 Members of the House of Representatives, Republicans and Democrats alike, stand ready to act on similar legislation.

169. STAR TRIBUNE. 1 August, 1996, p. 17A. "This is special-interest legislation at its worst," Bryan said. "The nuclear industry and its phalanx of lobbyists who quiet these halls every day with enormous amounts of money, power and influence, they're the ones that are driving this debate by creating a contrived and fabricated crisis."

170. John Cantlon. DISPOSAL AND STORAGE OF SPENT NUCLEAR FUEL—FINDING THE RIGHT BALANCE. March 1996, p. 30. Deciding to develop centralized storage most likely will create public opposition at the proposed storage site or even on a broader level. Some important general public acceptance issues include geographic equity, perceived transportation risks, the fear of creating de facto disposal sites, and the possibility of reprocessing spent fuel. General public acceptance of a centralized storage facility is probably unlikely. However, some choices about storage could influence the *intensity* of public opinion while others could shift opposition to different population groups, as discussed below.

171. Kathleen Reilly. 70 INDIANA LAW JOURNAL 679. Spring 1995, p. 683. See Dean Takahashi, Signs of Revival Buoy Nuclear Power Industry, L. A. TIMES, May 30, 1991, at D5 (pointing out that polls sponsored by Time Magazine and CNN showed a surge in support for nuclear power, and that one poll by the U.S. Council for Energy Awareness showed that 66% of Americans strongly favor nuclear reactor construction).

172. J. Bennett Johnson. CONGRESSIONAL PRESS RELEASES. 16 July 1996, Lexis/Nexis. Nuclear waste has never been a partisan issue. While the current law was signed by a Republican President, it has its roots in the Carter Administration. It was passed by a Democratic House and a Republican Senate and amended by a Democratic House and a Democratic Senate, with broad bipartisan support. It would be a terrible, terrible mistake to make it a partisan issue now.

173. John Ross. SMITHONIAN. November 1995, p. 42. For starters, they found that the public responds differently to voluntary and involuntary risks. You and I are willing to tolerate far greater risks when it is our own doing, such as smoking cigarettes or climbing mountains. But if the risk is something we can't control, such as pesticides on food or radiation from a nuclear power plant, we protest, even if the threat is minimal.

174. Robert Hirsch. PUBLIC UTILITIES FORTNIGHTLY. 1 March 1996, p. 13. With little fanfare, most aspects of the U.S. energy system seem to have settled into a fairly stable, predictable pattern. To my mind, we have reached an "energy plateau" likely to persist for maybe a decade or more into the future.

Energy is not now high on the radar screen of the general public, so there is little public pressure for significant change in the U.S. energy system. (Electric utility deregulation, which is government driven, I discuss below.) The low level of public interest reflects the availability of abundant, reliable, and low-cost energy supplies. In addition, the general public feels that the environment is in fairly good condition in most places, so energy related environmental concerns have abated significantly. All in all, the U.S. public seems generally happy with its energy system and has focused its attention on much more pressing issues.

175. Kathleen Reilly. 70 INDIANA LAW JOURNAL 679. Spring 1995, p. 683. Many argue that no other energy source carries the potential for such vast economic and environmental benefits as nuclear power, which currently provides twenty-two percent of the nation's electricity. Even President Clinton, who vowed upon entering office to halt an increase in nuclear power development, now supports a new generation of nuclear power plants.

176. WASHINGTON POST. 31 December 1996, p. A1. For dozens of states and utility companies, the most pressing issues are money and fairness. They want DOE to take the spent fuel off their property because,

they argue, U.S. law calls for the government to do so by 1998, because it's costing millions each year to store on site and because utility users have paid billions into a federal fund for a permanent disposal facility. DOE says it isn't obliged to take the waste because no long-term dump site has been approved. Next month, the fight goes to federal appeals court, where lawyers will attempt to sort out who's responsible for what.

In the interim, Congress is considering bills that would mandate a "temporary" waste repository in southern Nevada. That plan is supported by many influential Republicans, opposed by the White House, and certain to add yet another contentious chapter to the volatile and confusing history of U.S. nuclear waste policy.

177. John Cantlon. DISPOSAL AND STORAGE OF SPENT NUCLEAR FUEL—FINDING THE RIGHT BALANCE. March 1996, p. 32. Opposition to spent fuel storage often reflects underlying concerns about a variety of issues and can be expected to some degree around any radioactive waste management facility. Opposition to storing spent nuclear fuel has occurred primarily at the local level in some communities where spent fuel is stored or where utilities are adding additional storage capacity. Concerns seem to reflect the belief that the sites might become de facto disposal sites or, in a few cases, they reflect unease about storing spent fuel at shutdown reactors.

178. John Cantlon. DISPOSAL AND STORAGE OF SPENT NUCLEAR FUEL—FINDING THE RIGHT BALANCE. March 1996, p. 33. Recently there has been opposition at some operating reactor sites where utilities tried to add on-site dry-storage capacity. This happened at sites in Minnesota, Michigan, and Wisconsin. Many of the opponents to dry storage seem worried about the risks associated with the continued accumulation of spent fuel and about the lack of assurance that the fuel will ever leave the plant, especially in light of delays in the repository schedule. In general, plans to develop a centralized storage facility could reduce public opposition at reactor sites. The large the capacity of such a centralized facility, the smaller the amount of spent fuel that would have to stay at reactors, thus lessening the basis for public opposition at affected reactor sites.

179. John Cantlon. DISPOSAL AND STORAGE OF SPENT NUCLEAR FUEL—FINDING THE RIGHT BALANCE. March 1996, p. 14. Although developing dry storage at reactors generally has proceeded smoothly, installing dry-storage systems has produced public opposition at several locations recently. At least some of this opposition has been caused by concern that development of a repository for permanent disposal is not proceeding fast enough. Consequently, people fear that a temporary storage site could become a de facto disposal site. For example, in Michigan in 1993, the attorney general and an environmental group challenged the validity of the Nuclear Regulatory Commission's (NRC) generic licensing rule when Consumers Power Co., sought to use dry-storage casks at its Palisades reactor. Although litigated for several years, the suit was unsuccessful. In Minnesota in 1994, the Minnesota state legislature imposed several conditions for dry-storage development on Northern States Power Co., including that some storage capacity had to be secured elsewhere. Although, every attempt to develop at-reactor dry storage has succeeded so far, utilities are concerned about opponents' political, economic, regulatory, or legal challenges to on-site storage.

180. John Cantlon. DISPOSAL AND STORAGE OF SPENT NUCLEAR FUEL—FINDING THE RIGHT BALANCE. March 1996, p. 33. Although the risk of storing spent fuel at shutdown reactors is low, there is some uneasiness in that regard. So far, community response to spent fuel storage at shutdown reactors has been mixed. For example, in a survey of residents living near the shutdown Humboldt reactor in northern California, respondents believed that the risk of on-site dry storage of spent fuel is higher than the risk of continuing operation of the power plant itself (Pasqualetti 1993). On the other hand, the owners of Rancho Seco in central California encountered little if any opposition to storing spent fuel on site after the plant was shut down. However, the level of concern about having spent fuel at shutdown reactors could change in response to increasing uncertainties in the changing electric power industry, or as more reactors begin shutting down in a decade or so.

181. WASHINGTON POST. 31 December 1995, p. A1. Some DOE officials, however, feel that some definitive short-term action is essential in order to encourage public confidence in government in general and DOE in particular. "Part of what the political system is frustrated at," said Grumbly, "is that no matter how much money we throw at this, nothing happens. Well, we have to stay the course three or four more years."

The only way the department can repair its damaged credibility, Grumbly argued, is to proceed methodically, opening up the data to independent outside review and proving that DOE is not ignoring important information—even information that might imperil the Yucca sites' chances. That would leave the department open to more charges of spending billions without a result.

"But the only way we're going to get the public's trust," Grumbly said, "is to put our cards on the table face up all the time. Even if it's painful."

182. Mike Zapler. STATES NEWS SERVICE. 18 January 1996, Lexis/Nexis. Americans overwhelmingly believe Congress should establish a commission to study nuclear waste disposal before shipping the highly radioactive material to Nevada, environmental groups said today.

The groups based their claim on a survey which found that 70 percent of respondents reject the waste strategy of two proposals in Congress. The bills would create a temporary nuclear waste facility at the Nevada Test Site, 65 miles northwest of Las Vegas, until a permanent dump in Yucca Mountain is completed and deemed safe.

183. WASHINGTON POST. 31 December 1995, p. A1. Meanwhile, bills mandating "temporary" above-ground storage are moving through the House and Senate. As a result, even as taxpayers are spending $1 million a day to evaluate the Yucca site, legislation is being debated that would—critics say—make an underground repository effectively irrelevant and turn the less expensive "interim" storage area into a de facto permanent dump.

184. John Cantlon. DISPOSAL AND STORAGE OF SPENT NUCLEAR FUEL—FINDING THE RIGHT BALANCE. March 1996, p. 35. Finally, most spent fuel is still stored at reactor sites, despite past initiatives to develop centralized storage. In part this is because of community opposition created during these past attempts to site a storage facility. As in the past, much opposition will result from concern that a proposed storage site could become a de facto disposal site.

185. John Cantlon. DISPOSAL AND STORAGE OF SPENT NUCLEAR FUEL—FINDING THE RIGHT BALANCE. March 1996, p. 5. In its draft mission plan, the DOE's Office of Civilian Radioactive Waste Management (OCRWM) proposed a centralized storage facility to provide backup storage to utilities who needed it (DOE 1984). A year later, the DOE foresaw a much expanded role for the MRS facility. It would function as a pass-through facility located in the East near the majority of reactors, where eastern spent fuel could be collected and prepared for disposal at a yet-to-be determined repository site. The DOE identified one preferred site (at Oak Ridge on the Clinch River) and two alternative sites for the MRS facility—all in the state of Tennessee (DOE 1985). The plan proposed that the facility begin operating in 1998 and provided for the storage of 15,000 metric tons of spent fuel. This capacity limit was intended to allay concerns that the storage facility would become a de facto disposal site. In 1987 the DOE finally submitted its MRS proposal to Congress. but by then, Tennessee state opposition to a centralized storage facility was so great that Congress annulled and revoked the Clinch River MRS facility proposal in the 1987 amendments to the NWPA.

186. Kent Hansen. THE CHRISTIAN SCIENCE MONITOR. 30 May 1996, p. 16. Even as Congress considers legislation to fix the government's nuclear-waste program, it must be acknowledged that the program still has a very long way to go. It imposes an enormous burden on electricity customers and thus undermines public faith in our country's ability to dispose of spent fuel in a deep underground repository.

Repeated efforts to keep the program on course have failed. In 1982, when the Nuclear Waste Policy Act was adopted, the law called for electricity customers to pay a fee of 1/10th of a cent for each kilowatt-hour of nuclear-generated electricity. But today the government has little to show for the more than $12 billion it has collected, in large part because more than half of the money has been used to help reduce the federal budget deficit.

187. Larry Craig. ROLL CALL. 15 July 1996, Lexis/Nexis. The United States Senate will soon have a rare opportunity to pass bipartisan legislation to protect America's environment and preserve a vital source of US energy—an energy source that accounts for 20 percent of our nation's electricity production with relatively minor amounts of waste byproducts.

For all the right reasons, our nation has been a generator of radioactive material for nearly five decades. Most of this material is a byproduct of two principal sources: national defense and commercial nuclear power plants.

These two important activities have benefited every American. Managing the resulting waste and spent fuel should therefore be a national priority. We cannot and must not walk away from this responsibility.

188. John Barry. HERITAGE FOUNDATION REPORTS. 9 November 1995, p. 1. And yet, the department remains. In fact, DOE has grown in tax dollars spent and functions performed—the result of 15 years of searching for something to do. As Victor Rezendes of the General Accounting Office has testified, "DOE's mission and priorities have changed drastically over time so that the Department is now very different from what it was in 1977. While energy research, conservation and policy-making dominated early DOE priorities, weapons production and now environments cleanup overshadow its budget." Thanks to this continual empire-building, the department's budget has increased by 235 percent, and 85 percent today is spent on activities other than energy resources. For instance, nearly $12 billion is budgeted annually for environmental quality and nuclear waste disposal, with close to another $3 billion earmarked for fundamental science research.

189. Larry Craig. FDCH TESTIMONY. Congressional Press Releases, 15 July 1996, Lexis/Nexis. To pay for constructing and operating a permanent facility, the law established the Nuclear Waste Fund in the U.S. Treasury. It currently receives about $600 million per year from collections of a fee paid by ratepayers of nuclear-generated electricity. The fund currently contains over $11 billion in tax and interest. In addition, defense funds are annually appropriated to cover the cost of storing defense spent fuel and waste. [Note that S. 1936 would convert the fee paid by electricity consumers into a user fee that would be

assessed based on the level of appropriations for the year beginning in Fiscal Year 2003. (Sec. 401).

190.   Larry Craig. ROLL CALL. 15 July 1996, Lexis/Nexis. Storage and disposal facilities for commercial spent fuels created by S. 1936 are paid for lock, stock, and barrel by the people who use electricity from nuclear power plants. The American taxpayer will pay nothing for commercial fuel disposal.

Since 1983, consumers of electricity have been paying one-tenth of a cent for each kilowatt-hour of electricity produced at our nuclear power plants. Since then, the federal government has collected almost $12 billion from ratepayers and continues to collect more than $600 million every year—more than enough to finance construction of an interim storage facility, continued scientific investigations at Yucca Mountain, and the development of a transportation system to move nuclear waste to the storage and disposal facilities.

191.   Noah Sachs. RISKY RELAPSE INTO REPROCESSING.
http://www.ieer.org/ieer/reports/risky.html)

5. *Electrometallurgical processing is not an appropriate waste management technology, and its continued development keeps the door open to a return of civilian reprocessing in the United States.* Argonne National Laboratory has touted the potential applications of the technology to commercial nuclear power plant spent fuel, and the technology has been tested with a small amount of commercial spent fuel. In addition, DOE plans to apply it to commercial spent fuel under future research and development efforts. Reprocessing commercial spent fuel would reverse long-standing U.S. practice and undermine U.S. authority to discourage commercial reprocessing in other countries.

6. *Reprocessing small amounts of spent fuel or nuclear material pursuant to one EIS with a short time-frame raises the incentives to use the reprocessing plants for larger amounts of spent fuel or nuclear material over the long-term.* It is a ripple effect that could result in eventually reprocessing all of the aluminum-clad spent fuel in the United States, once it is consolidated at the Savannah River Site under DOE's preferred alternative in the SNF/INEL EIS.

192.   Paul Leventhal. SEPARATING MYTHS FROM REALITY. 16 April 1996, http://www. nci.org/nci/ spr14696a.htm
Myth #4: Plutonium can be Effectively Safeguarded against Diversions or Thefts.
In fact, even the most modern plutonium fuel cycle facilities cannot be safeguarded to the extent that diversions of bomb-quantities of plutonium can be detected with high confidence. Even with the best measurement and accounting instruments available, there must be some margin of error because of unavoidable measurement uncertainties. As a result, in large, commercial-scale plutonium facilities, the allowable discrepancies between book and physical inventories amount to hundreds of kilograms of plutonium a year—equivalent to dozens of nuclear weapons–leaving ample room for large diversions of atom bomb material. "Up to three percent of Japan's plutonium pile is 'unaccounted for' at any one time," according to a recent account in *The Economist*.

193.   Paul Leventhal. SEPARATING MYTHS FROM REALITY. 16 April 1996, http://www.nci.org/nci/spr1696a.htm
Myth #2: Reactor-Grade Plutonium Cannot be Made into Nuclear Bombs.
Japan's advocates of recovering and recycling plutonium have made false or misleading claims about the weapons potential of "reactor-grade plutonium," the type that Japan will recover from its spent fuel. Ryukichi Imai, former Japanese ambassador for non-proliferation, has written that the reactor-grade plutonium shipped in 1992 from France to Japan "is quite unfit to make a bomb." Hiroyoshi Kurihara, executive director of PNC, stated that "many Japanese experts express the opinion that reactor-grade plutonium could not be used for workable nuclear weapons." He speculated it "can be merely a nuclear fireworks, namely it produces glare and a big noise, but would not cause disastrous effects of nuclear bombs. . . .", and claimed that such a weapon would "fizzle like a firecracker."

PNC has also distributed a video in which "Pluto Boy," a cartoon character representing plutonium, reassures the audience that a workable bomb cannot really be made from reactor-grade plutonium—the same video in which this cute character demonstrates that plutonium is safe enough to drink.

In fact, the ability to construct a weapon from plutonium separated from the spent fuel of nuclear power plants was settled long ago. It is dangerous even to consider it an open question. Hans Blix, director-general of the International Atomic Energy Agency (IAEA), informed my institute that there is "no debate" on this point at IAEA, and that the agency considers virtually all isotopes of plutonium, including those that comprise reactor-grade plutonium, to be usable in nuclear weapons. Back in 1976, the U.S. government first declassified the information that reactor-grade plutonium could be used to make weapons and could even be the basis for a national military program. The following year, it declassified the fact that the U.S. successfully detonated a nuclear bomb made from reactor-grade plutonium at the Nevada Test Site in 1962.

In January 1994, a study by the U.S. National Academy of Sciences concluded that "it would be quite possible for a potential proliferator to make a nuclear explosive from reactor-grade plutonium using a simple design that would be assured of having a yield in the range

of one to a few kilotons, and more using an advanced design. Theft of separated plutonium, whether weapon-grade or reactor-grade, would pose a grave security risk. [ellipses in original]

194. THE NIKKEI WEEKLY. 27 May 1996, p. 1.
Plutonium produced by commercial nuclear reactors already far exceeds that produced for military use. About 800 metric tons of plutonium now exist in commercial spent fuel, while 130 additional tons are in separated form.
The U.S. has avoided reprocessing, instead storing spent fuel at reactor sites and eventually in underground repositories. The U.S. has urged other countries to adopt its so-called once-through cycle to help reduce the risk of plutonium being diverted for bombs.

195. THE ENERGY DAILY. 31 July 1996, Lexis/Nexis. The lawsuit comes at a sensitive time for DOE as it prepares to begin shipments of up to 22,700 foreign spent fuel rods to the nuclear weapons facility near Aiken, S.C. The lawsuit could disrupt a high-priority DOE program designed to address nuclear proliferation concerns posed by the foreign spent fuel, which contains weapons-usable high-enriched uranium (HEU) initially provided by the U.S. to its allies under peaceful nuclear cooperation programs.

196. THE ENERGY DAILY. 19 July 1996, Lexis/Nexis. But for the second year in a row, a coalition of moderate Republicans and Democrats overrode the wishes of the House Republican leadership and decisively approved a major increase in funding for solar and renewable energy programs. Further, the margin of victory for renewable energy proponents was much larger than last year, which environmentalists said reflects the recognition by many Republicans that renewable energy is politically popular.
Lawmakers voted 279-135 in favor of an amendment by Rep. Dan Schaefer (R-Colo.), chairman of the House Commerce energy and power subcommittee, providing an additional $30 million, including $22.5 million more for wind energy programs, which had been targeted for extinction.

197. Kenneth Baker. FY 1996 APPROPRIATIONS HEARINGS STATEMENT. www.doe.gov)
However, with all of these accomplishments, there is still much more to do. In fiscal year 1996, we must accelerate our efforts to protect fissile materials and redirect nuclear expertise in the former Soviet Union to peaceful projects. The Department will expand its efforts to end the civilian production and use of weapons-usable fissile materials through promotion of alternative energy sources, the Reduced Enrichment for Research and Test Reactors (RERTR) Program, and nuclear material purchases. Also in fiscal year 1996, the Department will continue its efforts to monitor U.S. and Russian inventories of plutonium and highly enriched uranium from weapons dismantlement through inspections and other activities that make dismantlement transparent and irreversible.
Among other measures, we will advance nonproliferation: by our efforts in North Korea; by continuing support of negotiations of a Comprehensive Test Ban Treaty and an international fissile material cutoff convention; and by facilitating International Atomic Energy Agency inspections in the United States. Reflecting this need to accelerate our nonproliferation efforts, our budget for fiscal year 1996 is increased by more than $86 million over fiscal year 1995.

198. Steven Dolley. COMMENTS ON THE DOE STORAGE AND DISPOSITION OF WEAPONS USABLE FISSILE MATERIALS DRAFT PROGRAM EIS. http://www.nci.org/nci/ib6796a.htm A further danger is that the MOX option would undercut U.S. non-proliferation diplomacy directed at so-called "rogue states." With the U.S. actively pursuing the MOX option, it becomes far more difficult to deny nations of proliferation concern their "right to civil use of plutonium as members of the NPT. North Korea claimed, albeit with little credibility, that its reprocessing plant at Yongbyon was intended to separate plutonium for use in MOX fuel for civilian nuclear power reactors. Though it agreed in October 1994 to abandon plans for indigenous reprocessing facilities, the ultimate disposition of plutonium contained in the remaining MAGNOX spent fuel and in future LWR spent fuel has yet to be determined. Overseas reprocessing has not been ruled out; nor has recycling of plutonium as MOX fuel in North Korea been specifically precluded.
At the third NPT PrepComm meeting, Iran threatened to withdraw from the NPT because, it charged, Iran and other NPT non-nuclear-weapon states were being denied nuclear technology that was their due under the terms of Article IV. India and Pakistan, though not NPT members, pursue plutonium programs that they justify as a legitimate part of their civil nuclear programs, and China anticipates reprocessing the spent fuel from its nuclear-power plants.
The only credible way to oppose the separation and use of plutonium as well as acquisition and use of HEU in nations of proliferation concern is to oppose it comprehensively—that is, to oppose such use in any nation for any purpose. This approach is effectively precluded if the U.S. insists upon retaining the right to use MOX fuel in civilian reactors, even if only for the purpose of weapons plutonium disposition.

199. NUCLEONICS WEEK. 13 June 1995, p. 1.
Graham thinks tritium production will take a back seat to other missions at the site, including the disposition of plutonium from dismantled nuclear warheads, the related production of mixed-oxide (MOX) fuel for commercial reactors, reprocessing of spent fuel from overseas

research reactors, and various environmental cleanup initiatives.

200. Noah Sachs. RISKY RELAPSE INTO REPROCESSING. January 1996, http://www.ieer.org/ieer/reports/risky.html The best alternative to reprocessing is to store spent fuel for an interim period in dry storage facilities. DOE's own data show that interim storage poses far fewer safety, environmental, and health risks than reprocessing, and interim storage would allow DOE to conduct research on engineered barriers and non-separative processing options. In addition, interim storage would allow DOE to gain more information about a repository before making spent fuel stabilization decisions. Keeping corroding spent fuel in current storage facilities while new ones are built does carry risks, but reprocessing involves much greater risks. Further, current wet storage could be improved in some cases by putting the spent fuel in sealed containers, as was done at Hanford for some spent fuel well over a decade ago.

201. Noah Sachs. RISKY RELAPSE INTO REPROCESSING. January 1996, http://www.ieer.org/ieer/reports/risky.html Perhaps the greatest flaw in DOE's current reprocessing policy is its open-ended nature, especially given the fact that DOE decided in 1992 to phase-out reprocessing operations. As long as DOE views reprocessing as a sound method for managing spent fuel, reprocessing will continue to be justified as the solution when additional types of spent fuel are found to be corroding or unstable in the future, and funds will not be devoted in a serious way to developing alternatives. In a November 15, 1995 letter, the Defense Nuclear Facilities Safety Board recommended that both reprocessing plants at the Savannah River Site be kept open indefinitely, stating that "the Department of Energy will always need to have available a capability for chemical processing of spent nuclear fuel . . ." This is just one example of the commitment to reprocessing in some quarters.

A recent DOE study recommended against re-starting one of the reprocessing plants at the Savannah River Site and proposed consolidation operations in the other. While this is a positive development, DOE is also considering several possible "future missions" for reprocessing at the Savannah River Site that could involve reprocessing through 2012 in the single reprocessing plant. Because spent fuel management is such a long-term project, the current signs of a favorable attitude toward reprocessing provide a very real possibility that the United States will still be reprocessing in a decade or more from now. This is a remarkable retreat from DOE's 1992 position of phasing reprocessing out. [ellipses in original]

202. Daniel Horner. NUCLEAR CONTROL INSTITUTE. Press Releases, 23 July 1996, http://www.nci.org/nci/pr72396.htm A coalition of non-proliferation and environment organizations praised a recently completed Department of Energy (DOE) report that establishes a very low ceiling on the amount of nuclear material that is eligible for reprocessing at DOE's Savan-nah River Site in South Carolina. But the coalition also faulted the report for having failed to "follow its findings to their logical conclusion and recommend that DOE make a clean break from its dirty and dangerous reprocessing habit."

Reprocessing is a Cold-War technology to separate plutonium and uranium from spent (irradiated) nuclear fuel for weapons and other purposes. DOE is now seeking to justify continued operation of its 40-year-old reprocessing facilities at the Savannah River Site, invoking a new mission of nuclear-waste management.

The coalition groups and many other environmental and non-proliferation advocates have rejected reprocessing as a waste-management solution since it produces nuclear-bomb material and enormous quantities of radioactive waste. As the report makes clear, there are other, less dangerous ways to manage spent fuel.

203. Daniel Horner. NUCLEAR CONTROL INSTITUTE. Press Releases, 23 July 1996, http://www.nci.org/nci/pr72396.htm) "Since it shows that at least 97 percent of DOE's aluminum-based spent fuel will not need to be reprocessed, it should establish the basis for DOE to broaden and accelerate its pursuit of alternatives to reprocessing. We commend the report for providing the first technical analysis of alternatives to reprocessing, and for presenting a plan for moving forward with one or more of those alternatives.

204. Arjun Makhijani. FISSILE MATERIALS IN A GLASS, DARKLY. 1996, http://www/ ieer.org/ieer/pubs/fissmats.html The difficulties of disposition of surplus plutonium from dismantled nuclear weapons are compounded by continued reprocessing of civilian spent fuel in Russia, France, Japan, Britain, and India. The governments of these countries are wedded to civilian plutonium separation as an important long-term component of energy programs. They are very unlikely in the near-term to give up these programs unless their energy concerns are addressed. Yet, if reprocessing, whether military or civilian, continues, disposition decisions on U.S. surplus military plutonium alone will not fundamentally change the global security picture. The separation and circulation of civilian plutonium will, in the coming decades, far exceed the approximately 250 metric tons of military plutonium in the world. Moreover, reprocessing civilian spent fuel is continuing in Russia; until it is halted, the security concerns in relation to weapons-usable materials associated with the state of the economy and society there cannot be resolved. Therefore, policies directed at achieving a *universal but interim halt* to reprocessing are essential so that the plutonium problem is not being aggravated while long-term energy and security issues are sorted out.

No country now engaged in civilian plutonium production is likely to stop even on an interim basis without

vigorous U.S. leadership. A clear and formal declaration by the U.S. government that plutonium is a security, environmental, and economic liability should be the starting point of such leadership. The text of a letter sent by 43 organizations and individuals to President Clinton on October 19, 1994 requesting such a declaration is attached as Appendix A to this report.

205. Arjun Makhijani. FISSILE MATERIALS IN A GLASS, DARKLY. 1996, http://www/ieer.org/ieer/pubs/fissmats.html The U.S. should address disposition of plutonium scrap and residues as part of its overall plutonium disposition plan. Because of proliferation concerns, it should rule out all options for processing of residues that, in practice, promote development of reprocessing technologies, such as pyroprocessing. The inclusion of residues in disposition policy will also be very important for non-proliferation and materials accounting in Russia. The U.S. should stop funding the development of pyroprocessing even as a plutonium disposition option. One pilot plant for plutonium vitrification should be devoted to the problem of processing scrap residues. According to our preliminary evaluation the use of a new technology for direct vitrification of residues, developed by Oak Ridge National Laboratory, appears to be a promising choice for this plant.

206. Arjun Makhijani. FISSILE MATERIALS IN A GLASS, DARKLY. 1996, http://www/ieer.org/ieer/pubs/fissmats.html)

Major Findings and Recommendations
Putting plutonium and HEU into forms not easily usable for making nuclear weapons is one of the most urgent security problems facing the world today. A great deal of the urgency derives from the severe economic decline that has occurred in the former Soviet Union since the late 1980s. Several political upheavals have accompanied that decline and the time-scale for these political changes has been on the order of a year or two. Further upheavals are possible and, if economic decline is not reversed soon, likely.

Despite the progress that has occurred between the United States and Russia on many nuclear-weapons-related issues, neither country has a coherent policy for disposition of nuclear materials. Russia is unlikely to act without U.S. leadership and reciprocity, especially given the rising nationalist sentiment that has accompanied economic decline in Russia in the last two to three years. There are already signs that such sentiment may take the form of Russian government policies favoring of preserving large stores of weapon-usable fissile materials and nuclear weapons, rather than reducing them. Thus, the U.S. must develop its disposition policy with an eye to its effects in Russia. Given the danger that a global black market in weapons-usable fissile materials originating in Russia may develop, it is imperative that the United States choose a disposition policy and persuade Russia to do the same.

207. Arjun Makhijani. FISSILE MATERIALS IN A GLASS, DARKLY. 1996, http://www/ieer.org/ieer/pubs/fissmats.html Russia has even less reason to stick with civilian plutonium production because it has huge reserves of various forms of energy, including fossil fuels and uranium. There is also immense room for improving energy efficiency in Russia. Further, Russia has been the scene of the worst civilian and military accidents of the nuclear era, namely the fire in one of the reactors at Chernobyl in 1986 and an explosion in a high-level radioactive waste tank at the Chelyabinsk-65 nuclear weapons plant in 1957. The frequency of accidents in recent years as well as the past record of despoliation of the environment are further reasons for Russia to reconsider its nuclear policies; many people in Russia are working toward that end. Britain also has plentiful fossil fuel reserves, and is an oil exporter.

208. Arjun Makhijani. FISSILE MATERIALS IN A GLASS, DARKLY. 1996, http://www/ieer.org/ieer/pubs/fissmats.html The modest theoretical merit of such arguments is overwhelmed by a number of realities. First, the danger of plutonium diversion is very real, especially in the context of continued economic, political and military instability and uncertainty in the former Soviet Union. Continued arguments that some countries need plutonium separation now for potential use in some distant future only encourages further plutonium separation and development of ancillary facilities in Russia.

The risk of diversion exists in all countries, though it is now most acute in Russia. The large-scale use of plutonium in the civilian sector will create new opportunities for diversion and for involvement of organized criminal elements in the traffic. Finally, the use of civilian plutonium in Western Europe and Japan creates obstacles to the stopping of reprocessing in Russia by depriving the United States of important leverage in dealing with Russia. The U.S. can hardly turn a blind eye to reprocessing in Western Europe and Japan while persuading Russia to stop.

209. Paul Leventhal. SEPARATING MYTHS FROM REALITY. http://www.nci.org/nci/spr41696a.htm Even Great Britain and France, which rely on the provision of reprocessing services to Japan, Germany and other nations as a major source of foreign exchange, are less than enthusiastic about using plutonium themselves. Indeed, Great Britain has no plans to use plutonium in its domestic reactors, and will have a civilian plutonium surplus totaling some 50 metric tons by the turn of the century, with no plans for its disposition. Great Britain also canceled its own breeder reactor project and withdrew support for development of a European FBR.

210. Paul Leventhal. SEPARATING MYTHS FROM REALITY. http://www.nci.org/nci/spr41696a.htm France, despite a long-standing commitment to close its fuel cycle, has been extremely slow to introduce plutonium

fuel into its own reactors. In fact, its electricity, Electricite de France (EDF), recently changed its bookkeeping practices to assign an economic value of zero to its plutonium stocks. And plans to breed plutonium at the troubled Superphenix fast reactor have been abandoned for safety reasons in favor of small-scale research projects, after a series of technical problems, shutdowns and, yes, sodium coolant leaks.

211. NUCLEAR FUEL. 12 August 1996, p. 3.

Immediately after the meeting with research reactor operators in Bonn in June, officials said, DOE sent draft contracts for spent fuel takeback to all German operators. Commercial contracts between DOE and GKSS and PTB were signed; both German institutions face regulatory difficulties if a solution to their spent fuel problem is not found soon. The BMFB official said it will be up to each research reactor operator to decide when and under what commercial terms it would agree to have DOE take back the spent fuel.

212. NUCLEAR FUEL. 12 August 1996, p. 3.

A senior German government official said last week that after considering the alternative of reprocessing the country's spent fuel at a UKAEA facility in Dounreay, German research reactors have instead accepted an offer from U.S. DOE to take back the spent fuel.

The official said that, thus far, three German reactors have signed, or are about to sign, contracts with DOE for takeback of their spent fuel. The three are GKSS in Geesthacht, the German Standards Agency (PTB) in Brunswick, and the Hahn-Meitner Institute (HMI) in Berlin.

According to officials involved, German reactors were convinced by a presentation made by DOE in Bonn two months ago that the U.S. agency would make a concerted effort to implemented the takeback (NF, 17 June, 11). At that time, a legal expert from DOE demonstrated to the German side that DOE "stood a better chance of making the takeback stand up in court than (the Germans) believed at first on the basis of (their) encounters" with the German justice system in nuclear affairs, one participant in U.S.-German talks said.

213. Paul Leventhal. SEPARATING MYTHS FROM REALITY. http://www.nci.org/nci/spr41696a.htm
Germany, once committed to an ambitious plutonium program, cancelled its nearly-completed Kalkar breeder reactor in 1991. Plans for the Wackersdorf reprocessing plant were also abandoned. In 1995, Germany's nuclear waste law was revised to allow direct disposal of spent fuel without reprocessing, and a number of German utilities are now considering cancellation of their long-term reprocessing contracts.

214. NUCLEAR FUEL. 12 August 1996, p. 3.

According to an official at the Federal Ministry of Education & Research (BMBF), analysis of comparative costs between DOE takeback and reprocessing "clearly indicated that the U.S. alternative was cheaper." In addition, he said, had reactor operators opted for Dounreay, the reprocessing waste "would have had to come back to Germany, but we don't have a final repository."

Since the mid-1970s, Germany has spent about DM 2-billion ($1.4 billion) to develop a repository at Gorleben. But the project has lost utility support behind the scenes and has been openly challenged by an internal debate in the German administration on a proposed one-repository policy that could sacrifice Gorleben in favor of the former Konrad iron mine, which until now was seen only as a host to low- and medium-level waste.

215. THE NIKKEI WEEKLY. 27 May 1996, p. 1.

Those on both sides of the issue agree that plutonium is less readily usable for bombs if it rains in spent fuel than if separated out. U.S. officials say it is easier to steal the 4kg of separated plutonium needed to make one bomb than to steal the 400kg of spent fuel needed.

216. THE NIKKEI WEEKLY. 27 May 1996, p. 1.

Even so, the distinction between separated plutonium and plutonium in spent fuel is still important, said Richard Garwin, an IBM fellow emeritus at the Thomas J. Watson Research Center in New York. If countries want to recycle plutonium, they should develop a new method in which plutonium is never completely separated from uranium or other elements, he said in an interview. Separated plutonium in Japan and Europe should be as carefully protected as nuclear weapons are in the U.S., and this makes recycling less economical, he said.

217. Paul Leventhal. SEPARATING MYTHS FROM REALITY. http://www.nci.org/nci/spr41696a.htm
The 1994 NAS study emphasized the importance of the "Fuel Cycle Policy Signal":

> [P]olicymakers will have to take into account the fact that choosing to use weapons plutonium in reactors would be perceived by some as representing generalized U.S. approval of separated plutonium fuel cycles, thereby compromising the ability of the U.S. government to oppose such fuel cycles elsewhere. Conversely, choosing to dispose of weapons plutonium without extracting any energy from it could be interpreted as reflecting a generalized U.S. government opposition to plutonium recycle. Either choice could have an impact on fuel cycle debates now underway in Japan, Europe, and Russia.

The Nuclear Control Institute believes that the vitrification option would send the *right* fuel cycle policy signal to the civilian nuclear sector and would be fully consistent with the Clinton Administration's September 1993 non-proliferation policy statement. The statement declared that "the United States does not encourage the civil use of plu-

tonium and, accordingly, does not itself engage in plutonium reprocessing for either nuclear power or nuclear explosive purposes.

218. Paul Leventhal. BURY IT, DON'T BURN IT: A NON PROLIFERATION PERSPECTIVE ON WARHEAD PLUTONIUM DISPOSAL. 12 December 1995, http://www.nci.org/nci/spr1696a.htm Most experts agree with the conclusion of this year's NAS study that the two most viable options for disposing of plutonium recovered from retired nuclear warheads are the irradiation in rectors of mixed-oxide fuel made from this plutonium (the MOX option), and direct disposal of warhead plutonium by means of vitrifying it with high-level radioactive waste (the VHLW option). The Nuclear Control Institute regards the vitrification approach as posing fewer risks than the MOX approach with regard to diversion or theft of warhead material, reversal of the disarmament process, and other adverse effects on international arms control and non-proliferation efforts. Proposals for transferring warhead plutonium MOX fuel to third countries not now processing nuclear weapons pose additional risks.

To minimize proliferation and terrorism risks in the post Cold War world, we advocate that a symmetrical regime be developed to address the dual threat of military and civilian plutonium. Such a regime would place comparable obligations on nuclear-weapon and non-nuclear weapon states to meet the spent-fuel standard and to avoid production and use of separated plutonium in any form. A decision to dispose of warhead plutonium by means of vitrification or other immobilization technology would be an essential step toward achievement of such a regime.

219. Paul Leventhal. LETTER TO SECRETARY O'LEARY. 20 December 1995, http://www.nci.org/nci/122095.htm Disposing of research reactor spent fuel via the CIC approach could be carried out with only a minor impact on DWPF operations. The total volume of foreign HEU spent fuel which will be eligible for acceptance by the management program by 2006 is 61 m3 (11,200 fuel elements); this would displace the equivalent of 90 DWPF canisters, corresponding to less than 3 months' operation of the facility.

The total number of DWPF canisters that will be required to contain the entire inventory of spent fuel depends on the number of fuel elements which can be loaded into each canister. This, in turn, will be determined by both near- and long-term criticality considerations. However, it is unlikely that criticality concerns will pose a major obstacle to this approach. The fissile content of a typical research reactor spent fuel element is low (less than 200 g 235U), which is well below the single-parameter limit for criticality in the most unfavorable configuration. The long-term criticality risk due to leaching and reconcentration of uranium can be made extremely low by limiting the number of cans per DWPF canister to fewer than four, or by packing the cans with a neutron absorber with chemical properties similar to those of uranium.

This strategy may also facilitate resolution of the most challenging stakeholder controversy surrounding this issue: the opposition of the State of South Carolina to acceptance of foreign spent fuel because of concerns about the safety of long-term storage and the absence of a firm plan for eventual removal from the state. The CIC option would solve the first problem by immobilizing the spent fuel in an inert matrix, and the second by directly linking its fate with the disposition of the much larger volume of vitrified high-level waste. Also, uncertainties associated with the direct disposal of research reactor spent fuel would be minimized, as the fuel would be isolated from groundwater by a matrix which has been thoroughly qualified as a waste form for geologic disposal.

To summarize, our preliminary analysis of extending the CIC approach to disposition of HEU spent fuels indicates that it is a technically viable politically acceptable alternative to reprocessing and merits careful consideration. We hope that this suggestion will make a useful contribution to a difficult decision.

220. Arjun Makhijani. FISSILE MATERIALS IN A GLASS, DARKLY. 1996, http://www/ieer.org/ieer/pubs/fissmats.html One way of achieving the spent fuel standard and still having a disposition policy that is compatible with policies needed for an interim halt to reprocessing would be to vitrify plutonium with rare earths or actnides first and add a gamma-emitting fission product, such as cesium-137, to the canister (instead of adding fission products to the glass). This would provide the same high resistance to theft as spent fuel and also greatly reduce the amount of fission products for achieving it compared to the option of mixing the fission products in the glass itself. As a result worker exposures and other health and environmental risks may be lower compared to other spent fuel standard options. A feasibility study and laboratory experiments should be initiated to examine this option. *This option appears to be the most promising of all the options that we have examined for achieving the principal disposition goals to the maximum feasible extent.*

221. Noah Sachs. RISKY RELAPSE INTO REPROCESSING. 1995, (http://www.ieer.org/ ieer/reports/risky.html) Disposing of the fissile materials from dismantled nuclear weapons is one of the most vexing problems the United States faces today. The Department of Energy is dismantling about 1,500 warheads per year at its Pantex facility and is currently evaluating disposition methods for the tens of tons of plutonium and highly enriched uranium (HEU) that will become surplus. These materials were produced during the Cold War, but no contingency plan was developed for a time when the U.S. would no longer want or need them. Now we face the challenge of preventing their re-use in nuclear weapons and isolating them from the environment.

IEER recently released a report on fissile materials disposition, entitled *Fissile Materials in a Glass, Darkly*. The report provides an independent analysis of the disposition problem and proposes some concrete solutions. It is the first study of this issue to detail a concrete plan that could put all excess plutonium into non-weapons-usable form in about ten years.

The report recognizes that all disposition methods have drawbacks. No existing technology can completely eliminate fissile materials, and the U.S. must choose from a menu of difficult disposition options. After a careful assessment, the report concludes that the most promising method for plutonium disposition is vitrification, that is, mixing plutonium with molten glass to form glass logs. Vitrification accords with U.S. non-proliferation goals (see below) and is technically feasible. The report suggests that the Department of Energy (DOE) build three or four vitrification plants within the next two years to test various vitrification methods.

222. Samuel Skinner. FEDERAL NEWS SERVICE.
12 July 1996, Lexis/Nexis. Notwithstanding the financial cost to customers, a number of utilities already have been forced to build additional storage capacity for spent nuclear fuel. In some cases, this has subjected those companies to unacceptably high political and financial costs. In 1993–94, Northern States Power Co. was forced into a bitter battle in the Minnesota legislature over a request to add dry cask storage at its Prairie Island station, one of the best performing, lowest-cost power plants in the country and the source of 20 percent of that state's electricity. Although the utility obtained the necessary regulatory approval for the facility, a small but vocal group of anti-nuclear activists exploited the utility's effort to store spent fuel safely to hold a hostile referendum on the future of nuclear power in that state. In this case, and others, state government authorities approved less additional storage capacity than requested by the utility, because they did not trust the Department of Energy to meet its obligations.

Precipitating these battles on a state-by-state basis is not what federal policy intended. It is not fair to expose utility customers and state legislatures to this kind of highly charged conflict, and it is avoidable. As difficult as they are today, these issues will only get more complex. Legislation required to correct problems that have been evolving for more than 10 years must be aggressive and demanding. The schedules and milestones necessary to site, build and license a central storage facility by 1998 are tight, and do not allow for slippage, delay or further excuses.

223. Samuel Skinner. FEDERAL NEWS SERVICE.
12 July 1996, Lexis/Nexis. At its summer meeting in July 1994, NARUC reaffirmed its earlier position in a resolution urging DOE to take steps to meet its responsibility to accept spent nuclear fuel starting in 1998. The state regulators also resolved that no nuclear power plant should be forced out of service, or prevented from returning to service, because of the federal government's inability to begin accepting fuel in 1998.

224. Susan Clark. FEDERAL NEWS SERVICE.
28 June 1995, Lexis/Nexis. Based on current projections, a permanent repository will not be available until the year 2010. State governments, utilities, and the public have acted in reliance of the federal government's promise that spent nuclear fuel would be removed from power plant sites beginning in 1998. If utilities cannot dispose of their spent fuel in a timely manner, then utility ratepayers and shareholders will face expensive choices, such as expanding on-site storage capabilities or shutting down economically viable power plants. In addition, the lack of an operational storage facility may delay the decommissioning process of plants whose licenses have expired. Utility ratepayers should not have to pay twice for the disposal of nuclear fuel: once in the form of payments into the Nuclear Waste Fund, and a second time in the form of costs incurred to manage the spent fuel until it is removed from the power plant site. Furthermore, the need for ratepayers to pay trice can be mitigated through corrective legislation.

225. Samuel Skinner. FEDERAL NEWS SERVICE.
12 July 1996, Lexis/Nexis. Without passage of new legislation—and particularly without a central storage facility to accept spent fuel starting in 1998—more and more nuclear power plants will run out of on-site storage space for spent fuel. Just 15 years from now, 80 of America's 109 nuclear power units will be out of space. Without passage of this legislation, more and more nuclear units will need to build additional on-site storage. More and more companies will be forced to explain to their state regulators why consumers should pay twice for spent fuel management, and why anyone should trust the federal government to keep its word in this area when its track record is so dismal.

226. THE VIRGINIAN-PILOT. 8 December 1995,
p. C12. Congress is slowly beginning to deal with a problem that successive administrations have sidestepped for more than 20 years—the safe and long-term disposal of highly radioactive spent fuel from nuclear-power plants around the country.

Congress may not be acting soon enough to satisfy most state governments and state utility commissions that want the spent fuel to be placed in a central storage facility by 1998. They have a right to expect action since the Department of Energy made a commitment to take this material more than 12 years ago.

Meanwhile, the spent fuel keeps accumulating, and by 2010, more than 70 nuclear-generating stations will have no more storage space for this material. All along, this problem has been a political rather than a technical one; key members of the House Commerce Committee have

publicly acknowledged that the government needs to stop procrastinating on this issue.

Twenty-seven states, including Virginia, have sued the DOE for failing to keep its commitment, and recently the National Association of Regulatory Utility Commissioners urged Congress to "take immediate action." They point out that "centralized, off-site spent fuel isolation is far preferable to on-site storage at reactor sites throughout the country."

227. Kathleen Reilly. 70 INDIANA LAW JOURNAL 679. Spring 1995, p. 685. When analyzing the "full objectives" of Congress behind the Atomic Energy Act, one must look to the language of the Act itself. Congress passed the Act in 1954 "to encourage widespread participation in the development and utilization of atomic energy for peaceful purposes to the maximum extent consistent with the common defense and security and with the health and safety of the public." Originally, the Federal Government had complete control over peacetime uses of nuclear power. Congress passed the AEA to eliminate this federal monopoly. Thus, Congress' purpose in the Act was clearly to promote nationwide commercial nuclear development. Since Congress also saw the need for expert supervision of such development, however, it granted the Commission expansive nuclear regulatory authority. Congress determined that federal regulation of private nuclear development would be necessary for "optimum progress, efficiency, and economy in this area of atomic endeavor."

228. Kathleen Reilly. 70 INDIANA LAW JOURNAL 679. Spring 1995, p. 686. The Atomic Energy Act designated the Commission as the federal nuclear regulatory agency, and directed it to regulate as it "may deem necessary or desirable to promote the common defense and security or to protect health or to minimize danger to life or property." In this provision of the Act, Congress' "full objective" is to give the Commission great authority to evaluate the costs and benefits that the Commission should consider economic costs in its promotion of nuclear energy. However, nothing in the Act expressly provides that economic costs should be excluded from these equations.

229. Margaret Kriz. THE NATIONAL JOURNAL. 6 April 1996, p. 763. California state officials and nuclear industry executives blasted the Interior Department's decision, contending that for political reasons, President Clinton approved the delay to placate California environmentalists and his liberal Hollywood supporters.

California's Republican Gov. Pete Wilson argued that Clinton is undermining the 1980 Low-Level Radioactive Waste Policy Act, which made the states responsible for disposal of commercial low-level radioactive waste. That law was actively supported by Clinton when he was governor of Arkansas and by Babbitt when he was governor of Arizona. Wilson wrote in a letter to Congress asking it to deed the Ward Valley lands to California.

"If this impasse between federal mandate and obstruction of that mandate by a federal agency is not resolved within the next six months, the (law) will, for all practical purposes, have been invalidated, and the Congress will be forced to make disposal of commercial low-level radioactive waste a federal responsibility," Wilson added.

230. Mike Zapler. STATES NEWS SERVICE. 2 May 1996, Lexis/Nexis. The bill, which the Senate may take up this week (EDS: WEEK OF MAY 6–10), would preempt most federal and state environmental laws—including the Safe Drinking Water Act and the Clean Air Act—in order to expedite the opening of a nuclear waste repository at Yucca Mountain, 90 miles northwest of Las Vegas.

Environmentalists and scientists say the legislation would set a dangerous precedent, inviting lawmakers to ignore air and water protections when they interfere with legislative initiatives. They also say the bill would increase the risk of cancer for people living near Yucca Mountain.

"These standards could cause quite a bit of environmental damage if they are adopted as a precedent," said Arjun Makhijani, president of the Institute for Energy and Environmental Research, an independent group that produces technical reports on environmental issues.

"People in Las Vegas could well be affected depending on how leaky the storage facility turns out to be, especially if Las Vegas continues to expand," added Bill Magavern, who specializes in energy issues at Public Citizen, a consumer rights group.

231. Bertram Wolfe. ISSUES IN SCIENCE AND TECHNOLOGY. Summer 1996, p. 56. This is the wrong time for the nation or the world to ignore nuclear power. Demand for energy will grow, and our options are limited. Ironically, environmentalists, who have opposed nuclear power since the 1970s, should have the strongest rationale for promoting nuclear energy. Like almost all large endeavors, nuclear power has its problems and its risks. But the problems of nuclear power do not look so bad when compared with the air pollution, global warming, and the supply limitations associated with fossil fuels. Besides, the major drawbacks of nuclear power—from cost to waste disposal–are due more to institutional impediments than to technological difficulties. Considering the growth in energy demand and the risks associated with other energy sources, the benefit-risk ratio for nuclear power is very attractive. Indeed, the welfare of our future generations and the environment may depend on maintaining the viability of nuclear power.

232. Bertram Wolfe. ISSUES IN SCIENCE AND TECHNOLOGY. Summer 1996, pp. 57–58. The added nuclear capacity allowed for the shutdown of oil-

fired plants, permitting the utilities to reduce oil imports by some 100 million barrels per year and thus lower the trade deficit by over a billion dollars per year. The substitution of nuclear for fossil-fueled plants has reduced present CO_2 atmospheric emissions by more than 130 million metric tons of carbon per year, roughly 10 percent of total U.S. CO_2 production. Nevertheless, the United States still needs to reduce carbon production by an additional 10 percent to reach its goal of returning to the 1990 production level. In addition, replacement of fossil-fuel plants with nuclear power has reduced nitrogen oxide emissions to the air by over 2 million tons annually, meeting the goal set by the Clean Air Act for the year 2000, and has reduced sulfur dioxide emissions by almost 5 million tons per year, half the goal for the year 2000. Both nitrogen oxide and sulfur dioxide are harmful to human health and the environment.

233. Lester Thurow. WORLD BUSINESS. July/August 1996, p. 7. When something has gone for a long time, human beings have a tendency to act as if it could go on forever—even when they know that it cannot. Consider the triangular trading pattern between the United States, Japan, and most of the rest of East Asia. Japan runs a huge trade surplus with the United States and now an even larger one with the other countries of East Asia, most of which pay for their enormous trade deficits with Japan by running even bigger trade surpluses with the United States The United States ends up with a current account deficit (more than $150 billion in 1995) that is mostly attributable to its unfavorable balance of trade with Japan and most of the rest of East Asia.

Yet if there is one thing that we know about international trade, it is that no country, not even one as big as the United States, can run a trade deficit forever. Money must be borrowed to pay for the deficit, but money must also be borrowed to pay interest on previous borrowing. Even if the annual trade deficit does not grow, interest payments do until they are so large that they can no longer be financed. Americans can also sell their assets (land, companies, buildings) to foreigners to finance deficits, but that approach is also limited since eventually there will be nothing of value left to sell.

At some point the world's capital markets will quit lending to Americans (the risks of default and of being paid back in a currency of much lower value are simply too great), just as they have quit lending to everyone else. The question is not *whether* the end will come. It will. The question is when and how fast. Will it come as one big shock or as a series of smaller shocks that do less damage?

234. Robert Vijuk. FEDERAL NEWS SERVICE. 29 February 1996, Lexis/Nexis. Congress's Energy Policy Act of 1992 defines several imperatives which must still be addressed. Key among these is the need to keep the nuclear option open for the United States. Keeping in the forefront of nuclear is also vital from an international perspective. The United States has long been a world leader in commercial nuclear technology. Without a strong program, our influence in shaping international nuclear, nonproliferation would be greatly diminished.

235. Evelyne Bertel. IAEA BULLETIN. Vol. 37, No. 4, 1995, http://www.iaea.or.at/worldatom./inforesource/bulletin)
Environmental Impacts. With regard to environmental impacts, nuclear power offers specific benefits. In routine operation, nuclear power plants and the fuel cycle facilities do release small quantities of radioactive materials. However, the rules developed and implemented several decades ago for limiting radioactive emissions satisfy criteria for protecting human health and are more than adequate to protect the environment. The other emissions, residuals, and burdens from nuclear power plants and fuel cycle facilities are lower than those arising from fossil-fuel electricity generation chains and comparable or lower than those from renewable energy systems. Taking into account the entire up-stream and down-stream energy chains for electricity generation, unclear power emits 40 to 100 times less carbon dioxide than currently used fossil-fuel chains. Greenhouse gas emissions from the chain are due mainly to the use of fossil fuels in the extraction, processing, and enrichment of uranium and to fuels used in the production of steel and cement for the construction of reactors and fuel cycle facilities. These emissions, which are negligible relative to those from the direct use of fossil fuels for electricity generation, can be reduced even further by energy efficiency improvements. Such improvements at the enrichment step include, for example, replacing the gaseous diffusion process by less energy-intensive processes such as centrifugation or laser isotope separation.

The role that nuclear electricity already plays in alleviating the risk of global climate change is notable. It is illustrated by the fact that if the nuclear power plants in operation worldwide would be substituted by fossil-fuelled power plants, the CO_2 emissions from the energy sector would increase by more than 8%. This level which almost equals the avoidance of emissions by hydropower has been achieved in a number of countries in about two decades of nuclear power development.

The analysis of statistical data in different countries over the last 20 years shows that countries which implemented large nuclear programmes, such as Belgium, France, and Sweden, achieved simultaneously significant reductions of their CO_2 emissions. In France, for example, both CO_2 and sulphur dioxide emissions were reduced by more than three between 1982 and 1992, although electricity production nearly doubled, owing to the share of nuclear power in electricity supply. In the United States, if nuclear energy would not have been used between 1973 and 1994, some additional 1750 million metric tons of CO_2 would have been released in the atmosphere. Countries and regions which do not deploy nuclear power on a large scale

for example, developing countries had a relatively high increase rate of CO_2 emissions.

236. Phillip Bayne. FEDERAL NEWS SERVICE. 29 February 1996, Lexis/Nexis. To help maximize NRC resources, the industry recommends that the NRC revise its approach to regulation and allow utilities to meet regulatory requirements without step-by-step prescriptive regulation. In the area of advance nuclear research and development, the industry supports appropriating $64 million in FY '97. The Advance Reactor Corporation, a consortium of 12 utilities, participates in the development of new reactor designs and represents one of the largest sources of private sector cost-sharing at DOE's energy supply research and development activities.

In recent years, the nuclear industry through ARC has front-loaded its cost share with the ALWR program. Now, with the program on the verge of producing market-ready nuclear designs, the industry expects the federal government to fulfill its funding obligation with the same commitment. These reactor technologies are between 50 and 90 percent complete. Furthermore, payback provisions in the program will provide a return to the US Treasury when these advance reactor designs are sold.

This new generation of reactor technology represents a growing high-tech industry and trade opportunities that will strengthen both the US economy and the nation's leadership.

237. NEI PRESS RELEASES. 25 July 1996, hhpi://www:nei.org/main/ pressrm/release "We are pleased that the House defeated Representative Obey's amendment to eliminate funding for the advanced light water reactor program. We are deeply concerned that funding cuts to the program seriously compromise our nation's ability to meet future energy needs and environmental challenges. U.S. electricity demand is expected to grow 21 percent by the year 2010.

"Americans recognize the value of preserving the option to build advanced nuclear plants. Nearly three-quarters say that advanced plants should be considered for their areas of the country when new generating capacity is needed. We urge the Senate to restore the funds needed for this critical industry/government program.

"The Department of Energy's (DOE) $40 million budget request for the ALWR program is just 2 percent of the DOE's energy supply R&D budget. If Congress terminates the ALWR program prematurely, or fails to provide adequate funding, it will essentially eliminate all federally funded nuclear energy R&D, while sustaining significant funding levels in all other energy supply technologies. Such a decision would also overturn the mandate of the Energy Policy Act of 1992.

"Without new designs, the U.S. may be forced to relinquish its global leadership position in peaceful nuclear technology and could lose American jobs and revenues to foreign countries.

"Nuclear energy is a vital investment in our nation's energy future—a minimum investment of dollars, but a rich reward of clean energy for America. The ALWR program is an excellent example of a cost-shared industry/government partnership. The industry has invested 62 percent toward this important effort. We hope the Senate will continue support for the program during its consideration of the Energy and Water Development Appropriations Act."

238. NEI PRESS RELEASES. 9 April 1996, hhpi://www:nei.org/main/ pressrm/release).The federal government has invested $3.7 billion in nuclear power since 1950. In return, we've received a safe, clean energy source that meets 20 percent of America's electricity needs—enough for 65 million homes.

- Today's joint industry-government R&D programs are the best examples of partnerships between the private and public sectors. For every federal dollar that goes toward the development of new advanced nuclear plant designs, utilities contribute $1.75 to this important effort.

- The advanced plant designs now being developed hold the potential for generating thousands of high-tech American jobs, and will maintain a U.S. leadership role in the peaceful uses of nuclear technology worldwide.

239. Evelyne Bertel. IAEA BULLETIN. Vol. 37, No. 4, 1995, http://www.iaea.or.at/worldatom./ inforesource/bulletin The technical constraints taken into account in estimating potential nuclear capacity growth rates include construction lead times and industrial capabilities for building nuclear power plants and fuel cycle facilities. The availability of sites for nuclear installations, including radioactive waste repositories, was also considered by region, taking into account seismicity, cooling water requirements and the need to build nuclear facilities in areas with relatively low population density. The availability of natural resources for nuclear fuel would not place any major constraint on the development of nuclear power, taking into account known uranium and thorium resources and expected technological progress in fissile material utilization. This scenario would require the deployment of breeder reactors by 2025 in order to support nuclear electricity generation over the period up to 2100 with the presently known uranium resources. However, within that time frame, additional uranium resources would likely become available whenever necessary. Moreover, other types of nuclear power plants, such as thorium fuelled reactors, hybrid systems, and even fusion reactors, might be developed and commercially deployed.

The implementation of this nuclear scenario would allow reductions in carbon dioxide emissions worldwide by a factor of three as compared to the present level. A similar reduction would be feasible without nuclear power

only if renewable energy sources, which have not yet reached the level of commercial development, would enter into the market early in the next century and would be deployed at very high rates throughout the next century.

240. Evelyne Bertel. IAEA BULLETIN. Vol. 37, 1995,No. 4, http://www.iaea.or.at/worldatom./inforesource/bulletin) Nuclear power is proven technology available today that can contribute significantly to reducing greenhouse gas emissions and other environmental burdens from the energy sector and to meeting environmental protection objectives. In the long term as the executive summary of the SAR chapter on energy supply mitigation options states nuclear energy could replace baseload fossil fuel electricity generation in most parts of the world, if generally acceptable responses can be found to concerns about reactor safety, radioactive waste disposal, and proliferation.

The use of nuclear energy for electricity generation dates back to the late 1950s and it has reached a stage of industrial maturity. At the end of 1994, there were 432 nuclear units connected to the grid with a total installed capacity of some 340 gigawatts-electric (GWe) in consumption. The accumulated operating experience of nuclear power plants is now over 7200 reactor-years and the average operating performance is improving continuously with an energy availability factor above 70% since the mid-1980s. This experience places nuclear power among the technologies that decision-makers can consider for sustainable electricity system expansion in the coming years and decades.

241. NEI PRESS RELEASES. 9 April 1996, http://www:nei.org/main/ pressrm/release
America's nuclear power plants are performing at world-class levels of safety, efficiency and power generation, and the proof is in the numbers:

- At an average production cost of 2 cents per kilowatt hour, nuclear power is competitive with coal (1.92 cents), and cheaper than gas (2.90 cents), oil (3.21 cents) and renewables (3 cents).
- Capacity factor—the key measure of efficiency—is projected to be 78 percent for 1995—the highest ever.
- Safety—nuclear power plant unplanned shutdowns have declined by 90 percent since 1980.
- Power production—U.S. nuclear power plants produced a record of 673 billion kilowatt-hours of clean electricity in 1995, up five percent from 1994.
- Without nuclear power, CO_2 emissions from power production would be 30 percent higher—the equivalent of 94 million more cars on the road.

242. Evelyne Bertel. IAEA BULLETIN. Vol. 37, No. 4, 1995,http://www.iaea.or.at/worldatom./inforesource/bulletin) Environmental issues are high on international agendas. Governments, interest groups, and citizens are increasingly aware of the need to limit environmental impacts from human activities. In the energy sector, one focus has been on greenhouse gas emissions which could lead to global climate change. The issue is likely to be a driving factor in choices about energy options for electricity generation during the coming decades. Nuclear power's future will undoubtedly be influenced by this debate, and its potential role in reducing environmental impacts from the electricity sector will be of central importance.

Scientifically there is little doubt that increasing atmospheric levels of greenhouse gases, such as carbon dioxide (CO_2) and methane, will cause climate change on a global scale. However, the natural climate variability is still larger than the estimated anthropogenic contributions to climate change.

Despite uncertainties, the threat of climate change remains a serious long-term global risk. Scenarios with time horizons of 2100 and beyond have to be developed, requiring insight into long-term development of life-styles, socio-economics, and technology. Such scenarios are of a normative character and therefore are inherently subjective. What is known is that energy consumption is one of the major sources of greenhouse gases, and nuclear power nowadays avoids more than 8% of the worldwide CO_2 emissions.

243. Hans Blix. IAEA BULLETIN. Vol. 37, No. 3, 1995, http://www.iaea.or.at/worldatom/inforesource/bulletin
Nuclear power and energy needs. Greater attention to issues of global safety should not mask the overall nuclear record, which is excellent. The world's 432 nuclear power plants, for example, generate about 17% of the world's total electricity, and far higher shares in many countries. Their normal operation has little environmental impact. As the environmentally conscious Club of Rome has noted and many States have realized in practice, nuclear power is a greener option than those emitting carbon dioxide and other gases as waste products threatening the atmosphere.

As sustainable development brings better living conditions to a growing world population, greater use of energy, especially electricity, will be demanded. Where will it come from? Extensive analyses of energy options are needed to factually frame answers. The IAEA and several other international organizations are assisting in comparative assessments of the benefits and problems of different electrical power options, including nuclear energy.

244. Poong Eli Juhn. IAEA BULLETIN. Vol. 38, No. 1, 1994, (http://www.iaea.or.at/worldatom/inforesource/bulletin) Today, energy consumption trends reflect the important role that electricity plays in modernization efforts and in total energy use and efficiency improvements. Also increasingly evident is that full participation in the information and communications age requires reliable sources of electricity. Studies have shown

that there is a distinct correlation between the trends of electricity consumption and national economic output in a wide range of countries. From 1960 to 1990, the share of electricity in global energy consumption has grown from 17% to 30% and the annual consumption of electricity *per capita* has almost tripled (from 765 to 2225 kWh per person). Still, two billion people in the world do not have access to electricity in their homes.

For years ahead, it is quite obvious that *per capita* consumption of electricity in developing countries will have to increase substantially to sustain economic growth and improve standards-of-living. The accelerating movement toward urbanization, which allows easier access to electrical distribution systems, together with the electrification of rural areas, will also contribute to a steadily increasing role for electrical power. In the Republic of Korea, where nuclear power already is a major producer, *per capita* electricity consumption has grown from 70 kWh per year in 1960 to almost 3200 kWh per year in 1992.

Over the next decades, the projected growth in the Earth's population, mostly in developing countries, will place higher demands on energy and electricity supplies. According to the World Energy Council, global consumption of electricity can be expected to increase between 50% to 75% by the year 2020. The potential clearly exists for nuclear power to play an important role as a safe and clean source of electricity to help countries meet their future energy needs.

245. Roger Seitz. IAEA BULLETIN. Vol. 38, No. 2, 1996, (http://www.iaea.or.at/worldatom/inforesource/bulletin) Health and environmental impacts that may result from disposal of waste are a growing concern for sustainable development of human society. Waste posing potential hazards to human health and the environment is generated in a number of industry sectors (mining/quarrying, agriculture, manufacturing, electricity generation, medical, etc.). When properly managed, this waste will pose minimal risks to human health and the environment.

However, environmental concerns arise from the fact that the quantity of waste being generated is growing (and expected to keep growing) as a result of increases in the world population, industrialization, and urbanization. Thus, one challenge in developing a strategy for sustainable development is to provide the services necessary to support economic growth and improving quality of life, while limiting the waste generated in terms of potential hazards and quantities and its health and environmental impacts.

As sustainable development brings better living conditions to a growing world population, greater use of energy, especially electricity, will be demanded. Until a suitable alternative capable of meeting the growing demand for electricity is developed, the vast majority of future electricity demand will need to be met by conventional fuels such as coal, natural gas, oil, and uranium/thorium. Thus, sustainable development strategies must include consideration of the waste that is generated throughout energy chains based on these fuels.

This article provides an overview of the initial stages of an IAEA project to compare wastes and disposal methods from different electricity generation systems and to review approaches used to assess and compare the health and environmental impacts resulting from disposal of such waste. The role of nuclear power in a strategy for sustainable development of human society is emphasized. In this respect, the article highlights the small mass of waste generated as a result of nuclear power when compared to the total mass of waste from all energy chains and other common activities. Selected waste and respective disposal methods from all steps in the energy chains for electricity generation are discussed. (Liquid and gaseous effluents that are discharged directly to air or natural water bodies are not included in this article.) Emphasis is placed on the importance of considering all steps in the energy chains, which yields information on large quantities of waste poisoning potential long-term impacts from electricity generation systems that are often considered "clean". Radionuclides that are present in many non-nuclear wastes are also discussed.

246. NEI PRESS RELEASES. 6 June 1996, http://www:nei.org/main/ pressrm/release WASHINGTON, D.C., June 6, 1996—Worldwide, nuclear power generated more electricity in 1995 than the entire world got from all sources of power in 1958, according to the International Atomic Energy Agency (IAEA). Seventeen countries relied on nuclear power for at least a quarter of their total energy needs.

Nuclear plants generate electricity in 31 countries and provide 17 percent of the world's electricity. They also provide one of the largest reductions in carbon dioxide emissions—445 million metric tons of carbon through 1993, the most recent data available.

In the United States, nuclear power plants generated more than 22 percent of the nation's electricity.

Four new nuclear power plants were connected to the electricity grids in India, South Korea, the United Kingdom and Ukraine in 1995, bringing the total number of operating reactors worldwide to 437.

Many nations expect to meet increased energy needs with nuclear power, but perhaps no countries' plans are as aggressive as those in Asia. Long-term planning for energy needs in Japan, South Korea, China, Taiwan, Indonesia, and Thailand calls for new nuclear power plants to fuel economic expansion.

In its annual worldwide survey of nuclear power, IAEA said eight plants are under construction in Japan and South Korea. South Korea's long-range plan calls for 19 new nuclear plants by 2010, including four 1,300-megawatt advanced pressurized water reactors. Japan's nuclear capacity is expected to reach 70.5 gigawatts by 2010, or 42 percent of total electricity supplies.

Last year, Japan started generating electricity at the world's first advanced-design nuclear power plant, which was designed in America. The 1,356-megawatt General Electric advanced boiling water reactor began operation in late 1995 and dispatching electricity in January 1996—10 weeks early and $6 million under budget.

Indonesia and Thailand are also considering building their first nuclear power plants to support growing population and industry. In Indonesia, electricity demand is growing by more than 7 percent a year, while forecasts for Thailand show a 10 percent increase in demand during the next 20 years.

247. THE ENERGY DAILY. 11 September 1996, Lexis/Nexis. Given the uncertainty of global energy supply and demand in the next few decades and the role renewables will play, as well as the current limited understanding of global warming, it is "prudent to keep open the option for future use of nuclear energy," Hafele says.

"Such future use could be large, far larger than the present 17 percent of our world electricity generation. This implies the use of large amounts of plutonium. Therefore the existing (plutonium) should be saved and eventually more will need to be produced."

248. Poong Eli Juhn. IAEA BULLETIN. Vol. 38, No. 1, 1996, (http://www.iaea.or.at/worldatom/inforesource/bulletin) Generally in Europe, the Chernobyl accident is one of the reasons that increases in nuclear deployment have been brought more or less to a complete halt; politicians do not dare to promote nuclear power as a safe and clean energy source, and utilities are afraid of the financial risks with respect to the economic consequences of a major nuclear accident. Besides, the economies of the European countries have been growing at a very slow rate, if growing at all, and the rate of growth in overall energy demand has been very small. On the other hand, recent studies have shown that the picture is likely to change. Electricity consumption continues to rise with economic growth, even when total energy consumption is decreasing. In a few years time more generating capacity will be needed in many countries, and chances are that some of this additional capacity will be nuclear power plants.

In regions of Asia, the situation is different; the rate of growth in energy demand is very high and nuclear is seen as an attractive option. Many countries in these regions have ambitious programmes for deployment of nuclear power in the next decade, and vendors worldwide are anticipating a revitalization of the nuclear market.

249. Poong Eli Juhn. IAEA BULLETIN. Vol. 38, No. 1, 1996, (http://www.iaea.or.at/worldatom/inforesource/bulletin) Projecting nuclear power development is a somewhat difficult exercise. A number of factors can influence policies and decision-making, and the implementation of programmes cannot be assessed with certainty.

Up to the year 2000, the installed nuclear capacity worldwide will grow to between 367 GWe and 375 GWe, compared to 340 GWe in 1994, according to IAEA estimates. Since all the units to be commissioned by the turn of the century are already under construction, the range of uncertainty reflects potential delays in construction and licensing. New nuclear units will be connected to the grid mainly in Asia, while in Western Europe and North America the installed nuclear capacity will remain practically unchanged. In Eastern Europe, although some of the units under construction will be completed, the economic transition will delay significantly the implementation of nuclear programmes in most countries.

250. Poong Eli Juhn. IAEA BULLETIN. Vol. 38, No. 1, 1996, (http://www.iaea.or.at/worldatom/inforesource/bulletin) Statistics show that nuclear power has become an important energy source in many countries since its introduction four decades ago. In the 1960s, the industry was convinced that nuclear power represented the solution to cheap and reliable energy supply, and programmes were eagerly pursued worldwide. As a result, the construction of nuclear power plants expanded rapidly through the 1970s. According to data in the IAEA's Power Reactor Information System (PRIS), by 1980 nuclear power production had increased to 692.1 terawatt-hours (TWh), contributing 8.4% of total electricity generation. *(See figures, next page.)* This represented an almost ninefold increase since 1970, and an average annual growth of 24% over the decade.

Since the 1980s, the expansion rate has dropped markedly even though significant new nuclear electricity capacity is still added each year. In the period 1980–85, nuclear electricity generation increased to 1402 TWh, corresponding to an average annual growth of 15.2% in generation; in the next half decade, during 1985–90, it increased to 1913 TWh, yielding an annual growth of 6.4%. In the period 1990–94, it increased to 2130 TWh, corresponding to a growth of about 2.8% per annum. The five largest producers in 1994 were the United States (639.4 TWh), France (341.8 TWh), Japan (258.3 TWh), Germany (143.0 TWh), and Canada (101.7 TWh).

Worldwide, the nuclear option today accounts for about 17% of the total electricity produced. In altogether 14 countries (as well as in Taiwan, China), one-quarter or more of the electricity is generated by nuclear power plants. Nuclear power accounts for some 40% or more in eight countries: Belgium, Bulgaria, France, Hungary, Lithuania, Slovakia, Sweden, and Switzerland. *(See the International Data File section, page 53.)*

By the end of 1995, a total of 437 units with a generating capacity of over 343-gigawatts-electric (GWe) were in operation in the world, and 39 units with a generating capacity of over 32 GWe were under construction. The total accumulated operating experience of the operating

plants amounts to more than 6637 reactor years, or an average operating period per plant of about 15 years. Substantial operating experience has also been gained by plants that have now been shut down, and factoring in these plants yields a total of more than 7700 reactor years. In other words, by now, nuclear energy has acquired a strong position in the electricity generation sector as a mature technology.

251. NEI PRESS RELEASES. 29 May 1996, (www:nei.org/main/ pressrm/release). "The United States has an important interest in seeing to it that these technologies are available to meet growing world energy needs," Curtis said. "These needs are especially significant in Asia, and much of the demand will be met by nuclear. This is a huge export opportunity for the United States. We have a responsibility to see that countries that want nuclear power build the safest plants—and those are ours."

DOE is supporting certification of three advanced nuclear plant designs and is partnering with a consortium of nuclear utilities to perform detailed engineering work on two of the designs. These efforts must be completed, Curtis said, to maintain U.S. influence in the peaceful uses of nuclear technology worldwide and to keep nuclear power a "viable option for the United States."

Along with completing advanced plant designs, Curtis said, "the future of nuclear power in this country is dependent on resolution of the high-level radioactive waste disposal issue. Scientific studies of the proposed nuclear waste repository at Yucca Mountain, Nev., have made "significant progress," and DOE is more than five months ahead of schedule on building an underground laboratory at the site.

252. Poong Eli Juhn. IAEA BULLETIN. Vol. 38, No. 1, 1996, (http://www.iaea.or.at/worldatom/inforesource/bulletin) Prospects for nuclear power should be assessed in the context of growing electricity demand and greater awareness of environmental issues. Nuclear power alone will not solve all the problems, but it will form part of the answers.

Over the near term, nuclear power projects will be pursued mostly in Asian countries, including China, Japan, and the Republic of Korea. In many other regions of the world, safe and reliable operation of nuclear power plants, convincing solutions to high-level nuclear waste storage and disposal, and a predictable licensing process are essential prerequisites for the revival and expansion of nuclear power.

253. LOS ANGELES TIMES. 4 July 1996, p. A1. Public unease is spreading about the nuclear power industry in Asia—the one region of the world where the promise of atomic energy has never until now lost any of its luster. For booming Asian nations with limited natural resources, burgeoning populations and polluted skies, nuclear power always has held special appeal. Some portray it as the only choice for fulfilling soaring power needs. And unstated but never forgotten, for countries hoping to quietly build a nuclear arms program, material for reactors can be converted for weapons use.

That enthusiasm for atomic power is pushed along by European and North American power-plant builders who see Asia as their greatest potential market–and after having no new orders to build U.S. plants in 23 years, perhaps their last hope.

254. Shirley Jackson. FDCH TESTIMONY. 15 May 1996, Lexis/Nexis. On the fuel cycle issues, the Koreans view these somewhat differently from the U.S. position. They dominated the discussions in my various meetings. As you know, since 1979 it has been U.S. policy not to reprocess spent fuel, because of concerns over non-proliferation and terrorism, and we have urged other countries to adopt the same approach. In 1991, the South Korean government made a formal commitment not to develop reprocessing or enrichment facilities. This unilateral declaration formed a cornerstone of the North Korea-South Korea non-nuclear declaration. This policy recognized the destabilizing effect within the Korean Peninsula of unirradiated nuclear material directly usable for nuclear weapons. The U.S. strongly endorses this exercise of self-restraint on South Korea's part. At the same time, however, the U.S. recognizes South Korea's desire to extend its uranium resources and better manage wastes and, as an advancing nuclear country, to engage in fuel cycle research.

255. Melvin Green. FDCH TESTIMONY, 4 June 1996, Lexis/Nexis. Worldwatch staff researcher Megan Ryan said power-plant manufacturers are facing a crisis, given the move in the US and Europe to restack energy priorities by incorporating renewables, efficiency, and a dramatically different mix of fossil-fuel options.

She said: "Nuclear vendors are trying to hook the Third World on nuclear power. South-east Asia may be one of the industry's last hopes.

Amidst such a controversy, planners in the region face a challenge in making rational policy decisions on energy without being co-opted by either the anti- or pro-nuclear lobby.

256. FINANCIAL TIMES. 26 June 1996, p. A1. In nuclear energy, the prospects are even bleaker. The last three plants to be built in western Europe and north America have recently come into production, leaving empty order books. The only realistic prospect of selling a new nuclear power station is in east Asia, where Japan, South Korea, and Taiwan all have construction plans.

257. FINANCIAL TIMES. 19 July 1996, p. A1. The government fears that a further rise in the anti-nuclear climate will hit the country's nuclear industry, con-

sisting of plant makers, general contractors, heavy industry companies, and leading electronics companies. Many have already started to restructure their nuclear plant divisions and are looking for business opportunities overseas, especially in Asia.

258. THE CHRISTIAN SCIENCE MONITOR. 17 April 1996, p. 1. These issues are now looming largest in Asia, the region anticipating the largest increase in new nuclear plants. The Pacific Rim is experiencing rapid economic growth, says Wilfrid Kohl, director of Johns Hopkins University's International Energy and Environmental Program in Washington.

"Electricity has been the fastest-growing fuel sector in the industrial and developing countries" in Asia because most of their economies are shifting from heavy industries to service and high-tech companies," says Mr. Kohl. "High-tech oriented societies are willing to place their bet on nuclear" to make up for a lack of indigenous energy sources and to ensure energy security.

259. FINANCIAL TIMES. 19 July 1996, p. A1. For the companies, the grass looks greener in Asia. The region's rapidly growing economies are creating demand for electricity, and about 20 nuclear plant projects are being planned in Asia, including China and Indonesia. Plant and equipment-makers, including Mitsubishi Heavy Industries, Hitachi and Toshiba, are eager to enter the expanding market.

260. LOS ANGELES TIMES. 4 July 1996, p. A1. The ability to learn and adapt in Asia is outpaced only by the region's economic growth. As China buys plants from overseas firms like Westinghouse, the government will demand technology transfers that speed Beijing's ability to build and sell plants of its own, analysts say. And though U.S. laws ban the export of militarily sensitive technology, China—which has reactors that run on bomb-grade enriched uranium—won't necessarily be hindered.

261. THE ASSOCIATED PRESS. 24 June 1986, p. 1. Her break from gloomy lectures came just before Chernobyl, the kind of nuclear disaster she had warned about for years. At least 26 people have died from radiation sickness and injuries suffered in the April 26 explosion at the Soviet power plant.

"It was inevitable," she said. "This is the beginning of many meltdowns. The Northern Hemisphere is in a certain way doomed." It was this vision of doom that first drew Caldicott to her work and that her critics attack as hyperbolic and traumatizing.

262. THE VANCOUVER SUN. 23 May 1992, p. A1. Caldicott also endorses shutting down all nuclear power plants, distributing contraceptives to every citizen on earth, making abortions available to all women and having the United States pass an edict limiting all families to one child.

"It's absurd to even discuss abortion when there are 5.4 billion of us," she says, adding that the world population is projected to reach 14 billion in less than a century.

As for the overthrow of governments, particularly the government of the U.S., Caldicott says there is really no alternative.

"It has to happen or we will all die."

263. SAN DIEGO UNION-TRIBUNE. 30 March 1996, p. 12. It's a sad tale, exhaustively documented. But at times Caldicott's clear and well-informed voice turns shrill.

Just as she once insisted that nuclear war was "inevitable" if Reagan was elected president, she writes that the cooling water in a nuclear power plant will "inevitably" become radioactive (it is designed to stay separate from the radioactive core and generally does.) She also describes transnational corporations as "sinister"—in other words, evil.

264. THE VANCOUVER SUN. 23 May 1992, p. 3. In the book, she singles out villains: trans-national corporations, the banks (quick to loan Third World countries money, with those countries now destroying their wilderness areas to pay it back) and governments that allow themselves to be manipulated by big business.

Corporations, she says, are the source of most ecological evils. Governments need to rein them in, but governments are controlled by them. So, she says, those governments need to be overthrown.

"You change things by getting into government yourself," says Caldicott, who ran, unsuccessfully, for the Australian parliament in 1990. "You remove people who are corporate prostitutes, so you can rein in the corporations. Corporations are aberrations, they're obsolete. They're killing the planet."

265. THE ASSOCIATED PRESS. 24 June 1986, p. 1. Harvard social scientist Robert Coles called her statements the "exaggerated rhetoric of the polemicist." Sanford Gottlieb, a leader of the United Campuses to Prevent Nuclear War, warned that students were "becoming numbed by the emphasis on nuclear blast, fire and radiation." Caldicott replied that her speeches have drawn millions into the anti-nuclear movement.

266. THE ASSOCIATED PRESS. 18 May 1986, Sunday AM cycle, p. 1. Dr. Helen Caldicott, a leading voice in the anti-nuclear movement, said Sunday she wishes the Chernobyl nuclear meltdown had happened in the United States because that would have ended the arms race.

"Chernobyl should have happened here. Maybe then people would stop this insane desire to build weapons and

atomic power plants," she said, referring to the recent accident at a Soviet nuclear power plant.

267. SAN DIEGO UNION-TRIBUNE. 30 March 1996, p. 12. Among the solutions she proposes is an end to political ads on radio and TV. The costly ads can only be supported with money from corporate donors, perpetuating the cycle of politicians who are beholden to large financial interests, said Caldicott.

"True democracy is representing the people, not the corporations that are chopping down the forests and producing the radioactive waste," she said.

Beyond such legislative measures, people must connect with nature by hiking, gardening, or other pursuits, Caldicott says. And they must gain confidence that they can effect the change necessary to heal the planet.

268. UNITED PRESS INTERNATIONAL. 18 May 1986, p. 1. Caldicott, president emeritus of Physicians for Social Responsibility, told 4,000 graduates and guests at the Renssalaer Polytechnic Institute commencement exercises what would happen if 20-megaton bomb fell on the school fieldhouse where they were sitting.

"It would dig a hole three-quarters of a mile wide," she said. "Buildings and people would turn into fallout shot up in a radioactive cloud.

"For six miles around everybody would be vaporized. In a 20 mile radius, everybody would be killed or lethally injured. Winds of 500 miles-per-hour would pick people up and turn them into 100 miles-per-hour missiles," she said.

Caldicott said she believes nuclear war is imminent and it will probably occur "by mistake" during the next ten years.

269. SAN DIEGO UNION-TRIBUNE. 5 April 1992, p. 12. It's good to know Helen Caldicott is looking after us all. With excruciating modesty, she deigns to take credit for the end of the Cold War, thereby giving the planet Earth another chance at survival.

Caldicott places all the blame for the world's ills on America. How does she reconcile the ecological devastation that grips the ex-Soviet empire, with her anti-nuclear principles?

But Caldicott is too busy scolding America for being wasteful and decadently rich—as if advanced technology and too many TV sets were the cause of the world's ills.

What technology would she find to her taste? That of 1900? 1800? The Middle Ages?

Does she want us to employ the kind of slash-and-burn agriculture and nomadic goatherds that helped to spread the Sahara Desert across the face of Africa? What would she say to the Brazilian farmer who chopped down an acre of rain forest to plant crops he needs for survival?

Somehow, she will find a way to blame the United States for all the fluorocarbons in use in Zambia, or all the whales killed by Japanese fishing fleets.

Helen Caldicott is in so far over her head that her activism self-destructs.

270. THE CHRISTIAN SCIENCE MONITOR. 3 June 1988, p. 3. Variations on this set of images form the basis of legend, Grade-B monster movies, and countless science-fiction novels. It is also (except for the happy ending) the substance of public debate on nuclear energy. For, Spencer Weart argues, much public discussion of nuclear power is not about nuclear physics. It is about nuclear fear.

"Nuclear cliches," he argues, distort rational evaluation of "real facts." Public understanding of nuclear issues is caught up in a "tangle of imagery," the bright and the somber together, and the images are crowding out the facts.

This book offers no analysis of superpower relations, force-ratio calculations, conflict resolution strategies, or calls for ethical rearmament. Instead, it sifts through elements of popular culture—comic books, pulp magazine stories, Saturday movie serials, science fiction, science journalism—and traces the origins and development of the images that surround nuclear questions.

- The "good" atom; radioactive rays with powers to heal; the ocean liner that encircles the glob, "fueled by a single gram of radium", the "white city" of hope.
- The "bad" atom: the atomic scientist who glows in the dark, killing everything he touches; radioactive mutant monsters "who in their anguish stamp around smashing everything"; radioactive rays with powers to destroy.

For Weart, these images of fantastic hopes and fears have a history. They are "encrusted with complex and emotive meanings." And they are powerful. Nuclear images short-circuit though, he says, allowing the public to set aside the actual experience of power plants and bombs, in favor of comfortable, plausible images.

271. THE CHRISTIAN SCIENCE MONITOR. 3 June 1988, p. 3. Beyond this core of faux-physicists is a larger group of nuclear whistleblowers, with enough technical training to be taken seriously by news media hungry for sensational images. Weart dismisses antinuclear activist Helen Caldicott, for example, as a "lively and tireless pediatrician." These pseudo-scientists, he says, feed the fantasies of larger groups ("students, housewives, independent tradesmen, farmers") who judge nuclear policy questions on the basis of feelings. They tend to be, he says, "allergic to authority," intemperate, and susceptible to "attacks of imagery."

272. Roger Seitz. IAEA BULLETIN. Vol. 38, No. 2, 1996, http://www.iaea.or.at/worldatom/inforesource/bulletin) The continuing growth in the quantities of waste being generated and the need for appropriate disposal facilities which protect human health and the envi-

ronment has led to the increased involvement of a number of United Nations organizations in waste management issues. The United Nations Conference for Environment and Development (UNCED), held in Rio de Janeiro in June 1992, provided an international forum for discussing sustainable development strategies related to waste management in addition to a number of other environmental issues. Agenda 21, the programme of action for sustainable development agreed upon by the governments participating in UNCED, reflects the significance of waste-related concerns. It includes three chapters specifically directed at waste management and references to waste management issues in a number of other chapters.

Through UNCED and Agenda 21, the United Nations and world governments have called global attention to the need for a comprehensive strategy for sustainable development of human society. Agenda 21 includes a number of statements emphasizing that reduction of the amount of wastes being generated is a necessary part of any such strategy. It also includes recognition that regardless of he success of efforts for cleaner production, waste is a consequence of development and will continue to be generated, and thus disposal options capable of protecting health and the environment must continue to be available. The data available support the argument that the minimal amounts of waste generated through nuclear power may help make it a beneficial contributor to a global strategy for cleaner production and sustainable development.

273. Roger Seitz. IAEA BULLETIN. Vol. 38, No. 2, 1996, (http://www.iaea.or.at/worldatom/inforesource/bulletin) Waste from the operation of nuclear power plants is probably the most studied waste in the world, especially spent nuclear fuel. However, data show that the amount of waste generated by a nuclear power plant is very small compared to the waste generated by electricity generation systems as a whole. The main concern for nuclear waste is the high levels of radioactivity in the much smaller quantities of high-level waste. Reprocessing of the spent fuel is conducted in several countries, which reduces the long-term hazards associated with the waste that must be disposed. Low- and intermediate-level waste also results from nuclear power plant operations. This waste includes various trash, piping, and used equipment that are contaminated by radionuclides with relatively short half-lives.

Decommissioning of closed power plants is the last step in the generic energy chain. For coal, oil, and gas-fired plants, decommissioning waste would include building rubble, old equipment from the facility, and contaminated soil resulting from operations. These materials would be contaminated by combustion byproducts and other substances associated with plant operation. Wastes from nuclear power plant coolant may require special handling due to elevated levels of primarily short-lived radionuclides. Decommissioning of solar cells, dams, and wind machines would also result in wastes that must be managed. The solar cells, in particular, will contain hazardous compounds posing potential long-term health hazards.

274. Roger Seitz. IAEA BULLETIN. Vol. 38, No. 2, 1996, (http://www.iaea.or.at/worldatom/inforesource/bulletin) The second step in the generic energy chain is fuel preparation. Fuel preparation can also be a large source of wastes. For conventional fuels, fuel preparation includes cleaning raw coal to remove impurities, refining petroleum products, and milling and fuel fabrication for nuclear power. These wastes from post-mining activities include tailings, water, and solids contaminated with similar materials to the mining wastes (e.g., trace metals, salts, metals and NORM). Refineries generate waste oil and water, a variety of sludges contaminated with NORM, hydrocarbons, trace metals, PCBs, and other contaminants. Fuel fabrication for nuclear power plants results in waste including ashes and sludges contaminated with NORM and trace metals. The manufacture of solar cells (photvoltaics) can be considered an analogue to fuel preparation in the context of solar power. Manufacture of photovoltaics for solar cells results in a variety of toxic or hazardous waste contaminated with arsenic, copper, cadmium, gallium, and zinc compounds.

275. Roger Seitz. IAEA BULLETIN. Vol. 38, No. 2, 1996, (http://www.iaea.or.at/worldatom/inforesource/bulletin) The third step in the energy chain includes waste generated during operation of the electricity generation plant. These are the most recognized wastes as they tend to receive the most attention. Coal-fired plants generate large quantities of combustion waste including fly ash (airborne combustion products) and bottom ash (heavier combustion products) that result from the burned fuel, as well as gypsum and sludges from different FGD techniques. These wastes are contaminated with NORM and trace metals. It is somewhat ironic that the use of flue gas desulfurization to reduce the greenhouse gases emitted from fossil fuel facilities results in more waste being produced than the ash from the burned fuel. Recycling of fly ash and FGD wastes is being heavily promoted and large quantities are being used for other purposes (e.g., cement additive, backfill, gypsum in construction materials, and many other uses). However, even with recycling, the enormous quantities of ash and FGD waste that are generated far exceed the demand (it is estimated that more than 450 million tonnes of these waste are unused in the world each year). Oil-fired plants generate lesser quantities of ashes, but can be a large source of FGD waste. Furthermore, some of the boiler cleaning and waste water treatment wastes also contain hazardous materials.

276. Roger Seitz. IAEA BULLETIN. Vol. 38, No. 2, 1996, (http://www.iaea.or.at/worldatom/inforesource/bulletin)
Waste management and sustainable development
The Organization for Economic Co-operation and

Development (OECD) estimates that roughly nine billion tonnes of solid wastes were generated by its Member States in 1990. In spite of waste minimization efforts implemented in recent years in nuclear and other industries, this total is continuing to rise. Reviews by the United Nations Environment Programme (UNEP) suggest that the mining/quarrying and agricultural (manure, crop residues, etc.) sectors generate the largest amounts of wastes. Data covering OCED countries and data from the United Nations Statistical Commission and Economic Commission for Europe (UNSC/ECE) support the general UNEP conclusion. They also suggest that in some countries the industrial, communal, and energy production sectors can account for a large proportion of the solid wastes generated. *(See graph.)* It is interesting to note that the mass of radioactive waste from nuclear power plants is a small fraction of that resulting from all energy production.

277. Roger Seitz. IAEA BULLETIN. Vol. 38, No. 2, 1996, (http://www.iaea.or.at/worldatom/inforesource/bulletin) Fuel preparation waste from coal and nuclear energy chains includes large amounts of liquids that are often disposed of in impoundments (man-made ponds or lagoons). Solid waste from fuel preparation (e.g., tailings and evaporation residues) are often covered with an engineered soil cap to minimize infiltration into the waste and to limit the release of gases from the waste. Oil refinery waste is often disposed of using land farming or pit disposal. Hazardous waste from the refineries or the manufacture of photovoltaics for solar power is generally sent to a licensed facility. A typical hazardous waste disposal facility includes a lined trench with leachate collection systems and an engineered soil cover to limit water contact with the waste. Other waste associated with fuel preparation will typically be disposed of in landfills or, in the case of some nuclear waste, engineered trenches, or concrete vaults.

Operational waste from coal and oil-fired power plants such as fly ash and FGD waste are typically disposed of in ponds, landfills, mine cavities, or surface waste piles. After evaporation and draining of the water, the sludge remaining at the base of disposal ponds is typically covered with soil. Boiler wash waste for coal, oil, and gas plants may have to be treated as a hazardous waste, which would require disposal in a licensed facility. Low- and intermediate-level waste from nuclear power plants is often disposed of in engineered trenches, concrete vaults, or mined cavities. This waste is also typically packaged prior to disposal. The high-level waste, including spent fuel, is planned to be disposed of in deep geologic formations or stored in a retrievable form.

278. Roger Seitz. IAEA BULLETIN. Vol. 38, No. 2, 1996, (http://www.iaea.or.at/worldatom/inforesource/bulletin) A number of wastes are generated as a result of construction, maintenance, transportation, and waste treatment processes in each of the steps of the energy chain. General construction, maintenance, and transportation wastes would, for the most part, be typical for all of the energy chains, although the quantities and types and levels of contamination will be different depending on the energy chain. For example, waste associated with transportation can be very significant for coal-fired plants due to the enormous volumes of fuel and resulting ash and waste that must be transported on a daily basis. It has been estimated that fifty 40-tonne trucks per day would be required to transport the fly ash and FGD wastes from a typical 1000-MWe coal-fired power plant to a disposal site (rail or other transport can also be used when available). A complete life-cycle analysis would need to include the waste generated while producing the fuel for the trucks or trains and the waste associated with maintaining the vehicles. Disposal of secondary waste resulting from treatment processes that are used for many of the wastes will also need to be considered in a comprehensive comparison.

279. ST. LOUIS POST-DISPATCH. 1 April 1996, p. 1. One important area for cost reduction is the regulatory process. The industry is working to shift to performance-based regulation. This would focus regulation where it is needed and make it more objective and efficient, while maintaining a high-level of safety.

Currently most Nuclear Regulatory Commission regulations are prescriptive. They tell a nuclear plant operator precisely what measures to take to meet certain objectives, leaving little room for alternative or innovative approaches which are sometimes both better and less costly.

For example, every system and component that is identified as "safety-related" or "important to safety" is given the same level of attention, whether it plays a big role in safety or a small one. This type of regulation focuses on processes and procedures instead of on results and is both more costly and less effective than it should be.

In contrast, a performance-based rule establishes basic requirements and sets overall performance goals, and then allows the nuclear plant operator to decide how best to meet these goals . For well performing plants, it promises substantial cost savings.

As an example, the Nuclear Regulatory Commission recently revised its rule on containment-leakage testing to allow plants with good performance on the test to extend the test interval to once in 10 years. Savings across the industry are expected to exceed $1 billion over the next 20 years, primarily in reduced outage time with no compromise in safety.

280. Shirley Jackson. FDCH TESTIMONY. 27 March 1996, Lexis/Nexis. Ensuring regulatory effectiveness has many facets. It means keeping a primary focus on adequate protection of public health and safety and minimizing risk at reasonable cost as a basis for any new regulation or change to an existing one. It also means considering ease of implementation, consistency with other applicable statutes and regulations and internal guidelines,

fairness, and the fit of the regulation into the entire regulatory program. An important element of regulatory effectiveness is the use of risk analysis insights. The movement to risk-informed, performance-based regulation, allows both the NRC and nuclear licensees to focus their resources on the most safety significant aspects of nuclear operations, while maintaining safety defense in-depth.

281. ELECTRICITY JOURNAL. May 1996, p. 3. Two major opportunities lie ahead for large-scale improvement in nuclear productivity. First, move the industry away from prescriptive regulation toward performance-based regulation, using quantifiable goals instead of mandated activities. The potential payoff can best be judged by looking to Europe, where nuclear plant outages are shorter, despite one-half the staffing we use in the United States. The Nuclear Energy Institute has the responsibility for working with the NRC to establish performance-based standards. EPRI has the responsibility for developing the technical basis for the regulations and the tools and procedures to reach these goals.

282. John Barry. HERITAGE FOUNDATION REPORTS. 9 November 1995, p. 1. Much of this DOE-funded research is already conducted by private firms or simply irrelevant to market needs. Electric utilities, for instance, voluntarily fund the $240-million-per-year Electric Power Research Institute at no government expense. Moreover, "the major new technologies for enhanced oil recovery . . . have come from private industry, not DOE," according to the Congressional Budget Office, and DOE's $9 billion investment in nuclear fission research has gone to waste because the private sector has no interest in building new nuclear power plants. Thus Energy "has little in the way of commercial applications to show for its investment." All the federal government's attempts to outguess the energy market have produced are such expensive failures as the Synthetic Fuels Corporation and the Clinch River Breeder Reactor. Federal funding for renewable and non-renewable energy research and development should be halted immediately. [ellipses in original]

283. John Barry. HERITAGE FOUNDATION REPORTS. 9 November 1995, p. 1. Second, the program's organizational structure reflects many of the inefficiencies that haunt the Department of Energy in general: excessively high expenses, scheduling lapses, poor contractor management, and a politically charged environment. Numerous studies over the past decade have commented on these problems and suggested that their cause is OCRWM's position within the Department of Energy. As early as 1984, only two years after the office was created, a report to the Secretary of Energy found that "Location of OCRWM within the Department of Energy makes it vulnerable to changes of policy and senior management as Administrations come and go." Studies generally have suggested that OCRWM be moved out of the Department of Energy and given the freedom to operate without political constraints.

284. John Barry. HERITAGE FOUNDATION REPORTS. 9 November 1995, p. 3. DOE's newest initiative, the fourth change in mission, is the creation of Cooperative Research and Development Agreements (CRADAs). These "agreements" are contracts whereby the Department of Energy allows individual companies to use federal laboratories, and even conducts research and development, all at taxpayer expense. CRADAs are meant to increase the competitiveness of American companies and support quality jobs at a time when much of the defense-related work completed at DOE's laboratories in the past is no longer seen as necessary. CRADAs thus offer the dual advantage of providing private companies with free research while preventing the closing of federal facilities—or so proponents argue.

The Department of Energy's CRADA initiatives are among the most egregious examples of corporate welfare that benefits individual companies at tremendous expense to American taxpayers. According to a recent series on "High-Tech Handouts" in the *Philadelphia Inquirer*, researchers who once worked for a private company are now employed by DOE but continue to do the same work. The only thing that has changed is that taxpayers are picking up the tab. Moreover, when this research is successful, individual companies can earn protected trade secrets or patents that often are not available even to the federal government. Millions of American taxpayers thus may be underwriting a single company's market position.

Gilbert M. Gaul and Susan Q. Stranahan, "How Billions in Taxes Failed to Cream Jobs," *The Philadelphia Inquirer*, June 4, 1995, p. 1. The series dramatically demonstrates the inefficiency of government-sponsored research and industry cooperatives.

The results of most efforts are both disappointing and expensive. For example, *The Philadelphia Inquirer* notes that DOE has spent some $792 million on CRADAs between 1992 and 1995. The result has been 46 new companies at a cost of $17 million per company. Even more disappointing, many recipients of DOE assistance are reducing their research and development budgets and trimming their workforces. General Electric, which earned $4.7 billion in profits in 1994, received $25.4 million in federal assistance between 1990 and 1994. Over the same period, GE cut 80,000 positions from its payroll. Clearly, DOE's CRADA program is not meeting its stated goal of job creation.

285. John Barry. HERITAGE FOUNDATION REPORTS. 9 November 1995, p. 6. Rationale: Many of the laboratories now owned by the Department of Energy were established during World War II in conjunction with the Manhattan Project, which resulted in the world's first atomic bomb. since then, they have "expanded their missions to encompass civilian research

and development in many disciplines—from high-energy physics to advanced computing," reports the GAO; but because of DOE mismanagement, "the multi-program laboratories—both individually and as a group—do not have eider clearly defined missions or specific implementation strategies that bring together labor resources to focus on accomplishing departmental objectives or nation goals." With no defined mission, other analysts report, these laboratories often compete directly with the private sector.

286. Arthur Coulter. NEWS AND RECORD. 14 July 1996, p. 1. Electric utilities have the expertise to handle radioactive material. Considering the enormous amount of radioactive waste they produce they have done well keeping it out of the environment. In North Carolina they are now storing high-level waste on site and recently–since South Carolina's Barnwell dump was closed to North Carolina–they are also storing LLRW. Reactor site storage leaves ownership and liability for the waste with those primarily responsible for producing it. This would include liability for leaks. Leaks have been common from LLRW dumps.

287. Frank Munger. KNOXVILLE NEWS-SENTINEL. 20 May 1996, p. 1.
OAK RIDGE—During Cold War production of nuclear weapons, the U.S. Department of Energy had a virtually unlimited bank account and earned a reputation for abusing the environment and wasting taxpayer money.
   The critics were legion.
   In the post-Cold War era, DOE is committed to cleaning up the legacy of pollution, and agency officials insist they're trying to save money, which no longer is so plentiful.
   Critics still besiege the federal agency.
   The latest example of DOE's plan to use private companies to treat an estimated 60 million pound of hazardous and nuclear wastes.

288. Frank Munger. KNOXVILLE NEWS-SENTINEL. 20 May 1996, p. 1. The current timetable calls for DOE to complete its environmental assessment sometime this summer, but Korynta said that schedule could be altered if necessary.
   He also denied any contracts have been awarded or promised to any waste companies, although some critics believe otherwise. Once the assessment process is completed, there will be a competition for the waste contracts, he said.
   Asked if there is a possibility DOE will back away completely from the privatization proposal and do the waste treatment with its in-house contractors, Korynta said, "It's unlikely."

289. Roger Seitz. IAEA BULLETIN. Vol. 38, No. 2, 1996, (http://www.iaea.or.at/worldatom/inforesource/bulletin) Although natural gas is often considered a "clean" energy source, exploration and drilling for natural gas and oil are large sources of waste. Waste from these operations include radioactive scale that accumulates on the inside of pipes, drilling muds, and soil contaminated by spills of oil and treatment of produced water. Scale that accumulates on the inside of pipes can contain significant amounts of radionuclides *(see table)* and can require disposal as a radioactive waste. Drilling muds can be contaminated with salts, trace metals (selenium, arsenic, magnesium, curium, zinc, chromium, nickel, aluminum, and iron), and oils and other lubricants. Extraction of oil and gas also includes large quantities of "produced water" from the gas or oil bearing formation (up to 3,000,000 L/day). Such produced water contains a variety of contaminants including NORM (specifically radium), trace metals, ammonia, salts, aliphatic and aromatic petroleum hydrocarbons, phenols, and naphthalenes. Sludges resulting from ponds of such water are contaminated with elevated concentrations of the metals, hazardous substances, and radionuclides that were present in the water. Drilling operations also result in a variety of hazardous wastes including asbestos, pesticides, PCBs, and trichloroethylene.

290. Bertram Wolfe. ISSUES IN SCIENCE AND TECHNOLOGY. Summer 1996, p. 56. The United States has been a leader in the development of nuclear power technology and the adoption of stringent safety standards. Not a single member of the public has been harmed by the operation of any of the world's nuclear plants that meet U.S. standards. (The Chernobyl reactor, which lacked a containment structure, did not meet U.S. standards.) The United States has also been successful in using its peaceful nuclear power leadership to limit the worldwide spread of nuclear weapons.

291. Bertram Wolfe. ISSUES IN SCIENCE AND TECHNOLOGY. Summer 1996, p. 38. U.S. nuclear power plants themselves have an admirable environmental and public health record. Safety has been a critical consideration in plant design from the beginning. Standard operation of a nuclear plant produces no ill effects, and even in the case of a major malfunction or accident, the use of a containment structure that surrounds the plant prevents the release of significant amounts of radioactive material. The wisdom of the U.S. approach is evident in a comparison of the accidents at the Three Mile Island plant in Pennsyl-vania and the Soviet Union's Chernobyl reactor. Thanks to the containment structure, not a single member of the public was injured by nuclear radiation from the Three Mile Island accident. In fact, a person standing outside the plant would have received less radiation exposure from it than from a two-week vacation in high-altitude Denver with its uranium-rich soil. Significant harm to humans resulted from the accident at the Chernobyl plant, which lacked a containment structure.

292. Nuclear Energy Institute. KEEPING IT SAFE. 1996, http://www.nei.org/main/safe/scte.htm)

Keeping It Safe

How safe are nuclear power plants? They do, after all, contain radioactive materials—the stuff left over after uranium atoms are split to create heat to generate electricity. To prevent any harmful release of **radiation** into the environment, U.S. nuclear power plants follow a philosophy of "safety in depth." This philosophy begins with the **design** of nuclear plants, which takes advantage of the laws of nature and incorporates backup safety systems and protective barriers, including a thick steel and concrete containment building. For example, it is physically impossible for a nuclear power plant to explode because of the low concentration of fissionable uranium 235 in the fuel. The philosophy is carried forward in plant **operations** through extensive and continuous **training** of plant personnel.

An additional level of safety is provided through **plant regulation** by the Nuclear Regulatory Commission, an independent federal agency. As a result of this safety-in-depth philosophy, even the 1979 **Three Mile Island** accident—the only major accident in the history of U.S. commercial nuclear energy—caused no injuries, deaths or discernible direct health effects to the population of the vicinity of the plant. The 1986 **Chernobyl** accident in the former Soviet Union was the product of a badly flawed plant reactor design which never would have been licensed in the United States.

The nuclear industry exercises the same degree of vigilance for safety in its handling, transportation, and disposal of the relatively small and manageable amounts of high-level waste and low-level waste produced at nuclear power plants.

293. Poong Eli Juhn. IAEA BULLETIN. Vol. 38, No. 1, 1996, (http://www.iaea.or.at/worldatom/inforesource/bulletin) In a number of countries, great emphasis is being placed on the development of advanced nuclear power plant designs. These new generations of nuclear power plants have been, or are being, developed by building upon experience and applying lessons learned from existing plants. Hence, the new, advanced designs are anticipated to become even more safe, economic, and reliable than their predecessors.

The advanced designs generally incorporate improvements of the safety concepts, including, among others, features that will allow operators more time to perform safety actions, and that will provide even more protection against any possible releases of radioactivity to the environment. Great attention is also paid to making new plants simpler to operate, inspect, maintain, and repair, thus increasing their overall reliability and economy.

Advanced designs comprise two sub-categories: evolutionary and developmental designs. The first encompasses direct descendants from predecessors (existing plant designs) that feature improvements and modifications based on feedback of experience and adoption of new technological achievements. they also take into account possible introduction of some new features, e.g., by incorporating passive safety functions. Evolutionary designs are characterized by requiring, at the most, engineering and confirmatory testing prior to commercial development. Developmental designs consist of those that deviate more significantly from existing designs, and that consequently need substantially more testing and verification, probably including also construction of a demonstration plant and/or prototype plant before large-scale commercial deployment.

294. NEI PRESS RELEASES, 15 April 1996, (http://www.nei.org/main/pressrm)

WASHINGTON, D.C., April 15, 1996—U.S. nuclear power plants are exceeding ambitious performance goals for safety and reliability, according to new information released by the Institute of Nuclear Power Operations (INPO).

Based in Atlanta, INPO was established after the Three Mile Island accident in 1979 to promote the highest levels of safety and reliability—to promote excellence—in the operation of nuclear power plants. Since 1983, the industry-sponsored organization has been collecting, monitoring and reporting data on America's nuclear power plants to measure their performance against industry-established goals.

"This year's outstanding performance by U.S. nuclear plants is the result of 15 years of dedication and hard work by the people who run them," said Phillip Bayne, chief executive officer of the Nuclear Energy Institute (NEI). "These professionals have demonstrated a commitment to continually seeking excellence in plant performance, and the proof of their success is in the 1995 numbers."

The World Organization of Nuclear Operators (WANO), the nuclear industry's international safety organization founded after the 1986 accident at the Chernobyl nuclear power plant, adopted these performance indicators in 1989 for use by nuclear plants worldwide.

Nuclear plants with good performance, as measured by such performance indicators, are more reliable have higher margins of safety and are generally recognized as well-managed plants. Utilities use performance indicator data to set both annual and long-term performance goals.

295. NEI PRESS RELEASES, 15 April 1996, (http://www.nei.org/main/pressrm) In 1995, the U.S. nuclear industry met a number of challenging five-year goals set in 1990. Highlights include:

- **Unit Capability Factor**—Expressed as a percentage of maximum electricity a plant is capable of producing limited only by factors within the control of plant management. Last year, U.S. plants again achieved the highest capability factor ever—a median value of 82.6 percent—exceeding the 1995 goal of 80 percent.

- **Unplanned Automatic Shutdowns**—Nuclear plant "scrams," or unplanned automatic shutdowns, have

declined by almost 90 percent since 1980. The 1995 median of 0.9 unplanned shutdowns is better than the 1995 goal for the third consecutive year.
- **Safety System Performance**—This indicator measures the availability of three key plant safety systems. Last year, 94 percent of safety systems met challenging industry goals for availability. The percentage of safety systems meeting industry goals for availability has improved since tracking began in 1989 and has been better than the 1995 goal since 1992.
- **Low-level Solid Radioactive Waste**—Minimizing the volume of tools, rags, clothing and other materials that come in contact with radiation reduces the need for storage, transportation, and disposal. Last year, pressurized water reactors produced a median value of only 30 cubic meters of solid low-level waste—well under the 1995 target of 110 cubic meters. Boiling water reactors produced a median of 107 cubic meters—also well under their 1995 target of 245 cubic meters.

296. Nuclear Energy Institute. THE CHERNOBYL DESIGN.1996,(http://www.nei.org/main/pressrm/special/differ.htm)
Why a Chernobyl-Type Explosion Couldn't Happen in the West
No Western nuclear power plant has a positive void coefficient–the most significant of the design flaws that led to the Chernobyl accident. In addition, Western light water plants have both a thick reactor vessel containing the nuclear fuel, and a massive containment structure around the reactor vessel and the reactor coolant system with its associated piping.

In addition, Western plants are designed to handle a complete rupture of the largest pipe in the coolant system. The Chernobyl plant was designed to accommodate an accident involving the rupture of no more than about four pressure tubes out of a total of approximately 1,660. In the Chernobyl accident, a substantial number of tubes ruptured, which led to the rupture of all of them. As a result, the reactor core was exposed to the environment. Such core exposure could not happen in a Western plant.

297. Nuclear Energy Institute. THE CHERNOBYL DESIGN. 1996, http://www.nei.org/main/pressrm/special/differ.utm) The Chernobyl plant's RBMK design uses graphite in place of water as a moderator to control the nuclear chain reaction. The graphite surrounds vertical pressure tubes that hold the nuclear fuel and the water that will be boiled to steam. Unlike Western-designed nuclear power plants, the RBMK's nuclear chain reaction and power output increase when cooling water is lost. This design flaw–called a "positive void coefficient"—was a major cause of the uncontrollable power surge that led to the Chernobyl accident.

298. James Rhodes. FDCH CONGRESSIONAL TESTIMONY. 5 February 1997, Lexis/Nexis. To be clear to the new members of this committee who may not be familiar with the operation of a commercial nuclear facility, used fuel from these plants consists of solid ceramic-coated uranium pellets encased in zirconium alloy tubes called fuel rods. A typical power plant produces about 20 metric tons of used fuel each year. All the used fuel produced by the nuclear energy industry in more than 30 years of operation—if stacked end to end—would cover an area the size of a football field to a depth of about four yards.

299. Bertram Wolfe. ISSUES IN SCIENCE AND TECHNOLOGY. Summer 1996, p. 758. One commonly cited drawback of nuclear power is that it creates radioactive waste that must be contained for thousands of years. Nuclear waste is a serious concern, but one that can be successfully managed and is less worrisome than the emissions from fossil-fuel plants. Coal, gas, wood, and oil plants emit greenhouse gases and other undesirable materials to the environment. Of course, radioactive waste can represent a serious hazard if it is not properly maintained, but its small volume allows very high expenditures and great care per unit volume. If all the country's high-level nuclear waste from over three decades of plant operations were collected on a football field, it would be only 9 feet deep. Nuclear power plant wastes have been carefully maintained at the plants for decades without harm to the environment or the public. Because high-level waste, compose largely of spent nuclear fuel, remains radioactive for thousands of years, the plan is to seal this waste in sturdy containers and bury it in underground geological structures that have remained stable for millions of years. The feasibility of this approach has been supported by a large number of national and international studies.

300. Ivan Selin. FEDERAL NEWS SERVICE. 16 May 1995, Lexis/Nexis. This improved license renewal option comes at a key time for industry, when the prospect of open competition has forced utilities to face the prospect of large financial unknowns. These financial pressures are leading our reactor licensees to face major licensing decisions that most thought would not have to be made for decades. Among their hard choices are: whether to shut down plants prematurely due to high operating costs or the high costs of needed capital improvements, or whether to consider license renewal in order to spread the cost of capital investments. The NRC must be, and is, ready to respond to this changing situation by working efficiently with industry to process license renewal applications, or to oversee plant decommissionings.

CHAPTER IV
CLIMATE
OUTLINE

I. CLIMATE CRISIS NOW

  A. MUST ACT NOW
1. Long lead times for climate change necessitates gradual actions now. (1–4)
2. Technological incentives now are key. Long lead time for development means we must act. (5–6)
3. Delay increases the costs of the transition. (7)
4. Delay risks runaway warming. Positive feedbacks will snowball after a certain point. (8)
5. Each delay makes the shift harder. (9)
6. Next decade is the key to solving warming. (10)
7. Winners and losers impact action in the future. (11)
8. Ozone proves delay is disastrous. (12)
9. Technologies may fail. Action now is a safer bet. (13)
10. Small long-run impact calculus is irrelevant. Any additional solvency justifies action now. (14)

  B. IPCC MODELS AND STUDIES ARE BEST
1. IPCC on balance most qualified. (15)
2. Methodology of IPCC modes solid. (16)
3. IPCC documents are the most informed sources. (17)
4. IPCC is the best science available. (18)
5. IPCC second assessment was a scientific milestone. (19)
7. General consensus exists on man-made warming. (20)
8. New IPCC reports are extremely certain. (21)
9. Scientists always express uncertainty. They speak in the language of probability. (22)

  C. NEGATIVE STUDIES FLAWED
1. Negatives authors are paid off by the energy industry. (23)
2. Media gives climate skeptics more credibility than justified by their science. (24)
3. Fossil fuel industry massively outstrips media funding to deny climate change. (25)

  D. ADAPTATION AND MIGRATION FAIL
1. Adaptation policy risks irreversible damage. Prevention is essential. (27)
2. Adaptation possible but difficult. (28)
3. Simultaneous disasters from warming overwhelm ability to adapt. (29)
4. Population growth outstrips ability of science and technology to cushion effects. (30)
5. Adaptation impossible. Society is fixed. (31)
6. Small countries cannot adapt. (32)
6. Status quo is not pursuing adaptive measures. (33)
7. Human activity impacts migration of animals. (34)
8. Geographical and man-made constraints impact species migration. (35)
9. Natural barriers impact migration. (36)
10. Increased flood size overwhelms levees. (37)

  E. UNCERTAINTY OF CLIMATE CHANGE DOES NOT DEJUSTIFY ACTION
1. Climate action independently justified. (38)
2. Barriers impact even "no regrets" actions. (39)
3. Regulations will encourage research, greater certainty. (40)
4. Delay in the name of uncertainty increases the chance of irreversible harm. Ozone proves. (41)
5. Uncertainty cuts both ways. Emissions reductions now are prudent. (42)
6. No regrets policy ignores risk of climate change. (43)
7. Uncertainty irrelevant to decision-making. (44)
8. Uncertainty does not deny consensus. (45)
9. Uncertainty irrelevant. Positive feedbacks and alternative advantages justify action. (46)
10. Precautionary principle justifies action. (47)
11. Scientific certainty is increasing. Action now is imperative. (48)
12. Precautionary principle mandates quick action. (49)
13. Even low end probability of impact justifies action. (50)

F. ANSWERS TO COMMONLY ARGUED BENEFITS OF GLOBAL WARMING
1. Warming increases disease outbreaks. (51)
2. Warming causes significant heat deaths. (52)
3. Benefits from warming occur only with stability. Must halt emissions. (53)
4. Rate of change kills benefits of warming. (54)
5. Spastic nature of change magnifies the impact of warming. (55)
6. Continuous change prevents benefits. (56)
7. Rate of change impacts benefits. (57)
8. Even slow warming is disastrous. (58)
9. Slow warming causes monsoons. (59)
10. Warming will reduce rainfall where it is needed the most. (60)
11. Equity concerns impact consideration of warming benefits. (61)
12. Night warming hurts crops. (62)

G. MISCELLANEOUS AFFIRMATIVE WARMING ARGUMENTS
1. Move towards energy efficiency changes social structure. Solves inevitable resource wars. (63)
2. China will cooperate with the U.S. on alternative energy development to reduce warming. (64)
3. Increased use of alternative energy decreases start-up costs. This permits export of technologies and solves global warming. (65)

II. NO WARMING CRISIS

A. STATUS QUO WARMING IS SLOW
1. Observational evidence supports slow warming. (100)
2. Warming will be at low end on GCM scale. (101)
3. Models overestimate rate of warming. (102)
4. Rapid warming increasingly unlikely. (103)
5. Positive feedbacks are massively uncertain. Warming will be slow. (167)
6. Runaway warming cuts both ways. Just as likely to be disastrous to cut CO_2. (168)
7. At least an equal chance that runaway warming effect is negative. (169)

B. MODELS PREDICTING GLOBAL WARMING ARE FLAWED
1. Even new GCMs overestimate warming. (104)
2. Satellite data deny warming. (105)
3. Cloud feedback solves warming. (106)
4. Models are more and more uncertain. Dire predictions unwarranted. (107)
5. Models ignore feedbacks. (108, 141)
6. Urban heat islands invalidate models. (109, 147)
7. Models do not allow for oceans. (110)
8. Northern hemisphere vortex contradiction invalidates GCMs. (111)
9. Models impossible. Soil moisture errors. (112)
10. CFCs prove models are fallible. (113)
11. Models are not supported by observational data. (114, 140)
12. GCMs misanalyze polar warming. (145)
13. GCMs misanalyze storm activity. (146)
14. Tropospheric ozone models uncertain. (148–149)
15. Models inadequate for policy-making. (150–153)

C. SATELLITE DATA DISPROVE WARMING HYPOTHESIS
1. Satellite data deny warming. (115, 125)
2. No warming over natural fluctuations. (116)
3. Satellite measurement is comparatively best source. (117–118)

D. CLOUDS SOLVE WARMING
1. Clouds prevent warming. (118–119)
2. Clouds are a key feedback mechanism. (120)
3. Clouds, not CO_2, are the key element in global warming. (121)
4. Low level cloudiness is increasing. (122)
5. GCMs ignore clouds. (123–124, 142–143)

E. ON BALANCE, NO GLOBAL WARMING
1. No observational evidence. (132)
2. No warming over natural fluctuations. (126)
3. Temperature not rising. (127)
4. No real world proof of warming. (128-129)
5. No warming after heat island corrections. (130)
6. Oceans are cooling. (131)
7. Warming insignificant. Changes in southern hemisphere contradict predictions. (133)
8. No empirical proof distinguishes warming from natural variability. (134)

F. WARMING STUDIES ARE HYPE
1. No observational data. Scientists hype problem to get grants. (135)

2. Scientists and political elites create warming crisis to feed their own interests. (136)
3. Warming crisis created by government. (137)
4. Environmentalists hype warming for political reasons. (138)
5. Media pushes aside negative warming evidence. (139)

G. IPCC SCIENCE IS BAD
1. Leipzig Declaration denies consensus. (154)
2. IPCC artificially erased uncertainty. (155)
3. No adequate review of IPCC. (156)
4. Government support of IPCC twists truth seeking. (157)
5. IPCC conference survey denies consensus. (158)
6. Majority of IPCC scientists disagree with executive summary. (159)
7. No scientific consensus. (160)
8. Gallup poll denies consensus. (161)
9. IPCC ignores clouds. (162)
10. IPCC empirically wrong. (163)
11. IPCC twisted by self-interest. (164)
12. IPCC and other scientists exaggerate problem to get research money. (165–166)

III. IMPOSSIBLE TO PREVENT GLOBAL WARMING

A. ENVIRONMENTAL LINKGES PREVENT SOLVENCY (170, 173)

B. POPULATION GROWTH, DEFORESTATION, ALTERNATE GASES IMPACT SOLVENCY (171, 184)

C. MOMENTUM PREVENTS SOLVENCY (172)

D. DESERTIFICATION CAUSES WARMING (174)

E. CO_2 WARMING LINKS EMPIRICALLY DENIED (175)

F. CO_2 IMPACTS HAVE DIMINISHING SIGNIFICANCE AS RATES INCREASE (176)

G. NON-CO_2 GASES WILL SOON BE THE KEY TO WARMING (177, 199)

H. URBAN HEAT ISLANDS ARE THE KEY TO WARMING (178–179)

I. NATURAL VARIABILITY SWAMPS TEMPERATURE EFFECTS OF CO_2 (180)

J. VOLCANOES SOLVE CO_2 SOURCE (181)

K. LUNAR CYCLE KEY TO WARMING (182)

L. WATER VAPOR IS KEY TO GREENHOUSE GAS (183)

M. POPULATION GROWTH IS KEY TO WARMING (185–186)

N. WARMING OCCURRED BEFORE CO_2 (187)

O. OBSERVATIONAL EVIDENCE DENIES WARMING/CO_2 LINK (188)

P. WARMING EFFECTS NOT A CORRECT TIME TO BE HUMAN INFLUENCED (189)

Q. TROPICAL RECORD SHOWS SLOW WARMING (190)

R. CO_2 IMPACTS ON CLIMATE HAVE DIMINISHING RETURNS AS LEVELS INCREASE (191)

S. SUNSPOTS ARE KEY TO WARMING (192–193)

T. LUNAR ACTIVITYIS KEY TO CLIMATE (194)

U. CO_2 IS A MINOR SOURCE OF WARMING (195–197)

V. WATER VAPOR IS KEY TO WARMING (198)

W. LONG LIFETIME OF GASES IMPACTS QUICK SOLVENCY (200)

X. INERTIA ENSURES WARMING (201)

Y. OCEANS ARE THE KEY (202)

IV. WARMING GOOD

A. ADAPTATION SOLVES ADVERSE CONSEQUENCES
1. Year to year variations swamp warming effects. (203)
2. Rate and intensity irrelevant. Technology and adaptation solves impacts. (204)
3. Human adaptation solves the impact. (212–213)
4. Ice age proves that humans adapt. (214)
5. Slow warming makes adaptation the best alternative. (215)
6. Slow warming on balance beneficial. (216–217)
7. Warming will be slow. (218–221)
8. Must provide adequate resources for adaptation. (222)
9. Must prioritize limited resources and concentrate on adaptation. (223)

B. NIGHT WARMING GOOD
1. Warming will occur only over nighttime and winter. (205)
2. Winter/night warming provides a more benign climate. (206)

C. WORST CASE PLANNING BAD
1. Worst case assessments increase polarization and prevent real action. (207)
2. Action now is premature. Waiting will provide the best benefits. (207)
3. Worst case assessments increase polarization and prevent real action. (288)
4. Doomsaying subverts rational public progress. Most successful alternatives will be overlooked. (289)
5. Warming hype diverts funds from true environmental problems. (290)
6. Extreme positions cause polarization and distrust. (291)
7. Doomsaying and harsh action impacts international action. Cautious course is best. (292)

D. ACTION NOW CAUSES ECONOMIC CALAMITY
1. Warming action decreases economic resources. (208)
2. Must save economic resources to allow realistic adaptation in the future. (209)

E. REGIONAL ENVIRONMENTS
1. Climate action impedes action on other environmental issues. (210)
2. Regional impacts more pressing. They occur now. (211)

F. ICE CAPS ANSWERS
1. Ice caps thickening. (224)
2. Ice caps theory is ridiculous and a long way off. (225)
3. No impact to arctic melting: it just divides into ice cubes. (226)
4. No sea level rise. (227)

G. SPECIES ANSWERS
1. Nature is in constant flux. Warming effects are insignificant. (228)
2. Migration solves. (229)
3. Ice age denies climate/extinction link. (230)
4. Crabs prove beneficieal species effect. (231)
5. CO_2 causes speciation. (232)
6. Empirical evidence supports species turns. (233)
7. CO_2 increases biodiversity. (234)
8. CO_2 increases plants. (235)
9. Earthworms prove beneficial effects. (236)

H. DISEASE ANSWERS
1. World health improving. (237)
2. Tropics are cooling. No disease spread. (238)
3. Disease spread historically false. (239)
4. Sanitation and infrastructure are key to solving warming. (240)
5. Technology solves disease spread. (241)

I. AGRICULTURE BENEFITS FROM GLOBAL WARMING
1. CO_2 is the savior of world food production. (243)
2. CO_2 increases crop yields. (245)
3. Thousand studies confirm positive effects of CO_2. (246)
4. High quality scientists support CO_2 positive agriculture effects (247)
5. CO_2 offsets crop losses due to warming. (248)
6. CO_2 increases root growth, solving water stress. (249)
7. Slow warming ensures plants will overcome weeds. (250)
8. Rapid CO_2-induced growth reduces pest damage. (251)
9. Consensus of scientists see beneficial CO_2 effects. (252)
10. Media suppresses CO_2 agricultural benefits. (253)
11. Rising population requires increased agriculture productivity. (242)
12. Increasing output key to solving world starvation. (244)

J. RESTRICTING CO_2 HAS DISASTROUS ECONOMIC CONSEQUENCES
1. Unexpected imposition of new restrictions on CO_2 will produce an economic shock. (254)
2. Decreased energy use causes a world depression. (255)
3. Greenhouse reductions cause U.S. imports and crushes LDC economies. (256)
4. Energy reductions cause an economic disaster. (257)
5. Greenhouse control causes worldwide poverty. (258)
6. Drastic regulations devastate economic growth. (259)
7. Oil shocks prove economic harms of greenhouse cuts. (260)
8. Loss of thousands of jobs outweighs value of climate insurance. (261)
9. Warming solvency is expensive. (262–263)

10. Lower consumer demand for electricity impacts productivity growth and capital formation. (265)
11. International greenhouse plans decrease U.S. competitiveness due to higher population growth. (266)
12. Greenhouse reductions put U.S. at a competitive disadvantage. (267)
13. Stricter domestic regulations decrease international competitiveness. (268–269)
14. More stringent reductions increase cost. (270)
15. Quick reductions cause depression. (271)

K. AID FATIGUE DISADVANTAGE
1. Developing nations will use climate negotiations to gain substantial compensation. (276)
2. Developing nations will use climate issue to wrest concessions from North (277)
3. Massive U.S. aid is necessary to solve warming. (278)
4. Treaty aid is major political issue in the U.S. (279–280)
5. Start-up costs swamp savings. Extra aid is necessary to set up new institutions. (281)

L. AUTOMOBILES
1. Emissions controls are unpopular. (293)
2. Government partnership with big business overlooks more innovative small business. (294)
3. Tradition impacts solvency. (295)
4. Renewables are not feasible with transportation. (296)
5. Methanol is expensive and inefficient. (297)
6. Methanol is an environmental hazard. (298)
7. CNG diverts focus from true solutions. (299)
8. Status quo is too fast. Consumers backlash versus shift. (300)

M. MISCELLANEOUS ARGUMENTS CONCERNING THE ENERGY/CLIMATE LINK
1. Energy efficiency underfunded due to tight budget climate. Status quo is a great start. (272)
2. Enterprises won't buy new energy technologies. (273)
3. Consumer patterns impact use of new technologies. (274)
4. Consumer habit impacts the use of new technologies. (275)
5. Decreased U.S. energy demand increases LDC consumption and ensures MNC shift. (282)
6. Decreased U.S. fossil fuel consumption decreases prices, increasing world consumption. (283)
7. LDC non-participation ensures MNC flight. Results in net increase in CO_2. (284)
8. Increased electricity costs cause a shift to dirtier gasoline engines. (285)
9. Technological solutions only shift pollution to another medium. (286)
10. Exclusive focus on CO_2 decreases flexibility and risks increasing greenhouse emissions. (287)

CHAPTER IV

EVIDENCE

1. Alan Miller and Mack McFarland. FORUM FOR APPLIED RESEARCH AND PUBLIC POLICY. Summer 1996, p. 61. The infrastructure associated with energy use requires longer lead times. For example cars typically remain on the road for a decade, refrigerators and other equipment remains in use for two decades, and buildings remain standing for several decades. This implies a time lag between implementation of policies and realization of reductions in greenhouse-gas emissions. The costs of rapid changes in infrastructure associated with transportation systems and power generation, two significant sources of carbon dioxide, would be enormous.

This time lag, along with the long-lived nature of greenhouse gases and the slow response of the climate system, provides a case for gradually changing the infrastructure in advance of potential adverse effects that may precipitate rapid and expensive actions.

2. Alan Miller and Mack McFarland. FORUM FOR APPLIED RESEARCH AND PUBLIC POLICY. Summer 1996, p. 58. CFCs and some other ozone-depleting substances have atmospheric lifetimes of about 100 years. Such long lifetimes lead to a slow increase in atmospheric concentrations after emissions start and a slow decrease of concentrations after emissions cease. Thus, the effects of ozone depletion could persist for many decades.

Similarly, carbon dioxide, the major contributor to global climate change, has a long atmospheric residence time. Experts estimate emissions reductions of more than 60 percent would be required to stabilize concentrations at today's level. As with CFCs, if anthropogenic emissions were reduced significantly or even phased out entirely, carbon-dioxide concentrations would decrease slowly.

Due primarily to the heat-absorbing capacity of the oceans, the climate system does not respond quickly to changes caused by increases or decreases in concentrations of greenhouse gases. In contrast, because of the rapid rate of production and destruction of stratospheric ozone, changes in ozone levels will closely follow changes in the concentrations of ozone-depleting substances.

The long lifetime of carbon dioxide and the prolonged response of the climate system means the climate system would recover more slowly than the ozone layer after emissions were curbed. In short, once any negative impacts of climate change become apparent, they will persist for decades—or even centuries.

3. Thomas C. Roberts. GLOBAL WARMING— PART 2. Y4. En 2/3: 103-92, 1993, p. 80. At the other extreme if technological evolution drives the emissions reduction, implementing a policy today may give technology a head start that will enable it to be more advanced than it would otherwise be for many decades. Thus, cutting emissions this decade could lead to lower emissions in all succeeding decades than what would occur if a policy was delayed. In such a case, global warming by the year 2100 could prove to be very sensitive to the timing of emissions reductions. A moderate sensitivity to timing would occur to the extent that cutting emissions now leads us to make permanent changes that we would have otherwise made later, but not to a continual evolution in our ability to conserve.

4. Rosina M. Bierbaum. CCAP AND ASSESSMENT. Y4. Sci 2:103/111, 1993, p. 1. Uncertainty is often cited as the reason for not investing in research on adaptation/mitigation for climate change. However, uncertainty is in fact why we must prepare. With some forethought, the Nation can both position itself better to cope with the broad range of impacts possible under climate change and protect itself against some particularly costly future outcomes. Delay in responding may, in fact, leave the Nation poorly prepared to deal with the changes that do occur and may increase the possibility of impacts that are irreversible and very costly. As Congress is aware, there can be considerable delays between the time policy options are identified and the time those policies are legislated and then, ultimately, implemented.

5. Christopher Flavin and Odil Tunali. CLIMATE OF HOPE: NEW STRATEGIES FOR STABILIZING THE WORLD'S ATMOSPHERE. Ed. Jane A. Peterson, June 1996, p. 53. One critical priority is accelerating the commercial development of new, low-carbon energy technologies since it takes time for new systems to displace a significant portion of the devices currently in use. To achieve any reasonably safe emissions trajectory, the market for fuel cells, photovoltaics, electric vehicles, and other critical technologies will need to be accelerated soon. The highest priorities are the facilities that last the longest: a large office building or coal-fired power plant built today could be contributing to global warming a half century from now. During the next three decades, most of today's power plants, refineries, and factories will be replaced, while automobiles and some appliances will turn over two to three times. These turnovers are opportunities to lower greenhouse gas emissions that should not be missed.

6. Christopher Flavin and Odil Tunali. CLIMATE OF HOPE: NEW STRATEGIES FOR STABILIZING THE WORLD'S ATMOSPHERE. Ed. Jane A. Peterson, June 1996, p. 10. The time for action is at hand. With the global economy booming, particularly in developing nations, industries and governments are currently installing large numbers of factories, power plants, roads,

and buildings that could contribute to greenhouse gas emissions for many decades to come. Unless the world soon shifts to a new and more sustainable path of energy development, later efforts to stabilize the climate will be far more difficult and expensive—and too late to prevent some of the most wrenching damage. Industrial countries, which unwittingly created the climate problem in the first place and control the technologies that ultimately can solve it, have a clear responsibility to lead the way forward.

7. Christopher Flavin and Odil Tunali. CLIMATE OF HOPE: NEW STRATEGIES FOR STABILIZING THE WORLD'S ATMOSPHERE. Ed. Jane A. Peterson, June 1996, p. 52. It is correct, indeed obvious, that there are several different emission trajectories that can get the world to a low-carbon energy system; but the more important question is how to minimize the risks we now face. Since scientists do not yet know exactly what level of greenhouse gas concentrations we and future generations can live with, there is a real danger of overshooting critical limits. Delay could be dangerous since it will inevitably take time to prepare the groundwork for the kind of rapid change in energy and transport systems that may soon be needed. And as Figure 7 indicates, the longer action is delayed, the sharper and more disruptive the eventual shift in emission trends will have to be.

8. David A. Wirth and Daniel A. Lashof. GLOBAL WARMING FORUM: A SCIENTIFIC, ECONOMIC AND LEGAL OVERVIEW. 1993, p. 512. The long atmospheric lifetimes of GHGs necessitate major reductions in emissions from current levels. Even after these reductions, atmospheric concentrations of GHGs will fall only very slowly. The heat capacity of the oceans will further delay the climatic response by decades more. Indeed, temperatures might continue to rise for many years even after the elimination of all anthropogenic GHG emissions. Moreover, the required changes in utility and transportation infrastructure will take years or even decades to accomplish following decisions to embark on a GHG emissions reduction program. The likelihood of positive feedbacks, through which the warming itself further accelerates GHG emissions rates, raises the further frightening possibility that human efforts to reduce emissions could be overwhelmed by natural processes such as the release of methane from Arctic regions. Once such a crisis has been reached, it will be too late to act.

9. Justin Lancaster. A GLOBAL WARMING FORUM: SCIENTIFIC, ECONOMIC AND LEGAL OVERVIEW. Ed. R. Geyer, 1993, p. 535. First, as a global society, eventually we must grip the fact that our luxury fuels, i.e., oil and gas, which represent a billion-year heritage of ancestral solar energy storage, are finite resources. A shift away from fossil fuels may become more difficult each decade the shift is postponed, because we will be forced to burn increasing amounts of coal until alternative sources fill the gap. We will need a good portion of these fossil energies just to build the alternative energy technologies and infrastructure, i.e., nuclear, solar, wind, tidal, and biomass. The amount of our shrinking, finite resource that is needed for a transition to a long-term energy supply has not been calculated. Future generations will shake their heads in disbelief at the ease and rapidity with which we will have spent our fossil treasure. We should take every step we can to conserve it.

10. Christopher Flavin and Odil Tunali. CLIMATE OF HOPE: NEW STRATEGIES FOR STABILIZING THE WORLD'S ATMOSPHERE. Ed. Jane A. Peterson, June 1996, p. 45. The next few decades may be a particularly dangerous period. Continuing rapid increases in carbon dioxide and methane concentrations, combined with a leveling off or even a decline in sulfate aerosols—as a result of stringing sulfur emission standards enacted in industrial countries in the early 1990s—will tend to accelerate the pace of charcoal. Consequently, action to limit emissions of the shorter-life greenhouse gases—mainly methane and HCFCs—is essential during this period, along with strengthened efforts to lighten CO_2 emissions.

11. INFORMATION UNIT ON CLIMATE CHANGE. May 1, 1993, UNEP, http://www. unep.ch/iucc/fs107.html. **Climate change poses an unprecedented challenge to human society.** Not only is the challenge unique and of enormous scope, but tackling it will require a political consensus and determination never before seen at the global level. What's more, the predicted impacts of climate change may lead to economic and political instability, making it increasingly difficult for countries to respond to these impacts. At the same time, growing social and economic burdens may divert attention away from the need to address the underlying *causes* of climate change as well.
Climate change would probably produce "winners" and "losers." Winning may be relative, in the sense of losing less than others, but it may also be absolute. Nations at higher latitudes may actually benefit from a warmer climate, while a northward shift in the precipitation belts associated with the Intertropical Convergence Zone could benefit at least some of the Sahelian countries. Winners and losers may be entire continents, countries, or social groups. Some losers may have effective means for protecting themselves, but others would not. The resulting cleavages between winners and losers would make concerted action on climate change even more difficult.

12. David A. Wirth and Daniel A. Lashof. GLOBAL WARMING FORUM: A SCIENTIFIC, ECONOMIC AND LEGAL OVERVIEW. 1993, p. 523. Unfortunately, the U.S., a leader on the stratospheric ozone issue, has dragged its feet on the global warming agreement, Indeed, the moniker "no regrets" adopted by the Bush Administration to describe its current policy demon-

strates the inherently flawed bias of that position. The only "regrets" contemplated by the Administration's policy are those that result from environmental protection measures that subsequently prove unnecessary. The costs of uncertainty and delay in the event of unanticipated catastrophes like the ozone hole or major losses of stratospheric ozone over North America—or even of widely-accepted, if uncertain, predictions of climate disruption—are dismissed altogether.

"Delays have dangerous ends," wrote Shakespeare. This prudent advice is nowhere more relevant than for global warming. Procrastination today will cost dearly—perhaps not tomorrow, but certainly for the tomorrows of our children.

13. Dave Alden. AUSTRALIAN ECONOMIC REVIEW. First Quarter, 1995, p. 130. It may be argued that the best option for taking the heat out of global warming, to either prevent or reduce the likely level of global warming, is to allow technology to solve the problem after it has occurred. Technology has saved those in industrialised countries, in the past from natural and human-made limitations to our endeavours, so surely it will in the future? So why not adopt a research and development program with this aim? This is not necessarily the answer for three reasons. First, a new technology to do this may not come along. Second, it may come along, but not quickly enough. Third, it may be more expensive to clean up greenhouse gases from the atmosphere, such as shooting these gases into space, than the costs associated with preventing the build-up of these gases in the first place.

14. Thomas C. Roberts. GLOBAL WARMING—PART 2. Y4, En 2/3: 103-92, 1993, p. 82. Published studies have not thoroughly investigated the sensitivity of benefits to the timing of emissions reductions. A 1982 report in *Nature* (Schlesinger et al.) seemed to suggest that delaying a policy for 10 years might be rational; emissions expected during the next 10 years account for a small fraction of the emissions expected over the next century, so delay would only increase the total warming a few percent by the year 2100. EPA's view, however, is that it is not necessarily rational to delay action simply because doing so has a small impact in the very long run. Otherwise, we would rarely implement policies to address long-run problems, because it is almost always true that taking action today is only slightly better than taking action tomorrow (e.g., reducing federal deficits, quitting smoking, etc.)

15. Henry Lee. SHAPING NATIONAL RESPONSES TO CLIMATE CHANGE: A POST RIO GUIDE. Ed. Henry Lee, 1995, p. 2. Our point of departure is the report of the scientific working group of the Intergovernmental Panel on Climate Change (IPCC), which represented the best judgment of the world's scientific community at the time this book was written. The thrust of the report is that over the next hundred years, the global mean temperature is likely to increase at rates faster than the world has experienced in the past and that this will affect both climate and climate-dependent ecosystems in ways that are uncertain but, potentially, very harmful. Some reputable scientists disagree with these findings, and future research may prove them to be correct. Enormous scientific uncertainty surrounds this issue, and this uncertainty is not likely to evaporate any time soon.

16. 1992 SUPPLEMENT, GLOBAL WARMING—PART 2. Y4.En 2/3: 103-92, 1992, p. 315. New IPCC 1992 emissions scenarios (IS92a-f; see Section 4) have been derived in the light of new information and international agreements. In order to provide an initial assessment of the effect of the new scenarios, the change in the surface temperature has been estimated with the simple climate model used in the IPCC 1990 report which has been calibrated against the more comprehensive coupled ocean-atmosphere models (see box on models). These calculations include, in the same way as did the 1990 calculations, the direct radiative forcing effects of all the greenhouse gases included in the scenarios. The effect of stratospheric ozone depletion and of sulphate aerosols have not been included, which again parallels the 1990 calculations. The accompanying diagrams show (i) the temporal evolution of surface temperature for IS92a, assuming the high, "best estimate" and low climate sensitivities (4.5, 2.5 and 1.5°C), and (ii) the temperature changes for the six 1992 IPCC scenarios and the 1990 Scenario A, assuming the "best estimate" of climate sensitivity (see earlier box on "What tools do we use. . . .?" for the definition of climate sensitivity). [ellipses in original]

17. J. D. Mahlman, D. Albritton, and R. T. Watson. CCAP AND ASSESSMENT. Y4.Sci 2:103/111, 1993, p. 77. While much uncertainty remains in our ability to determine in detail the timing, magnitude and regional patterns of climate change, it is equally true that much is already known scientifically about the greenhouse effect and how it is likely to play out under various scenarios. The scientific uncertainties have been used by some to fuel the current emotional intensity of the public debates on this problem.

We strongly recommend that the 1990 Intergovernmental Panel on Climate Change (IPCC) scientific assessment and its supplement in 1992 be used as the current-best foundation for the scientific understanding of the greenhouse-warming issue. The report preparation for these documents involved nearly two years and nearly 200 scientists from scores of countries, followed by a comparable number of scientific reviewers. These IPCC documents are generally regarded to be the most authoritative and widely accepted statements on climate change that have ever been made by the scientific community.

18. Ned Nanda. DENVER POST. 4 August 1996, p. D-4. In Geneva, Burt Bolan, a Swedish scientist and chairman of the IPCC, responded: "This is the work of 1,000 scientists, reviewed, revised and reviewed again. There is no compromise in one or other direction. . . . We utterly reject accusations and allegations made against us. We have completely and carefully done our work; it is the best science on climate the world has to offer." [ellipses in original]

19. David Gardiner. FEDERAL DOCUMENT CLEARING HOUSE. 19 June 1996. Three prominent factors have intensified the Administration's recent focus on the problem of human-induced climate change, which has also been labeled "global warming" and the "greenhouse effect." First, just this past December, the most comprehensive assessment of climate change ever undertaken was completed and released under the auspices of the Intergovernmental Panel on Climate Change (IPCC), a scientific body drawing on the work of 2,500 scientists that was set up by the United Nations to advise world leaders on this critical issue facing our civilization. The IPCC's first assessment in 1990 provided the basis for negotiation of the 1992 Framework Convention on Climate Change, which was signed by over 150 nations at the Earth Summit in Rio de Janeiro. The IPCC's Second Assessment Report constituted a major scientific milestone that advanced the scientific understanding of climate change and reinforced the urgency of further action.

20. Kenan Malik. THE SPECTATOR. 11 November 1995, p. 24. As we know, there still remains a debate about the exact nature, cause and consequence of global warming; the effect of some industrial pollutants, such as sulphur dioxide aerosols, for example, is to cool the earth, not warm it. There is, however, a general consensus now that the principal cause is the increase in the amount of greenhouse gases, in particular carbon dioxide, in the atmosphere as the result of the burning of fossil fuels. Greenhouse gases form themselves into a kind of glass roof over the earth, soaking up the sun's radiation, just as a greenhouse traps the sun's rays. To create a greenhouse gasses warm up the earth by absorbing sunlight.

21. Ross Gelbspan. HARPER'S MAGAZINE. December, 1995, p. 31. The question became more pointed in September, when the 2,500 climate scientists serving on the Intergovernmental Panel on Climate Change issued a new statement on the prospect of forthcoming catastrophe. Never before had the IPCC (called into existence in 1988) come to so unambiguous a conclusion. Always in past years there had been people saying that we didn't yet know enough, or that the evidence was problematical, or our system of computer simulation was subject to too many uncertainties. Not this year. The panel flatly announced that the earth had entered a period of climatic instability likely to cause "widespread economic, social and environmental dislocation over the next century." The continuing emission of greenhouse gases would create protracted, crop-destroying droughts in continental interiors, a host of new and recurring diseases, hurricanes of extraordinary malevolence, and rising sea levels that could inundate island nations and low-tying coastal rims on the continents.

22. Ross Gelbspan. HARPER'S MAGAZINE. December, 1995, p. 31. I came across the report in *The New York Times* during the same week that the island of St. Thomas was blasted to shambles by one of thirteen hurricanes that roiled the Caribbean this fall. Scientists speak the language of probability. They prefer to avoid making statements that cannot be further corrected, reinterpreted, modified, or proven wrong. If its September announcement was uncharacteristically bold, possibly it was because the IPCC scientists understood that they were addressing their remarks to people profoundly unwilling to hear what they had to say.

23. Ross Gelbspan. HARPER'S MAGAZINE. December, 1995, p. 31. But while the skeptics portray themselves as besieged truth-seekers fending off irresponsible environmental doomsayers, their testimony in St. Paul and elsewhere revealed the source and scope of their funding for the first time. Michaels has received more than $115,000 over the last four years from coal and energy interests. *World Climate Review*, a quarterly he founded that routinely debunks climate concerns, was funded by Western Fuels. Over the last six years, either alone or with colleagues, Balling has received more than $200,000 from coal and oil interests in Great Britain, Germany, and elsewhere. Balling (along with Sherwood Idso) has also taken money from Cyprus Minerals, a mining company that has been a major funder of People for the West—a militantly anti-environmental "Wise Use" group. Lindzen, for his part, charges oil and coal interests $2,500 a day for his consulting services; his 1991 trip to testify before a Senate committee was paid for by Western Fuels, and a speech he wrote, entitled "Global Warming: the Origin and Nature of Alleged Scientific Consensus," was underwritten by OPEC. Singer, who last winter proposed a $95,000 publicity project to "stem the tide towards ever more onerous controls on energy use," has received consulting fees from Exxon, Shell, Unocal, ARCO, and Sun Oil, and has warned them that they face the same threat as the chemical firms that produced chlorofluorocarbons (CFCs), a class of chemicals found to be depleting atmospheric ozone. "It took only five years to go from . . . a simple freeze of production of CFCs," Singer has written," . . . to the 1992 decision of a complete production phase-out all on the basis of quite insubstantial science." (2)

The skeptics assert flatly that their science is untainted by funding. Nevertheless, in this persistent and well-funded campaign of denial they have become interchangeable ornaments on the hood of a high-powered engine of

disinformation. Their dissenting opinions are amplified beyond all proportion through the media while the concerns of the dominant majority of the world's scientific establishment are marginalized. (3) By keeping the discussion focused on whether there is a problem in the first place, they have effectively silenced the debate over what to do about it. [ellipses in original]

24. Christopher Flavin and Odil Tunali. CLIMATE OF HOPE: NEW STRATEGIES FOR STABILIZING THE WORLD'S ATMOSPHERE. Ed. Jane A. Peterson, June 1996, p. 16. These conclusions have gained wide acceptance among scientists and policymakers, but have failed to persuade so-called climate skeptics, who argue that computer-generated models are too flimsy to be predictive, that the temperature record shows a slower rate of warming than the models suggest, or that "negative feedbacks" will protect us from climate change. Scientists such as Patrick Michaels of the University of Virginia, Richard Lindzen of the Massachusetts Institute of Technology, and Robert Balling of Arizona State University have shifted frequently from one argument to another in an effort to minimize the risks of climate change. Many of their arguments are hardly reflected in the scientific literature, and their credibility is further undermined by the fact that they have been heavily funded by oil and coal industries, and even by the government of Kuwait. Although their arguments are not regarded as plausible by most scientists, the skeptics have received considerable media coverage, partly because of an attempt to create "balance," which has confused the public about the state of climate science and helped slow international climate negotiations.

25. Ross Gelbspan. HARPER'S MAGAZINE. December, 1995, p. 31. Capital keeps its nose to the wind. The people who run the world's oil and coal companies know that the march of science, and of political action, may be slowed by disinformation. In the last year and a half, one of the leading oil industry public relations outlets, the Global Climate Coalition, has spent more than a million dollars to downplay the threat of climate change. It expects to spend another $850,000 on the issue next year. Similarly, the National Coal Association spent more than $700,000 on the global climate issue in 1992 and 1993. In 1993 alone, the American Petroleum Institute, just one of fifty-four industry members of the GCC, paid $1.8 million to the public relations firm of Burson-Marsteller partly in an effort to defeat a proposed tax on fossil fuels. For perspective, this is only slightly less than the combined yearly expenditures on global warming of the five major environmental groups that focus on climate issues—about $2.1 million, according to officials of the Environmental Defense Fund, the Natural Resources Defense Council, the Sierra Club, the Union of Concerned Scientists, and the World Wildlife Fund.

26. Marit P. Assenza. FORUM FOR APPLIED RESEARCH AND PUBLIC POLICY. Summer 1996, p. 82. Adaptive policies may seem an alluring option because they postpone the costs of taking action, and because more scientific and economic knowledge could be gathered in the meantime. Most experts agree, however, that policy makers should resist the temptation to depend exclusively on adaptive policies.

Such policies are risky because the cumulative effects of greenhouse gases may cause irreversible damages. Moreover, society may not adjust to a warmer climate as well as some observers theorize; thus, the associated costs may be higher.

A society that is risk averse wants to protect itself against the possibility of incurring losses caused by climate change. Therefore, it would be willing to pay an insurance premium—in the form of investments in the preventive policies—to reduce the change of future damages caused by climate change.

Furthermore, if society adheres to the principle of intergenerational equity, as expressed in the concept of sustainable development, preventive policies would be necessary to avoid the risk of shifting the burden of a warmer climate to future generations.

27. William J. Taylor III. GLOBAL WARMING—PART 2. Y4. En2/3: 103-93, 1993, p. 107. The National Academy of Sciences study, "Policy Implications of Greenhouse Warming," discussed the adaptive capacities of both manmade and natural systems. These systems include agriculture, forests, grasslands, natural landscapes, and marine and coastal environments. Energy, industrial production, agriculture, and human health have demonstrated adaptability to different climate conditions. The possible effects of climate change on sea-level and water resources could pose significant adaptation challenges; yet the NAS study ranks the potential adaptability of human activities and nature, concluding that energy, industry and health have low sensitivity, but natural landscapes and marine ecosystems would be more sensitive to climate change. All other activities would possess intermediate sensitivity, with adaptation measures being available at some cost.

28. Christopher Flavin and Odil Tunali. CLIMATE OF HOPE: NEW STRATEGIES FOR STABILIZING THE WORLD'S ATMOSPHERE. Ed. Jane A. Peterson, June 1996, p. 21. In an age when many people live in air-conditioned homes and eat fresh food grown thousands of kilometers away, it is easy to ignore our dependence on the climate. Most cities are located near adequate water supplies, and their nutritional and material needs are met via agriculture, forestry, and fishery systems that require particular temperature, rainfall, and humidity levels. We can cope with isolated droughts, storms, or floods by bringing in relief supplies, but the widespread,

simultaneous disruptions that may result from greenhouse warming would be unmanageable.

29. Robert G. Feagle, GLOBAL ENVIRONMENTAL CHANGE: INTERACTIONS OF SCIENCE, POLICY, AND POLITICS IN THE UNITED STATES. 1994, p. 221. A joint statement of the U.S. National Academy of Sciences and the Royal Society of London, issued in 1992, began with the words, "World population is growing at the unprecedented rate of almost 100 million people every year, and human activities are producing major changes in the global environment. If current predictions of population growth prove accurate and patterns of human activity on the planet remain unchanged, science and technology may not be able to prevent either irreversible degradation of the environment or continued poverty for much of the world." These two prestigious science organizations had never before issued a joint statement, so it merits more than ordinary attention. The statement reflects the seriousness of environmental threats as well as the uncertainties inherent in projections into the future. It reflects also the recognition that more than science and technology will be needed to avoid a slide into an intolerable future.

30. Sylvan Wittwer. FOOD, CLIMATE AND CO_2: THE GLOBAL ENVIRONMENT AND WORLD FOOD PRODUCTION. 1995, p. 15. The difference today is that the human economy has become more vulnerable to climate variations. For centuries, pastoral nomads such as the Tuareg (Berber) and Fudani endured each Sahelian drought by migration to better grasslands with their flocks and herds, and in numbers of livestock. By endurance, Indian and Chinese peasants somehow survived countless failures of the monsoon. In North America, there was an adaptation to the ferocious winters of the late nineteenth century.

Today, human numbers have increased, especially in areas where food supplies are marginal. Old social and economic customs, notably nomadism, have largely disappeared. This is especially true in advanced Western societies, where there is the rising cost of energy, and a huge increase in the consumption of water. Demands for greater comfort, such as air-conditioning, are evidence of an increased sensitivity to climate anomalies (Hare, 1980).

31. Robert G. Feagle. GLOBAL ENVIRONMENTAL CHANGE: INTERACTIONS OF SCIENCE, POLICY, AND POLITICS IN THE UNITED STATES. 1994, p. 70. Effects of weather and climate on agriculture have been studied extensively, and there is a large literature on how weather and climate affect the growth of specific crops and the spread of pests and pathogens, which have a secondary effect on agriculture. In general, studies have shown that agriculture is adaptable to changes in climate elements, though such changes as the droughts in the U.S. midwest in the 1930s and the USSR in the early 1970s can be devastating to the regions directly affected and damaging to their national economies. But long-term changes of climate probably can be coped with by altering farming practices and developing new strains of seeds. Adaptation should be relatively easy for large, technologically advanced countries whose agricultural regions extend over a variety of climate zones, whereas smaller countries or those that lack sophisticated agricultural research and development could find adaptation difficult.

32. Thomas C. Roberts. GLOBAL WARMING—PART 2. Y4, En 2/3: 103-92, 1993, p. 80. Numerous studies by EPA and others have analyzed policies that would minimize adverse impacts of climate change. But the federal government is not currently implementing such policies. A few state governments have: For example, the State of Maine has developed a policy to minimize the impact of sea level rise on sand dunes and other coastal ecosystems.

33. Thomas C. Roberts. GLOBAL WARMING—PART 2. Y4, En 2/3: 103-92, 1993, p. 80. Terrestrial Ecosystems. Human activities do not appear to be substantially diminishing the ability of forests to adapt to climate change. In urban areas, air pollution may increase vulnerability; the decline of the passenger pigeon will make it more difficult for beech trees to migrate north; on the other hand, people can transport seeds into cooler regions.

On the other hand, human activities will impair the ability of terrestrial animals to migrate to cooler regions. For some species, highways and development present a virtually impenetrable barrier to migration. Moreover, the areas that would otherwise become suitable habitat as climate warms may already be developed.

34. Vera Noest, Eddy van der Maarel, and Frank van der Meulen. CLIMATE CHANGE: IMPACT ON COASTAL HABITATION. Ed. Doeke Eisma. 1995, p. 201. The geographical distribution of vegetation types on a global scale is mainly defined by climatic variables. An obvious response to a temperature rise would be a general migration of plant species, including dominants, to higher latitudes and/or altitudes. However, past migration rates, especially of trees, have been much slower than the expected shift of isotherms (100 to 200 km poleward per degree of warming). Inadequate seed dispersal, a lag in soil development, and competition from persisting vegetation would force species to adapt to a different climatic regime. Dispersal rates and adaptive abilities are extremely variable between species. Natural or man-made barriers to dispersal might cause local or even global extinction of some species (polar, montane, and island communities; isolated nature reserves), leading to a decrease in global species diversity.

35. Paul Brown. THE GUARDIAN. 5 December 1995, p. 10. Forests and plants suited to cooler climes could, in theory, migrate north over time as the once frozen land became more temperate. The southern flank of forests would die as the atmosphere became too hot and dry. But this takes no account of natural barriers such as mountain ranges or seas which might trap species and force them into extinction. Even without these problems, long-lived species of trees might not be able to breed quickly enough and distribute their seeds far enough in response to rapid climate change.

36. Thomas C. Roberts. GLOBAL WARMING—PART 2. Y4, En 2/3: 103-92, 1993, p. 79. River Flooding. Development, destruction of wetlands, and deforestation all decrease the recharge of aquifers and increase runoff during rainstorms. Therefore, these activities increase the vulnerability to floods and drought. Levees and dams, on the other hand, can decrease vulnerability during modest floods; but if the flood is worse than the system was designed for, the dams and levees are overtopped and hence provide no protection.

The recent events along the Mississippi and Missouri rivers, while not attributable to climate change, illustrate the potential for human activities to exacerbate or ameliorate the impacts of unexpected climatic events. In some areas, levees provided important protection; in other areas, overtopping of levees flooded communities built under the expectation that the levees would hold. Another example is reported by an EPA Report to Congress (Smith and Tirpak 1989), which showed that river flooding might increase in the Tennessee Valley, and the same principals apply elsewhere.

37. Robert Sussman. GLOBAL CC AND AIR POLLUTANTS. Y4.En2/3: 103-71, 1993, p. 2. Preventing air pollution and reducing greenhouse gas emissions go hand in hand, and the benefits of taking actions today to address both of these issues will accomplish the goals of cleaner air and a decreased threat of global warming. In many cases, these environmental goals can be achieved while simultaneously strengthening and increasing the efficiency of our economy. For example, a lighting upgrade in one of Hewlett-Packard's facilities in California reduced lighting energy use by 75%, and will save $33,000 in annual energy costs. These improvements create jobs by contributing to the demand for lighting installation and services, while freeing up capital that was spent on energy bills to be invested in productivity and competitiveness. As international competitiveness becomes more fierce, a much more efficient and productive private sector will ensure the United States' leadership in the world economy.

38. Rosina M. Bierbaum. CCAP AND ASSESSMENT. Y4. Sci 2:103/111, 1993, p. 49. Many of OTA's options would begin to remedy these already-perceived problems and additionally buy some added insurance against climate change impacts. They will also help assure use of the Nation's resources will be sustainable. Such options are often termed no regrets or low regrets because they make sense to pursue now, even assuming no climate change. Why, therefore, are actions that are supposed to be prudent anyway, even without the added impetus of climate change, being pursued in such a limited way? Actions that appear reasonable for protecting resources cannot be considered in a vacuum. In reality, barriers of many types exist—in information, institutions and process—that impede the implementation of no regrets or low regrets options. OTA's policy analysis focused on these barriers and tried to identify ways to overcome them.

39. Robert G. Feagle. GLOBAL ENVIRONMENTAL CHANGE: INTERACTIONS OF SCIENCE, POLICY, AND POLITICS IN THE UNITED STATES. 1994, p. 208. A few compensating positive factors can be identified: (1) A regulatory mission serving an important or powerful constituency can provide assurance of stable research support, thus helping the agency to build a capable research program and staff. Examples include NOAA's Marine Fisheries Laboratories and the Department of Energy's (DOE's) laboratories devoted to atmospheric boundary layer research. (2) Legislation that mandates an effort to solve an environmental problem can provide multiyear support for relatively free, unrestricted research. Examples include the FAA program of stratospheric research of the early 1970s (see Chapters 5 and 6), the National Acid Precipitation Assessment Program (NAPAP) of the 12980s (see Chapter 6), and the several Clean Air Acts. (3) The isolation of research from the regulatory process, illustrated by EPA research programs in the early years of the Reagan administration, can provide a protected sanctuary from the political storms buffeting the rest of the agency and, in this way, can permit research to proceed.

40. Seth Cagin and Philip Dray. BETWEEN EARTH AND SKY. 1993, p. 362. To Sherry Rowland or Mario Molina or Susan Solomon or James Anderson, however, what's happening in the atmosphere is completely tangible. Rising rates of chlorine in the stratosphere, and methane in the troposphere, are powerful evidence that humankind has fundamentally altered the chemistry of the atmosphere, probably irrevocably. They know that the harm that might accompany global warming is far less easy to predict or quantify than the harm associated with increased ultraviolet light reaching the earth's surface; and that monitoring ozone depletion was simple compared to the difficulties they face in obtaining conclusive proof of climate change. Knowing the immense damage that CFCs have caused, and the enormous obstacles that were overcome in regulating them, they recognize that international action to alleviate global warming will be even harder to achieve in the face of scientific uncertainty about its causes, and that it will be far more costly and complex to

regulate greenhouse gasses than it has been to regulate CFCs. Every year of delay in implementing change compounds the problems associated with chemical contamination of the atmosphere. Ultimately, waiting for an "ozone hole"—like aberration to force swift action increases the likelihood of irreversible harm.

41. Justin Lancaster. A GLOBAL WARMING FORUM: SCIENTIFIC, ECONOMIC AND LEGAL OVERVIEW. Ed. R. Geyer, 1993, p. 551. Human society as it faces global warming might be analogous to the prehistoric organism developing an eye. Most likely, survival initially became favored to the organism who followed its very first visual warnings only with a humble and hasty retreat. It would be later organisms, after investing much more time and energy into their ability to see, who would be able to derive advantage through responses specifically tailored to the changing scene. As we pride ourselves with our growing perception of a chemical system that was beyond our senses until only recently, we must not neglect the possibility that the greenhouse effect could be more complex and more rapidly destabilizing ecologically than we have estimated. Given that our biogeochemical understanding still may be in its infancy, immediately reducing our energy use is a humble and wise precaution.

42. Henry Lee. SHAPING NATIONAL RESPONSES TO CLIMATE CHANGE: A POST RIO GUIDE. Ed. Henry Lee. 1995, p. 25. Some people argue that since there is so much uncertainty surrounding the science of climate change, the United States should adopt a no-regrets policy. That is, it should only pursue actions that can be justified on grounds other than the reduction of greenhouse gases. Such a policy might include incentives for increased energy efficiency, but only if the costs could be supported by benefits such as cleaner air, greater energy security, or some other indices unrelated to climate change. This type of policy only makes sense if one believes that the probability for global climate change is very low or that the resulting dislocations and harm will be minor. Otherwise, taking some precautions against the possibility of global warming may be a prudent course to take.

43. Robert G. Feagle. GLOBAL ENVIRONMENTAL CHANGE: INTERACTIONS OF SCIENCE, POLICY, AND POLITICS IN THE UNITED STATES. 1994, p. 79. As we have seen, uncertainties regarding the future are large. But if we wait for unequivocal, quantitative evidence of the effects of global change, it may be too late to avoid serious consequences; we will have waited too long. Responses must be based on uncertain knowledge, as indeed are most important decisions ranging from private investments to national budget priorities.

44. Justin Lancaster. A GLOBAL WARMING FORUM: SCIENTIFIC, ECONOMIC AND LEGAL OVERVIEW. Ed. R. Geyer, 1993, p. 536. Countries will be reluctant to limit economic expansion because of uncertain warnings about climate change. Recently, as they struggle to evaluate the appropriateness of various response strategies, economists have grappled with the problem that impacts from global warming are uncertain. There is no solid basis for believing, some argue, that a modest degree of warming would be disadvantageous for the world as a whole. Uncertainty does not mean that there is not a stable scientific consensus. The estimates of the extent and rate of global warming have not fluctuated significantly over the past decade, remaining in the range of 1.5 to 4.5°C by roughly the year 2035, with a "best guess" increase of about 2.5 degrees. Most scientists agree that over the past 100 years global climate has warmed by about 0.5°C. However, a lively controversy has emerged about whether this warming can be connected to the theoretical greenhouse forcing. Also, computer modeling of future climate cannot provide reliable predictions about regional impacts. These uncertainties are being urged as support for a policy of deferring responses to limit CO_2 emissions.

45. Justin Lancaster. A GLOBAL WARMING FORUM: SCIENTIFIC, ECONOMIC AND LEGAL OVERVIEW. Ed. R. Geyer, 1993, p. 535. We know that complex feedbacks in the Earth's system can produce unexpected and potent responses. Possible biogeochemical feedbacks could exacerbate the effects of forest loss and ozone depletion, and this could result in a more rapid global warming than is predicted by purely physical models. When these feedbacks are incorporated into climate models, the most likely global mean temperature increase (resulting from an initial forcing equivalent to a doubling of CO_2) becomes greater than 6°C. It already appears that global warming could outpace the ability for some tree species to survive in some regions, and more complex interactions with the problem of ozone depletion may emerge. A potential contribution to the stratospheric ozone problem alone could warrant controlling atmospheric CO_2 emissions. CFCs directly affect global warming as absorbers of infrared energy, but it is possible for systemic feedbacks to provide indirect effects, as well. It is too soon to tell what feedbacks will occur, but governments might sensibly choose to avoid the risk that feedbacks could be more detrimental than we now estimate. If we need not move blindly into a doubled CO_2 future, as those who urge energy-efficient technologies claim, then we should not want to, even with uncertainty about global warming impacts, because significant reasons other than just the threat of climate change urge our restraint. Such burning, for instance, has created air pollution and acid deposition problems that need solutions before being made worse.

46. Christopher Flavin and Odil Tunali. CLIMATE OF HOPE: NEW STRATEGIES FOR STABILIZING THE WORLD'S ATMOSPHERE. Ed. Jane A. Peterson, June 1996, p. 20. For policymakers, this uncertainty presents a major challenge: how to respond to a problem whose precise dimensions are unclear. Yet in many other fields, governments and individuals frequently take action in the face of similar, or even greater, lack of certitude. Homeowner investments in insurance and government investments in military armament are obvious examples of people taking action to reduce risks despite large unknowns. A concept known as the "precautionary principle" suggests that with a problem such as global climate change it would be illogical to wait for absolute certainty in the form of a catastrophic climatic event before slowing greenhouse gas emissions.

A wait-and-see approach would likely commit the world to even faster climate change that would take centuries to reverse—extending the unplanned and unregulated experiment with the atmosphere that is already underway. As most governments implicitly acknowledged by signing the convention itself, the world cannot afford to continue ignoring the climate problem.

47. Dean Abrahamson, STAR TRIBUNE. 8 January 1996, p. 9A. The International Panel on Climate Change (IPCC), the international panel of scientists charged by the United Nations with reassessing the causes and consequences of climatic change, has concluded that global heating is real and that, unless we make large reductions in greenhouse gas emissions, major climatic change is expected within the next few decades that could cause flooding of coastal areas, heating and drying of the world's major agricultural areas, disappearance of entire forest types, more and stronger hurricanes, the spread of tropical diseases, an increase in heat-related deaths of humans and livestock—in sum, widespread suffering and economic loss.

We have long known that the burning of fossil fuels to run our vehicles, generate electricity and heat our buildings is the single largest contributor to the increase of atmospheric carbon dioxide, that carbon dioxide is a powerful greenhouse gas, that atmospheric greenhouse gases cause global warming, and that the atmospheric concentration of carbon dioxide is increasing rapidly.

What had been missing, until now, was proof that the global warming which has been observed over the past century can be attributed to human-produced greenhouse gases.

The reality and risks of ozone depletion have been recognized and measures taken to stop the pollution that is causing it. The science underlying climatic change now justifies, many say demands, comparable action.

The growing scientific certainty about the nature and impacts of climatic change and the availability of inexpensive alternatives to fossil fuels has paved the way for the world community, including the United States, to agree that emissions in the year 200 should not exceed those of 1990—as a first step toward the 60 to 80 percent reductions that will be necessary to halt rapid climatic change.

The new IPCC study shows that the longer we wait, the more painful will be the necessary actions.

48. Paul Ehrlich and Anne Ehrlich. HEALING THE PLANET: STRATEGIES FOR REDUCING THE ENVIRONMENTAL CRISIS. 1991, p. 108. British writer John Gribbin has suggested that sources of serious environmental hazards, unlike people accused of wrongdoing, should be considered guilty until proven innocent. He calls this approach the "precautionary principle." Considering the severity of the probable consequences of global warming, and the relatively small cost of taking out some "insurance," waiting for final "proof" rather than acting on 50–50 odds would be the height of imprudence.

49. Paul Ehrlich and Anne Ehrlich. HEALING THE PLANET: STRATEGIES FOR REDUCING THE ENVIRONMENTAL CRISIS. 1991, p. 108. The United States so far has not initiated any measures specifically designed to slow global warming (although the EPA has pointed out that the tougher regulations of the 1990 Clean Air Act will have a positive effect). In part, this lack of action has been due to the failure of scientists to communicate effectively with each other and with decision makers. Statistician Andrew Solow testified in 1990 that he considered the probability of unprecedented climatic change in the next century to be "low." Stephen Schneider, a distinguished climatologist from the National Center for Atmospheric Research, considered it "uncomfortably high." It turned out that they agreed that the chances were at least 50 percent! Schneider understood that to decision makers a 50-percent chance of catastrophe is high; Solow was using the arbitrary (and in this case, utterly inappropriate) standard of routine statistical analysis that the result of an experiment is "significant" if the odds are at least 95 percent that it was not caused by random chance alone.

This sort of miscommunication has maintained the impression that vast disagreement exists among scientists over the seriousness of the threat of global warming. The *lowest* probability we have heard any competent scientist attach to the chance of unprecedented climatic change in the next century is 25 percent. The Scientists Working Group of the IPCC, a panel of 200 atmospheric scientists whose work was reviewed by 200 others, reached a consensus in 1990 that there were at least even odds that a warming of 3° to 10° F will occur by the middle of the next century. Their "best guess" was about 5° F by 2050. They also estimated a sea-level rise of between 3 and 12 inches by 2030, and noted that continental interiors are likely to experience drier conditions in summer.

Most of us, of course, would consider a 10 or 25 percent chance of such events more than high enough to take action unless the costs of so doing were absolutely prohibitive. After all, safety-belt use in automobiles is mandated

to reduce the odds of an individual being killed in an accident—a chance without the belt of less than 0.01 percent per year. And you probably would skip lunch rather than eat in a restaurant where you knew there was "only" a 10 percent chance of fatal food poisoning.

50. Ned Nanda. DENVER POST. 4 August 1996, p. D-4. The World Health Organization reported along with the World Meteorological Organization and the U.N. Environmental Program that rising worldwide temperatures would increase the global incidence of malaria and other killer tropical diseases into current temperate zones. The range of illnesses such as rover blindness and Lyme disease could be extended to new parts of the world. The report added, "Current models indicate that by around 2050, many major cities around the world could be experiencing up to several thousand extra heat-related deaths annually." Insurance groups, which have incurred severe losses because of natural disasters blamed in part on global warming also called in Geneva for urgent measures to check emissions. Measures such as carbon taxes to reduce emissions after 2000 may have to be introduced.

51. Paul Ehrlich and Anne Ehrlich. HEALING THE PLANET: STRATEGIES FOR REDUCING THE ENVIRONMENTAL CRISIS. 1991, p. 95. Threats to health also may arise with global warming. One of the most serious potential hazards is the extension of the ranges of disease-carrying vectors, such as mosquitoes that carry diseases like malaria, dengue fever, encephalitis, and yellow fever, into temperate regions. One tropical mosquito species associated with dengue fever has already been found in the United States. Of course, heat itself will be a problem. The frequency of extremely hot days will almost certainly rise dramatically, especially in the warmer areas of the temperate zones, such as the southern and central United States. The annual number of summertime heat-related deaths (mostly among the elderly) in the United States could more than double in a world with the equivalent of twice the preindustrial concentration of atmospheric CO_2. The additional discomfort (and loss of productive work time) would also be great.

52. Tjeerd Deelstra. CLIMATE CHANGE: IMPACT ON COASTAL HABITATION. Ed. Doeke Eisma, 1995, p. 165. The microclimate of bigger cities is already different from rural areas (Table 8–2). Cities are warmer, have more dust and pollution in the air above the city, and are drier. The "urban heat island" causes heat stress from which older people in particular suffer. The hotter the city, the more these problems will increase. It has been calculated that a 4°C temperature increase in the San Francisco Bay area would increase maximum ozone concentrations by 20% and the number of exposure hours for people would triple. A rise in temperature also leads to smog, which has immediate impacts on people's well being, causing lung abnormalities in the long term.

Research has shown that ozone stimulates lung tumors. In warmer cities new types of vector-based diseases such as malaria could occur. When the need for cooling in hotter cities is met by using fossil fueled installations, the air above cities will be polluted with more photochemical oxidants and aerosols, changing visibility and albedo. This and increased acid rain will cause the quality of life to deteriorate. Urban trees and vegetation will also be affected.

53. David Wirth. WORLD SECURITY: TRENDS AND CHALLENGES AT THE CENTURY'S END. Ed. Michael Klare, 1991, p. 381. The odds are strongly stacked against every country in the game of climate roulette. Contrary to some speculation, it is very unlikely that any region of the world will be a net "winner" from climate change. The very concept of "winning" implies the existence of a stable warmer climate, which will not occur unless the warming trend is halted. There is no natural endpoint to climate disruption from the greenhouse effect. Even the limited goal of a steady-state warmer climate will require major policy reform. Otherwise, greenhouse gas concentrations and global temperatures will continue to increase indefinitely, nullifying any short-term benefits. Moreover, no single country will be able to guarantee that the phenomenon is arrested at an optimal point for that country. The only way to ensure that there will be any winners is to guarantee that all countries are winners by reversing the global buildup of greenhouse gases in the atmosphere.

54. David Wirth. WORLD SECURITY: TRENDS AND CHALLENGES AT THE CENTURY'S END. Ed. Michael Klare, 1991, p. 381. Even if a stable warmer climate were identified as a policy goal, the rate of climate change resulting from greenhouse gases already in the atmosphere would be faster than ever experienced in human history. This climate alteration would undoubtedly result in decades of destruction from an inability to alter human behavior, such as agricultural techniques, fast enough to take advantage of new weather patterns. The transition to warmer climates is expected to be turbulent, accompanied by an increase in the frequency, intensity, duration, and geographic extent of extreme weather events like droughts and storms. Moreover, sea-level rise would be certain to entail net harm the world over. No region or individual country should place the health and well-being of its public and environment at stake in what amounts to a crapshoot.

55. Robert G. Feagle. GLOBAL ENVIRONMENTAL CHANGE: INTERACTIONS OF SCIENCE, POLICY, AND POLITICS IN THE UNITED STATES. 1994, p. 64. It should be noted, also, that the climate projections discussed in Chapter 3 apply to a future climate implicitly expected to vary smoothly and uniformly as greenhouse gas concentrations increase. We

know that climate does not change in this way and that, in fact, climate is defined as the aggregate of weather that undergoes substantial fluctuations on many scales. Impacts of future climate change are likely to reflect spatial and temporal variations of the changing climate much more than they do the smoothly varying changes projected by averaging. This important consideration is easily overlooked; to do so may lead to underestimate of the consequences of climate change. Similar generalizations apply also to the problems of stratospheric ozone and acid precipitation.

56. David A. Wirth and Daniel A. Lashof. GLOBAL WARMING FORUM: A SCIENTIFIC, ECONOMIC AND LEGAL OVERVIEW. 1993, p. 512. Although the buildup of GHG concentrations is uniform around the globe, the impacts of the resulting climate change will vary from region to region. This has led to the erroneous suggestion that there will be "winners" and "losers" from global warming. So long as this notion persists, there is a serious risk that broad international agreement on environmentally meaningful reductions in GHG emissions will be stymied. The assumption that there will be winners from global warming is often based on a comparison of current GHG levels with a future, hypothetical climate regime in equilibrium, with CO_2 concentrations at double their preindustrial levels. This arbitrary and totally unrealistic scenario was developed solely for the convenience of climate modelers, who needed simple assumptions for their calculations.

The very concept of "winning" implies the existence of a stable warmer climate, which will not occur unless the current warming trend is halted. There is no natural endpoint to climate disruption from the greenhouse effect. Moreover, no single country will be able to guarantee that the phenomenon is arrested at an optimal point for that country. The only way to ensure that there will be any winners is to guarantee that all countries are winners by reversing the global buildup of GHGs in the atmosphere.

57. Jennifer Woodward. AMERICAN UNIVERSITY LAW REVIEW 203. Fall, 1989, p. 213. Although scientists speculate that some nations might benefit from the effects of a warmer climate—for example, the climate conditions that currently make the mid-region of the United States the world's "bread basket" could shift northward to southern Canada—they also caution that the pace of change could be disruptively rapid.

58. James Gustave Speth. THE RIGHT CLIMATE FOR CARBON TAXES: CREATING ECONOMIC INCENTIVES TO PROTECT THE ATMOSPHERE. August 1992, p. xii. Make no mistake about it, the threat is real. For scientists, the question is no longer *whether* the earth will heat up, but how much it will heat up, and how soon. The latest estimate by the Intergovernmental Panel on Climate Change is that the global average temperature will rise by from 1.5 to 3 degrees Centigrade over 1900 levels by the middle of the next century. That may not sound like much, but a few degrees can spell enormous change: during the depths of the last ice age, for instance, the earth was only 5 degrees Centigrade colder than it is now. Temperatures at the midlatitudes are expected to rise about twice as much as the global average temperature does. Together with declining rainfall, this could bring dust-bowl conditions to the American Great Plains and some Eurasian farmlands. The expected 48-centimeter rise in sea levels would devastate low-lying coasts and islands, especially if tropical storms become more fearsome in a climate-altered world. All these changes may outpace the ability of species to move to new habitat or adapt to changing conditions, exacerbating the species extinction crisis already in progress.

59. Christopher Flavin and Odil Tunali. CLIMATE OF HOPE: NEW STRATEGIES FOR STABILIZING THE WORLD'S ATMOSPHERE. Ed. Jane A. Peterson, June 1996, p. 43. Many scientists believe that anything more than an increase of 0.1 degrees Celsius (0.18 degrees Fahrenheit) per decade—1.00 degrees Celsius (1.8 degrees Fahrenheit) over the next century—would present unacceptable risks to natural systems as well as the human economy. For example, ice core data extracted from Greenland indicate that where climate change occasionally has exceeded this rate in the past, northern hemisphere forests have been devastated.

The global average temperature at the earth's surface has risen at more than twice the acceptable rate in recent decades, and so far shows no sign of letting up. Climate models indicate that to slow the rate of change to 0.1 degrees Celsius (0.2 degrees Fahrenheit) per decade, the total concentration of all greenhouse gases will have to be held to a maximum of less than 550 parts per million CO_2-equivalent at mid-century, compared to the current level of roughly 430 ppm CO_2-equivalent. This suggests that the concentration of carbon dioxide alone will have to level off at between 450 and 500 parts per million—compared to the current level of 360 ppm.

60. Norman Myers. ULTIMATE SECURITY: THE ENVIRONMENTAL BASIS FOR POLITICAL STABILITY. 1993, p. 199. On top of all this there is the prospect of other greenhouse effects, such as shifts in monsoon systems and the arrival of continuous droughts—with all that both entail for our capacity to feed ourselves. A temperature rise of only 1.8 degrees Fahrenheit, entirely likely by early next century, could affect monsoon patterns to an extent that would dwarf the direct drought effects of such a temperature rise. The area most vulnerable to monsoon dislocations is the Indian subcontinent, projected to feature 1.4 billion people by the year 2030. India depends upon the monsoon for 70 percent of its rainfall. In broader terms, the entire Asia-Pacific region is unduly vulnerable to monsoonal changes if only because it contains well over

half the world's population today, projected to become a still larger proportion by 2030.

61. INFORMATION UNIT ON CLIMATE CHANGE. May 1, 1993, UNEP, http://www.unep.ch/iucc/fs101.html **The impact on yields of low-latitude crops is ore difficult to predict.** While scientists are relatively confident that climate change will lead to higher temperatures, they are less sure of how it will affect precipitation—the key constraint on low-latitude and tropical agriculture. Climate models do suggest, however, that the intertropical convergence zones may migrate poleward, bringing the monsoon rains with them. The greatest risks for low-latitude countries, then, are that reduced rainfall and soil moisture will damage crops in semi-arid regions, and that additional heat stress will damage crops and especially livestock in humid tropical regions.

62. Dave Alden. AUSTRALIAN ECONOMIC REVIEW. First Quarter, 1995, p. 130. There are equity, or fairness, implications that arise when considering the optimal level of global warming. For example, what if Australians benefit from global warming by the same amount as costs incurred by those in Bangladesh. Should these cancel each other out, given that the costs in Bangladesh may be measured in terms of lives lost? It is likely that society would judge this to be unfair. If an adjustment for this were included in the calculations by Nordhaus, the revised optimal level of global warming would require a larger reduction in projected emissions of greenhouse gases, and may suggest a complete prevention of global warming is optimal.

63. Sylvan Wittwer. FOOD, CLIMATE AND CO_2: THE GLOBAL ENVIRONMENT AND WORLD FOOD PRODUCTION. 1995, p. 46. A recent report (Karl et al., 1993) pertinent to lengths of growing seasons states that although global average temperatures have been increasing, the warming has been primarily due to nighttime rather than daytime temperature increases, appearing as a decrease in the day-night temperature difference over land for the past 40 years. The potential benefits of nighttime warming, such as a longer growing season and fewer killing frosts, may be offset, however, by increased pest infestations, a reduced cropping area, and higher human heat-related mortality.

64. Justin Lancaster. A GLOBAL WARMING FORUM: SCIENTIFIC, ECONOMIC AND LEGAL OVERVIEW. Ed. R. Geyer, 1993, p. 536. Second, our momentous growth affects social stability. The rapid intensification of energy use in human society has changed social structure, and it might well create further change faster than our ability to cope. A structure dependent on steadily increasing energy flow becomes vulnerable to major disruptions in energy supply. We must establish throttles on the increasing flows of energy through society that are designed not for profit, which favors acceleration, but for long-term stability and public benefit.

Society's rampant energy use is caused by a fundamentally natural growth principle—an energy-technology feedback. Our environmental problems are intimately intertwined with this mutually reinforcing dynamic between energy acquisition, industrial metabolism and technological power. This feedback, which bounds social evolution and governmental control, impacts political systems and stability more directly than through its environmental manifestations alone. Wars over energy, for instance, are a more immediate threat to human health and welfare than global warming. It is not just the uncertain extent of climate change, then, that presents risks that are difficult to assess. It is important to realize that changing energy use and perturbing global biogeochemical cycles pose their own serious concerns for humanity, separate from their relevance to global warming. Even change within predicted trends could trigger shifts in biogeochemical and energy processes that could, particularly if compounded by internal societal stresses, present us with severe challenges in the long term.

On balance, it is time to work for international agreement on response strategies. The question then becomes, "how great is the unknown risk, to warrant how great a response cost"? The dilemma seems to be that at least the first part of the question cannot yet be answered.

65. Susan F. Tierney. FEDERAL DOCUMENT CLEARING HOUSE. 21 March 1995, FED. In February 1995, the Secretary led a Presidential Mission on Sustainable Energy and Trade (February 15–25, 1995) to Hong Kong, Shanghai, and Beijing, China. An official delegation of over 100 American business executives, environmental leaders and government officials accompanied Secretary O'Leary on the mission. The mission highlighted our two countries' shared belief that sustainable energy development is critical to economic progress and building mutually beneficial economic ties is critical to stable long term relationships.

Of the seven agreements signed or witnessed by the Secretary's mission, four specifically focus on the global need to improve the environment:

- protocol on Energy Efficiency Improvement and Renewable Energy Technology. Development and Utilization, which provides for cooperative activities with the State Science and Technology Commission in energy conservation, civil transportation (automotive) technology, and solar, wind, and geothermal energy.

- Annex on Cooperation in Coalbed Methane Recovery and Utilization with the Ministry of Coal Industry, which will help China reduce greenhouse gas emissions.

- Letter of Intent for a Renewable Energy Annex to the

Renewable Energy Protocol, which will promote cooperation with the Ministry of Agriculture on renewable energy in rural areas.

- Annex on Regional Climate Research to establish with the China Meteorological Administration a program of joint research to document, predict, and identify impacts of regional climate change.

The mission culminated in the signing of 34 business agreements totaling more than $6 billion ($4.6 billion of U.S. content). These included approximately $630 million in renewable energy projects, $428 million in energy efficiency, $31.5 million in oil and gas projects, $1.4 billion in advanced coal projects $2 billion in electric power generation, and $20 million in environmental technologies. These agreements will deliver clean, affordable energy to fuel China's economic growth while promoting the use of U.S. skills, personnel, technology, and investment funds, creating jobs in the U.S. and China.

66. David Gardiner. FEDERAL DOCUMENT CLEARING HOUSE. 19 June 1996, Lexis/Nexis. A second factor intensifying the Administration's attention to climate change is the accelerating pace of the international negotiation process. The Parties to the 1992 Convention decided in March of 1995 in Berlin to launch a process to define actions to be taken by the industrialized nations–designated in the Convention as "Annex I" nations—in the post-2000 period, and to promote implementation of commitments by all nations. This decision, known as the "Berlin Mandate," initiated international negotiations to develop a new legal instrument for adoption in 1997. The U.S. successfully urged the Parties to include in the Berlin Mandate a call for an initial phase of analysis and assessment to inform the negotiation of post-2000 obligations. It is the early lessons of that analytic phase that we are prepared to discuss today.

67. David Wirth. WORLD SECURITY: TRENDS AND CHALLENGES AT THE CENTURY'S END. Ed. Michael Klare, 1991, p. 381. Coordinating policies on the international level to fight greenhouse warming will maximize environmental and foreign policy benefits. Unilateral reductions in releases of greenhouse gases by large emitters such as the United States and the Soviet Union will go a long way toward arresting global climate disruption. However, a multilateral consensus strategy will further the crucial goals of creating incentives for universal participation and establishing an equitable balancing of responsibility for solving the problem. These and other international political, legal, and institutional challenges are likely to endure over time, even as the science of the global warming issue evolves.

Existing international mechanisms are an important part of such a strategy. A reassessment of the Montreal Protocol, a process that is provided for by the document itself, is the most expeditious way to eliminate the contributions CFCs and halons make to the global warming problem. The World Bank's institutional structure includes mechanisms for member countries to redirect priorities in the critical energy and forest sectors. But existing mechanisms by themselves are highly unlikely to be adequate for the task of crafting a comprehensive greenhouse gas regime.

The remainder of the greenhouse problem could be handled most effectively through a mulilateral treaty, with standards binding under international law that would require each country to take prescribed actions to reduce and halt greenhouse warming. An international agreement designed to arrest global climate change should satisfy several basic requirements. First, it must require reductions in releases of greenhouse gases of a magnitude and speed sufficient to stabilize the earth's climate. The most important gas to control is CO_2, for which global reductions of at least 60 percent are necessary. Participating countries should accomplish these reductions by means of environmentally and economically sound technologies that do not present unacceptable risks to public health or world security. The creation of new forested areas might be encouraged by allowing credits against reductions of CO_2 emissions that would otherwise be required and by provisions establishing or promoting forestry programs. Because the agreement could be expected to cover a large number of emissions sources, it should require strict mechanisms for enforcement through reporting of emissions, on-site audits, and internationally controlled remote sensing.

68. Dale E. Heydlauff. GLOBAL WARMING—PART 3. Y4.En2/3: 103–167), 6 October 1994, p. 95. We are at a very early and fragile stage in the implementation of the Convention and are without essential information on what action the Parties will be taking. Instead, national action plans from the various parties were only required to be submitted in late September, and the First Conference of the Parties will not take place until March. It is premature and potentially counterproductive to do anything more than develop a process for conducting a review of the adequacy of commitments in the future to see what, if any, appropriate action is needed in the post-2000 period.

69. Justin Lancaster, A GLOBAL WARMING FORUM: SCIENTIFIC, ECONOMIC AND LEGAL OVERVIEW. ed. R. Geyer, 1993, p. 537. In one sense, present scientific uncertainty may encourage early intergovernmental agreement, because it is likely to make negotiations less biased now, with individual nations unable to exactly calculate their positions. This could make it more likely for countries to consent to common goals. The more that CO_2 levels increase, and the closer we come to seeing which countries lose and which benefit, the more difficult it will be to get those who benefit to enter negotiations; the political trade-offs will become increasingly costly. A sensible approach would be to use the uncertainty

that exists now to push for a limiting scheme that will spread the cost and the impact over many nations who later might be less willing to join an agreement. The choice and timing of these responses must, to a large degree, be based on social value judgment, rather than any objective measure of scientific certainty. The debate is really over how, given all the uncertainties, we should respond to the possibility of climate change. Beyond establishing the facts and assessing the uncertainties, science can contribute relatively little.

70. Henry Lee, SHAPING NATIONAL RESPONSES TO CLIMATE CHANGE: A POST RIO GUIDE. Ed. Henry Lee, 1995, p. 52. Many environmental advocates believe that the opposition of key White House aides in the Bush administration was responsible for weakening the framework convention and for undercutting the negotiations leading to the Rio summit. The truth is that, currently, support for significant reductions in greenhouse gases among the nations of the world ranges from slight to nonexistent. The global climate change issue involves both scientific uncertainty and large economic costs. Mitigation responses must be paid for in the near term by identifiable entities, while their benefits are uncertain and dispersed and will be realized far in the future. For these reasons, a sophisticated and concerted political effort is needed to build a constituency both within the United States and abroad.

71. David Wirth. WORLD SECURITY: TRENDS AND CHALLENGES AT THE CENTURY'S END. Ed. Michael Klare, 1991, p. 398. Optimism about the prospects for a treaty to limit emissions of greenhouse gases through a global convention arises to a great extent from progress on CFCs and halons in the Montreal Protocol. However, most of the existing mandates for the negotiation of a global warming treaty consciously limit action to a "framework" or "umbrella" instrument analogous not to the 1987 Montreal Protocol, but to the 1985 Vienna convention. Unfortunately, this precursor to the Montreal Protocol establishes only a structure for cooperation in research and exchange of information. It does not mandate reductions in emissions of CFCs or halons, nor does it require any other measures that directly benefit the environment.

There is a serious risk that precious time will be lost in protracted negotiation over these "framework" functions, which in any event have already been largely performed by the IPCC process. Instead, a greenhouse-gas convention should contain minimum global coals commensurate with environmental necessity, such as a commitment to reduce global emissions of CO_2 by at least 60 percent on a specified timetable. The implementation of these goals could be taken up in negotiations on ancillary agreements similar to the Montreal Protocol, when complicated issues such as the distribution of national obligations for reductions could be addressed.

72. Justin Lancaster, A GLOBAL WARMING FORUM: SCIENTIFIC, ECONOMIC AND LEGAL OVERVIEW. Ed. R. Geyer, 1993, p. 534. Early agreement to reduce greenhouse gas emissions is urged by at least three considerations. First, more rapid and complex change could occur than has been previously anticipated by the "consensus" view. Second, current scientific uncertainty prevents states from yet knowing who will be "winners"; thus, consensus may be more easily achieved today than decades hence. Third, a fundamental need exists to regulate energy.

73. James K. Sebenius. SHAPING NATIONAL RESPONSES TO CLIMATE CHANGE: A POST RIO GUIDE. Ed. Henry Lee, 1995, p. 84. Reacting against the LOS approach (that is, a comprehensive agenda with the requirement of a package deal), climate change negotiators aimed for a framework convention to be followed by specific protocols. In line with the CFC experience, this approach retained the aims of universality and consensus, but dropped comprehensiveness and the goal of a package deal in favor of single, separable protocols on limited subjects. This alternative has attractive negotiating features, but it is worth noting that it was the failure of precisely this approach—negotiation of separate miniconventions, which are analogous to protocols—in earlier LOS conferences (in 1958 and 1960) that indirectly led back to the comprehensive package approach of the 1973 LOS conference.

74. Ned Nanda. DENVER POST. 4 August 1996, p. D-4. The threat of global warming can no longer be ignored. The controversy flared up again in mid-July when the U.N. Climate Change Convention concluded a two-week conference with publication of the Geneva Declaration, calling upon the developed countries to commit to "legally-binding objectives for [greenhouse gas] emission limitations and significant overall reduction within specified time frames."

The United States, previously unwilling to go beyond undertaking voluntary efforts, changed its stand to opt for mandatory limits on carbon dioxide emissions, caused mainly by the burning of coal, oil and gas. Undersecretary of State Tim Wirth presented the administration's position, committing itself to negotiating "realistic, verifiable and binding" targets to reverse global warming. The final decisions will be taken at a meeting of the 150 signatories to the Climate Change treaty, scheduled for the end of 1997 in Kyoto, Japan.

75. "Inside Energy/with Federal Lands," ENERGY. 8 July 1996, Lexis/Nexis. In addition, the group is concerned over calls by some countries for "common policies and measures" that could include such things as international fuel efficiency standards for automobiles, carbon taxes and efficiency standards.

Although administration officials have said they will not make any commitments at the Geneva meeting, GCC

members say the two-week session is part of a process that will eventually lead to reduction commitments. "Each one of these meetings takes you further down the road," Holmes said. "The gist of the conversations at the meeting begins to point in the same direction and it's hard to make good decisions when you lack information."

O'Keefe said the group wants the U.S. delegation to state at Geneva that all of the reduction proposals made thus far would do "significant damage to our economy. The United States should make clear that the current proposals are not acceptable."

76. FINANCIAL TIMES. 31 July 1996, p. 16. The US has returned to the driving seat in international climate change negotiations after it helped push through a pledge by developed nations earlier this month to seek legally binding reductions in greenhouse gas emissions after 2000.

This is the conclusion of Michael Grubb, a fellow at the Royal Institute for International Affairs, the London-based think tank, and a lead author for the Intergovernmental Panel on Climate Change, which studies the phenomenon for governments.

The US is widely recognised to have a poor record in responding to the existing voluntary cuts on greenhouse gases—despite accounting for 25 per cent of emissions.

The curbs are supposed to bring emissions back to 1990 levels by 2000—a goal which will be met only by Britain and Germany and only then for reasons not connected to climate change.

Some sceptics in Europe also point out that the fine print of the Geneva agreement still leaves room for countries to wriggle out of a legally binding deal when they meet in Kyoto next December because it refers to legally binding reductions "or another legal instrument."

But Grubb said the "brutal" US attack at the meeting in Geneva earlier this month on energy lobby groups which had questioned the integrity of the IPCC, together with its support for legally binding commitments, showed that the Clinton administration saw climate change as a "serious problem."

"It suits the United States system better to get a binding international commitment and go back to Congress and say 'Now you must agree measures to implement this.'"

77. James K. Sebenius. SHAPING NATIONAL RESPONSES TO CLIMATE CHANGE: A POST RIO GUIDE. Ed. Henry Lee, 1995, p. 60. In the greenhouse case, a wise course of action may be to proceed sequentially with protocols to avoid the creation of a potent unified opposing coalition. Not entirely tongue in cheek, it may be best to pick "easy" subjects first to generate momentum—protocols directed at greenhouse contributors that are politically weak, morally suspect, and concentrated in highly "green" countries—with later protocols strategically chosen to build on early success.

78. James K. Sebenius. SHAPING NATIONAL RESPONSES TO CLIMATE CHANGE: A POST RIO GUIDE. Ed. Henry Lee, 1995, p. 60. Beyond measures to prevent the formation of blocking coalitions in the first place, a number of other approaches can be characterized as incremental. The idea behind them is to gain agreement on a relatively weak or nonspecific treaty or plan of action in the expectation that, over time, it will progressively be strengthened. This approach may be a conscious initial choice or it may simply reflect the strength of opposing forces in the early negotiations. Advocates may settle for what they can get in the hope that they have set the stage for another round that will conclude more in line with their preferences.

79. Peter M. Morrisette. FORUM FOR APPLIED RESEARCH AND PUBLIC POLICY. Summer 1996, p. 5. The fact remains, however, that both the Montreal Protocol and the Climate Convention were the result of an incremental process in which the scientific and political debate evolved over decades.

Such an incremental process provides a scientific and political context, which in turn gives agreements like the Montreal Protocol and Climate Convention meaning and credibility. In addition, the openness and shared knowledge that the process fosters add credibility to the decisions that are made.

80. Henry Lee, SHAPING NATIONAL RESPONSES TO CLIMATE CHANGE: A POST RIO GUIDE. Ed. Henry Lee, 1995, p. 17. Both the Sebenius and the Parson and Zeckhauser chapters emphasize the importance of incremental approaches. The latter argues for the establishment of progressive obligations that are definable in terms of annual incremental improvements in trend lines for national emissions and carbon intensity, or even in terms of specific efforts, such as expenditures in emissions control or progressively increased tax levels for energy. Such an approach would respect nations' concerns about a precedent being set that might prove untenable and would also provide countries with maximum flexibility to review and adjust the annual incremental targets as they learn more about the costs and effectiveness of various policies and programs.

81. Christopher Flavin and Odil Tunali. CLIMATE OF HOPE: NEW STRATEGIES FOR STABILIZING THE WORLD'S ATMOSPHERE. Ed. Jane A. Peterson, June 1996, p. 60. Even as these approaches are being developed, the climate convention can be used to encourage narrower but still important policy reforms that will lower greenhouse gas emissions, particularly from major contributors such as automobiles, home appliances, power plants, and landfills. This sector-by-sector approach is sometimes criticized as inadequate to the scale of the climate problem, but it may be the most effective means of lowering the market barriers to new technologies—some of

which would not be much affected by a carbon tax or emissions trading system. Some argue that such measures should be left entirely to national governments to decide, but the treaty process could help standardize such norms internationally, reducing the cost to industry of compliance, particularly for those products that are heavily traded.

82. James K. Sebenius. SHAPING NATIONAL RESPONSES TO CLIMATE CHANGE: A POST RIO GUIDE. Ed. Henry Lee, 1995, p. 63. In the face of environmental challenges, a successful accord on climate change calls for a process designed to achieve results that can be sustained over time and modified as appropriate. In particular, a process like the ones that took place in Vienna and Montreal, with independent protocols to be negotiated on a step-by-step basis, was thought to have the advantage of speed and relative simplicity over a comprehensive LOS-like approach. This raises the more general question of how to deal with greenhouse issues (or protocols): singly, comprehensively, or in intermediate-sized linked packages. The answer, most usefully explored in the LOS context, has a direct implication for ensuring enough gain in an agreement to attract a winning coalition.

83. James K. Sebenius. SHAPING NATIONAL RESPONSES TO CLIMATE CHANGE: A POST RIO GUIDE. Ed. Henry Lee, 1995, p. 63. With respect to climate change negotiations in particular, it is quite likely that public concern will be cyclic, in part as a result of natural climate variability as well as unrelated environmental events (such as medical waste on beaches and the Exxon *Valdez* oil spill). Arguably, a naturally occurring period of climate calm, including milder summers and normal rainfall, will lead to reduced public concern and pressure for action. Moreover, scientific understanding will change over time. These prospects argue for more limited agreements, with analogues to the ratchet mechanism in the Montreal Protocol, if and when more stringent action appears warranted. Such agreements could constitute a "rolling process of intermediate or self-adjusting agreements that respond quickly to growing scientific understanding." In addition, an even more fundamentally adaptive institution might be envisioned that better matched the rapidly changing science and politics of this set of issues.

84. Henry Lee, SHAPING NATIONAL RESPONSES TO CLIMATE CHANGE: A POST RIO GUIDE. Ed. Henry Lee, 1995, p. 18. Sebenius suggests the use of ratchet mechanisms, whereby negotiators establish modest targets for greenhouse gas reductions and use these to institutionalize monitoring, enforcement, and implementation at both the international and domestic levels. Once the foundation is built, negotiators can simply ratchet up the targets if new information warrants more stringent reductions. Such a mechanism was a vital component of the Montreal Protocol for reducing CFC emissions. A process characterised by many small steps rather than a few large ones has a much better chance of success.

85. Edward Parson and Richard Zeckhauser. SHAPING NATIONAL RESPONSES TO CLIMATE CHANGE: A POST RIO GUIDE. Ed. Henry Lee, 1995, p. 101. This type of agreement simply provides a shift in focus from fixed-year goals to annual incremental ones. From a negotiating standpoint, this change in view offers several important advantages. First, progressive obligations integrate the understanding that no one knows how tight emissions limits should be, hence respecting nations' concerns about setting a precedent that might be wrong. Second, progressive obligations integrate the observation that since little is known about the effectiveness of emissions-control programs, nations are truly unable to pledge to meet a fixed target. By avoiding fixed targets (for good reasons), this approach also denies parties the obstructive rhetorical device of criticizing a proposed form of obligations as if they will extend unchanged forever.

Since attainment of targets over a period as short as a year is for the most part a consequence of random fluctuations, this approach basically amounts to establishing a continuing negotiation over the appropriateness and effectiveness of national emissions-reduction programs. The essence of this approach lies in its review process—both periodic negotiated review of the appropriateness of the annual incremental targets and review of national measures—which should strive both to give national policymakers the right incentives (to try hard, be flexible, and report progress honestly) and to maximize learning through international comparison of experiences. Such a system presupposes a very different forum from broad, comprehensive, diplomatic negotiations. Rather, this system seems to require a small, functional, professional group that would develop good working relations as its members continually negotiate, thus acquiring both skill at knowing what works and the ability to consult effectively enough with national officials to apply the pressure needed to keep plans effective and targets honest.

86. FINANCIAL TIMES. 20 March 1996, p. 15. Governments have undertaken to agree targets for carbon emission reductions into the next century at the Tokyo meeting. But very few, possibly including Sweden, will meet the existing target for industrialised countries to stabilise emissions by 2000 at 1990 levels.

Part of the problem facing the country as a whole is the same one encountered by its companies. Having done much already to reduce emissions, it becomes more difficult to take the process a stage further.

Unlike most of western industry, which argues that it is premature to set any targets for reducing carbon emissions until more is known about the exact nature of the climate change threat, Swedish industrialists do not object to such action. They do object, however, to being alone

(except for other Scandinavian countries) in taking such measures.

87. Henry Lee, SHAPING NATIONAL RESPONSES TO CLIMATE CHANGE: A POST RIO GUIDE. Ed. Henry Lee, 1995, p. 3. We start from the premise that the U.S. government, by signing the United Nations (UN) Framework Convention on Climate Change (FCCC), has concluded that it is in its national interests to reduce world greenhouse gas emissions. Our question is not whether the United States should act, but rather how an enforceable strategy can be fashioned that will be acceptable both domestically and internationally. We step back from the immediate debate and present a strategic framework to assess and compare specific alternatives. Many technological and programmatic options to reduce the threat of climate change have been suggested, but without political consensus of a strategy for building and institutionalizing such a consensus, no option will achieve its potential. Whether a nation believes that solar energy, energy conservation, or aggressive tree planting is the right answer, its first step must be the design of an overall strategy to realize its international and domestic objectives. Without a coherent strategy, mitigation efforts are the strategic equivalent of shooting stars—spectacularly bright, but short-lived, political phenomena in the global political firmament.

88. Christopher Flavin and Odil Tunali. CLIMATE OF HOPE: NEW STRATEGIES FOR STABILIZING THE WORLD'S ATMOSPHERE. Ed. Jane A. Peterson, June 1996, p. 63. None of these policies are likely to be adopted unless heavy political resistance is overcome. Fossil-fuel dependent industries, oil-exporting nations, and hired-gun scientists have confused the public and thrown up a host of procedural barriers. These interests, particularly those based in the United States, have spent millions of dollars lobbying against tightening the climate convention in recent years; their position papers have found their way into official Saudi and Kuwaiti statements and, in one embarrassing episode, a speech by a U.S. government official. Although obstructionist, they have slowed a process that normally operates by consensus and is easily bogged down by even one hold-out nation. It is Kuwait and Saudi Arabia, for example, that have prevented treaty members from even adopting rules of procedure.

89. James K. Sebenius. SHAPING NATIONAL RESPONSES TO CLIMATE CHANGE: A POST RIO GUIDE. Ed. Henry Lee, 1995, p. 64. The LOS experiences in 1958 and 1960 suggest that sometimes issues must be linked. By 1958, for the First UN LOS Conference, the International Law Commission had suggested a negotiating structure with four separate conventions concerning different issues, such as the breadth of the territorial sea and the extent of continental margin. With respect to the comprehensive agenda of the 1973 LOS talks, Conference President Tommy Koh observed that

> A disadvantage of adopting several conventions is that states will choose to adhere only to those which seem advantageous and not to others, leaving the door open to disagreement and confrontations. The rationale for this [comprehensive] approach was to avoid the situation that resulted from the 1958 conference which concluded four [separate] conventions.

Such an uneven pattern might also result from a framework-protocol structure on climate change. Imagine Libya signing a forestry convention while Nepal agreed to a transportation and automotive protocol. For individual countries or groups of similar ones, a single issue often represents either a clear gain or a clear loss. As with the early LOS conferences (with independent miniconventions), countries sign the gainers and shun the losers. In a climate context, for example, China may resist a specific fossil fuel protocol that would place restrictions on the development of its extensive coal resources.

90. Peter M. Morrisette. FORUM FOR APPLIED RESEARCH AND PUBLIC POLICY. Summer 1996, p. 68. An incremental process, however, does not guarantee success. It can easily be stalled by parties interested in seeing that agreement is not reached. In addition, an incremental process may not move fast enough for those who feel that there is a need to take immediate action. Yet it is a process that tends to foster compromise and agreement.

91. Dale E. Heydlauff. GLOBAL WARMING— PART 3, Y4.En2/3: 103–167), 5 October 1994, p. 96. At the recent meeting of the INC-10 in Geneva, we found substantial interest in the Climate Challenge Program and in the voluntary approach that it represents. As I mentioned before, both the electric utility industry and DOE hope to showcase the actions to be taken under the Climate Challenge Program at INC-11 next year. The U.S. government has taken a world leadership role by including voluntary programs such as Climate Challenge as a centerpiece of the mitigation portion of its national action plan. We applaud its efforts to promote voluntary, cost-effective programs in lieu of command-and-control actions, and we therefore urge the Administration not to make premature judgments on the issues and not to commit itself to actions to reduce greenhouse gas emissions in the post-2000 time frame that go beyond the current commitments of the Framework Convention.

92. Henry Lee, SHAPING NATIONAL RESPONSES TO CLIMATE CHANGE: A POST RIO GUIDE. Ed. Henry Lee, 1995, p. 16. An effective strategy is one that creates incentives, rather than requirements, for domestic political coalitions to act. An effective

strategy will be one that does not trigger the aggressive opposition of powerful blocking coalitions or threaten national priorities. Until there is a stronger popular consensus about the urgency of the need to respond to the problem of climate change, carrots will work better than sticks.

93. Christopher Flavin and Odil Tunali. CLIMATE OF HOPE: NEW STRATEGIES FOR STABILIZING THE WORLD'S ATMOSPHERE. Ed. Jane A. Peterson, June 1996, p. 54. One of the greatest challenges is the fact that climate policy cuts across so many lines of governmental jurisdiction. While it is one of the few issues that is truly global in scope—a kilogram of carbon dioxide emitted in Canada has exactly the same consequence as one released in Sri Lanka—essentially policy changes need to occur at the national, state, and local levels. Energy taxes, for example, are generally levied nationally, electric utilities are often regulated by states, and road and building codes are usually set by cities and towns. Effectively slowing the pace of climate change will therefore require cooperation and coordination among all levels of government, with broad guidelines and goals set at the international level, and much of the policy implementation occurring closer to where people live and work.

94. Ronald Mitchell and Abram Chayes. SHAPING NATIONAL RESPONSES TO CLIMATE CHANGE: A POST RIO GUIDE. Ed. Henry Lee, 1995, p. 115. Most of the analytic and negotiating energy surrounding the development of a climate change treaty has focused on substantive limitations on net emissions of greenhouse gases, whether through targets and timetables, emission permits, taxes, or technological standards. But no matter how stringent these commitments are, the treaty will not succeed unless the parties comply with them. The compliance problem must be addressed from the outset and a compliance system must be designed into the treaty from the beginning. In the climate change arena, the costs and magnitude of required behavioral changes, and the regulatory breadth and complexity pose especially difficult compliance problems. Moreover, unlike most international agreements, which seek to affect only the actions of governments, a successful climate change treaty must alter the behavior of ordinary individuals and business firms whose activities account for the emission of greenhouse gases. The treaty must encourage national governments not only to comply with its provisions by adopting legislation and other appropriate policies, but also to take action to facilitate compliance and condemn violations of private actors within their own borders.

95. Marc W. Chupka. FEDERAL DOCUMENT CLEARING HOUSE. 19 June 1996. Global Trends: During the first half of the 1990s, greenhouse gas emissions grew in the U.S. and most of the rest of the world. However, due to the restructuring of the former Soviet Union and Eastern Europe, global emissions were roughly stable. Nearly all of the countries of the Organization for Economic Cooperation and Development (OECD) are experiencing rising greenhouse gas emissions as illustrated in Figure 1. The important exceptions are the U.K, where energy sector deregulation is resulting in shifts away from coal and toward natural gas; and Germany, whose emissions have fallen due to dramatic emissions reductions in former East Germany. Projected U.S. emissions growth through the first decade of the next century is expected to be slightly higher than the OECD average. Emissions are expected to grow somewhat more slowly in OECD-Europe and somewhat faster in OECD-Asia, when compared to the U.S.

Over the next century, the bulk of emissions growth will occur in the non-OECD developing countries. While OECD carbon emissions are projected to grow by about 25% over the next 20 years, developing country emissions are projected to more than double. This observation underscores why addressing the climate change problem will require both developed country leadership and international cooperation.

96. Hazel R. O'Leary. GLOBAL WARMING— PART 3. Y4.En2/3:103–167, 6 October 1994, p. 14. The International Energy Agency has also studied the potential for growth in energy use and related emissions. It concludes that most of the growth occurs in Asia. China's emissions, for example are forecast to grow from 2.4 billion metric tones in 1990 to 5 billion metric tons in 2010—doubling its emissions in only 20 years. Other East Asian countries together will more than double their emissions over that same period, from 1 billion metric tons in 1990 to 2.6 billion metric tons in 2010. South Asia is also a major growth region. The International Energy Agency forecasts a growth in that region's emissions from 0.7 billion metric tons in 1990 to 1.7 billion metric tons by 2010. The remaining countries of the developing world also experience emissions growth, about a doubling—from 2.4 billion metric tons in 1990 to 4.7 billion metric tons in 2010.

In annual percentage terms, this growth in the developing word means that by 2010 the developing world will be responsible for 44 percent of global energy related CO_2 emissions, while the OECD is responsible for 42 percent in that year. This contrasts sharply with the picture in 1971 when the OECD countries produced 61 percent of these emissions while the developing world generated only 17 percent of the emissions.

97. James K. Sebenius. SHAPING NATIONAL RESPONSES TO CLIMATE CHANGE: A POST RIO GUIDE. Ed. Henry Lee, 1995, p. 58. Absent a natural climate catastrophe, it is unlikely that future climate negotiations will have anything like the public salience of the Earth Summit, which brought together more than 150 nations, 1,4000 NGOs, and 8,000 journalists. Nevertheless, future talks should explicitly seek to build on this wide-

spread public exposure. Given the potential of global communications technologies and the efforts of concerned governments and interested NGOs, future climate negotiations themselves and the public awareness they stimulate can help to spur informal control regimes, in part by building on and influencing domestic opinion, which is often led by the actions of NGOs. (The most striking example of this phenomenon probably occurs in the area of human rights.) The national reporting requirements contained in the current climate convention—if beefed up and properly funded—provide a natural vehicle for involving and mobilizing citizens and advocates. In turn, stronger informal regimes may come to be embodied in more potent formal instruments that might earlier have been blocked by opposing coalitions.

98.   Henry Lee, SHAPING NATIONAL RESPONSES TO CLIMATE CHANGE: A POST RIO GUIDE. Ed. Henry Lee, 1995, p. 8. Vernon makes the point that domestic policy cannot get too far in front of, or lag too far behind, positions taken internationally without undercutting the credibility and bargaining position of the U.S. negotiators. This linkage between international and domestic policies is often ignored or not fully appreciated. International policy is formulated by the executive branch, with only moderate involvement by special interests, whereas domestic policy is a product of a pluralistic, interactive process involving the executive branch, the Congress, state and local governments, and a spectrum of interest groups. Thus, ideological and specific economic concerns, such as the impact on regions or on certain industries, tend to surface with more emotion during the formation of domestic policy. If the United States is to craft an effective global strategy for greenhouse gas mitigation, the processes by which it formulates domestic and international policies must be closely linked.

99.   Henry Lee, SHAPING NATIONAL RESPONSES TO CLIMATE CHANGE: A POST RIO GUIDE. Ed. Henry Lee, 1995, p. 24. In designing domestic initiatives, U.S. policymakers should, first and foremost, remember Vernon's admonition: the development of international negotiation strategies and the design of domestic policy initiatives are inextricably linked. To push for protocols and policies at the international level that are unlikely to be acceptable to the U.S. Congress, or the domestic body politic, could result in a serious loss of credibility. Conversely, to design a domestic program that is out of symmetry with the United States's international position could seriously undermine the ability of U.S. negotiators to reach meaningful agreements. A domestic program that exceeds the targets and requirements acceptable to a majority of other nations encourages massive free riding and could result in negligible net reductions in carbon dioxide. Since the climate change negotiations will last several decades, the long-term credibility of U.S. negotiation positions is a significant concern.

100.   Sylvan Wittwer. FOOD, CLIMATE AND CO_2: THE GLOBAL ENVIRONMENT AND WORLD FOOD PRODUCTION. 1995, p. 39. Not all agree with such extreme levels or warming (Balling, 1992; Idso, 1989; Lindzen, 1993; Michaels, 1991; Michaels et al. 1993; Seitz et al., 1989a; Strommen, 1992) and project much more tolerable, and even favorable changes when coupled with the direct benefits of rising levels of atmospheric CO_2. They predict that there would be a net benefit for food production. The position of a very modest temperature change related to the lower levels of increase is supported by the record of the past 100 years, where the average global temperature has increased only between 0.3 and 0.6°C (0.5 to 1.1°F). During this time the CO_2 level has increased up to 30%, and has been added to by even more rapid increases in other so-called greenhouse gases. With these rather remarkable increases in atmospheric CO_2, it is difficult to explain such low increments in temperature change, when in reality with the increases in not only CO_2, but other greenhouse gases as well, we should already be approaching the halfway mark of temperature increases projected for a doubling of CO_2 by the GCMs. In other words, because of the combined effect of these so-called greenhouse gases, we have already gone over halfway to an equivalent doubling of CO_2 (Balling, 1992; Hansen et al., 1989; Michaels, 1990; Wigley, 1987). Even so, there has been less than a half a degree of warming in the past 100 years. This is not all. The minor temperature increases (0.3 to 0.6°C) that have occurred during the past century could be attributed either to greenhouse warning, heat island effects at weather stations, inadequate sampling, or natural climate variability. None can be ruled out.

101.   Sylvan Wittwer. FOOD, CLIMATE AND CO_2: THE GLOBAL ENVIRONMENT AND WORLD FOOD PRODUCTION. 1995, p. 56. Finally, there is a scenario that offers that there has not been, and will likely not be a significant global warming, and if there is, it may be overall more desirable than the climate we now have. Also, that any climate change (warming) will be at the very lower levels of general circulation coming from the interpretation of projections of general circulation climate modeling. This scenario is based on the reality of what has happened in the earth's atmosphere thus far. It is emphasized that the computer projections of global warming are of known computational errors and do not include many variables, among which the most important is the effects of clouds, which may exceed by severalfold any anthropomorphic inputs of greenhouse gases, the inputs of oceans, and the biological sequestering by green plants and soils.

The rationale is that the models have predicted much more warming than has already occurred. The GCMs, with their known flaws, form the only basis for predicting catastrophic warming, from a doubling of greenhouse gases, and particularly CO_2, during the next century. If all the global warming of 0.4 to 0.6°C, which has occurred during the past century could be attributed to an increased green-

house effect, it would still be only one-third to one-half of the lowest now being predicted by current models.

102. Sylvan Wittwer. FOOD, CLIMATE AND CO_2: THE GLOBAL ENVIRONMENT AND WORLD FOOD PRODUCTION. 1995, p. 50. Predictions of the effects of global warming on agriculture have not factored in the use of plant and crop models nor the now well-known effects of elevated levels of atmospheric CO_2, on increased photosynthetic and water use efficiencies. As indicated earlier, the 1992 IPCC supplement (1993) now prefers the temperature range 1.5 to 4.5°C rather than the 3.0 to 5.5°C as indicated by Schneider (1989). Reference is now being made to even a lower range of 1 to 4°C (Guilderson et al., 1994). Also, if warming occurs, it will likely be within the lower end of this range. General circulation models are even less adapted to estimating potential changes in precipitation than potential temperature change. Further, for a comparison of the estimated precipitation, the models output for key agricultural regions in the United States did not even agree on the sign of change (Strommen, 1992). Landsberg (1984) has stated that neither the picture of the climate system nor its representation by numerical models has been fully grasped by many who have drawn far-reaching conclusions from on going climate research. *Precipitation changes are the most import as far as agriculture is concerned. It is how much, how soon, and with what regional distribution it occurs.* Here the atmospheric scientists relying on their mathematical models are noticeably silent. As Landsberg (1984) has reported, 28 model experiments based on these groups of models scatter temperature changes from less than 1°C to nearly 10°C. Most cluster around the 1.5 to 4.5°C above current levels, agreeing with the latest IPCC (1992) assessment of consensus. A substantial minority, approximating 25%, suggest 1°C or lower. We expect that future clustering will be around the lower range of the 1.5 to 4.5°C.

103. Patrick J. Michaels and Robert C. Balling. A GLOBAL WARMING FORUM: SCIENTIFIC, ECONOMIC AND LEGAL OVERVIEW. Ed. R. Geyer, 1993, p. 314. One vision of the world influenced by an enhanced greenhouse is that of ecological disaster resulting from rising temperature, evaporation rates, and sea level. Several lines of observational and model evidence now suggest that this scenario is becoming increasingly improbable.

The Northern Hemisphere, which should warm first and most, shows no significant change in trend over the last half-century. Repeated measurements now show relative warming at night, which may in fact be beneficial. The amount of global warming is clearly less than it should be according to the earlier GCMs, given the fact that we are already halfway to an effective doubling of CO_2. The only high-latitude signal that seems consistent with a greenhouse alteration is a 2.0°C warming of air masses whose average surface temperature is approximately -40°C, which is a slight modification of the air mass type that is most inhospitable in North America. Other anthropogenerated compounds may be mitigating the warming. New climate models partition almost all of the warming of more than 4°C to polar twilight or night, which will have a minimal effect on ice melting.

If this is the course the Earth has embarked upon in response to human insults of the atmosphere, that response is primarily benign, and possibly beneficial. However, whether indeed this is the true response remains to be seen.

104. Patrick J. Michaels and Robert C. Balling. A GLOBAL WARMING FORUM: SCIENTIFIC, ECONOMIC AND LEGAL OVERVIEW. Ed. R. Geyer, 1993, p. 313. It is tempting to view the unanimity of these newer models as a sign of reliability, but that is not the case. In a version of the NCAR model that increases greenhouse gases by a realistic 1%/year (as opposed to the "shock doubling" used in the model cited above), calculations based upon its results show that the current global temperature should be 0.7C above the 1950 mean; in fact the rise has been on the order of 0.3°C. Thus, even the most conservative climate model appears to have substantially overestimated global warming. This version also predicts very unrealistic regional temperatures for the current greenhouse enhancement. Figure 12 details the distribution of December to February temperature in the 5-year average of years 26 to 30 in the 30 year transient run. Roughly speaking, the map should be analogous to a 5-year aggregate in the period 1975 to 1985 (years 26 to 30 beginning in 1950). The most significant project anomalies are the 2 to 4°C warming of the northern half of North America and the 3 to 6°C cooling of the North Atlantic. Neither occurred, although the Atlantic temperatures may have dropped some (Rogers, 1989). This projected cold anomaly does not appear in later years of this transient model. On 5-year scales, anomalies of this magnitude over such large areas are very rare.

105. Robert C. Balling. STATE OF THE CLIMATE REPORT. 1995, p. 27. The bottom line is inescapable. The satellite data are extremely accurate, and the satellite record does not show the warming signal that should be easily identifiable since 1979. Even when the potential effects of volcanism and El Niño/La Niña are taken into account, only a small warming of 0.08°C/decade can be extorted from the satellite records (Christy and McNider, 1994). We are now in the 18th year of the satellite database, and to make matters even more confusing, December 1995 and January 1996 have been cold, particularly so, given the lack of any major volcanic eruption.

106. Dasheng Yang and Weiyu Yang. GLOBAL WARMING FORUM: A SCIENTIFIC, ECONOMIC AND LEGAL OVERVIEW. 1993, p. 21. IPCC

has reached the conclusion that the effect of 2% change of low cloud amount is equivalent to that of double CO_2 concentration, which is in accordance with our results. The percentage variation of the global total marine-based cloud amounts is depicted in Figure 12, which indicates that within the recent 100 years, percentage variation of the total cloud amount largely exceeds 2%. Hence, the feedback effect of the cloud is very fundamental, which has an effect much stronger than that of the variation of CO_2. In a climate-predictive model, it is certain that the cloud feedback effect must be correctly dealt with, otherwise a slight error in its formulation may lead to large distortion in the results.

107. Robert C. Balling. TRUE STATE OF THE PLANET. 1995, p. 90. So although the numerical models for testing the sensitivity of climate to various changes in atmospheric chemistry are powerful and complex, they are crude and deficient in many of the critical representations of the climate system. The models are getting better all the time, and advances in computing technology, along with advances in the atmospheric sciences, will produce the next generation of models. Therefore, I suspect climate model outputs of the immediate future will look quite different from today's results as this critical coupling of ocean, cloud, and other effects is achieved.

These existing models may be viewed as crystal balls for peering into the climatic future, and we are free to believe in as much or as little of their predictions as we like. I believe that as we learn more about the models, less confidence will be placed on their projections for the future. Many of the best modelers are ill at ease with any reference to their model outputs as forecasts or projections; they are far more comfortable using the phrase sensitivity experiments to describe their work. They understand that their climate models are not very good crystal balls. Nevertheless, many greenhouse proponents place remarkable confidence in the model outputs and demand policy changes based on their climatic crystal balls.

108. Sylvan Wittwer. FOOD, CLIMATE AND CO_2: THE GLOBAL ENVIRONMENT AND WORLD FOOD PRODUCTION. 1995, p. 53. Out of the above simulations of climate for the future, resulting primarily from current versions of models that indicate that global surface air temperatures will rise by 1.5 to 4.5°C when the CO_2 concentration is doubled in the atmosphere, have come an array of scenarios, most of them seemingly designed to frighten the public. The effects of clouds, and the way that models treat them, of oceans, the sequestering of atmospheric CO_2 by plants, losses from the soil, volcanoes, and the direct effects of rising levels of CO_2 on plant photosynthesis, dry matter accumulation, and water use efficiency are not factored into such projections.

109. Robert C. Balling. TRUE STATE OF THE PLANET. 1995, p. 93. The problems already described cannot be overlooked in the search for any greenhouse signal, yet the potential impact on the temperature record caused by the urban heat island effect represents a major contaminant to many of the temperature records. Recognizing that cities tend to warm their local environments, a number of scientists have attempted to quantify the urban heat island effect in the historical land-based temperature records of the globe. A variety of schemes have been used in these analyses, and from this research, it would appear that the global temperature data set has a global urban warming bias somewhere between 0.01°C and 0.10°C per century, with the most likely value near 0.05°C.

110. Robert C. Balling. TRUE STATE OF THE PLANET. 1995, p. 89. Many people have argued that the best models in the world are basically all predicting something of a climate disaster for a doubling of the atmospheric concentration of greenhouse gases. Given that these models represent the best tools for peering into the climate future, they argue that we should pay close attention to the warning. They very fairly asked how we can neglect this warning, or they asked how will we ever explain to future generations that we knew about the coming disaster and did nothing to stop it. If we place much faith in the models, then global warming must be viewed as a very real threat.

The same scientists who work on these models, however, are among the first to point out that the models are far from perfect representations of reality and probably are not advanced enough for direct use in policy implementation. Major weaknesses remain in the models; in particular, the role of the ocean in absorbing CO_2 and storing and transporting heat is not adequately included. Coupling the best ocean models to the best climate models is very tricky business computationally, but a necessary step in building more reliable forecast tools.

111. Robert E. Davis, STATE OF THE CLIMATE REPORT. 1996, p. 17. Emissions of anthropogenic greenhouse gases continue to increase rapidly. Atmospheric trace gas concentrations have gone halfway toward a doubling of pre-industrial concentrations. It is not only logical, but imperative, that the climate record be scrutinized to find evidence of the impact of these gases.

Careful examination of the atmospheric circulation record shows no evidence of a greenhouse-warming signal. In January, when the signal should be strongest, the Northern Hemisphere vortex has expanded—exactly the opposite of what models and theory indicate should occur. This particular expansion was associated with a more meridional flow pattern and a related increase in extremes—more cold air masses extending farther south over the eastern half of the United Sates, more warm air in Alaska, and an increase in the strength or nor'easters. Furthermore, this meridional flow pattern is not new; a very similar situ-

ation existed prior to the 1930s when the cause could not have been greenhouse gases.

Climatic change is natural and (at present) unpredictable. We are currently in a general pattern that produces more extreme conditions. There is substantial evidence that circulation patterns shift abruptly from one regime into another, and these sudden changes can occur at any time. Until theory and models advance to the point that they meld with historic observations in a systematic fashion, it is unwise to attribute any particular weather event, series of weather events, or weather in a year or series of years, to global warming.

112. Robert C. Balling. STATE OF THE CLIMATE REPORT. 1995, p. 24. Prediction of future temperature is obviously a complex undertaking in which an infinite number of feedbacks can destroy model accuracy. For example, if the models do not adequately stimulate precipitation patterns, soil moisture errors are created immediately. If the soil moisture is not properly represented, the surface energy balance is miscalculated, and the near-surface air temperature error increases. This new error initiates feedbacks in to evaporation processes and infrared radiative transfers, and on and on. As this simple example shows, small errors in the model are compounded by numerical representations of feedbacks that underlie the climate system, and these relatively small, but compounding errors seriously compromise the predictions of temperatures long into the future.

113. Ronald Bailey. ECOSCAM. 1993, p. 146. One telling example of how the models go awry is the case of chlorofluorocarbons. CFCs, which are implicated in destroying stratospheric ozone (see Chapter Eight), are also greenhouse gases. In fact, 20 to 25 percent of the greenhouse effect had been attributed to the buildup of CFCs. However, ozone is also a potent greenhouse gas. Taking the rise of CFCs into account along with the simultaneous reduction in ozone, CFCs produce no net increase in global warming. "It's a net wash to zero," says Dan Albritton of the National Oceanic and Atmospheric Administration. Getting the net effect of CFCs wrong means that climate model projections of future greenhouse warming were greatly exaggerated. Projected global temperatures plunged once the models correctly accounted for CFCs.

114. Sylvan Wittwer. FOOD, CLIMATE AND CO_2: THE GLOBAL ENVIRONMENT AND WORLD FOOD PRODUCTION. 1995, p. 51. *Models have consistently predicted a greater warming in the northern than southern hemisphere because of the greater extent of land in the north, which responds to radiative forcing.* The opposite, however, has occurred. There was a 0.3°C increase between 1955 and 1985 in the south with no warming in the north. Strangely, the IPCC supplementary report (1992) to the 1990 report states that "the size of the warming is broadly consistent with the predictions of climate models, but it is also of the same magnitude as natural climate variability." The IPCC supplementary report (1992) continues (emphasis is the author's): "There has been some clarifications of the nature of water vapor feedback, although the radiative effects of clouds and related process continue to be the major source of uncertainty and there remains uncertainties in the predicted changes in the upper tropospheric water vapor in the tropics. *Biological feedbacks have not been taken into account in simulations of climate change." Computer modeling of temperature for future climates say nothing about nonclimatic or direct effects.*

115. Roy W. Spencer. STATE OF THE CLIMATE REPORT. 1996, p. 11. In the context of global warming, the average trend in the satellite record during the 1979–1995 is a cooling of -0.05°C/decade, with an uncertainty of about ±0.03°C/decade. If the effects of volcanoes, El Niño and La Niña are removed, the adjusted trend becomes about +0.07°C/decade. This is just beneath the range of the latest global warming projections of +0.08°C to +0.30°C, and far beneath the range for the model that served as the basis for the Rio Treaty on Climate Change. When will the satellite measurements match global warming rates predicted by computer climate models? Maybe when those models are improved to the point where they contain all of the temperature-stabilizing process that exist in nature.

116. S. Fred Singer. SAN DIEGO UNION TRIBUNE. 1995, p. G-4. Weather satellites, however, have been measuring temperature with great precision on a truly global scale since 1979 with a single, well-calibrated microwave radiometer instrument. As it is widely known, but rarely mentioned, the satellite record of the last 15 years shows no significant global temperature increase. While the mathematical models predict a "best" temperature rise of 0.3 degrees C (0.5 F) per decade, the satellite data indicate, at most, one-fifth of that. If extrapolated to the next century, the feared global warming may not even be detectable above the "noise" of natural climate fluctuations.

117. Patrick J. Michaels. STATE OF THE CLIMATE REPORT. 1996, p. 6. The difference between ground-measured and satellite-sensed temperatures continues to grow. Which is "right"?

It is now apparent that the two histories oscillate in unison: When the ground-measured temperature is warm, so is the satellite, and *vice versa*. But there's a warming trend in the ground temperatures that simply doesn't show up in the satellites (Figure 5). In fact, a plot of the difference between the two is statistically significant; scientifically demonstrating that one is warming while the other is not (Figure 6).

Scientists at NASA's Goddard Institute for Space Studies believe that the discrepancy is caused by stratospheric ozone depletion "cooling" the satellite data. Is this true?

Ozone depletion is greatest in the far southern latitudes and moderate (at best) in the north polar region. There is no stratospheric depletion in the tropics. Therefore the satellite and ground-based temperatures should match up best in the tropics and worst in the high latitudes. As shown in Figure 7, the opposite is true. Verdict: The satellite temperatures are correct. Moreover, because they match up perfectly with weather balloon-measured temperatures between 5,000 and 30,000 feet, the global warming models are wrong. They all predict this area should show a smooth and statistically significant warming over the last 18 years—since the satellite data began in 1979.

Amid all the hoopla about 1995 as the warmest year in the ground record, the original news stories only included data from January through November. In the Northern Hemisphere, the satellite temperature departure from normal took the biggest drop in history between November and December.

118. S. Fred Singer. SAN DIEGO UNION TRIBUNE. 1995, p. G-4. Experts agree that the mathematical models do not simulate the atmosphere well enough, with most attention focusing on the inadequate way the models deal with atmospheric water vapor and clouds. By reflecting about 30 percent of solar radiation, clouds reduce the energy input from the sun. However, they also contribute significantly to trapping the heat radiation from the earth. The net effect of the clouds is a cooling of about five times the warming effect that is supposed to arise from a doubling of greenhouse gases. Clearly, therefore, even a small long-term change in cloudiness—or in their spatial and temporal distribution, or in cloud altitude and water content—can have a significant impact on climate.

119. Dasheng Yang and Weiyu Yang. GLOBAL WARMING FORUM: A SCIENTIFIC, ECONOMIC AND LEGAL OVERVIEW. 1993, p. 4. The global ocean total cloud amounts in recent 126 years (1854 to 1979) were analyzed and it was found that the cloud cover had a tendency toward increasing. This tendency is in agreement with the global warming and intensification of the CO_2 greenhouse effect. By a one-dimensional radiative equilibrium model, we demonstrate that within the last 100 years, the cooling of the air temperature due to the variance of the cloud amount largely exceeds the warming arising from the greenhouse effect by the intensification of the CO_2 concentration.

120. Dasheng Yang and Weiyu Yang. GLOBAL WARMING FORUM: A SCIENTIFIC, ECONOMIC AND LEGAL OVERVIEW. 1993, p. 18. Cloud has following feedback effect to climate. When the globe gets warm, the evaporation on the underlying surface strengthens, which in turn induces the increase of atmospheric humidity and then the global cloud amount. As noted before, cloud has both positive and negative feedback effects to climate. The water vapor in clouds is the constituent of the greenhouse gases, which will induce global warming. On the other hand, cloud augments the albedo of the Earth atmosphere system, so it will lower the global temperature. Both of these effects may finally result in a diminution of the tendency of the global warming.

The feedback mechanism depends on the cloud height. The high cloud is biased towards greenhouse effect, whereas the middle and low clouds contribute chiefly to the negative feedback effect. Based on the report of IPCC, the warming effect by doubling atmospheric CO_2 concentration is cancelled by increasing 2% of the low cloud cover.

Climate is very sensitive to the cloud feedback effect. In order to understand the inner connection and the mutual restrictive mechanism of the changing climate system, plus to provide basis for improving the climatic model and also to eliminate the uncertainty in climate prediction, it is necessary to have a correct grasp of the variation of the global cloud amount.

121. Dasheng Yang and Weiyu Yang. GLOBAL WARMING FORUM: A SCIENTIFIC, ECONOMIC AND LEGAL OVERVIEW. 1993, p. 21. IPCC has reached the conclusion that the effect of 2% change of low cloud amount is equivalent to that of double CO_2 concentration, which is in accordance with our results. The percentage variation of the global total marine-based cloud amounts is depicted in Figure 12, which indicates that within the recent 100 years, percentage variation of the total cloud amount largely exceeds 2%. Hence, the feedback effect of the cloud is very fundamental, which has an effect much stronger than that of the variation of CO_2. In a climate-predictive model, it is certain that the cloud feedback effect must be correctly dealt with, otherwise a slight error in its formation may lead to large distortion in the results.

122. Patrick J. Michaels and Robert C. Balling. A GLOBAL WARMING FORUM: SCIENTIFIC, ECONOMIC AND LEGAL OVERVIEW. Ed. R. Geyer, 1993, p. 303. This behavior of the HCN is consistent with an enhanced greenhouse combined with the increases in cloudiness (of 3.5%) and reduced sunshine that have been documented across the coterminous U.S. (Angell, 1990), Weber (1990b) has also documented a decline in sunshine in Germany, and notes that the effect is enhanced in the mountains, which implicates stratocumulus (Sc)—the low altitude cloud type most effective at surface cooling—as the cause. Warren et al. (1988) have also found an increase in global cloudiness (with increases greater in the Northern Hemisphere) at most marine locations around the globe, although the shipboard observations used are subject to substantial observer and scale

biases. Nonetheless, the cloud type that shows the most increase is again the low-level Sc, and again in the Northern Hemisphere. The finding of a decline in UV radiation at low elevations (Scotto et al., 1988) in combination with increased values at heights >10 km (Bruhl and Crutzen, 1989) is also consistent with an increase in low level cloudiness.

123. Sylvan Wittwer. FOOD, CLIMATE AND CO_2: THE GLOBAL ENVIRONMENT AND WORLD FOOD PRODUCTION. 1995, p. 54. Scenarios based upon the results of three-dimensional mathematical GCMs may be the best climatic predictive models available, but they are not well adapted for estimating potential changes in either global or regional temperatures or precipitation. No definite trends have yet been shown to exist in either. They are deficient in that they do not adequately incorporate the effects of clouds, which are expected to increase if there is a warming.

124. Sylvan Wittwer. FOOD, CLIMATE AND CO_2: THE GLOBAL ENVIRONMENT AND WORLD FOOD PRODUCTION. 1995, p. 54. As already indicated, clouds may have both negative and positive impacts on temperature. The negative effect comes from reflecting sunlight into space. The trapping of heat from below increases the temperature (Abelson, 1990). The water vapor content in the atmosphere, which is highly variable with time and season, is the atmosphere's most abundant greenhouse gas. A small increase in the earth's albedo related to increased cloudiness expected with global warming would produce a net cooling 5 to 10 times greater than the potential contribution for warming from all greenhouse gas increases (Strommen, 1992). When the water content and other characteristics of clouds were incorporated in one model, the predicted global warming from the doubling of CO_2 dropped from 5.2 to 19.°C (Mitchell et al., 1990).

125. S. Fred Singer. WASHINGTON TIMES.
21 March 1996, p. A11. But climate warming is turning out to be a scientific non-problem, despite attempts to picture it as the "greatest global challenge facing mankind." Contrary to all forecasts from computer models, weather satellite data over the last 18 years show no warming whatsoever.

To overcome this embarrassing disagreement, the United Nations' scientific advisers, in a series of assessments published since 1990, have resorted to convoluted language, and recently even went so far as to alter an underlying scientific report to make it agree with a politically negotiated "policy-makers' summary." This change was only discovered when their just published 1996 report was compared with the approved final draft. The surreptitious alterations were brought to wide attention in a June 12 editorial essay in the Wall Street Journal, authored by Frederic Seitz, a former president of the U.S. National Academy of Sciences and holder of the National Medal of Science.

126. S. Fred Singer. SAN DIEGO UNION TRIBUNE.
1995, p. G-4. Weather satellites, however, have been measuring temperature with great precision on a truly global scale since 1979 with a single, well-calibrated microwave radiometer instrument. As is widely known, but rarely mentioned, the satellite record of the last 15 years shows no significant global temperature increase. While the mathematical models predict a "best" temperature rise of 0.3 degrees C (0.5 F) per decade, the satellite data indicate, at most, one-fifth of that. If extrapolated to the next century, the feared global warming may not even be detectable above the "noise" of natural climate fluctuations.

127. Thomas Gale Moore. ECONOMIC TIMES.
March 1995, p. 4. In fact, the evidence supporting the claim that the earth has grown warmer is shaky. The theory is weak and the models on which the conclusions are based cannot even replicate the current climate. It is asserted, for example, that over the last 100 years, the average temperature of the earth's surface has gone up by about 1° F (Fahrenheit). However, given the evidence that in the United States temperatures have failed to rise, and British naval records that find no significant change in temperatures at sea since the mid -1800s, the public is entitled to be wary.

128. Sylvan Wittwer. FOOD, CLIMATE AND CO_2: THE GLOBAL ENVIRONMENT AND WORLD FOOD PRODUCTION. 1995, p. 131. As to global warming, society is responding to a threat produced only in the world of admittedly flawed mathematical models. Any threat of climate change has been projected as a disadvantageous expectation. No real world data have been addressed to demonstrate any threat to human, animal, or plant welfare response to a CO_2 and other greenhouse gas warming.

129. C. J. E. Schuurmans. CLIMATE CHANGE: IMPACT ON COASTAL HABITATION. Ed. Docke Eisma, 1995, p. 9. Several reasons have been put forward to explain the apparently lower observed rate of warming over the past century. We mention them here, without a detailed discussion. First, our theoretical estimate of warming might be too large. We mentioned earlier the uncertainties in regard to the feedback of clouds. This and other reasons lead to a large range of uncertainty of $(\Delta T_s)_c$ at $2 \times CO_2$, namely $1.5 < (\Delta T_s)_c < 4.5°C$. Second, factors other than the greenhouse effect might have contributed to the observed record of T_s. The most important of these are changes of solar radiation, volcanic dust, and also man-made aerosols (particles) in the atmosphere. Another important factor is the internal interaction within the cli-

mate system, e.g., between the atmosphere and the oceans (see the next section).

Because of this large number of intervening factors, it is impossible to conclude what part of the observed heating is caused by the greenhouse effect, at least not from a comparison of global mean estimates alone. This attribution problem can only be overcome when observed warming rates are larger than might be expected from any of the factors, except the greenhouse effect.

130. S. Fred Singer et al. A GLOBAL WARMING FORUM: SCIENTIFIC, ECONOMIC AND LEGAL OVERVIEW. Ed. R. Geyer, 1993, p. 349. K. Hanson, T. Karl, and G. Maul, scientists of the National Oceanic and Atmospheric Administration (NOAA), find no overall warming in the U.S. temperature record, contrary to the global record assembled by J. Hansen, of the National Aeronautics and Space Administration (NASA). Using a technique that eliminates urban "heat islands" and other local distorting effects, the NOAA scientists confirm the temperature rise before 1940, followed, however, by a general decline. Newell and colleagues at the Massachusetts Institute of Technology (MIT) report no substantial change in the global sea-surface temperature in the South Indian Ocean and global network.

131. Richard A. Geyer. A GLOBAL WARMING FORUM: SCIENTIFIC, ECONOMIC AND LEGAL OVERVIEWS, Ed. R. Geyer, 1993, Introduction. Drs. John Moen and Tony Michaels, research oceanographers at the Bermuda Biological Station, have analyzed results of temperature data collected for the last 37 years. Their results indicated that the adjacent oceanic waters have cooled during the past 40 years. This is contrary to the greenhouse effect and the results of climate-change model studies. In addition, the temperatures dropped dramatically between 1950 to 1970. Also, over the past 20 years, the adjoining ocean waters have become steadily warmer, although some depths show even lower temperatures than those recorded in 1954. Additional examples of this difficulty to various degrees of intensity can be found from time to time in some of the chapters in this book, as well as in some of the references.

132. Patrick J. Michaels and Robert C. Balling. A GLOBAL WARMING FORUM: SCIENTIFIC, ECONOMIC AND LEGAL OVERVIEW. Ed. R. Geyer, 1993, p. 303. The very negative vision of future climate is not supported by recent studies on daily temperature regimes. Karl et al. (1988a) examined maximum and minimum values from the U.S. HCN and found that the daily range (difference between the two) has declined precipitously since 1950, and is now two standard deviations below the mean for the century. In that record, maximum values have actually declined, while minimum values have risen (Figure 5).

133. Sylvan Wittwer. FOOD, CLIMATE AND CO_2: THE GLOBAL ENVIRONMENT AND WORLD FOOD PRODUCTION. 1995, p. 40. The water vapor content in the atmosphere, although highly variable with time and season, is the most abundant greenhouse gas. It is only because water vapor is able to selectively trap outgoing longwave radiation that life on our earth is possible. Otherwise, it would be too cold to sustain life. Small increases in the earth's albedo relating to increased cloudiness expected from global warming will produce a net cooling effect far greater than the potential contribution of warming from all greenhouse gas increases including that of CO_2 (Strommen, 1992). It has been suggested that the world's clouds may have an effect on the earth's energy budget that is up to 10 times greater than would be produced by a doubling of the amount of CO_2 in the atmosphere (Wildavsky, 1992). A slight change in cloud cover could either negate global warming induced by greenhouse gases or greatly intensify it. It is concluded that until clouds can consistently find their way into GMCs, they will be of little value in such predictions (Abelson, 1990; Anonymous 1993b). These GMC projections, in turn, have also been hotly contested and seem not to be supported thus far by real earth observations. The National Academy of Sciences (1991; 1992) stated that increases in atmospheric greenhouse gas concentrations will probably be followed by increases in the average global atmospheric temperature. But we cannot accurately predict how rapidly such changes may occur, or how intense they will be for any given atmospheric concentration or location. More important will be what regional changes in temperature, precipitation, wind speed, frost occurrence, and length of the growing season can be expected. Thus far, no large or rapid increases in temperature have occurred, and there is no evidence yet of a CO_2-induced temperature or climate change. One projection of the GCMs is that warming would start first and grow fastest at the poles. Yet a major study of satellite data has shown a slight cooling trend above the Arctic, with warming only at night, and only in the southern hemisphere (Karl et al., 1993). Both the benefits of longer growing seasons and fewer killing frosts would have to be factored against liabilities of increased pest infestation. Decreased cropping areas would have to be considered if such conditions prevailed.

134. Patrick J. Michaels and Robert C. Balling. A GLOBAL WARMING FORUM: SCIENTIFIC, ECONOMIC AND LEGAL OVERVIEW. Ed. R. Geyer, 1993, p. 300. While the citation is often made that the history of global temperature as "at least not contradictory to" (MacCracken and Kukla, 1985) the mid-1980s climate model projections, those models produce an expected equilibrium warming for this trace gas change of over 2.0°C, assuming the climate temperature sensitivity is 1°C/W/m², while a least-squares fit of the updated set of Jones et al. (1990) data of the U.S. Department of Energy (1990) global data gives a rise of 0.45 ± 0.10°C in the last

100 years (Balling and Idso, 1990). Further, Figure 1 demonstrates that virtually all of the warming of the Northern Hemisphere was prior to the major postwar emissions of the trace gases; a linear trend through the date since 1940 is statistically indistinguishable from zero (p = .05). In addition, the "water" (Southern) Hemisphere, which should warm up slower, in fact shows the more "greenhouse-like" signal. While most of the Southern Hemisphere stations are on land, the fact that so much of the Southern Hemisphere is water results in more oceanic influence on the record than occurs in the Northern Hemisphere. Least-squares analysis of this record demonstrates, in a statistical sense, that much of the observed warming was already realized prior to 1945 (Balling and Idso, 1990); see Figure 2. At this time the actual CO_2 concentration was in the range of 307 ppm and the effective concentration was approximately 322 ppm. This is only one third of the total change in forcing to date, and implies that there has been very little additional warming during the period in which the majority of the radiative forcing has taken place.

135. Richard A. Geyer. A GLOBAL WARMING FORUM: SCIENTIFIC, ECONOMIC AND LEGAL OVERVIEWS. Ed. R. Geyer, 1993, Introduction. The Phoenix (or comparison) Group consisted primarily of academicians from the University of Virginia, M.I.T., and other well-known institutions. They concluded that, "it is not possible to attribute all, or even a large part of the observed global warming since 1890 to an increased greenhouse effect based on currently available data." Similarly, on future warming the various theoretical models used to predict climate have not been adequately validated by the existing climatic record. Thus, in theory if this is applied retroactively to 1890, it fails to predict the observed climatic patterns that have actually occurred. Also, global mean-surface air temperatures have increased between 0.3 to 0.6°C over the last 100 years. However, these values are of the same magnitude as those observed for natural climate variability.

136. S. Fred Singer. WASHINGTON TIMES.
21 March 1996, Lexis/Nexis. Global warming is rapidly becoming a non-problem. The latest calculations predict a much reduced temperature rise for the next century; at its low end it will be barely detectable and certainly inconsequential. The problem for the promoters of warming is how to keep the good news from leaking out to the public.

Activists, politicians, and some scientists have by now acquired a vested interest in a climate catastrophe. They react strongly to any skepticism, often with personal attacks that verge on desperation. Vice President Al Gore, for example, has even invented a group label—the "know-nothing society"—a "tiny minority of scientists" that regard global warming as the "empirical equivalent of the Easter Bunny." (His words, not mine—but well said.)

137. William F. O'Keefe. VITAL SPEECHES.
15 August 1995, p. 654. I will tell you at the outset that I intend to be provocative and to challenge what is portrayed as the prevailing wisdom, not to discount the potential of future climate change or the possibility of serious consequences, but to challenge, the current public policy progress.

That process is in fact not consistent with the view that mankind is confronted with a potential catastrophe. Instead, global warming appears to be yet another example of predicting catastrophe to reshape political decisions and economic outcomes. This is an overused strategy.

Indeed over five decades ago the political satirist H. L. Mencken observed, "The whole aim of popular politics is to keep the populace alarmed (and hence clamorous to be led to safety) by menacing it with an endless series of hobgoblins, all of them imaginary."

Dr. Richard Lindzen of MIT recently sent me a paper that compared global warming to the debate over heredity, eugenics and immigration laws in the 1920s. Fears of genetic degradation led to an oversimplification of the science and pseudo-science became the basis for political advocacy. The reluctance of the scientific community to become visibly involved in the policy debate made it easier for Congress to pass the 1924 Immigration Restriction Act—closing America's doors until the 1960s—long after the pseudo-science had been discredited. The parallels here with the "forged consensus" on global warming are many and instructive.

138. WASHINGTON POST. 16 July 1995, p. W17.
Several years ago, John Harte, an ecologist at the University of California at Berkeley, was among those invited by the Library of Congress to brief some Congress members on global climate change. A few senators drifted in and out of the presentation, recalls Harte, whose own study area has recently been the effect of warmer temperatures on a plot of forbs and sagebrush. After listening to the earnest academics detail the possibilities and perils of warming, Sen. Bill Bradley (D-N.J.) asked: "But where are the rabbits?"

If the academics meant to teach the politicians a lesson in science, Bradley meant to give the scientists a lesson in politics. Without a threat to something cute and furry, or at least to something immediate in the lives of his constituents, the issue was going to be an extremely hard sell.

And that political fact of life, argue the skeptics, has led environmentalists to hype the threat of global warming.

Those environmentalists most concerned about it have found the threat hard to sell even to others in the movement. Michael Oppenheimer of the Environmental Defense Fund, whose training in atmospheric physics alerted him early on to the potential for climate change, remembers the early 1980s, when he couldn't drum up support from other groups. They found global warming too complicated and the threat too far in the future. Environmentalists didn't

think they could interest, let alone rally, the public, telling Oppenheimer: "It's not in their face."

139. J. Sebastian Sinisi. DENVER POST. 24 March 1996, Lexis/Nexis. "Regardless of where you stand on global warming," said Robert Balling, director of climatology at Arizona State University in Tempe, "there's no question that starting to plan for a future with less fossil fuel makes sense. But to run out and impose a carbon tax tomorrow makes no sense to me.

"I'm in the skeptical camp and don't think there's a climate mandate demanding policy action at this time. In my view, the evidence isn't conclusive enough for much deviation from 'business as usual' over the next 20 years."

His 1992 book, "Greenhouse Prediction Versus Climate Reality," challenged the accuracy of computer models that forecast dire greenhouse effects.

"The computer models that point to catastrophe are still primitive," he said. "And they don't accurately measure many of the key processes that are relevant in the greenhouse debate."

The debate is egged on by the media, which seek to set up polarized controversy, he said.

"Stephen Schneider (climate scientist at Stanford University) and I were at a recent climate-change retreat outside Washington, D.C.," said Balling.

"The cameras were rolling, and all they seemed interested in was setting up a 'crossfire' debate. The reality isn't that simple, but the media demands simple answers.

"I don't think any credible scientist would say the Blizzard of '96 was the result of global warming—but that's the hook seized by the mainstream media.

"Last December was one of the coldest months on record in this hemisphere. Where were the headlines?

"Nobody covers all the caveats in the global warming theory because it isn't consistent with the story that the media has become part of in recent years.

"The answers aren't conclusive," Balling said.

"The world is warming—but not in a pattern consistent with the official global warming theory. I can buy into another half-degree of warming during the next century-along with cloudier and wetter weather.

"But that doesn't add up to catastrophe. In Phoenix, a wetter world doesn't seem that bad."

140. Sylvan Wittwer. FOOD, CLIMATE AND CO_2: THE GLOBAL ENVIRONMENT AND WORLD FOOD PRODUCTION. 1995, p. 51. *Models have consistently predicted a greater warming in the northern than southern hemisphere because of the greater extent of land in the north, which responds to radiative forcing*. The opposite, however, has occurred. There was a 0.3°C increase between 1955 and 1985 in the south with no warming in the north. Strangely, the IPCC supplementary report (1992) to the 1990 report states that "the size of the warming is broadly consistent with the predictions of climate models, but it is also of the same magnitude as natural climate variability." The IPCC supplementary report (1992) continues (emphasis is the author's): "There has been some clarifications of the nature of water vapor feedback, although the radiative effects of clouds and related process continue to be the major source of uncertainty and there remains uncertainties in the predicted changes in the upper tropospheric water vapor in the tropics. *Biological feedbacks have not been taken into account in simulations of climate change." Computer modeling of temperature for future climates say nothing about nonclimatic or direct effects.*

141. Sylvan Wittwer. FOOD, CLIMATE AND CO_2: THE GLOBAL ENVIRONMENT AND WORLD FOOD PRODUCTION. 1995, p. 53. Out of the above simulations of climate for the future, resulting primarily from current versions of models that indicate that global surface air temperatures will rise by 1.5 to 4.5°C when the CO_2 concentration is doubled in the atmosphere, have come an array of scenarios, most of them seemingly designed to frighten the public. The effects of clouds, and the way that models treat them, of oceans, the sequestering of atmospheric CO_2 by plants, losses from the soil, volcanoes, and the direct effects of rising levels of CO_2 on plant photosynthesis, dry matter accumulation, and water use efficiency are not factored into such projections.

142. 1992 SUPPLEMENT, GLOBAL WARMING—PART 2. Y4.En 2/3: 103-92, 1992, p. 314. Confidence in regional climate patterns based directly on GCM output remains low and there is no consistent evidence regarding changes in variability or storminess. GCM results can be interpolated to smaller scales using statistical methods (correlating regional climate with the large-scale flow) or a nested approach (high-resolution, regional climate models driven by large-scale GCM results). Both methods show promise but an insufficient number of studies have yet been completed to give an improved global picture of regional climate change due to increases in greenhouse gases; in any event both interpolation methods depend critically on the quality of the large-scale flow in the GCM. Given our incomplete knowledge of climate, we cannot rule out the possibility of surprises.

143. 1992 SUPPLEMENT, GLOBAL WARMING—PART 2. Y4.En 2/3: 103-92, 1992, p. 314. There has been some clarification of the nature of water vapour feedback, although the radiative effects of clouds and related processes continue to be the major source of uncertainty and there remain uncertainties in the predicted changes in upper tropospheric water vapour in the tropics. Biological feedbacks have not yet been taken into account in simulations of climate change.

Increased confidence in the geographical patterns of climate change will require new simulations with improved coupled models and with radiative forcing scenarios that include aerosols.

144. Dr. Robert T. Watson. GLOBAL WARMING—PART 3. Y4.EN2/3: 103–167. 1994, p. 44.
Aerosols: While we have an improved understanding of: (i) the different sources of aerosols (emissions of sulfur dioxide arising from the combustion of fossil fuels, and biomass burning), and (ii) the influence of aerosols on radiative forcing, the implications of increased atmospheric concentrations of aerosols on changes in surface climate are not as well understood. Quantification of the regional and global effects of aerosols on surface climate will require fully coupled ocean-atmosphere models. Understanding the influence of aerosols on the Earth's climate is a high priority research area for the U.S. Global Change Research Program.

145. Patrick J. Michaels and Robert C. Balling. A GLOBAL WARMING FORUM: SCIENTIFIC, ECONOMIC AND LEGAL OVERVIEW. Ed. R. Geyer, 1993, p. 303. All GCMs suggest that the polar warming will be magnified in winter. Nonetheless, there has been a substantial secular decline in winter temperatures over the Atlantic Arctic since 1920 (Rogers, 1989), and there has been no change in polar night temperatures at the South Pole (Figure 4B)—a station that surely has no urban warming (Sansom, 1989). Kalkstein et al. (1990) have documented that wile there has been no net warming of the North American arctic, the coldest air masses, whose mean surface temperatures are approximately -40°, have warmed some 2°C. Thus, the most severe air mass in North America may be undergoing some mitigation that is consistent with an enhanced greenhouse; nonetheless, the lack of any observed warming in the Antarctic polar night data seems at variance with this finding. In their coupled ocean-atmosphere model. Nanabe et al. (1991) project very little Antarctic warming, and even some cooling where cloud cover may increase.

146. C. J. E. Schuurmans. CLIMATE CHANGE: IMPACT ON COASTAL HABITATION. Ed. Doeke Eisma, 1995, p. 14. As far as storms are concerned, again we have to rely on the GCM experiments. First, it must be remarked that until now simulations of storm tracks of extratropical cyclones for the present climate have not been completely successful. In the simulations, cyclone activity in general is somewhat lower than in the real atmosphere, while Atlantic depressions on average move too far into Europe before they disappear. The reasons for these systematic errors are being studied.

147. Patrick J. Michaels and Robert C. Balling. A GLOBAL WARMING FORUM: SCIENTIFIC, ECONOMIC AND LEGAL OVERVIEW. Ed. R. Geyer, 1993, p. 301. Localized warming related to urban buildup has long been recognized as a confounding factor in regional and global temperature analysis. While causes include changes in the surface energy balance, local wind flow, and amount of open sky, precise calculations on the various components of the "urban effect" are not available. However, Karl et al. (1988a) have found in the U.S. that the effect is statistically detectable at population levels of 2500 and up, and Balling and Idso (1989) note the effect with population as low as 500. The temperature record is further confounded by the fact that exponential populations increases suggest the urban warming bias should be of the same functional form as enhanced greenhouse forcing, and that it will therefore tend to contaminate the latest years of temperature measurement. In other words, it would be striking if the last years in the record *were not* the warmest, although the least-squares global warming in the Jones and Wigley record since 1970 (which, on a smoothed basis, is the lowest point since 1915) of 0.35°C is certainly greater than any urban warming. The global trend for the last half-century (since 1940) is 0.21°C.

148. 1992 SUPPLEMENT, GLOBAL WARMING—PART 2. Y4.En 2/3: 103-92, 1992, p. 318. Since the 1990 report there has been a greater appreciation of many of the uncertainties which affect our predictions of the timing, magnitude and regional patterns of climate change. These continue to be rooted in our inadequate understanding of:

- sources and sinks of greenhouse gases and aerosols and their atmospheric concentrations (including their indirect effects on global warming)
- clouds (particularly their feedback effect on greenhouse-gas-induced global warming, also the effect of aerosols on clouds and their radiative properties) and other elements of the atmospheric water budget, including the processes controlling upper-level water vapour
- oceans, which through their thermal inertia and possible changes in circulation, influence the timing and pattern of climate change
- polar ice sheets (whose response to climate change also affects predictions of sea level rise)
- land surface process and feedbacks, including hydrological and ecological processes which couple regional and global climates.

149. 1992 SUPPLEMENT, GLOBAL WARMING—PART 2. Y4.En 2/3: 103-92, 1992, p. 310.
Atmospheric Gas Phase Chemistry Models: Current tropospheric models exhibit substantial differences in their predictions of changes in O_3, in the hydroxyl radical (OH) and in other chemically active gases due to emissions of CH_4, CO, non-methane hydrocarbons and, in particular NO_x. These arise from uncertainties in the knowledge of background chemical composition and from our inability to represent small-scale processes occurring within the atmosphere. These deficiencies limit the accuracy of predicted changes in the abundance and distribution of tropospheric O_3, and in the lifetimes of a number of other greenhouse gases, including the HCFCs and HFCs, all of

which depend upon the abundance of the OH radical. Increases in CH_4, NMHC, and CO all lead to increases in O_3 and decreases in OH, thus leading to an increase in radiative forcing. On the other hand because increases in NO_x lead to an increase in both O_3 and OH, the net effect of NO_x on radiative forcing is uncertain.

150. Sylvan Wittwer. FOOD, CLIMATE AND CO_2: THE GLOBAL ENVIRONMENT AND WORLD FOOD PRODUCTION. 1995, p. 6. It is further recognized that predictive climate modeling, especially global circulation models, have severe limitations, and little credence is given them. Successful crop modeling with multiple factor inputs relating to productivity is in its infancy. Crop modeling even for single factor inputs, such as heat sums for predicting maturity and scheduling times of harvests and times for processing in sweet corn and peas, is of limited value. General circulation predictive climate modeling is one dimension. Modeling the productivity responses of plants and ecosystems is a much more formidable task. It requires a predictive understanding in the disciplines of physiology, genetics, organismic biology, population dynamics, community dynamics, soil physics and chemistry, and systems ecology. To add to this complexity, the direct benefits and possible hazards to plants of a rising level of atmospheric CO_2 as well as the development of the methodology to determine this effectively remains a still more formidable task in global and crop productivity research (U.S. Department of Energy, 1992).

151. 1992 SUPPLEMENT, GLOBAL WARMING—PART 2. Y4.En 2/3: 103-92, 1992, p. 362. The dire predictions of elevated temperatures, rising sea levels, increased storm activity and agricultural dislocation that are made by some are primarily based on the general circulation models (GCMs). These computer models, as sophisticated as they are, remain inadequate for setting national and international policy regarding future climate change. One example to prove this point is the timing of any "global warming". The Intergovernmental Panel on Climate Change (IPCC) indicated in 1992 that a doubling of total CO_2 in the atmosphere may indicate (based on the GCMs) a potential warming of 1.5° C to 4.5° C by the year 2100, <u>almost 50 years later</u> than the estimate reported in 1990. Indeed, current models that have not been able to replicate the past climate changes cannot be relied on for forecasting temperatures 50 to 100 years from now.

152. 1992 SUPPLEMENT, GLOBAL WARMING—PART 2. Y4.En 2/3: 103-92, 1992, p. 368. Future research on and incorporation into the GCMs of the role of water vapor, clouds, oceans, polar ice caps, solar activity and a host of other variables is vital to improving the capability of the models. The Administration should acknowledge that it will approach the need to resolve scientific uncertainty with the same level of urgency as it does its plans to monitor and evaluate progress on the Action Plan. Long term strategies for reducing greenhouse gas emissions, and particularly policies which contemplate legislative or regulatory mandated actions with adverse economic consequences to the U.S. and the world, cannot be prudently fashioned until this uncertainty is substantially reduced.

153. S. Fred Singer et al. A GLOBAL WARMING FORUM: SCIENTIFIC, ECONOMIC AND LEGAL OVERVIEW. Ed. R. Geyer, 1993, p. 351. The models are "tuned" to give the right mean temperature and seasonal temperature variation, but they fall short of modeling other important atmospheric processes, such as the poleward transport of energy via ocean currents and atmosphere from its source in the equatorial region. Nor do the models encompass longer-scale processes that involve the deep layers of the oceans or the ice and snow in the Earth's cryosphere, nor fine-scale processes that involve convection, cloud formation, boundary layers, or processes that depend on the Earth's detailed topography.

154 RICHMOND TIMES DISPATCH. 25 June 1996, p. A-12. In view of which, now might be the time to call attention to the Leipzig Declaration on Global Climate Change. Signed this month by more than 100 European and American scientists, it says the available evidence does not support global-warming theories, and that attempts to eliminate "what are perceived to be" the driving forces behind "potential climate change" would be "dangerously simplistic."

The Declaration is worth quoting at some length:

> As scientists, we . . . are intensely interested in the possibility that human activities may affect the global climate; indeed, land clearing and urban growth have been changing local climates for centuries. Historically, climate has always been a factor in human affairs—with warmer periods, such as the medieval "climate optimum," playing an important role in economic expansion.

Contrary to conventional wisdom . . . there does not exist today a general scientific consensus about the importance of greenhouse warming from rising levels of carbon dioxide. On the contrary, most scientists now accept the fact that actual observations from Earth satellites show no climate warming whatsoever. . . . [ellipses in original]

155. Michelle Malkin. SEATTLE TIMES. 16 July 1996, p. B4. The latest report from the United Nations' International Panel on Climate Change (IPCC) suggests that curtailing human-induced carbon emissions is the key to staying cool. But throw in additional natural factors such as volcanic eruptions, oceans, water vapor, phytoplankton, and vegetation that may influence climate change, and the idea of wielding carbon taxes as a primary weather-altering weapon suddenly seems quixotic. (Dissidents are now assailing the IPCC after discovering that

some members altered language in its report to expunge all signs of doubt about the role of industrial activity in global warming.

156. Dale E. Heydlauff. GLOBAL WARMING—PART 3. Y4.En2/3: 103–167), 6 October 1994, p. 97. Recent developments have raised serious concerns on our part regarding the IPCC process. We believe that the U.S. representatives to the IPCC should strive to ensure that the activities undertaken under the auspices of the IPCC promote a full, open and robust evaluation of the science of the global climate change issue. Even with the increasing base of knowledge about the global climate system, there is no consensus on and substantial uncertainty remains concerning many critical scientific issues. Therefore, the proceedings of the various working groups of the IPCC should accord due process to all commenters and should include all views in the published findings of these groups.

157. Peter M. Morrisette. FORUM FOR APPLIED RESEARCH AND PUBLIC POLICY. Summer 1996, p. 65. In the same year, the Intergovernmental Panel on Climate Change (IPCC), an international team of scientists sponsored by WMO and UNEP, became the principle instrument for assessing global climate change. This panel had the full support of participating governments.

Such support, however, proved a two-edged sword. On the one hand, it legitimized the panel's findings. On the other, it limited the scientists' ability to pursue other issues and agendas by placing their research under government supervision. The panel's initial findings were made available in time for the Second World Climate Conference in November 1990.

158. Richard A. Geyer. A GLOBAL WARMING FORUM: SCIENTIFIC, ECONOMIC AND LEGAL OVERVIEW. Ed. R. Geyer, 1993, Introduction. Another more recent example involves the reporting by the IPCC on the Scientific Assessment of Climate Change Conference held in 1990 in which 170 scientists from all over the world participated. No scientific consensus was actually reached on greenhouse effects as these relate to the global warming theory. It was only in the much publicized nontechnical summary, that this appeared to be so. It was widely distributed to politicians and the general public. They could easily interpret this summary as an authoritative analysis reached by the "international community."

Subsequently, a separate survey was conducted of the conference participants by Dr. Singer, Professor of Environmental Sciences at the University of Virginia and Mr. Winston, former Director of the Climate Analysis Center of the National Weather Service. Their objective was to determine the views of the conference participants independently. The response was sent to 126 scientists, 31 IPCC reviewers, and the 24 members of the Phoenix Group. The results were statistically significant. The response of the Phoenix Group was 60% and 21% for the IPCC contributors.

159. S. Fred Singer, SAN DIEGO UNION TRIBUNE. 1995, p. G-4. Trying to prop up a scientific rationale for global warming, Gore keeps referring to a "tiny minority" of scientific skeptics. At the same time, however, more and more independent researchers are raising their voices in the greenhouse debate and attacking the scientific basis of the Global Climate Treaty.

Even the U.N.-sponsored Intergovernmental Panel on Climate Change (IPCC) had to admit to the 1992 Earth Summit that there were dissenting views which the IPCC report editors "could not accommodate." As surveys published later revealed, a majority of the scientists who worked on the IPCC report disagreed with its executive summary.

160. William F. O'Keefe. VITAL SPEECHES. 15 August 1995, p. 654. Those who advocate deep emission reductions contend that a scientific consensus exists and that man-made emissions of greenhouse gases will lead to a dangerous level of global warming. That's simply wrong. For example, it's been demonstrated that uncertainty about variations in water vapor can overwhelm models that are used for policy recommendations. According to the various polls that have been taken of climate experts, no more than 40 percent support the premise that man-induced warming is now underway. That is hardly a consensus.

161. Michelle Malkin. SEATTLE TIMES. 16 July 1996, p. B4. National newspapers, magazines and nightly news reports have largely dismissed dissident scientists who dispute the validity of historic temperature data and question the role of industrial activity in global climate change. As a recent seven-year study by Robert Lichter of the Center for Media and Public Affairs found, four out of five scientists quoted in the news maintained that scientific evidence of global warming was sufficient enough to begin regulating human activity believed to enhance its effect.

Yet, a 1991 Gallup poll of 400 randomly selected members of the American Meteorological Society and the American Geophysical Union showed that only one in three attributed global temperature increases to human activities. Almost half attributed the increases instead to nonhuman causes, such as the natural range of the earth's temperature fluctuations. And more than half of those directly involved in global-warming research said they did not believe in a historical increase of any kind. The only scientific consensus about global warming is that there is no consensus—except in the media.

162. Dasheng Yang and Weiyu Yang. GLOBAL WARMING FORUM: A SCIENTIFIC, ECONOMIC AND LEGAL OVERVIEW. 1993, p. 18. As mentioned in the Introduction, in recent years greenhouse

effect has occupied so much of people's attention because it is believed that the greenhouse effect is responsible for global warming. According to the report of Intergovernmental Panel on Climate Change (IPCC), if the present situation keeps on developing, the global temperature will rise 0.3°C, and the sea surface will rise 6 cm every 10 years until the end of the next century. Thus, the temperature will increase 3°C, and the sea surface will rise 65 cm. This can give rise to serious consequences. However, this result may be questionable since in the predictive study of the climate change, the effect of clouds on the climate has yet to be defined accurately.

163. S. Fred Singer. WASHINGTON TIMES.
21 March 1996, Lexis/Nexis. The key IPCC conclusion is that global climate will warm between 0.8 and 3.5 degrees C by about 2100, according to the best current model calculations. (The lower limit was later raised to 1.0 C.) The summary does not reveal, however, that only three years ago in Rio the IPCC had predicted nearly double this temperature increase, 1.5 to 4.5 C, to occur within only half the time, by about 2050. That's a reduction in the rate of warming by nearly a factor of four! No doubt, the models have improved and become more realistic; but didn't the IPCC assure the assembled statesmen signing the climate treaty in 1992 that its predictions were the best and of the highest quality, and "broadly consistent" with the observations? Obviously, the claim was wrong—but this fact is never admitted.

164. REUTER. 20 December 1995 [reprint: GW
Online Today], Lexis/Nexis. Not all environmental scientists agree with the consensus reached by members of the UN Intergovernmental Panel on Climate Change (IPCC) recently that the Earth's climate is warming by dint of human activity. Dissenters say that interest groups with their own political agendas have been effective in pushing their views. The summarization methods of the masses of data by the IPCC, "lacked rigor" claimed some who disagree.

Dr. Frits Bottcher, chairman of the Global Institute for the Study of Natural Resources in the Hague explained, "Instead of scientific discussion we have groups of hundreds of scientists and civil servants, and by a majority of vote they decide. That's not how science should work, in my opinion." He derided the computer predictions used by the IPCC study groups saying they were loaded to make certain the desired results emerged.

"I totally disagree with the IPCC conclusions," said Bottcher. "They put wrong physical equations in their computer, the wrong figures and all kinds of tricks, but they have to defend the case of global warming because they get hundreds of millions of dollars. If they say global warming is not happening, they won't get their money."

165. WASHINGTON POST. 16 July 1995, p. W17.
Industry has looked with favor on the handful of scientists who do loudly disagree with the IPCC and its estimates. These scientists, generally called the contrarians, remain suspicious of the IPCC's internationalist motives and charge that the attempt to establish consensus is really meant to marginalize those who are out of the mainstream of opinion. "If you want to look for the truth, don't look at the center, look at the edges," says Patrick Michaels, a climatologist at the University of Virginia and one of the most voluble of those who argue that future climate change is nothing much to worry about. As for those researchers in the mainstream, Michaels suggests that they play the game to get more funding. "I can't tell you how many phone calls I get from scientists saying, 'We know it's exaggerated, but it's been good for research money,'" he says, though he declines to furnish any of these callers' names.

166. Sylvan Wittwer. FOOD, CLIMATE AND CO_2:
THE GLOBAL ENVIRONMENT AND WORLD FOOD PRODUCTION. 1995, p. 53. Kellogg and Schware (1981) have stated that the CO_2-climate problem, more correctly the CO_2-climate issue, is a cascade of uncertainty arising in part from a wide spectrum of scenarios. Those of Houghton and Woodwell (1989), for example, and those of Schneider (1989, 1992) predict rising sea levels inundating millions of hectares of land and coastal cities, of dislocations of agriculture, the desertification of what is now the United States breadbasket, of devastating droughts, changes in the locations of monsoons, disappearance of endangered species, and disruptions of forests and natural ecosystems, including the carbon balance of arctic ecosystems (Shaver et al., 1992), which may be unable to survive in their current localities. All these are well known, having been given great visibility in the press. Some scenarios, reviewed by Landsberg (1984), call for a drastic and rapid climate modification on a large scale with catastrophic possibilities never before witnessed in human history.

The most famous and widely referenced scenarios is that of James Hansen, director of the Goddard Institute for Space Studies, who declared before the U.S. Congress during the peak of the drought and heat wave of the summer of 1988, that the odds were 99 to 1 that the greenhouse effect or global warming was already here. As supporting evidence, it is further emphasized that there has been about 0.5°C of unexplained "real" warming over the past 100 years, and that in keeping with the trend, the 1980s appear to be the warmest decade on record, with 1988, 1987, and 1981 the warmest years in that order. Although the inaccuracies of such statements and conclusions have been well documented (Balling, 1992, 1993; Strommen, 1992), the fallacy still exists. It has given way to widespread glamorized press releases, the publication of best sellers, an advocacy for policy decisions involving billions of dollars to restrict the burning of fossil fuels, and ammunition for many press releases, some in prestigious scientific journals, and the issuing of annual reports by Worldwatch, World Resources Institute, and other earth organizations detailing the apocalyptic path we are following. Embellish-

ments of such reports, the clever use of caveats by Schneider (1989) and others, and the responsiveness of the press are discussed by Friedly (1993). Frightening the public does get support for more research.

166. Richard Lindzen. WORLD CLIMATE REPORT. 4 March 1996, p. A4. Gelbspan claims that I simply asserted that increasing carbon dioxide would increase global mean temperature by only 0.3°C. What I actually stated was that the warming due to doubling CO_2 alone was only about 0.3°C, and that predictions of greater warming required that nature greatly amplify the effect of CO_2 alone. Such amplification is found in models due to physical process handled so poorly as to involve errors far larger than the effect of doubling CO_2. . . . Several indirect inferences from data are consistent with the low value. . . . [ellipses in original]

168. Ronald Bailey. ECOSCAM. 1993, p. 146. The current GCMs depend on a destabilizing feedback loop in which higher surface temperatures increase upper tropospheric water vapor (five to twelve kilometers up), which is the major greenhouse gas. "Without this feedback, no current model would predict warming in excess of 1.7C (3F) degrees," says climatologist Lindzen. The physics of upper tropospheric water vapor is simply unknown, and some models actually produce regions of negative water vapor—a physical impossibility. He adds that each destabilizing feedback loop included in the models might actually be stabilizing feedback.

169. Robert C. Balling. THE HEATED DEBATE, 1992, p. 134. Decisionmakers must realize that there are legitimate reasons to believe in the apocalypse, and there is a chance that we may still be headed for the disaster. Scientists may be underestimating the thermal inertia of the oceans, the climate system may one day jump to a new, very much warmer equilibrium, or some cloud feedback or biological response may exacerbate global warming. There could be surprises, and we may fully realize the apocalypse. However, decision-makers must be aware that there is at least an equal chance, if not a much higher change, that many of the greenhouse effects will not be disastrous to social, biological, or economic systems.

170. Robert G. Feagle. GLOBAL ENVIRONMENTAL CHANGE: INTERACTIONS OF SCIENCE, POLICY, AND POLITICS IN THE UNITED STATES. 1994, p. 7. However, the linkages of these problems to those of desertification, biological diversity, and others mentioned above must not be overlooked. The linkages are both scientific—long-term biogeochemical cycles incorporate processes of the atmosphere, land surface, solid earth, ocean, and biosphere—and economic, social, and political—responses to global environmental problems will be temporary at best, and more likely futile, unless the problems of world population and economic development are addressed at the same time as the environmental problems.

171. Justin Lancaster. A GLOBAL WARMING FORUM: SCIENTIFIC, ECONOMIC AND LEGAL OVERVIEW. Ed. R. Geyer, 1993, p. 535. Global environmental change is human society altering the Earth's biogeochemical cycles. Deforestation is a key aspect of this change as it alters carbon and water cycles. Another geochemical perturbation, infusion of halocarbons into the stratosphere, threatens the ozone layer, an impact unforeseen before 1973. Spewing carbon dioxide into the atmosphere threatens global warming, while adding nitrogen and sulfur oxides into the troposphere threatens lakes and forests. Global environmental change is caused by a growing population rapidly consuming energy to drive an accelerating technology. Cutting forests, changing agricultural techniques, and manufacturing modern chemicals are merely a few of the multiple paths by which energy expenditure is stressing the global environment.

172. William D. Nordhaus. MANAGING THE GLOBAL COMMONS. 1994, p. 5. A complete analysis of the economics of climate change must recognize the extraordinarily long time lags involved in the reaction of the climate and economy to greenhouse gas emissions. Current scientific estimates indicate that the major GHGs have an atmospheric residence time of over 100 years; moreover, because of the thermal inertia of the oceans, the climate appears to lag perhaps a half century behind the changes in GHG concentrations; and there are long lags in the introduction of capital stocks and new technologies in human economies in response to changing economic conditions. Dynamics are therefore of the essence, and a study that overlooks the dynamics will produce misleading conclusions for the steps that we should take at the dawn of the age of greenhouse warming.

173. Sylvan Wittwer. FOOD, CLIMATE AND CO_2: THE GLOBAL ENVIRONMENT AND WORLD FOOD PRODUCTION. 1995, p. 51. Jones and Wigley (1990) have further warned that models cannot prove the emissions of greenhouse gases will significantly alter the earth's climate, and that the causes of global warming, if it exists, are less certain that trend itself. Although the earth has experienced an overall warming trend of half a degree since the late 19th century, there are many questions concerning reliability. The earlier averages were compiled from a much smaller set of stations comparable to those of today. Are biases related to urban warming (heat islands) removed? Finally, how well do land-based temperatures represent the climate of the earth, which is two thirds covered by water? Finally, in a recent report, when a doubling of atmospheric CO_2 was considered, the radiative forcing reported among 15 atmospheric general circulation models showed roughly a three fold variation in predicted increases in global mean surface temperature (Cess et al.,

1993). *The sensitivity of the world's climate system to rising levels of atmospheric CO_2 has not been resolved* (Hansen et al., 1993).

174. Robert C. Balling. TRUE STATE OF THE PLANET. 1995, p. 94. The urban effect creates a localized warming signal that is not representative of the surrounding area. Recently it has been discovered that overgrazing and desertification may be producing a large-scale warming signal that is clearly not related to the greenhouse gases. The role of desertification in changing the regional temperature was strongly debated following a landmark article suggesting that overgrazing in arid and semiarid lands would increase the albedo (reflectivity) by removing the dark-colored vegetation. The increased albedo would reflect more of the sun's energy, less solar energy would be absorbed by the surface, and surface and air temperatures would drop. Soon after the introduction of this hypothesis, others argued that removal of vegetation would reduce evapotranspiration rates; less solar energy would be consumed in evaporating and transpiring water, leaving more solar energy to warm the surface and the air. Most empirical data and recent theoretical findings support the notion that overgrazing and desertification would act to warm, not cool, the surface and air temperatures.

175. Sylvan Wittwer. FOOD, CLIMATE AND CO_2: THE GLOBAL ENVIRONMENT AND WORLD FOOD PRODUCTION. 1995, p. 51. As reported by Wittwer (1994), an extensive review of the CO_2-climate issue by a joint committee of the World Meteorological Organization and International Council of Scientific Union (ICSU) reached a most perceptive conclusion in 1983, that still holds (UNEP, 1985; WMO, 1983). It was stated that by simple comparison of the overall temperature and CO_2 increase for the last 100 years, one must conclude that the climate (temperature) sensitivity to CO_2 is at the lower limit of mathematical climate computer model prediction.

176. Chris Hope and Philip Maul. ENERGY POLICY. Vol. 1, No. 3, 1996, p. 217. The second effect, which would tend to make the impact of a pulse on top of BAU emissions smaller than the impact of a pulse on its own, is that the concentration of CO_2 in the atmosphere is sufficiently high that the radiative forcing effect, and therefore the eventual temperature rise, is not linear in CO_2 concentration, but logarithmic (IPCC, 1990). What this means is that the higher the concentration of CO_2 in the atmosphere, the lower the extra temperature rise caused by a given pulse of emissions. Superimposing the pulse on top of substantial business as usual emissions will cause a lower rise in the temperature than would occur from the pulse on its own.

177. S. Fred Singer et al. A GLOBAL WARMING FORUM: SCIENTIFIC, ECONOMIC AND LEGAL OVERVIEW. Ed. R. Geyer, 1993, p. 349. Precise measurements of the increase in atmospheric CO_2 date to the International Geophysical Year of 1957 to 1958. More recently, it has been discovered that other GHG, i.e., gases that absorb strongly in the IR have also been increasing—at least partly as a result of human activities. These gases currently produce a greenhouse effect nearly equal to that of CO_2, and could soon outdistance it.

178. Sylvan Wittwer. FOOD, CLIMATE AND CO_2: THE GLOBAL ENVIRONMENT AND WORLD FOOD PRODUCTION. 1995, p. 41. Changnon (1992) indicated that more than half of all North Americans live and work in anthropogenically generated urban climates that are drastically different than those of 100 to 150 years ago. Totally new climates have been created in cities and downwind from urban areas. Societies and individuals have successfully adapted to them. The magnitude and extremes of these climates in temperature and precipitation are at least similar to, if not greater than, those predicted by GCMs for the next 100 years. In fact, the average urban warming per 100 years ranges from 2 to 3°C in cities of the United States such as Boston, Cleveland, and Washington, D.C. According to Jones et al. (1990), global warming is inflated 2 to 20% by the heat island effect. Also, the urban-enhancing influence has been greatest in the increase of summer temperatures. Urban heat island effects have also been noted in Paris, Berlin, Vienna, London, Sao Paulo, Tokyo, and Shanghai. Changnon (1992) concludes that a study of the inadvertent weather and climate modification that has occurred over the past century can serve as a useful historical analog for analyzing many aspects of the global issue. The past does hold some guidance for the future. People in cities have adapted to these increases in temperature.

179. Sylvan Wittwer. FOOD, CLIMATE AND CO_2: THE GLOBAL ENVIRONMENT AND WORLD FOOD PRODUCTION. 1995, p. 41. There is one category of anthropogenic impact that has occurred over the past 150 years not entirely attributable to the rise in atmospheric greenhouse gases. It is inadvertent modification in urban areas or the so-called "heat island effect." These climate modifications related not only to temperature but to summer rainfall increases, especially downwind of cities, cloud cover, air composition, visibility, fog and smog, aided by lower wind speeds. Sunshine is notably attenuated. Abatement efforts for some of the latter have been partially successful in some localities but not others. Some are virtually hopeless because of orographic settings. Notable are Mexico City, Los Angeles, and more recently the Salt Lake City area during some winter months. If the heat island effect is eliminated by noninclusion of weather stations in cities and airports, there has been no temperature change in the lower 48 states of the United States during the past 100 years. Although the United States is not one of the largest land areas of the earth, it is one portion of the earth's surface in which greenhouse warming is predicted by some GCMs to be most apparent. It also has one

of the most sophisticated weather station records, and is a major food-producing country (Ruttan, 1990).

180. Robert C. Balling. THE HEATED DEBATE. 1992, p. 139. Because the warming in Figure 51 is the warming thought to be forced by the buildup of the greenhouse gases, we would have to assume that all of the warming of the past 100 years is caused by the greenhouse gases. But as we saw throughout the book, stratospheric aerosols and contaminants to the temperature record account for some of the warming. If we accept 0.2°C (0.4°F) as the amount of unexplained warming over the past decade (the global and United States temperature trends would make this value reasonable), then the temperature curves in Figure 51 would be reduced by a factor of four, and the effect of Scenario Be reduces to only 0.18°C (0.2°F) by the year 2040. It is crucial to recall that the natural variability in the climate system produces global temperature fluctuations much larger than this amount; thus, we may never be able to detect the temperature response of our policies.

181. Dixie Lee Ray. TRASHING THE PLANET. 1990, p. 37. Gigantic as it was, Mount St. Helens was not a large volcanic eruption. It was dwarfed by Mount St. Augustine and Mount Redoubt in Alaska in 1976 and 1989 and El Chicon in Mexico in 1982. El Chicon was an exceptionally sulfurous eruption. The violence of its explosion sent more than 100 million tons of sulfur gases high into the stratosphere. Droplets of sulfuric acid formed; these continue to rain down onto the earth's surface. The earth, at present, appears to be in a period of active volcanism, with volcanic eruptions occurring at a rate of about 100 per year. Most of these are in remote locations, where accurate measurement of the gaseous emissions is not possible, but they must be considerable. Some estimates from large volcanic eruptions in the past suggest that all of the air polluting materials produced by man since the beginning of the industrial revolution do not begin to equal the quantities of toxic materials, aerosols, and particulates spewed into the air from just three volcanoes: Krakatoa in Indonesia in 1883; Mount Katmai in Alaska in 1912, and Helka in Iceland in 1947. Despite these prodigious emissions, Krakatoa, for example, produced some chilly winters, spectacular sunsets, and a global temperature drop of 0.3 degrees Centigrade, but no climate change. From written records, we also know that the famous "year without a summer" that followed the eruption of Mount Tambora in 1816 meant that the summer temperature in Hartford, Connecticut, did not exceed 82 degrees Fahrenheit. No doom.

182. Robert C. Balling. STATE OF THE CLIMATE REPORT. 1995, p. A4. An interesting message also came from the analyses of lunar impacts on global temperature. The amount of energy redirected to earth from a full moon, as opposed to a new moon, is 0.014 Watts per square meter (W/m^2); this is dwarfed by the solar output of 1.370 W/m^2. However, the radiative effect of the unknown buildup of greenhouse gases from 1979 to the present is near 0.6 W/m^2. The anthropogenic greenhouse forcing should be approximately 50 times stronger than the lunar forcing, yet, the lunar effect is loud and clear in the satellite record while the greenhouse impact remains undetectable. Even when the cooling effects of the sulfates are included in the comparison, the warming signal should be very much larger than the oscillation in global temperature related to lunar phase.

183. Robert C. Cowen. THE CHRISTIAN SCIENCE MONITOR. 25 June 1996, p. 16. Meanwhile, Jeff Severighaus at the University of Rhode Island Graduate School of Oceanography at Narraganset notes that "one of the really big uncertainties" in global-warming forecasting is the role of water vapor. When it comes to warming, water vapor is the big gun of the climate system. It is the most abundant and powerful of the greenhouse gases. It plays the main role in keeping Earth warm enough to support life. Significant changes in atmospheric water-vapor content would overwhelm any effect of human pollution.

184. Justin Lancaster. A GLOBAL WARMING FORUM: SCIENTIFIC, ECONOMIC AND LEGAL OVERVIEW. Ed. R. Geyer, 1993, p. 535. Global environmental change is human society altering the Earth's biogeochemical cycle. Deforestation is a key aspect of this change as it alters carbon and water cycles. Another geochemical perturbation, infusion of halocarbons into the stratosphere, threatens the ozone layer, an impact unforeseen before 1973. Spewing carbon dioxide into the atmosphere threatens global warming, while adding nitrogen and sulfur oxides into the troposphere threatens lakes and forests. Global environmental change is caused by a growing population rapidly consuming energy to drive an accelerating technology. Cutting forests, changing agricultural techniques, and manufacturing modern chemicals are merely a few of the multiple paths by which energy expenditure is stressing the global environment.

185. Dasheng Yang and Weiyu Yang. GLOBAL WARMING FORUM: A SCIENTIFIC, ECONOMIC AND LEGAL OVERVIEW. 1993, p. 5. Weather is a chaotic system, and due to its sensitivity to initial conditions, it is unpredictable. Regional climate is also highly sensitive to man-made perturbations to the Earth's atmospheric system, thus, making it unpredictable. Man-made perturbations are increasing the climate variability. This is evidenced by the comparison of the previous climate changes with the current greenhouse gases forced changes, and the recently noted changes in ocean circulation as well as the probable thinning of the Arctic ice. Hence, within 20 years there may be many larger-than-expected regional and global climatic changes. This indicates that climatic change is driven partly by human activity. The relationship

between population and natural disasters is being recognized, and most scientists agree that population is important in global climatic change. The concept of the human bolide is suggested.

186. Ronald Mitchell and Abram Chayes. SHAPING NATIONAL RESPONSES TO CLIMATE CHANGE: A POST RIO GUIDE. Ed. Henry Lee. 1995, p. 138. Third, an increased effort to address population growth not only would have long-term benefits in decreasing one of the major sources of global stress underlying greenhouse emissions, but also would have significant benefits in terms of development and quality of life. Without addressing this issue (and doing so with a long-term perspective), many of the other actions discussed here may prove inadequate to forestall global warming.

187. Michelle Malkin. SEATTLE TIMES. 16 July 1996, p. A1. Between 1881 and 1993, the earth warmed roughly 0.54 degrees Celsius. But nearly 70 percent of that warming occurred during the first half of the century—that is, prior to World War II and long before the most dramatic increase in greenhouse gases took place. What the media further fail to emphasize is that the earth's climate is in constant flux regardless of human activity. Weather varies wildly from year to year, city to city, pole to pole. And it is simply not clear from computer climate models that projected warming would actually cause catastrophic conditions, or whether it might actually result in greater prosperity due to longer grower seasons and increased agricultural productivity.

188. Robert C. Balling. TRUE STATE OF THE PLANET. 1995, p. 92. Given the approximate 40 percent increase in atmospheric concentration of equivalent CO_2 over this 1881- to 1993 time period, the observed rise in temperature is low given numerical model predictions for a doubling of equivalent CO_2. Given this buildup in equivalent CO_2, the models suggest that we should have observed at least 1.0°C of warming over the same period. Even if all of the 0.54°C can be ascribed to the effect of greenhouse gases, the models appear to be off by a factor of two. Noted climate modeler Stephen Schneider acknowledged this point and concluded that "the twofold discrepancy . . . is still fairly small." However, alternative explanations for the observed warming, the timing of the warming, and/or the geography of the warming may increase the size of the discrepancy and cast even more doubt on the ability of the models to simulate changes in climate accurately. [ellipses in original]

189. Patrick J. Michaels. STATE OF THE CLIMATE REPORT. 1996, p. 5. While the news wires were ablaze with the story that the U.N. Intergovernmental Panel on Climate Change claims that the consensus of scientists is humans have changed the climate, the true story was how little people have affected it.

Examination of hemispheric ground-based temperatures shows that at least half of the half-degree warming of the last century occurred before the greenhouse effect changed very much (Figure 2). That leaves, at best, about a quarter of a degree, or several times less than what was predicted by the Framework Convention climate models.

Despite the very small change in temperature, there are other subtle signs, according to U.S. Department of Commerce climatologist Tom Karl, that are consistent with greenhouse changes. One is a slight (2%) increase in rainfall from summer storms that produce two or more inches of rain in 24 hours. In terms of stormy days, there is now, on the average, one more day in every 730 that has a rainstorm of more than two inches. In terms of public perception, no one (except perhaps the most dedicated weather nerd) could notice this change. In fact, when asked bout it on a nationally syndicated radio program, Karl said that, with respect to global warming, his findings were "no smoking gun," and that any climate change was "very subtle."

Indeed. On page 6 is the core graphic from Karl's paper (Figure 3), detailing the time-history of the change in rainfall. The most rapid increase is from 1930 to 1945—before the major changes in the greenhouse effect.

190. Sylvan Wittwer. FOOD, CLIMATE AND CO_2: THE GLOBAL ENVIRONMENT AND WORLD FOOD PRODUCTION. 1995, p. 51. As reported by Wittwer (1994), an extensive review of the CO_2-climate issue by a joint committee of the World Meteorological Organization and International Council of Scientific Union (ICSU) reached a most perceptive conclusion in 1983, that still holds (UNEP, 1985; WMO, 193). It was stated that by simple comparison of the overall temperature and CO_2 increase for the last 100 years, one must conclude that the climate (temperature) sensitivity to CO_2 is at the lower limit of mathematical climate computer model prediction.

191. Chris Hope and Philip Maul. ENERGY POLICY. Vol. 24, No. 3, 1996, p. 217. The second effect, which would tend to make the impact of a pulse on top of BAU emissions smaller than the impact of a pulse on its own, is that the concentration of CO_2 in the atmosphere is sufficiently high that the radiative forcing effect, and therefore the eventual temperature rise, is not linear in CO_2 concentration, but logarithmic (IPCC, 1990). What this means is that the higher the concentration of CO_2 in the atmosphere, the lower the extra temperature rise caused by a given pulse of emissions. Superimposing the pulse on top of substantial business as usual emissions will cause a lower rise in temperature than would occur from the pulse on its own.

192. Ronald Bailey. ECOSCAM. 1993, p. 165. In November 1991, two scientists, Eigil Friis-Christensen and Knud Lassen of the Danish Meteorological Institute, startled climatologists by showing a remarkably close correlation between the length of the sunspot cycle

and global temperature changes over the past 130 years. Sunspots are huge energetic storms that appear as dark blotches on the sun's surface. Their numbers rise and decline in regular nine to 13-year-long cycles. The Danes show that shorter nine-year cycles correlate with hotter global temperatures while longer eleven-year cycles signal cooler weather. For example, when global temperatures declined from 1940 to 1975, solar activity decreased and the sunspot cycles lasted longer. When temperatures began to rise in the 1970s, the sunspot cycle shortened to nine and a half years. The two scientists conclude that their findings may mean the greenhouse effect plays a smaller role in warming than asserted by environmentalists.

193. Robert C. Balling. TRUE STATE OF THE PLANET. 1995, p. 95. Obviously the total energy output of the sun could play a major role in governing the planetary temperature. For many years, some scientists have argued strongly in favor of this mechanism as a primary control of planetary temperature, while others have rejected the idea that small variations in solar output can explain much of the trend of the past century. Recently two researchers have found that the length of the solar sunspot cycle is related strongly to the fluctuations in temperatures on the earth. Although the physical mechanism responsible for the linkage remains elusive, it is noteworthy that over 75 percent of the observed global warming in this century can be statistically explained by the variations in the length of the solar sunspot cycle.

194. Dixie Lee Ray. TRASHING THE PLANET. 1990, p. 39. Coupled with the activity of the sun, there is the moon's gravitational force, to which the earth's waters respond daily and in 28-day cycles of tides. Also, there are 20-year and 60-year tidal cycles, as well as longer ones. Moreover, the solid land also responds to the moon's gravitational force, but because we move with the ground, we do not feel it. Recently, a 556-year variation in the moon's orbit around the earth was analyzed; some meteorologists believe that the occasional confluence of all these sun-and-moon cycles may trigger dramatic changes in ocean currents and temperatures. And it is now widely acknowledged that the oceans are a major influence on the climate. There is also a 500-to-600-year cycle in volcanic activity, which appears to be near a peak at the present time.

195. Michelle Malkin. SEATTLE TIMES. 16 July 1996, p. A1. The latest report from the United Nations' International Panel on Climate Change (IPCC) suggests that curtailing human-induced carbon emissions is the key to staying cool. But throw in additional natural factors such as volcanic eruptions, oceans, water vapor, phytoplankton, and vegetation that may influence climate change, and the idea of wielding carbon taxes as a primary weather-altering weapon suddenly seems quixotic. (Dissidents are now assailing the IPCC after discovering that some members altered language in its report to expunge all signs of doubt about the role of industrial activity in global warming.)

196. Ken Miller. GANNETT NEWS SERVICE. 12 April 1996, p. A1. Baliunas also said there is no proof of a human signature on global warming. "The science does not suggest dangerous global warming," she said. "If there is any trace at all of a greenhouse warming, it is too small to be seen in the climate record. That means future warming due to human activities will be quite small, well under 1 degree centigrade."

197. WORLD CLIMATE REPORT. 30 September 1996, p. A4. Additionally, Baliunas showed the relative uncertainties in climate predictions compared with the amount that human beings can perturb the greenhouse effect (Figure 2).

For doubling atmospheric carbon dioxide, the amount of warming radiation that rains down on the earth's surface increases by about 4 watts per square meter, or the equivalent of a very dim bulb. This is the right-hand bar in Figure 2. But the errors in our estimation of other important climate parameters are far greater. For example, 100 watts (25 times the greenhouse change) is used to "correct" the models' inaccurate northward movement of oceanic heat.

198. S. Fred Singer et al. A GLOBAL WARMING FORUM: SCIENTIFIC, ECONOMIC AND LEGAL OVERVIEW. Ed. R. Geyer, 1993, p. 350. **Water vapor (H_2O)**—Turns out to be the most effective GHG by far. It is not man-made, but is assumed to amplify the warming effects of the gases produced by human activities. We don't really know whether H_2O has increased in the atmosphere or whether it will increase in the future—although that is what all the model calculations assume. Indeed, predictions of future warming depend not only on the amount, but also on the horizontal and especially vertical distribution, of H_2O, and on whether it will be in the atmosphere in the form of a gas, or liquid cloud droplets, or ice particles. The current computer models are not complete enough to test these crucial points.

199. S. Fred Singer et al. A GLOBAL WARMING FORUM: SCIENTIFIC, ECONOMIC AND LEGAL OVERVIEW. Ed. R. Geyer, 1993, p. 349. Precise measurements of the increase in atmospheric CO_2 date to the International Geophysical Year of 1957 to 1958. More recently, it has been discovered that other GHG, i.e., gases that absorb strongly in the IR have also been increasing—at least partly as a result of human activities. These gases currently produce a greenhouse effect nearly equal to that of CO_2, and could soon outdistance it.

200. Henry Lee. SHAPING NATIONAL RESPONSES TO CLIMATE CHANGE: A POST RIO GUIDE. Ed. Henry Lee. 1995, p. 5. Although this

book does not examine the scientific debate, it is important to discuss a few aspects of the global climate change problem that will affect the magnitude and scope of governmental response. First, if the IPCC's scientific report is correct, it is unlikely that the world will be able to avoid some increase in average world temperature and perturbations in the form of changes in rainfall, storms, and other weather phenomena. Even if the greenhouse gas emissions were radically reduced, substantial momentum has already been built into the system. Greenhouse gases have long atmospheric lifetimes. Carbon dioxide remains in the atmosphere for 50–200 years. Nitrous oxide lasts for 150 years, and CFCs, 65-130 years. Only methane among the major man-made gases has a relatively short atmospheric lifetime, but even it lasts 10 years. The U.S. Environmental Protection Agency (EPA) calculated in a 1990 study that to stabilize atmospheric concentrations, a 75-percent reduction in greenhouse gas emissions would be required. In fact, the EPA estimated that carbon dioxide emissions would have to be reduced by 50 percent; nitrous oxides, by 80–85 percent; and methane by 10–20 percent; CFCs would have to be phased out altogether.

201. William D. Nordhaus. MANAGING THE GLOBAL COMMONS. 1994, p. 189. A fourth issue that requires emphasis is the unpleasant fact of inertia in climate change. Scientific estimates suggest that the stock of greenhouse gases in the atmosphere will lead to significant climate change *even if we take stringent steps to reduce emissions.* for example, we examined above a policy that caps emissions at 80 percent of the 1990 levels, producing emissions reductions of 70 percent of the baseline late in the next century, and costs that total $11 trillion in discounted consumption. Even so, global temperatures rise by 2.2°C by 2100 with this stringent set of controls as opposed to 3.1°C in the optima policy or 3.3°C in the uncontrolled run. These calculations show how, even with major technological breakthroughs and stringent controls, the momentum of past GHC emissions appears to be leading us to an inevitable rendezvous with massive climate change. Mitigate we might; adapt we must.

202. M. A. K. Khalil. GLOBAL WARMING FORUM: A SCIENTIFIC, ECONOMIC AND LEGAL OVERVIEW. 1993, p. 360. The second effect is the "thermal inertia" of the oceans. As trace gases increase radiative forcing, the oceans warm very slowly because of the large heat capacity of water. Heat that would warm the surface is therefore taken up to heat the water. This effect slows down and even buffers global warming from increasing concentrations of trace gases. Similarly, if circumstances change and the atmosphere begins to absorb less of the Earth's radiation each year, the warmed up oceans will delay global cooling.

203. S. Fred Singer et al. A GLOBAL WARMING FORUM: SCIENTIFIC, ECONOMIC AND LEGAL OVERVIEW. Ed. R. Geyer, 1993, p. 352. Keep in mind also that year-to-year changes at any location are far greater and more rapid that what might be expected from GHW: nature, crops, and people are already adapted to such changes. It is the extreme climate events that cause the great ecological and economic problems: crippling winters, persistent droughts, extreme heat spells, killer hurricanes, etc. There is no indication from modeling or actual experience that such extreme events would become more frequent if GHW becomes appreciable. The exception might be tropical cyclones, which—R. Balling and R. Cerveney argue—would be more frequent but weaker, would cool vast areas of the ocean surface, and increase annual rainfall. In summary climate models predict that global precipitation should increase by 10 to 15%, and polar temperatures should warm the most, thus reducing the driving forces for severe winter-weather conditions.

204. Ronald Bailey. ECOSCAM. 1993, p. 162. What if the prophets of doom turn out to be right about future temperature increases? Does greater warming necessarily spell disaster for humanity? No. Humanity can easily adapt to the projected changes in climate during the next century, according to the National Academy of Sciences (NAS). Humanity's ever-increasing store of knowledge and technology enables us to adjust to changing circumstances faster and more easily as the years pass. Human beings thrive in climates ranging from the Sahara to Sarawak and from the Amazon to Alaska. Therefore, adaptation to any future climate change will most likely be the same as for more familiar climate variations. "The technologies, small and large, that buffer human activity over the long-term will be the same ones that mollify the difference between daytime and nighttime temperatures, protect against normal variability between days, shield from storms and hail, adjust to seasons, and adapt to the wide range of climates where people already live," says Rockefeller University researcher Jesse Ausubel. He notes—and it should be obvious to anyone who considers the matter for a moment—that technology has enormously reduced humanity's vulnerability to climate variation and will continue to do so in the future. We've exchanged smoky campfires for central heating, precarious dependence on local harvests for the abundance of the world's grain markets, and the shaman's rattles for modern medicine.

205. Robert C. Balling. THE HEATED DEBATE. 1992, p. 96. As we saw in this chapter, six fundamental points emerge in the analysis of hemispheric and latitudinal climate trends. First, the amount of warming experienced over the past 100 years in either hemisphere is considerably less than expected given the observed increase in equivalent CO_2 levels. Second, despite model predictions to the contrary, the Southern Hemisphere has warmed more than the Northern Hemisphere, particularly in the most recent half-century. Third, some evidence shows that the long-term warming is greatest in the polar

regions and least in the tropical regions; this distribution is broadly consistent with the predictions of the models. Fourth, the models predict greatest warming in the low-sun season and the least warming in the high-sun season, and generally this is observed in the mid-latitude temperature records. Fifth, the Arctic region as a whole, where the greatest amount of warming should be observed, has shown cooling over the past 50 years. And sixth, the observed patterns for precipitation trends are only broadly consistent with the highly varied predictions of the climate models.

Therefore, we again see evidence that a doubling of CO_2 will produce a temperature response on the low end of the predictions, possibly near 1.0°C (1.8°F). The warming will likely occur more at night than during the day, the warming will probably occur most in the high-latitude locations, and the warming will be revealed largely during the winter season; cloudiness and precipitation will increase in most locations. this internally consistent change in the climate is supported by model predictions, analyses of the global climate records, and now by the analyses of the hemispheric and latitudinal records.

206. Jonathan Adler. WASHINGTON TIMES. 27 September 1994, Lexis/Nexis. The horrific vision of a scorched Earth and withering crops combined with monsoons of untold ferocity may help environmental lobbyists solicit donations, but this is not an accurate picture of what climate scientists expect a warmed world to be like. Most warming is likely to occur at winter and at night, creating a more benign climate, not a more hostile one. The reality is that a change in the Earth-s climate—human-induced or otherwise—will bring both positive and negative effects.

207. Alan Miller and Mack McFarland. FORUM FOR APPLIED RESEARCH AND PUBLIC POLICY. Summer 1996, p. 62. Many experts involved in the negotiations of the Montreal Protocol have acknowledged the key role played by environmental groups in enhancing public awareness of the problem and citizen support for action. Such groups also were instrumental in gaining support for the ban on CFCs used as aerosol propellants in the United States and Scandinavia in the 1970s and for building public support for recent national and international regulations to phase out CFC production. Grassroots activism increased consumer pressure and caused manufacturers to switch to other products.

However, some environmental groups tend to overstate the degree of scientific uncertainty and concentrate on worst-case risk scenarios to attract media interest and public support. Such tactics can cause a polarization of views and a surge in media interest with every new prediction of disaster.

More important, these tactics can undermine the credibility of the scientific community and the bases for taking action on environmental problems.

Conveying reliable climate-change information to the public will be critical in maintaining support for necessary policies as society begins to address this complicated environmental issue.

208. Michelle Malkin. SEATTLE TIMES. 16 July 1996, p. A1. There are those who argue that "better safe than sorry" is always the most prudent course of action. But prudence cannot be purchased for free. As economist Richard Stroup of the Montana-based Political Economy Research Center notes, "If 'insurance' against a particular risk, such as the threat of global warming, is bought at the cost of reduced economic growth, then a decline in the automatic insurance represented by wealth, and the social resilience it provides, is one of the costs borne by future generations."

209. S. Fred Singer et al. A GLOBAL WARMING FORUM: SCIENTIFIC, ECONOMIC AND LEGAL OVERVIEW. Ed. R. Geyer, 1993, p. 348. We can sum up our conclusions in a simple message: *The scientific base for a greenhouse warming is too uncertain to justify drastic action at this time.*

There is little risk in delaying policy responses to this century-old problem since there is every expectation that scientific understanding will be substantially improved within the next decade. Instead of premature and likely ineffective controls on fuel use that would only slow down and not stop the further growth of CO_2, we may prefer to use the same resources—trillions of dollars, by some estimates—to increase our economic and technological resilience so that we can then apply specific remedies as necessary to reduce climate change or to adapt to it.

210. S. Fred Singer. WASHINGTON TIMES. 21 March 1996, Lexis/Nexis. By now, more than a hundred countries are involved in these exercises, keeping thousands of well-paid functionaries busy and spending some tens of millions of dollars on meeting and planning—funds that could be earmarked for solving real environmental problems. Such waste is bad enough. But what is alarming is that the measures considered would certainly stifle economic growth and reduce living standards for a large part of the global population. They may even reduce funds paid out to bureaucrats. Now that's really scary!

211. Thomas Homer-Dixon. INTERNATIONAL SECURITY, Winter 1995–1996, p. 190. I also agree with Levy that ozone depletion and climate change could endanger core American values and are therefore direct threats to U.S. security interests. Unfortunately, though, Levy does not adequately acknowledge that these are unlikely to be near-term threats to the United States, whereas many regional environmental problems—including land scarcity, fuelwood scarcity, and depletion of water supplies and fish stocks—are today affecting the core val-

ues of hundreds of millions of people in the developing world.

212. Kenan Malik. THE SPECTATOR. 11 November 1995, p. 26. Global warming would certainly create problems for humanity, but it may also provide opportunities. The weather would worsen in certain areas, but it would improve in others. Of course, there would be winners and losers: Northern Europe, for one, would gain. We might have olive trees in England and wine in Wales. Italy would suffer, but as the development of agriculture in places like Israel or the Himalayas shows, human ingenuity can transform even the most inhospitable terrain given sufficient will and resources. On the other hand, as the desertification of Africa's Sahel region reveals, poverty and social neglect can have devastating impact on previously clement areas.

213. Thomas Gale Moore. ECONOMIC TIMES. March 1995, p. 5. The influence of climate on human activities has declined with the growth in wealth and resources. Primitive man and hunter-gatherer tribes were at the mercy of the weather, as are societies that are still bound to the soil. But since the time of the industrial revolution, climate has basically been confined to a minor role in human affairs.

Since its origins, the earth has experienced periods significantly warmer than the modern world—some epochs have been even hotter than the most extreme predictions of global warming. Today's cool temperatures are well below average for the globe in its more-than-four-billion-year history. Studies of climatic history show that sharp changes in temperatures over brief periods of time have occurred frequently without setting into motion any disastrous feedback systems that would lead either to a heating that would cook the earth or a freezing that would eliminate all life.

214. Kenan Malik. THE SPECTATOR. 11 November 1995, p. 25. Nor is there any evidence that a warmer climate in itself is harmful to humanity. Human beings have endured major climatic changes throughout their history without major problems. In fact, human culture began to develop some 30,000 years ago during the last major Ice Age. The end of that Ice Age, 10,000 years ago, marked the beginnings of agriculture. In more recent times, the transformation of the life of Europe during the Renaissance and the Enlightenment occurred during the little Ice Age. This is not to claim that climatic changes cause human progress, but simply that such changes need not be a barrier to social advance, so long as we approach them with the right attitude.

215. WORLD CLIMATE REPORT. 15 April 1996. CT: "Sir John points to the majority of scientists, who, armed with their computers . . . fear there is precious little time to take action."

FACT: Tom Wigley, of the U.S. National Center for Atmospheric Research recently published a paper in *Nature* that demonstrated that the IPCC itself is only projecting 1.3°C of additional warming even if effective concentrations of CO_2 are stabilized at around 700 parts per million (the current effective value is around 440ppm, and the background is usually given as 279ppm). Wigley et al. argue that it is more cost-effective and environmentally sound to consider initiation of mitigation technology about two decades from now, when the cost of that technology is much lower and more investment can be made. [ellipses in original]

216. Robert C. Balling. THE HEATED DEBATE. 1992, p. 134. The empirical evidence suggests that our future will see a rise in temperature of approximately 1.0°C (1.8°F), with most of that warming occurring at night, in higher latitudes, and in the winter season. Extreme high temperatures probably will not increase in the 2×CO_2 world, and the earth will probably be wetter, cloudier, with substantial increases in soil moisture; droughts may diminish in frequency, duration, and intensity. these predictions for the future are largely consistent with the models and largely consistent with how the climate system has already responded to a 40 percent increase in equivalent CO_2 over the past century. From a climate perspective, we have had a far more moderate greenhouse effect than what the public has been led to believe.

217. Robert C. Balling. THE HEATED DEBATE. 1992, p. 81. So how did the models score? In many respects, the models did remarkably well simulating the observed changes to the global climate system. The observed increases in cloud coverage and precipitation are generally consistent with model expectations given a 40 percent increase in equivalent CO_2; in fact, the models have underestimated the change in these moisture-related climate components. Although not conclusive, the data from glaciers, snow coverage, and sea ice extent certainly do not contradict the models. A rise in temperature was observed, but the increase appeared to be far less than the model predictions. Not surprisingly given the smaller than expected increase in temperature, the rise in sea level has been less severe than what would have been predicted from the various models. In addition, some evidence suggests that the diurnal temperature range was decreasing faster than the predictions by some of the models.

From this discussion alone, a new vision of the greenhouse world begins to emerge. This vision is consistent with the general predictions of the models and it is consistent with the observations of climate change over the past century. When compared to the climate of the past 100 years, the greenhouse world (at 600 ppm) is likely to receive more precipitation; an increase in precipitation near 10 percent may be reasonable. Cloud coverage is likely to increase about the same amount. Global temperatures will rise, but possibly by no more than 1.0°C (1.8°F); the

decrease in diurnal temperature range would suggest that the warming will largely occur at night. Because the afternoon temperatures do not rise significantly, evaporation rates do not increase (as seen by the "popular vision"), and soil moisture levels may increase. The moderate rise in global temperature would likely translate into only moderate rises in sea level. These global patterns are internally consistent with one another, they are consistent with climate changes observed during a time when equivalent CO_2 increased by 40 percent, and they are consistent generally with the predictions of the numerical models. We are already watching the greenhouse effect unfold before our eyes, and the observational evidence is not pointing to disaster.

218. Sylvan Wittwer. FOOD, CLIMATE AND CO_2: THE GLOBAL ENVIRONMENT AND WORLD FOOD PRODUCTION. 1995, p. 56. Finally, there is a scenario that offers that there has not been, and will likely not be a significant global warming, and if there is, it may be overall more desirable than the climate we now have. Also, that any climate change (warming) will be at the very lower levels of projections coming from the interpretation of projections of general circulation climate modeling. This scenario is based on the reality of what has happened in the earth's atmosphere thus far. It is emphasized that the computer projections of global warming are of known computational errors and do not include many variables, among which the most important is the effects of clouds, which may exceed by severalfold any anthropomorphic inputs of greenhouse gases, the inputs of oceans, and the biological sequestering by green plants and soils.

The rationale is that the models have predicted much more warming than has already occurred. The GCMs, with their known flaws, form the only basis for predicting catastrophic warming, from a doubling of greenhouse gases, and particularly CO_2, during the next century. If all the global warming of 0.4 to 0.6°C, which has occurred during the past century could be attributed to an increased greenhouse effect, it would still be only one-third to one-half of the lowest now being predicted by current models.

219. WORLD CLIMATE REPORT. 22 January 1996. Scientific "moderates' have maintained for decades that the United Nations position—which is labeled as the "consensus"—was not borne out by the data and that a net additional warming of around 1.3°C for doubling atmospheric carbon dioxide was much more likely than the "best estimate" of 3.2°C given by the U.N. It turns out that 1.3°C is only 40% of the 1990 U.N. estimate and is precisely what the data featured in the *Times* imply. The more facts come in, the more the U.N. "consensus" that built the Rio Treaty seems extreme.

220. Sylvan Wittwer. FOOD, CLIMATE AND CO_2: THE GLOBAL ENVIRONMENT AND WORLD FOOD PRODUCTION. 1995, p. 39. Not all agree with such extreme levels of warming (Balling, 1992; Idso, 1989; Lindzen, 1993; Michaels, 1991; Michaels et al., 1993; Seitz et al., 1989a; Strommen, 1992) and project much more tolerable, and even favorable changes when coupled with the direct benefits of rising levels of atmospheric CO_2. They predict that there would be a net benefit for food production. The position of a very modest temperature change related to the lower levels of increase is supported by the record of the past 100 years, where the average global temperature has increased only between 0.3 and 0.6°C (0.5 to 1.1°F). During this time the CO_2 level has increased up to 30%, and has been added to by even more rapid increases in other so-called greenhouse gases. With these rather remarkable increases in atmospheric CO_2, it is difficult to explain such low increments in temperature change, when in reality with the increases in not only CO_2, but other greenhouse gases as well, we should already be approaching the halfway mark of temperature increases projected for a doubling of CO_2 by the GCMs. In other words, because of the combined effect of these so-called greenhouse gases, we have already gone over halfway to an equivalent doubling of CO_2 (Balling, 1992; Hansen et al., 1989; Michaels, 1990; Wigley, 1987). Even so, there has been less than a half a degree of warming in the past 100 years. This is not all. the minor temperature increases (0.3 to 0.6°C) that have occurred during the past century could be attributed either to greenhouse warming, heat island effects at weather stations, inadequate sampling, or natural climate variability. None can be ruled out.

221. Ronald Bailey. ECOSCAM. 1993, p. 153. Many climatologists believe that there may eventually be some warming associated with increased atmospheric CO_2, but nothing like the climate disaster predicted by the apocalyptics. Lindzen thinks that over the next century "a virtually undetectable rise of a few tenths of a degree might occur." Robert Balling says, "the temperature record of the past century suggests that a doubling of carbon dioxide will produce a global temperature response at the lowest end of the model predictions, probably not more than 1.0C degrees." Patrick Michaels also believes global temperature might rise as much as 1.0C (1.8F) degrees.

222. Kenan Malik. THE SPECTATOR. 11 November 1995, p. 26. Whatever measures are taken now to limit emissions of carbon dioxide and other greenhouse gases, global warming is likely to occur to some degree thanks to the accumulation of gases already released. The IPPC estimates that, at 1991 prices, it would require $227 billion to provide a coastal defence of low-lying areas from rising seas. Where will such resources come from with reduced production and slowed-down development? Third World nations will find it particularly hard to cope with global warming. Over the next century, it is estimated that it will cost richer nations 1 percent of their GNP to cope with the change, while poorer nations will have to spend 2 per cent. Environmentalists' call for a slow-down in the

223. Justin Lancaster. A GLOBAL WARMING FORUM: SCIENTIFIC, ECONOMIC AND LEGAL OVERVIEW. Ed. R. Geyer, 1993, p. 550. The upcoming international agreement in 1992 should address continued research, a goal for atmospheric composition, emissions limitations, alternative energy sources and adaptive strategies. Even if not a formal "law of the atmosphere," the agreement must still specifically require enforcement of guidelines through national Clean Air Act mechanisms, or "harmonization" of legislation. It should link compliance with a share of the GEF to provide incentive for maintaining commitment. The time to begin applying the brakes on our energy use has clearly arrived. The momentum of our energy-consuming society threatens to make some measure of climate change unavoidable. The energetic cost of society reacting to the greenhouse effect has to be added to the energy we will need to react to other problems. We will likely have to set priorities, as our budget for response may not allow optimal responses to every looming ecological problem.

224. Ronald Bailey. ECOSCAM. 1993, p. 190. New Yorkers, Londoners, and Bangladeshis can relax. Recent evidence shows that instead of melting away, there has been a "significant increase" in the accumulation of snow and ice in Antarctica. In addition, the West Antarctic ice sheet is stable and still reacting to the changes that occurred at the end of the last ice age. The Greenland ice cap is also growing. So instead of sea levels rising, the accumulation of ice should lower them by a little over one millimeter per year. Another apocalypse avoided.

225. Dixie Lee Ray. TRASHING THE PLANET. 1990, p. 43. Some scientists postulate that the west Antarctic ice sheet, which is anchored on bedrock below sea level, could melt and add enough water to raise the world sea level by six or seven meters. this would be disastrous for most coastlines, but if it should happen, it would probably take several hundred years, and there is currently neither observational evidence nor scientific measurements to indicate that it is under way. In fact, new measurements show that the glaciers in Antarctica are growing, not melting.

226. Dixie Lee Ray. TRASHING THE PLANET. 1990, p. 43. Finally, let's suppose that a worst case scenario does develop and that global warming does occur. If the warming caused polar ice to melt, only that on land, as in the Antarctic continent (or the glaciers of Greenland), would materially affect global sea level. When ice floats, as in the Arctic ocean, it already displaces approximately the same amount of water that would result if it were to melt. (There would be some slight thermal expansion.) Whether Arctic ice stays solid or melts would no more cause the sea level to rise than ice cubes melting would cause a full glass of ice water to overflow.

227. Dixie Lee Ray. TRASHING THE PLANET. 1990, p. 43. Analysis of sea level data since 1900 indicates that the oceans may be rising at a rate of 10 to 25 centimeters per century (about 0.1 inch per year). the data are very sketchy and uncertain. The sea rise, if it is real, is not uniform and other phenomena, such as land subsidence or upthrust, the building and erosion of beaches by weather, and the variation of inshore currents, could all affect the few measurements that are available.

228. Kenan Malik. THE SPECTATOR. 11 November 1995, p. 25. But who is to say that global warming is not good for us? Global warming is certainly not a problem for nature. The planet has no preferred state, no ideal climate or temperature. Nor, for that matter is there an ideal or normal carbon dioxide level. The concept of ecological balance is a romantic Green myth. Nature is in a constant state of flux and the earth's history is characterized by continual change.

The dinosaurs lived in a world that was 10 to 20 degrees warmer than today, a world in which carbon dioxide levels were between five and 10 times higher. Around 8,000 years ago, the earth abruptly cooled by four degrees before rapidly warming again 200 years later. More recently still, in the warm period between the 12th and the 14th centuries, temperatures were on average one degree higher than today, while in the 'little Ice Age' that followed they fell significantly lower. In raising global temperature, therefore, humanity is not doing anything to nature that nature has not already done to itself, and often much more drastically than we could conceive.

229. Ronald Bailey. ECOSCAM. 1993, p. 162. Wild plants and animals dwell successfully in habitats as different as the Himalayas are from Death Valley, and Greenland from the Congo. The NAS recognized that some species might become extinct, but noted that species and entire ecosystems have the ability to migrate to more congenial climates. The NAS advised the creation and preservation of natural corridors to facilitate ecosystem migration.

230. WORLD CLIMATE REPORT. 2 September 1996, p. 4. One of the pillars of current climate hysteria is the argument that human-induced changes will result in massive extinctions, particularly in communities that have seen little climate change over the eons.

The crown jewel of that argument is the tropical rain forest, the most diverse ecosystem on earth. The argument goes like this: It's so diverse because the climate has been so stable that evolution exploited all those weird little niches. Temperatures there only budged a degree or so during the last ice age, while protoChicago lay under 5,000 feet of ice.

Now comes evidence that tropical temperatures were much lower during the ice age than was previously thought. According to Wallace Broeker, writing in *Science*, tropical temperatures at midelevations declined by 5°C, and possibly even 6°C to 8°C. Sea surface temperatures cooled a remarkable 1.5°C to 5.5°C.

It is doubtless that coolings like these would have severely restricted the tropical rain forest, probably confining it to microclimatic refugia in the variegated terrain of the Andean base. And when it warmed back up . . . it expanded again!

Does all of this mean that human-induced changes may in fact confine some biotic communities to smaller refuges? Possibly—but it also argues that biodiversity is a bit more insensitive to climate change than was thought before this hit the stands.

We'll bet that, when cooler heads ultimately prevail in the next century (as they must), the predominant ecological question will change from "How can we protect our fragile earth?" to "Why is it so resilient?" [ellipses in original]

231. WORLD CLIMATE REPORT. 14 October 1996, p.14. According to a recent study by *World Climate Report* (motto: Protecting Our Planet from Global Warmers), crab populations will swell because of global warming. "Changes in the timing of shorebird migration will cause them to miss the annual emergence of crab larvae into the Mid-Atlantic Bight," said an unidentified WCR spokesperson. "There's no doubt that the number of crabs will increase dramatically, benefiting the area's watermen and economy."

232. Sherwood Idso. CO_2 AND THE BIOSPHERE: THE INCREDIBLE LEGACY OF THE INDUSTRIAL REVOLUTION. 12 October 1995, p. 214. Scores of ecosystem studies reveal two other pertinent facts. First, with increasing productivity, ecosystems typically produce a more diverse flora that supports greater populations of herbivores and other animals. Second, with the increased plant turnover rates that appear to be a common consequence of vegetative stimulation, ecosystem detrital mass generally declines. Consequently, as the air's CO_2 content rises, enhancing plant biodiversity, productivity and turnover rates, there is a significant transfer of organic carbon from earth's non-living reservoir, where most of it currently resides, to the living portion that animates the planet.

233. Keith E. Idso. WORLD CLIMATE REVIEW. Spring 1995, p. 23. Evidence to support this hypothesis is voluminous. In a massive review of vegetative productivity in 51 terrestrial ecosystems, it was found that the biomass of plant-eating animals or herbivores increased linearly with net above-ground primary production (which id dramatically enhanced by atmospheric CO_2 enrichment) over an observed productivity range spanning two orders of magnitude. And in a similar study of paired measurements of herbivore biomass and net primary production in 22 aquatic ecosystems, it was found that herbivore biomass in watery habitats also rose linearly with increases in vegetative productivity.

234. Sherwood Idso. CO_2 AND THE BIOSPHERE: THE INCREDIBLE LEGACY OF THE INDUSTRIAL REVOLUTION. 12 October 1995, p. 31. Just as turnover rates are strongly correlated with ecosystem productivity, so are they also strongly correlated with ecosystem detrital mass. But whereas increasing turnover rates are indicative of increasing growth rates, they imply just the opposite with respect to an ecosystem's inventory of plant detritus. Consequently, as the air's CO_2 content has risen in the past, and as it continues to rise in the future, we have experienced, and will continue to experience, a positive net transfer of organic carbon from *dead* vegetative tissue to *living* vegetative tissue, which transfer increases the planet's abundance of living plants. And with that increase in plant abundance should come an impetus for increasing the species richness or *biodiversity* of earth's natural plant communities; for in a recent study of vascular plant diversity in 94 terrestrial ecosystems from every continent of the globe except Antarctica, it was found that ecosystem species richness is more positively correlated with ecosystem productivity than it is with anything else. What is more, a 1989 review of plant-animal interactions in 51 terrestrial ecosystems found that the biomass of plant-eating animals or herbivores was also an increasing function of aboveground primary production; while a 1993 review of 22 aquatic ecosystems found that the herbivore biomass of watery habitats increased in the same manner with a rise in underwater vegetative productivity. Consequently, as K. E. Idso has recently written, it is abundantly clear that earth's animal life will experience population responses to rising levels of atmospheric CO_2 that will parallel those of the plant kingdom; for the greater the food base, the greater the superstructure of life that can be supported.

235. Keith E. Idso. WORLD CLIMATE REVIEW. Spring 1995, p. 23. Contrary to allegations that increasing levels of atmospheric CO_2 may lead to decreases in biodiversity, the preceding experimental evidence suggests just the opposite. As pressures exerted by resource limitations and environmental stresses are alleviated by atmospheric CO_2 enrichment, plants of all types should find it easier to grow and reproduce, leading to an increased production of biomass at the most fundamental level of the planet's many food chains, which in their interlocking complexity comprise the biosphere's "web of life."

236. Sherwood Idso. CO_2 AND THE BIOSPHERE: THE INCREDIBLE LEGACY OF THE INDUSTRIAL REVOLUTION. 12 October 1995, p. 4. Creatures that live in the soil, such as earthworms, for example, are greatly stimulated by additions of organic matter, and an increase in their activity would likely lead to the creation

of much new soil, while at the same time improving the fertility, structure, aeration and drainage of existing soils. These improvements, in turn, would likely boost plant productivity higher yet, putting still more organic matter into the soil, and so on, with the several phenomena impacting each other in a positively reinforcing fashion that lifts the whole biosphere to a new level of activity.

237. WORLD CLIMATE REPORT. 10 June 1996, In reality, the health of the world's people is improving. Over the last 35 years, the world's death rate has been cut in half. Life expectancy continues to grow to unprecedented levels (Figure 1), and each year infant mortality falls to record lows (Figure 2). Today 86 percent of the world's population lives in a country where newborns enjoy a life expectancy greater than 60 years, compared with six out of 10 in 1980 (Figure 3).

238. WORLD CLIMATE REPORT. 29 April 1996, p. 14. *Tropical diseases will spread poleward from higher temperatures arising from greenhouse warming.*

The graph below shows decadal temperature changes (in degrees C per decade) within the Tropics based on the satellite record since 1979. Trends are computed across the globe (land and ocean) in 5-degree-wide latitude bands. In fact, the Tropics have been cooling for the last 17 years. So recent outbreaks of tropical diseases can't logically be attributed to global warming.

239. WORLD CLIMATE REPORT. 5 August 1996, p. 4. Why am I not surprised that U.N. agencies are hyping global warming as a threat to public health (*NYT*, July 8)? Since data from satellites point to the absence of a predicted temperature increase, U.N. agencies are getting desperate—grasping at any available scare tactic to influence the ongoing international negotiations on the Global Climate Treaty.

The U.N. report hypothesizes that tropic diseases, spread beyond the tropics by mosquitoes and similar vectors, will put "tens of millions of people" at risk for malaria, etc. But the report ignores crucial questions:

Why didn't malaria spread widely during the major warming in the decades around 1920?

... Why is there no mention of the human vector in the spread of infectious diseases, likely to be more important in this age of mega-conferences?

... The U.N. report now calls global warming one of the greatest health challenges facing the world in the 21st century. This strange emphasis on a phantom global threat at the expense of real local and national health problems strikes many as both unscientific and irresponsible. [ellipses in original]

240. WORLD CLIMATE REPORT. 30 September 1996, p. 41. If climate change didn't wipe out malaria in the United States, what did? Technology.

According to Reiter, air-conditioning and elimination of urban overpopulation (brought about by the development of suburbs and commuting by automobile) caused the decline. Reduce urban population density, wipe out malaria. Need proof? In 1995, a dengue pandemic (another mosquito-borne disease) hit the Caribbean and Mexico. More than 2,000 of the people infected were from Reynosa, Mexico (the city that borders Hidalgo, Texas). Yet only seven cases were reported in the entire state of Texas. Apparently those mosquitoes have a harder time crossing the border than some other species!

The environmental pests that keep bugging us with their ecological apocalypses have yet to provide a shred of hard (or even soft) evidence that minuscule temperature changes mean anything in the face of advanced sanitation and infrastructure.

241. WORLD CLIMATE REPORT. 29 April 1996, p. 27. The major factor missing from these dire predictions is technology. It is assumed that technology will remain constant–the people will sit back and wait for the impending disaster to strike. Fortunately, humans have an incredible capacity to both avoid problems and adapt to those that do occur. Current global agricultural production far exceeds predictions agriculturalists would have made only 50 years ago. Pandemic diseases that have caused incredible human death and suffering for centuries have been significantly reduced or completely eradicated. But technological advances can't be predicted. So people have a choice. They can either assume that we've run out of good ideas for solving the world's problems, or they can put their faith in the adaptability and ingenuity of human invention. If anything, the crux of the debate on climate change pits the pessimists against the optimists. Based on human history, the latter will win out.

242. Sylvan Wittwer. FOOD, CLIMATE AND CO$_2$: THE GLOBAL ENVIRONMENT AND WORLD FOOD PRODUCTION. 1995, p. 107. A recent report by Waggoner (1994a) emphasizes that advances in farming technologies, combined with changes in diets in response to health and price, will ensure that the world's population will use existing cropland and water more economically and efficiently and save more land for natural or wilderness use in the next 50 years, even with a population increase to 10 billion people. The global total of sun on land, CO$_2$ in the air, and fertilizer and water could produce far more food than 10 billion people need. Waggoner (1994a) goes on to affirm *"the silver lining in the growing cloud of atmospheric CO$_2$ that may warm the planet is more raw material for photosynthesis."* The CO$_2$ in the air is rising at the rate of 2ppm/year, which will raise the atmospheric concentration by approximately one third while the population of the earth grows to 10 billion. There will be an increase in food crop production of about 10%.

243. Sylvan Wittwer. FOOD, CLIMATE AND CO$_2$: THE GLOBAL ENVIRONMENT AND WORLD FOOD PRODUCTION. 1995, p. 189. The rising level of atmospheric CO$_2$ could be the one global natural resource that is progressively increasing food production and total biological output, in a world of otherwise diminishing natural resources of land, water, energy, minerals, and fertilizer. It is a means of inadvertently increasing the productivity of farming systems and other photosynthetically active ecosystems. The effects know no boundaries and both developing and developed countries are, and will be, sharing equally. Further, there is now strong evidence that the rising abundance of this resource is enabling greater efficiency in both land and water use and enhancing the value of crop fertilization—both biological and chemical. Crop and plant responses to CO$_2$ enrichment offer fertile grounds, with frontiers not yet explored for enormous strides in world food production and for the conservation of water, which is rapidly becoming the world's most critical natural resource.

244. Sylvan Wittwer. FOOD, CLIMATE AND CO$_2$: THE GLOBAL ENVIRONMENT AND WORLD FOOD PRODUCTION. 1995, p. 103. Globally, some 25 crops stand between people and starvation (Wittwer, 1981a). The production of these crops, their climatic adaptability, and their direct response to CO$_2$ holds priority in any assessment of the effects of currently rising levels of atmospheric CO$_2$ on agricultural output, world food production, and global food security (Table 5.1).

245. Sylvan Wittwer. FOOD, CLIMATE AND CO$_2$: THE GLOBAL ENVIRONMENT AND WORLD FOOD PRODUCTION. 1995, p. 63. There are important reasons for the increased crop productivity, dry matter increases, and higher yields from elevated levels of CO$_2$. First is the superior efficiency of photosynthesis, resulting in marked reductions of respiration. Net photosynthesis is the sum of gross photosynthesis minus respiration, including both photorespiration and dark respiration. The second is an increased water use efficiency resulting from sharp reductions in water loss per unit of leaf area. This relates to the partial closing of stomata associated with higher CO$_2$ levels. While stomata admit air and CO$_2$ into the leaf for photosynthesis, they are also the chief conduit for moisture loss. While higher CO$_2$ levels greatly reduce water loss, there is little if any impairment of CO$_2$ uptake. The mechanism by which the concentration of CO$_2$ in the atmosphere determines whether stomata tend to open or close is still largely unknown.

Other observed and possible benefits are substantial increases in root growth resulting in an increase in root/top ratios, reductions in injury from air pollutants, and a partial compensation for deficiencies of light, moisture, temperature, and some soil nutrients and greater resistance to soil salinity.

246. Sylvan Wittwer. FOOD, CLIMATE AND CO$_2$: THE GLOBAL ENVIRONMENT AND WORLD FOOD PRODUCTION. 1995, p. 59. Indeed, CO$_2$ fertilization of air has proven decidedly beneficial for crops grown in enclosed greenhouse structures, and also for plants and crops exposed to natural atmospheric conditions. Many experimental observations of the influence of elevated levels of atmospheric CO$_2$ on plants (Bazzaz, 1990; Bazzaz, et al., 1989; Dahlman, 1993; Eamus and Jarvis, 1989; Idso, 1989a; Kimball, 1983b; Rogers et al., 1981; Strain and Cure, 1986; Woodward, 1993; Wittwer, 1994) have found a number of repeatable effects across most species. These include growth stimulation, especially of roots, reduced stomatal conduction and leaf nitrogen, and an increase in water use efficiency. Lawlor and Mitchell (1991) report, in a summary of field studies, that elevated atmospheric levels of CO$_2$ increase photosynthesis, dry matter production, and yield, substantially in C$_3$ species and less in C$_4$, decrease stomatal conductance and transpiration in C$_3$ and C$_4$ species, and greatly improve water use efficiency in all plants. Increased productivity is related to greater leaf area. Stimulation of yield is more from an increase in the number of yield-forming structures than their size. Partitioning of dry matter among organs is mostly evident in root and tuber crops. Over 1000 experiments have now been conducted around the world reporting favorable growth improvement or responses, approximating an overall increase in yield of 33% for all mature and immature plants from a doubling of the CO$_2$ concentration from its current level of 360 ppm (Kimball, 1985). These data have been further extrapolated by Goudriaan and Unsworth (1990), who suggested that about 5 to 10% of the actual rate of increase of agricultural productivity worldwide during the past century can be ascribed to the fertilizing effect of rising atmospheric CO$_2$. Rogers et al., (1994a) correctly state that "Carbon dioxide is the first molecular link from atmosphere to biosphere. It is essential for photosynthesis which sustains plant life, the basis of the entire food chain. No substance is more pivotal for ecosystems, either natural or managed." The rising levels of atmospheric CO$_2$ represent a resource for agricultural and food production rather than a conventional pollutant.

247. Sylvan Wittwer. FOOD, CLIMATE AND CO$_2$: THE GLOBAL ENVIRONMENT AND WORLD FOOD PRODUCTION. 1995, p. 57. Finally, there would be the growth enhancement caused by CO$_2$. Carbon dioxide at its current atmospheric level, is a limiting nutrient for plants. A voluminous scientific literature has now demonstrated more growth through photosynthetic enhancement; improved CO$_2$ dark fixation, reductions in both photo and dark respiration; increased water use efficiency by plants, including food crops, and greater resistance to air pollutants, as the atmospheric CO$_2$ concentration increases. Except for the height of the ice ages, both global temperatures and CO$_2$ concentrations are currently near their lowest values for the past 100 million years. A

prestigious group of scientists, including meteorologists and those familiar with the direct biological effects of elevated levels of atmospheric CO_2 is supportive of this scenario (Balling, 1992, 1993; Ellsaesser, 1993; Idso, 1989a; Landsberg, 1984; Lindzen, 1990; Michaels, 1992, 1993; Seitz et al., 1989; Reifsnyder, 1989).

248. Hayri Onal. GLOBAL WARMING FORUM: A SCIENTIFIC, ECONOMIC AND LEGAL OVERVIEW. 1993, p. 411. If areas with increased production potential offset the production loss in other regions, a significant change in agricultural market conditions may not occur. However, an overall production loss for some crops would lead to new market equilibria where prices of disadvantaged crops would increase, while prices of other crops would decline or remain the same. Several studies incorporated this endogenous price adjustment mechanism while determining the economic impacts of global warming. Using a spatial equilibrium model, Dudek (1988) analyzed the economic effects upon California agriculture. His findings, based on the GISS and GFDL climatic scenarios, show drastic yield reductions (between 3 and 40%) especially in the southern and central parts of California, resulting in an overall welfare loss (between 14 to 17%). When the direct fertilization effect of CO_2 is incorporated, however, yields of some crops (especially vegetables) are recovered and statewide welfare even slightly increases (1.4%).

249. N. C. Bhattacharya. GLOBAL WARMING FORUM: A SCIENTIFIC, ECONOMIC AND LEGAL OVERVIEW. 1993, p. 493. In general a high CO_2 environment appears to alleviate water stress effects in several species, e.g., soybean (Rogers et al., 1984), sweet potato (Bhattacharya et al. 1990b), wheat (Sionit et al., 1980), and cotton (Kimball et al., 1991; Bhattacharya et al., 1992b). In a few cases, growth and biomass even increased under water-stressed conditions compared to that seen under well-water conditions (Kimball, 1985; Gifford, 1979; Paez et al. 1983, 1984; Tolley and Strain, 1984; Conroy et al., 1986, 1988; Wray and Strain, 1986). According to Kimball and Mauney (1992), the probable explanation for growth stimulation under high CO_2 and water stress may be through profuse root growth at high CO_2 which enables plants to more fully mine a volume of soil and thereby withstand moisture stress.

250. Sylvan Wittwer. FOOD, CLIMATE AND CO_2: THE GLOBAL ENVIRONMENT AND WORLD FOOD PRODUCTION. 1995, p. 86. For the future, it is conceivable the weeds, having a C_3 photosynthetic pathway, could, at higher levels of atmospheric CO_2 and no appreciable change in temperature, become more competitive with C_4 food plants, such as corn, sorghum, sugarcane, and millet. With a global warming accompanying elevated levels of atmospheric CO_2, the reverse could occur. Conversely, cotton, soybeans, cowpeas, field beans, mung beans, peanuts, wheat, rice, potatoes, sweet potatoes, cassava, sugar beets, bananas, coconuts, and most all horticultural crops and forest trees are C_3 plants and are now plagued by weeds, particularly the grasses, many of them C_4 plants. These important agricultural crops may become better competitors with weeds in a CO_2-enriched world (Table 4.1). With an accompanying global warming of some magnitude, the reverse could occur. Kimball (1984) has stated it well: "Even as C_4 species face stiffer direct weed competition, so will they face greater competition from C_3 crops for a share of the market."

251. Sylvan Wittwer. FOOD, CLIMATE AND CO_2: THE GLOBAL ENVIRONMENT AND WORLD FOOD PRODUCTION. 1995, p. 82. Idso (1989a), with a thorough review of the literature then available, suggests a tipping of the scales in favor of plants over foraging insects in a higher CO_2 world. This relates to the assumption that insect damage and attack may be partially or totally circumvented by the more rapid vegetative growth and earlier maturity of plants grown at elevated levels of atmospheric CO_2. Earlier maturity induced by elevated levels of atmospheric CO_2 should enable some crops to escape the ravages of insect invasions, as well as disease and weed infestations, and drought. The Chinese have been very adept at this approach. In the United States, this has already been partially achieved by genetically selecting cotton, corn, wheat, and soybeans for early maturity. Clearly, more laboratory and field experiments are required. It is unfortunate that the large-scale free air CO_2 experiments (FACE) underway with cotton and wheat include neither predator nor pollinator interactions, or even more importantly, competition with other crop species, and weeds (Kimball et al., 1993b).

252. Sylvan Wittwer. FOOD, CLIMATE AND CO_2: THE GLOBAL ENVIRONMENT AND WORLD FOOD PRODUCTION. 1995, p. 132. Virtually all investigators dealing with the direct effects of high levels of atmospheric CO_2 have concluded that the impact on food crops will be positive. There are those, however, that emphasize the few exceptions. Even the most vocal critics (Bazzaz and Fajer, 1992), however, admit with legumes, particularly the soybean, the crop will be positively impacted by elevated levels of atmospheric CO_2 and this will include the impact on ecosystem-wide changes, as well (Stulen and den Hertog, 1993).

253. E. M. Bridges and N. H. Bates. GEOGRAPHY. April 1996, p. 167. Climatic change can bring with it both benefits and problems, and neither can be predicted with confidence (Scharpenseel et al., 1990). It is mostly the negative aspects which have been stressed by the media, but there are some positive effects as well. In the first place, some plants grown in enhanced carbon dioxide atmospheres have shown improved growth and increased yields (Idso and Idso, 1994). Warmer tempera-

tures will further enhance this trend, but may also encourage pests. A secondary effect is that with higher carbon dioxide concentrations in the atmosphere plants do not need to open their stomata for so long, and do not develop as many stomata, with the result that the water demand for plants is less. This could be a very significant advantage in arid and semi-arid regions where water is scarce. The photosynthetic capacity of plants has also been reported to be greater under increased carbon dioxide concentrations (Bazzaz and Fajer, 1992). This increased carbon dioxide content of the atmosphere has the power to encourage plant growth in what has come to be called the 'CO_2 fertilization' effect, providing water and plant nutrients do not become a limiting factor.

254. David Gardiner. FEDERAL DOCUMENT CLEARING HOUSE. 19 June 1996. As noted earlier, it is not only model design that matters but also the selection of which policy instrument to use in pursuing a particular emissions reduction goal. Analysis to date suggests that a well-designed implementation policy can substantially reduce the costs of greenhouse gas mitigation. While this analysis is still incomplete, it appears that a well-designed policy would include some of the following features:

- Advance Signals: Policies that are announced well in advance and implemented gradually will be substantially less costly than policies that impose large controls unexpectedly.

- Flexibility: We know from our experience in implementing the Clean Air Act that control strategies that grant sources broad flexibility in meeting desired targets will dramatically reduce costs. A model for this flexibility is the tradable allowances system for sulfur dioxide.

- Broadly Shared Impacts Across the Economy: Our studies show that how the cost of emission reductions is shared among energy producers and consumers can be important, in terms of both the magnitude and distribution of overall economic impacts.

- Policies to Improve the Operation of Energy Markets: Given the magnitude of potential emissions reductions that can be realized at a net savings, one must ask why more of these technologies are not now being implemented even in the absence of a control strategy. The reasons include a lack of information on the performance and economics of new technologies, higher up-front costs for the equipment (although lower operating costs), and other institutional problems (e.g., landlords minimize up-front costs and load higher operating costs onto tenants). Many of the programs included in the Administration's current Climate Change Action Plan were designed to address these issues. Such programs help ensure that the vast potential of technology to address the climate problem will be realized.

255. Ross Gelbspan. HARPER'S MAGAZINE. December 1995, p. 31. That resistance is understandable, given the immensity of the stakes. The energy industries now constitute the largest single enterprise known to mankind. Moreover, they are indivisible from automobile, farming, shipping, air freight, and banking interests, as well as from the governments dependent on oil revenues for their very existence. With annual sales in excess of one trillion dollars and daily sales of more than two billion dollars, the oil industry alone supports the economies of the Middle East and large segments of the economies of Russia, Mexico, Venezuela, Nigeria, Indonesia, Norway, and Great Britain. Begin to enforce restriction on the consumption of oil and coal, and the effects on the global economy—unemployment, depression, social breakdown and war—might lay waste to what we have come to call civilization. It is no wonder that for the last five or six years many of the world's politicians and most of the world's news media have been promoting the perception that the worries about the weather are overwrought.

256. Steve Kidney. ENERGY DAILY. 11 January 1996, p. 14. Thorning says it would not do any good for the U.S. and other developed countries to ratchet their emissions down to 1990 levels, since "the greatest source over the next 50 years will be developing countries" such as India and China.

"These goals are politically determined," she says of the Berlin proposal. "All we will do with a carbon tax is slow our own growth rate and make it harder on developing countries, because we won't be importing nearly as many goods. And there won't be any material impact on carbon dioxide emissions."

257. S. Fred Singer. WASHINGTON TIMES. 21 March 1996, p. A 11. The Global Climate Treaty, signed at the 1992 U.N. "Earth Summit," is supposed to save the planet from global warming—a threat based entirely on unconfirmed theory. Lacking scientific justification, the treaty may turn into a giant U.N. scheme for taxing the use of energy—with the burden falling primarily on consumers in the United States and the other industrialized nations.

To be sure, the level of atmospheric carbon dioxide from the burning of coal, oil and gas has been rising since the beginning of the Industrial Revolution. The announced purpose of the treaty is to keep the level from increasing further—an unrealistic goal that spells economic disaster. To stabilize CO_2 levels, one has to reduce emissions by about 70 percent—effectively reducing energy use worldwide to less than one-third its present amount!

258. RICHMOND TIMES DISPATCH. 25 June 1996, p. A-12. The Declaration notes that the Global Climate Treaty arising from the 1992 "Earth Summit" in Rio de Janeiro calls for stabilization of atmospheric green-

house gases—and that this stabilization would require a 60 to 80 percent reduction in the use of fuel worldwide.

In a world in which poverty is the greatest social pollutant, any restriction on energy use that inhibits economic growth should be viewed with caution. For this reason, we consider "carbon taxes" and other drastic control policies—lacking credible support from underlying science—to be ill-advised, premature, wrought with economic danger, and likely to be counterproductive. The Declaration is not signed by loons. Signatories include scientists at such esteemed institutions as Yale, the University of Virginia, Johns Hopkins, Oxford, Lawrence Livermore Laboratory, the University of Stockholm, the University of Vienna, the University of Prague, and the Max Planck Institute in Munich. Perhaps President Clinton should take a look at what these experts have to say before launching the nation—and the world—on a costly campaign to solve a problem that might not (and probably doesn't) even exist.

259. S. Fred Singer et al. A GLOBAL WARMING FORUM: SCIENTIFIC, ECONOMIC AND LEGAL OVERVIEW. Ed. R. Geyer, 1993, p. 354. Drastic precipitous—and, especially, unilateral—steps to delay the putative greenhouse impacts can cost jobs and prosperity and increase the human costs of global poverty, without being effective. Stringent controls exacted now would be economically devastating—particularly for developing countries for whom reduced energy consumption would mean slower rates of economic growth—without being able to greatly delay the growth of GHG in the atmosphere. Yale economist, W. Nordhaus, one of the few who has been trying to deal quantitatively with the economics of the greenhouse effect, has pointed out that ". . . Those who argue for strong measures to show greenhouse warming have reached their conclusion without any discernible analysis of the costs and benefits. . . ." It would be prudent to complete the ongoing and recently expanded research so that we will know what we are doing before we act. "Look before you leap" may still be good advice. [ellipses in original]

260. Dale W. Jorgensen and Peter J. Wilcoxen. SHAPING NATIONAL RESPONSE TO CLIMATE CHANGE: A POST RIO GUIDE. Ed. Henry Lee. 1995, p. 238. Somewhat surprisingly, the United States has had an extended period of stable carbon dioxide emissions once before: from 1972 to 1985. During that period, high oil prices reduced energy demand and lowered carbon dioxide emissions substantially. The relationship between oil prices and carbon emissions can be seen by a comparison of historical oil prices, shown in figure 8.1, with the history of U.S. carbon emissions, shown in figure 8.2. The large increases in oil prices in 1974 and 1979 led to drops in the trend rate of emissions growth. However, this reduction came at a very high price: the oil price shocks reduced U.S. GNP growth by 0.2 percent per year from 1974 to 1985. The lesson from this episode is that a 0.2 percentage point of annual GNP growth is an upper bound on the cost of stabilizing U.S. carbon dioxide emissions.

261. Ned Nanda. DENVER POST. 4 August 1996, p. D4. Scientific uncertainties remain, and the IPCC's work has been challenged. The Global Climate Coalition, an energy lobbying group, has stated that efforts suggested by the Geneva Declaration "would eliminate millions of American jobs, reduce America's ability to compete and force Americans into second-class lifestyles." And a group of scientists, most of whom represent the fossil-fuel-based industries, proclaimed in their alternative Leipzig Declaration that the IPCC's conclusion is based upon dubious science and that global warming is not really occurring.

262. Henry Lee. SHAPING NATIONAL RESPONSES TO CLIMATE CHANGE: A POST RIO GUIDE. Ed. Henry Lee. 1995, p. 8. First, the design of U.S. domestic policy options to mitigate global warming will be a much more complex task than was the design of responses to past environmental problems. The cost of even stabilizing emissions is several orders of magnitude larger than is the cost of phasing out chemicals that destroy the ozone layer or of reducing the threat of acid rain.

263. Norman Myers. ULTIMATE SECURITY: THE ENVIRONMENTAL BASIS FOR POLITICAL STABILITY. 1993, p. 178. This is not to say that a global campaign to avoid global warming would come cheap in every respect. According to one exploratory estimate, it could cost $30 billion a year to counter the more prominent causes of climate debacle. But remember: this amount is less than the United States alone spends to subsidize fossil fuels, among other forms of energy. Another estimate proposes an outlay of $100 billion a year to achieve our switch not only away from carbon dioxide but from methane and other greenhouse gases. Yet that sum is only one-tenth of what we now spend to deter war, among other military activities. Wouldn't it be a solid investment to deter global insecurity of climacteric scale?

264. William D. Nordhaus. MANAGING THE GLOBAL COMMONS. 1994, p. 6. The need to address the potential issues raised by future climate change is one of the most challenging economic problems of today, and it is daunting for those who take policy analysis seriously. It raises formidable issues of data, modeling, uncertainty, international coordination, and institutional design. In addition, the economic stakes are enormous, involving investments on the order of hundreds of billions of dollars a year to slow or prevent climate change.

265. Dale W. Jorgensen and Peter J. Wilcoxen. SHAPING NATIONAL RESPONSE TO CLIMATE CHANGE: A POST RIO GUIDE. Ed. Henry Lee, 1995, p. 6. In this chapter, we use a detailed model of the U. S.

economy to calculate the magnitude of these and other effects. Our findings are roughly as follows. A carbon tax would raise fuel prices, particularly for coal. Coal demand would fall substantially, leading to a large drop in coal production. Oil and gas output would also decline, but by a much smaller percentage. Higher coal prices, in turn, would raise the cost of electricity. Consumers and firms would demand less electricity, which would slow productivity growth and capital formation. This, in turn, would tend to reduce GNP.

266. Russell O. Jones. FORUM FOR APPLIED RESEARCH AND PUBLIC POLICY. Summer 1996, p. 78. By contrast, the United States, Canada, Australia, and New Zealand have growing populations. According to the World Bank, populations in these countries will increase by 26 to 35 percent between 1990 and 2025. To achieve the EU target, per-capita emissions in the United States, Canada, Australia, and New Zealand would have to decrease by 20 to 25 percent.

267. William E. Rees. ALTERNATIVES. October 1995, p. 31. Finally, even without these problems, it would be difficult for any country to attempt the transition to sustainability on its own. By acting decisively to reduce energy and material throughput, individuals and countries alike risk putting themselves at a short-term disadvantage while freeing up common-pool assets that well might simply be used by someone else. In this sense, sustainability has the quality of a public good—the costs to the initiating country may be greater than its share of any benefit accruing to the common pool, and all will be lost if others don't follow suit.

268. Carlos J. Moorhead. GLOBAL WARMING—PART 2. Y4.En2/3: 103–92, 1993, p. 338. I also have other concerns with the plan. To begin with, the goal of the plan is tougher than the goal of the International Treaty on Climate Change. I recognize that it is the President's privilege to set goals for the country, but I just hope he is not setting climate change goals that will put the United States at a disadvantage in terms of international trade and economic growth.

269. GLOBAL WARMING—PART 2. Y4.En2/3: 103–92, 1993, p. 369. The Administration has correctly acknowledged that the President's "commitment" to return greenhouse gas emissions to their 1990 levels by the year 2000 is a "political commitment" as opposed to a legally binding requirement of the Framework Convention. Nevertheless, the GCC is concerned that this important distinction will be overlooked or forgotten when, as early as one year from now, the U.S. must submit its National Action Plan under the Framework Convention. The Climate Change Action Plan states that the Plan, or an "updated version" will be the cornerstone of the U.S. National Action Plan. If the targets and timetables of the "political commitment" become incorporated into the National Action Plan, the United States runs the risk of unilaterally placing U.S. industry at a competitive disadvantage in the world markets.

270. David Gardiner. FEDERAL DOCUMENT CLEARING HOUSE. 19 June 1996. Modest emissions reductions can be achieved with macroeconomic impacts ranging from mildly negative to slightly positive. All else held equal, more stringent emission reduction targets would generally impose higher costs. The range of analytic results depends on the modeling framework used as well as the type of policy intervention simulated.

271. William F. O'Keefe. VITAL SPEECHES. 15 August 1995, p. 654. First, the cost of mitigation is a function of time. Reducing emissions over a longer time period will be far less harmful to our economic well-being than forcing rapid reductions. I have no doubt that if we tried to reduce greenhouse gas emissions by 20 percent over the next few years we would bring on a full-scale depression. On the other hand, returning to 1990 levels over a period of several decades might not impose unreasonable economic burdens. So time is important. Accepting this reality helps us better understand the trade-offs worthy of consideration.

272. Hazel R. O'Leary. GLOBAL WARMING—PART 3. Y4.En2/3: 103–167, 6 October 1994, p. 11. The Motor Challenge is also off to a promising start. As you know, two-thirds of all the electricity used in the industrial sector is used to run motors, accounting for about 100 million metric tonnes of carbon. This partnership seeks to increase the market penetration of highly efficient industrial motor systems in order to reduce this level of emissions. We established a national technical assistance service, held workshops, and published a request for showcase demonstration proposals just this week in the *Federal Register*. To date, 105 partners are participating in this program—including such industrial giants as General Motors, Ford, Dow, and Johnson & Johnson.

Now that the first fiscal year has begun, we are poised for an early roll-out of many of the Plan's actions. While we did not fare as well in the appropriations process as we had hoped due to the tight budget climate, we believe that we have sufficient funding to demonstrate the viability of our approach with an aggressive start. We received $107 million out of a total of $208 million requested. We received virtually our entire request for continued deployment of renewable sources of energy from the Energy and Water Development Appropriations Subcommittee. These clean energy supplies should help displace carbon generating energy supplies in the longer term. We received an increase of about 15 percent over the last year from the Interior Subcommittee for energy efficiency programs, which represents the largest increase of any Interior account this year. A table (Attachment 1) summarizing fis-

cal year 1995 program funding is attached to my testimony. the Administration continues to support the goals of this important program. We will, of course, continue to request the funding that is needed to assure the success of the program in the future. We thank the Committee for your support.

273.   Vicki Norberg-Bohm and David Hart. SHAPING NATIONAL RESPONSES TO CLIMATE CHANGE: A POST RIO GUIDE. Ed. Henry Lee. 1995, p. 265. Private and state-owned enterprises are also potential adopters of many technologies that could help to cut greenhouse gas emissions, and they demonstrate many of the same tendencies. Enterprises often are risk-averse and lack access to capital and information. As with consumers, they tend to be slow to adopt energy-saving technologies.

274.   Vicki Norberg-Bohm and David Hart. SHAPING NATIONAL RESPONSES TO CLIMATE CHANGE: A POST RIO GUIDE. Ed. Henry Lee. 1995, p. 265. High perceived costs, lack of information, and embedded habits impede better choices. Consumers are very sensitive to the initial cost of a product. In practice, new technologies have a very high discount rate, ensuring future savings far less than present savings. Compact fluorescent lamps, for instance, use about one-quarter as much electricity as incandescent lamps, and they last ten or more times longer. However, they cost about ten to fifteen times as much. The time needed to make up the difference in first cost depends on the cost of electricity and the usage of the lamp. Yet even in industrialized countries, where the payback period is typically only two years or less and the price of a compact fluorescent lamp is a small fraction of a consumer's income, market penetration has been slow. In developing countries, where the payback period is likely to be longer (because electricity is subsidized and lamps may be used less), this problem is magnified.

Consumers may simply lack the capital to buy a product with a high first cost. Loans may not be available at reasonable rates of interest, or they may be viewed as risky. Consumers may also be so uncertain about their future earnings that they are reluctant to make the investment. In addition, they may be skeptical about savings that depend on factors outside their control, such as future energy prices.

275.   Vicki Norberg-Bohm and David Hart. SHAPING NATIONAL RESPONSES TO CLIMATE CHANGE: A POST RIO GUIDE. Ed. Henry Lee. 1995, p. 265. Perhaps more important, consumer decisions are not based solely on prices and financing. Consumers are largely creatures of habit, and for good reason: they have little money or time to spare. Many of the technologies that could contribute substantially to greenhouse mitigation if widely adopted provide essential services, such as cooking, heating, and lighting. Consumers can ill afford an interruption in them. Consumers value reliability and ease of operation and often lack the information or ability to evaluate new or complex products. Furthermore, the cost of a large appliance—or even of a compact fluorescent lamp—may be a substantial fraction of an average family's income in a developing country. The failure or breakage of such a product could be a catastrophic loss of a capital investment. Consumers thus tend to be risk-averse, choosing products with which they are familiar.

276.   Edward Parson and Richard Zeckhauser. SHAPING NATIONAL RESPONSES TO CLIMATE CHANGE: A POST RIO GUIDE. Ed. Henry Lee. 1995, p. 104. However, climate change negotiations will raise the question of transfer payments, in particular the question of how large they should be, in a far more difficult form. An agreement such as the Montreal Protocol, which pays precisely the developing countries' marginal costs of compliance, leaves the recipients no better off than without the agreement. In principle, it leaves them indifferent between joining and not joining the treaty (neglecting for the moment that the treaty's penalties against non-parties make not joining worse than the status quo). If the negotiated level of controls is even roughly right, the emissions controls to be realized in any country should be worth more to the rest of the world than they cost the country to implement, possibly much more. Unless the investment or project to control emissions brings external benefits to the country undertaking it (which may be the case), for the donors to pay only incremental cost is to award them the entire surplus realized in the transaction.

There is substantial evidence in the recent negotiating record that representatives from developing countries perceive their bargaining position to be much stronger than would warrant such a disadvantageous outcome. If the industrial world's concern about greenhouse emissions remains at present levels or grows substantially, then the range of different development plans of such nations as China, India, and Brazil may include some that impose substantial harm on the rest of the world. For them, this possibility can represent a threat, which can be used to extract transfers larger than the costs they should incur in changing paths. Such demands can be characterized morally in terms ranging from compensation for historical inequity and exploitation to claiming a share of a globally created surplus to blackmail.

The redistributive negotiations related to climate change have not yet been seriously engaged. Financial negotiations represented the most divisive item on the Earth Summit agenda, but the resolution was a standoff: developing countries accepted minimal new substantive environmental obligations, and industrial countries accepted minimal new financial obligations.

277.   James K. Sebenius. SHAPING NATIONAL RESPONSES TO CLIMATE CHANGE: A POST RIO GUIDE. Ed. Henry Lee. 1995, p. 51. Given the prevailing levels of distrust—not to mention the steep energy

requirements vital to development—a threat by key developing nations not to cooperate with an emerging climate regime could have a clear rationale and a measure of credibility, even if such steps are ultimately mutually destructive and even if their effects might be more severe in the developing world. No wonder that, in the words from a recent discussion of climate change and overall Third World concerns, "the problems presented by climate change also present opportunities to reexamine and correct many of the underlying problems of development that have led to the current dilemma . . . including trade issues, debt, technology transfer, technical assistance, and financial assistance." To Southern diplomats who hold this view, the climate change issue may prove to be a very potent bargaining lever, with applications well beyond the climate context. According to another observer, "this group sees environment as the same kind of issue in the 1990s that energy was in the 1970s. They hope that the developed countries' interest in the environment can be used over time to wring concessions on development issues from the North." [ellipses in original]

278. Henry Lee. SHAPING NATIONAL RESPONSES TO CLIMATE CHANGE: A POST RIO GUIDE. Ed. Henry Lee, 1995, p. 22. Inevitably, any workable, long-term reduction program will require large transfers of money and resources from developed to developing nations. In 1990 the United States provided foreign aid equivalent to only 0.28 percent of its GNP, much of it in the form of surplus agriculture, military equipment, and tied export loans (that is, the loans had to be used to purchase goods and services from the United States). In 1986 the World Bank calculated that only 8 percent of the budget of the U.S. Agency for International Development was for developmental aid to low-income countries. To make any difference, the United States would have to dramatically increase its level of foreign aid and channel all of the additional funds to projects that directly or indirectly reduce greenhouse gases. Furthermore, this program could not be a one-time arrangement. The United States would have to transfer this higher level of assistance every year for the foreseeable future, and U.S. industrialized allies would have to increase their assistance proportionately as well.

279. Edward Parson and Richard Zeckhauser. SHAPING NATIONAL RESPONSES TO CLIMATE CHANGE: A POST RIO GUIDE. Ed. Henry Lee. 1995, p. 103. The reality of environmental negotiations is far from this ideal. First, international side payments seem to be hard to negotiate and hard to deliver. In 1990 it was surprisingly difficult to obtain U.S. consent to an international obligation of only $13 million per year for three years to support international CFC reductions. The United States chronically fails to meet its long-established obligations to the UN and the development banks, even for such a widely supported activity as UN peacekeeping operations. Even more problematic than international payments to developing countries, for which some domestic constituency exists, would be side payments to other industrial countries—such as European countries or states of the former Soviet Union. Moreover, it seems that the explicit exchange of money among nations is even harder politically than is mutual accommodation on substantive issues intended to achieve the same result. It is ironic that the strongest example of a complex, asymmetric international environmental agreement, the EC's Large Combustion Plants directive described above, appears to use asymmetric emissions targets to achieve distributive goals at the cost of efficiency losses rather than use the distribution of targets to reduce costs with compensating financial transfers. The largest reductions are required of countries whose marginal costs are probably the highest.

280. Ronald Mitchell and Abram Chayes. SHAPING NATIONAL RESPONSES TO CLIMATE CHANGE: A POST RIO GUIDE. Ed. Henry Lee. 1995, p. 128. The most frequently discussed and most expensive means of encouraging states to comply is direct financial transfers conditioned on nations undertaking activities that will improve compliance. A modest compliance fund has been established under the Montreal Protocol. Funding might be provided for the technology transfers discussed above, for administrative programs to improve the effectiveness of human resources, or for capital projects. One frequent suggestion is that an international fund be established to which industrialized states would contribute and from which developing states would receive funds to cover the costs of compliance. Financing could also occur bilaterally, perhaps with an international organization as a clearinghouse, although this would probably mean fewer funds for developing countries. In any case, the magnitude of the required funding will be a major political issue in the legislatures of developed countries.

281. Vicki Norberg-Bohm and David Hart. SHAPING NATIONAL RESPONSES TO CLIMATE CHANGE: A POST RIO GUIDE. Ed. Henry Lee. 1995, p. 282. Many other technologies have lower life cycle costs than those currently planned for use, but their first costs are higher. this situation is particularly true of many technologies used by consumers. The transaction costs of diffusing such technologies can be high, and consumers often must be subsidized to adopt them. In the near term, funds are likely to be needed to finance the higher first cost to the consumer. In addition, for this category of technologies, more technical assistance and institutional development funds are likely to be required than in the previous category because of the difficulty of developing institutions that can take a long-term perspective and that have an interest in diffusion.

282. Marc W. Chupka. FEDERAL DOCUMENT CLEARING HOUSE. 19 June 1996. Carbon

Emission Reduction "Leakage": Some analysts argue that if the developed nations constrain the use of fossil fuels in order to reduce greenhouse gas emissions, two effects could lead to an offsetting increase in developing country fuel use and greenhouse gas emissions. First, reduced demand could cause world oil prices to decline, thereby potentially spurring increased oil demand in the developing world. Second, over the longer term, energy-intensive industries facing higher energy prices in developed countries might migrate to developing countries that are not controlling emissions. Taken together, these two influences are called the "leakage effect."

283. Gary Marchant. ENVIRONMENTAL LAW. Winter 1992, 22:623. In fact, unilateral action by a single major country such as the United States or a group of nations such as the Organization for Economic Cooperation and Development could result in an increase in greenhouse gas emissions in other nations. Action to reduce CO_2 emissions by only some countries would cause a substantial decrease in world demand for fossil fuels, which would cause the price of these fuels to drop and encourage greater fuel consumption by nonparticipating countries.

284. Dale W. Jorgensen and Peter J. Wilcoxen. SHAPING NATIONAL RESPONSE TO CLIMATE CHANGE: A POST RIO GUIDE. Ed. Henry Lee. 1995, p. 250. A tax with only partial international coverage could be vitiated by the movement of energy-intensive industries away from participating countries to other nations. In fact, Hoel (1991) has shown that if nonparticipating nations have inefficient energy technologies, it is theoretically possible for such a policy to result in a net increase in world carbon dioxide emissions. To date, however, only a modest amount of empirical research has been done on how an incomplete carbon policy would affect patterns of international trade. The principal study was conducted by Felder and Rutherford (1992), who found that the amount of redirected emissions could be considerable. An OCED carbon tax could reduce OCED oil demand enough to lower the world price of oil substantially. Lower world oil prices would lead, in the Felder-Rutherord model, to a large increase in oil demand by developing countries.

285. Bernard S. Black and Richard J. Pierce, Jr. COLUMBIA LAW REVIEW 1341. October 1993, p. 1341. Energy Source Substitution—Adders raise the cost of utility-supplied electric power relative to other energy sources. This gives consumers an incentive to switch from electric power to direct burning of fuel. This shift will generally reduce the environmental gains from adders. To take an extreme example, gasoline-powered lawnmowers, leaf-blowers, and hedge-trimmers compete with electric powered models. The gasoline-powered models use notoriously dirty, 2-stroke engines, which emit roughly 100 times more pollution than a power plant for the same amount of delivered power. Thus, even a 1% shift from electric to gasoline-powered models will increase emissions from this equipment class. A second example involves the choice between gasoline-powered and electric-powered motor vehicles. Electric vehicles can sharply reduce air pollution, but the more the utility charges for electricity, the fewer of them we are likely to see.

286. Arnold W. Reitze. ENVIRONMENTAL LAW. Summer 1991, p. 1549. Even when air pollution is reduced, the actual environmental effect may be difficult to ascertain. For example, air pollutants may be collected in a control device and then dumped into a waterway, or carcinogens can be removed from water by air stripping techniques, but either of these processes only transfers the pollution to another medium. What seems to be a reduction in pollution may in reality only be a shift of the pollution to another media. In the case of lead, the petroleum industry replaced lead with substitutes such as higher levels of benzene and other aromatic hydrocarbons to increase octane in gasoline. These additives often evaporate to become air pollutants. Now the petroleum industry is under pressure to produce less polluting, "reformulated" gasoline. The challenge today is to accomplish more effective air pollution control, and in such a way that further improvements do not involve unacceptably high marginal costs or unwanted governmental intrusions on life-style choices and the personal freedom of its citizens.

287. Gary Marchant. ENVIRONMENTAL LAW. Winter 1992, 22:623, p. 623. So far, the offset program has been described as applying only to CO_2 emissions. However, CO_2 is only responsible for about 55% of the current commitment to global warming. A regulatory program that attempts to control all greenhouse gases, not just CO_2, would have a larger impact on reducing the potential for global warming. Furthermore, a piecemeal approach that only addresses CO_2 may result in firms shifting to processes or fuels that result in reduced CO_2 emissions but increased emissions of other greenhouse gases. Thus, without a comprehensive program that includes all greenhouse gases, the benefits of CO_2 emissions reductions may be diminished by increased emissions of other gases. Finally, since there are likely to be significant differences in the marginal costs of reducing the various greenhouse gases, an offset program that allows trading between different gases will provide additional flexibility that will increase further cost-effectiveness of the program.

288. Alan Miller and Mack McFarland. FORUM FOR APPLIED RESEARCH AND PUBLIC POLICY. Summer 1996, p. 62. Many experts involved in the negotiation of the Montreal Protocol have acknowledged the key role played by environmental groups in enhancing public awareness of the problem and citizen support for action. Such groups also were instrumental in gaining support for the ban on CFCs used as aerosol propellants in the United States and Scandinavia in the 1970s and for building public

support for recent national and international regulations to phase out CFC production. Grassroots activism increased consumer pressure and caused manufacturers to switch to other products.

However, some environmental groups tend to overstate the degree of scientific uncertainty and concentrate on worst-case risk scenarios to attract media interest and public support. Such tactics can cause a polarization of views and a surge in media interest with every new prediction of disaster.

More important, these tactics can undermine the credibility of the scientific community and the basis for taking action on environmental problems.

Conveying reliable climate-change information to the public will be critical in maintaining support for necessary policies as society begins to address this complicated environmental issue.

289.   William F. O'Keefe. VITAL SPEECHES.
15 August 1995, p. 654. Unfortunately, such wooden-headed thinking persists. In the last twenty years, we've been assailed by one failed apocalyptic vision after another. All of this simply confirms that there is nothing so powerful as a plausible but false idea. Global warming may turn out to be a reality but right now there also are enough reasons to conclude that what is masquerading as the most serious of environmental threats may be just another hobgoblin being used to advance agendas that can't survive on their own merits.

I want to be absolutely clear that I am not asserting that the global warming threat is a hoax. I am challenging the process—the way it is being addressed. Instead of rational debate, what we have is advocacy driven by pseudo-science and hyped by the media. The current approach will eventually be seen as flawed because politicizing such an important environmental issue, and that's clearly what's been done, subverts rational policymaking and threatens to impose unreasonable burdens on the public. Ted Koppel made that point with Vice President Gore two years ago on this very issue, and I quote,

> The measure of good science is neither the politics of the scientist nor the people with whom the scientist associates. It is the immersion of hypothesis into the acid of truth. That's the hard way to do it, but it's the only way that works.

Right now, no one knows enough about global warming to advocate with certainty the kinds of actions that could jeopardize our economic well-being, and the economic aspirations of developing countries. That doesn't mean no action, which is usually described pejoratively and erroneously as "business as usual." It does mean actions must be based on facts, not misperceptions and myths. It does mean a mindset that reexamines, rethinks and changes course based on new knowledge. This is what former arms control negotiator Paul Nitze called, "working the problem," subjecting it to exhaustive examination, discussing it from many angles and being willing to go over it repeatedly until it yields a solution.

In short, I am advocating a reality check on the process based on the political, scientific and economic realities. Each of these realities has an important role in determining how we respond to the global warming threat. Our goal should be to identify actions that do the least damage to material well-being and that preserve the path to a better way of life, especially for the developing nations.

290.   S. Fred Singer. WASHINGTON TIMES.
21 March 1996, p. A11. By now, more than a hundred countries are involved in these exercises, keeping thousands of well-paid functionaries busy and spending some tens of millions of dollars on meeting and planning—funds that could be earmarked for solving real environmental problems. Such waste is bad enough. But what is alarming is that the measures considered would certainly stifle economic growth and reduce living standards for a large part of the global population. They may even reduce funds paid out to bureaucrats. Now that's really scary!

291.   Alan Miller and Mack McFarland. FORUM FOR APPLIED RESEARCH AND PUBLIC POLICY.
Summer 1996, p. 62. Because of the complexities and uncertainties of climate change, it is even more critical than was the case for ozone depletion that both sides avoid taking extreme positions and making overstatements. Selective use of evidence or promotion of theories divorced from scientific understanding may help promote a particular point of view, but such statements contribute to polarization and distrust.

The ability to achieve meaningful action on measures to address global climate change may hinge on continued diligence by all parties to ensure that the Montreal Protocol is successful. The public must receive reliable scientific information on the ozone issue to counteract a growing backlash against regulations imposed on ozone-depleting substances. Most important, governments in industrialized countries must demonstrate their commitment to assisting developing countries and avoid sending the message that they should develop their economies more slowly for the sake of a more sustainable and healthier global environment.

292.   William F. O'Keefe. VITAL SPEECHES.
15 August 1995, p. 654. Without a credible plan of action for a clearly defined problem—one that reflects our state of knowledge and that considers the implications of alternative courses of action—most Annex I countries will not in fact have the political will to implement the type of actions discussed at recent U.N. meetings. In the end, political rhetoric will not be sufficient to encourage citizens to forego economic growth or a rising standard of living without a well-documented threat. Look at the debate that has taken place in Europe over the European Union

proposal for a carbon tax. And, I doubt we will soon forget the ferocious resistance to the ill-conceived BTU tax.

Nonetheless, the Administration's climate change action plan so far represents a sensible course of action. It is precautionary and advocates actions that make sense in their own right. Unfortunately, the political will has been lacking to draw the line and reject additional arbitrary targets and timetables.

293. Alex Daniels. GOVERNING. February 1995, p. 37. "When you look outside on a summer day, you generally see clean air. Even when you look at the tailpipe of your car, you won't see much," says Dennis Keschl of the Maine Department of Environmental Protection. "It's hard to sell air toxicity."

Keschl speaks from experience. Last July, Maine became the first state on the East Coast to implement enhanced auto inspections, mandated by the U.S. Environmental Protection Agency. The high-tech, centralized emissions control program was designed to replace traditional tailpipe tests conducted at gas stations and satisfy requirements of the Clean Air Act. But things didn't go according to plan. The costly program proved to be overwhelmingly unpopular with the public, as well as politically explosive, and after only eight weeks, the operation was suspended.

294. Tom Arrandale. GOVERNING. January 1995, p. 42. That tension runs through a number of federal and state policies dealing with transportation fuels. Through its "clean-car initiative," for instance, the Clinton administration has begun working closely with Ford, Chrysler and General Motors to develop a new generation of lightweight vehicles capable of going 80 miles on a gallon of gasoline. But officials of Calstart, the state/business consortium trying to build an electric car industry in Southern California, worry that a federal research effort linked to the Big 3 automakers will overlook innovative start-up firms that could put alternatively fueled vehicles on the road more quickly than Detroit is willing to move.

295. Tom Arrandale. GOVERNING. January 1995, p. 41. By converting their own fleets, government agencies expect to set an example that millions of American car buyers will eventually follow. But government's experience so far suggests that major drawbacks must be overcome before most motorists will be willing to fill up with cheap and plentiful natural gas or drive battery-powered cars.

Even in Oklahoma, a natural gas-producing state, Smitherman says many state and local government employees often switch to gasoline tanks as soon as they leave government garages. New York City spent $948,000 last year to buy 184 light-duty cars and trucks that can run on alternative fuels, but a new CNG station in Brooklyn sits behind padlocked gates because so few city drivers bother refueling there. When drivers have a choice, "most of them are going to pick what they're used to," Smitherman observes. "That's the biggest obstacle to a successful alternative-fuels program."

296. World Resources Institute. A NEW GENERATION OF ENVIRONMENTAL LEADERSHIP: ACTION FOR THE ENVIRONMENT AND THE ECONOMY. 1993, p. 8. While renewable energy sources—with forms of energy storage—can directly replace fossil fuels in power generation and heating, the transportation sector—accounting for more than 25 percent of the total U.S. energy consumption—presents special problems. A number of energy sources have been promoted as potential replacements for oil in vehicles, including compressed natural gas and methanol. But many of these entail only marginal environmental benefits.

297. Tom Arrandale. GOVERNING. January 1995, p. 42. "It's not only expensive but you can't go very far," Spielberg says. Tom Cackette, the California Air Resources Board's deputy executive officer, says methanol was "a good stalking horse" that's now been passed by as stricter standards go into effect and fuel technology develops.

298. Tom Arrandale. GOVERNING. January 1995, p. 42. In the late 1980s, California officials and the oil industry were touting methanol as a likely gasoline replacement. But methanol poses its own environmental hazards, since it will dissolve in groundwater if it leaks and is highly toxic to humans. Its energy value is low, so mileage is limited, and its price was driven up when refiners began developing reformulated gasolines that use methyl tertiary butyl ether, an octane-raising additive made from methanol, to control carbon monoxide emissions.

299. Tom Arrandale. GOVERNING. January 1995, p. 43. Citing potential climate impacts, James J. McKenzie, a senior analyst for the World Resources Institute, argued before the California board last May that programs promoting CNG, methanol and ethanol fuels "represent a diversion of valuable time and resources and should not be pursued." McKenzie contends that governments should mount an all-out campaign to develop electric cars.

300. Tom Arrandale. GOVERNING. January 1995, p. 42. For now, however, much of the battle centers on clean-car mandates that state governments plan on imposing in the next few years to meet air quality goals. Over the Detroit auto industry's protests, the California Air Resources Board has approved low-emission-vehicle regulations that over 15 years will make it more and more difficult for the state's residents to keep driving gasoline-powered vehicles. In four stages, the rules will require the automakers to start selling progressively cleaner-running cars, culminating in 1988 with the begin-

ning of the phase-in of a "zero-emission" standard that will force them to bring electric vehicles onto the market.

New York and Massachusetts are following California's technology-forcing approach. This fall, the U.S. Environmental Protection Agency was weighing Detroit's objections to plans by 10 other Northeastern states to adopt the California car standards across the region. The auto industry maintains that it's doubtful that motorists will be ready to buy expensive electric cars that, given the current state of battery technology, are likely to offer limited range between rechargings. They want EPA to stretch out the transition so they can meet standards with the reformulated gasoline blends.

CHAPTER V
NEGATIVE APPROACHES
OUTLINE

DISADVANTAGE: TRADE

I. SHELL

   A. WTO GAINING IN POLITICAL EFFICACY AND CREDIBILITY (1)

   B. U.S. LEADERSHIP CRITICAL TO WTO SUCCESS (2)

   C. ENVIRONMENTAL PROTECTION CAUSES DOMESTIC INDUSTRY TO PRESSURE FOR PROTECTIONISM (3)

   D. THIS DESTROYS INTERNATIONAL COMMITMENT TO WTO (4)

   E. STRONG WTO IS CRITICAL TO PREVENTING GLOBAL CONFLICT (5)

II. LINK EXTENSIONS

   A. ENVIRONMENTALLY SAFE TECHNOLOGIES
1. Pressure for environmentally safe technologies induces business and related groups to petition for protection. (6)
2. Process standards—restrictions on the means of production—are a priority concern for business. (7, 32–35)
3. Business believes that requirements for environmentally safe technologies distort their international trade and investment. (8–9)
4. Industry will pressure to "Level the Field"—Protectionist trend will result and threaten demise of free trade. (11)
5. Environmental lobby will fear a lessening of environmental protections—They pressure for protection. (12)
6. Protectionists will align with environmentalists to use "clean energy" requirements as a platform for eco-duties. (13)
7. GATT interprets health exceptions narrowly. (62)

   B. CARBON REDUCTIONS
1. U.S. industries will pressure for eco-duties after carbon abatement policy. (14)
2. Taxes for abatement causes competitive losses. Industry will pressure for tariffs. (15)
3. Carbon tax increases pressure on the government for protection. (16)
4. Unilateral carbon restrictions increase trade distortions and pressure for compensation. (17)
5. Greenhouse gas abatement effects on manufacturing growth increases protectionist pressures. (10)
6. Using Montreal Protocol standards to force compliance violates GATT if non-parties are affected. (45)

   C. ENERGY USAGE RESTRICTIONS
1. Energy restrictions domestically give foreign consumers an advantage. (18)
2. Energy sustainability increases pressure on industry to increase productivity. Protectionism pressures result. (19)
3. New technology developments create incentives for protection. (24)
4. Conservation restrictions unfairly shut out foreign consumption. Violates GATT. (44)
5. Enhanced CAFE standards will be challenged under GATT. (45)

   D. EMISSION RESTRICTIONS
1. Emission taxes cause pressure for similar taxes on imports, destroying GATT. (20)
2. Equally applied taxes with differing effects violate GATT. (21)
3. Strict Emission Standards create a non-tariff barrier. (36)

   E. SUBSIDIES
1. Subsidies for environmentally safe technologies controversial under GATT. (22)
2. Strict Standards for subsidy agreement means diversion will be contested. (23)
3. Subsidies violate GATT. (42, 46)
4. Domestic incentives violate GATT. (43)

   F. REGULATION
1. Increased regulations on business cause pressure for protection. (29)
2. Enviro-sanctions destroy confidence in the free trade system. (57)
3. Standards insure retaliation and snowball to collapse of the free trade system. (58)
4. Enviro-tariffs will be exploited by protectionism and invoke a slippery slope of social trade restrictions. (59)

   G. NON-TARIFF BARRIERS
1. Environmental standards are a non-tariff barrier to trade. (30)

2. Regulations that ban or limit product use violate non-tariff barrier discrimination rules of GATT. (31)
3. Plan can't require foreign imports to meet domestic standards. Violates GATT. (41)

III. INTERNAL LINKS

A. PERCEPTION OF "UNFAIR" TRADE WILL TURN PUBLIC OPINION INWARD TO PROTECTIONISM (25)
B. DOMESTIC INDUSTRY CLAIMS PERSUASIVE TO GOVERNMENT. BARRIERS WILL BE ADOPTED (26)
C. FAIRNESS CLAIMS HIGHLY MANIPULATIVE. PROTECTIONISTS WILL SUCCESSFULLY PUSH FOR "LEVELING" (27)
D. "UNFAIRNESS" CLAIMS PERSUASIVE TO CONGRESS (28)
E. WTO BILL BLOCK IMPORT BANS RESULTANT OF HIGHER U.S. STANDARDS (37)
F. GATT WILL STRIKE DOWN "DISCRIMINATORY" STANDARDS (38)
G. THIRD WORLD HOLDS 85% OF WTO VOTES. WILL VOTE AGAINST U.S. STANDARDS THAT RESULT IN ECO-TARIFFS (39)
H. GATT/WTO TRUMPS INTERNATIONAL ENVIRONMENTAL PROTOCOLS (40)
I. GATT WILL CONSIDER THEORETICAL TRADE IMPACT—NOT THE ACTUAL EFFECTS OF THE PLAN (47–48)
J. GATT IGNORES THE ENVIRONMENT. TRADE IMPLICATIONS ARE THE ONLY RELEVANT CONCERN (49)

IV. UNIQUENESS AND BRINK EXTENSIONS

A. U.S. SOVEREIGNTY ISSUE PUTS WTO ON THE BRINK (50)
B. RISK OF U.S. WITHDRAWAL FROM WTO IF IT OVERRULES U.S. LAWS (51)
C. U.S. PLAYING BY THE RULES IS CRITICAL TO WTO CREDIBILITY (52–53)
D. DISCRIMINATORY TREATMENT WILL KILL GATT AND SPUR WORLD-WIDE PROTECTIONISM (54)
E. UNILATERAL STANDARD DETERMINATION CAUSES RETALIATION (55)
G. UNILATERAL ACTIONS SUSPECTED OF CHEATING FREE TRADE NORMS (50)
H. LIBERALIZATION BANDWAGON GAINING MOMENTUM (63)
I. WE ARE ON THE VERGE OF ELIMINATING ALL BARRIERS. NEXT BIG PUSH WILL BE KEY (64)

V. IMPACT EXTENSIONS

A. STRENGTH OF FREE TRADE FRAMEWORK IS CRITICAL TO PREVENTING DEPRESSION AND WAR (60)
B. FAITH TO INTERNATIONAL COMMITMENTS CRITICAL TO AVOIDING PROTECTIONISM REVERSION (61)
C. PROTECTIONISM FORCES SHIFT TO DOMESTIC DEFENSE PROCUREMENT (65)
D. PROTECTION INDUCED DOMESTIC PROCUREMENT CAUSES R&D AND MAINTENANCE CUTS, JEOPARDIZING MILITARY READINESS (66)
E. PROTECTIONISM PREVENTS COALITION BUILDING. KEY TO AID IN MILITARY CRISIS (67)
F. PROTECTIONISM DESTROYS COOPERATIVE TIES TO DEAL WITH FUTURE OIL SHOCKS (68)
G. U.S. PROTECTIONISM WILL SPREAD WORLDWIDE. THIS SHUTS OUT DEVELOPING ECONOMIES IN EASTERN EUROPE AND THREATENS EAST-WEST RELATIONS (69)
H. PROTECTIONISM THREATENS INTERNATIONAL STABILITY AND RISKS LONG-TERM DOMESTIC CRISIS (70)
I. PROTECTIONISM KILLS SPECIALIZATION AND INNOVATION. RESULTING RETALIATION KILLS GLOBAL ECONOMIC GROWTH (71)
J. PROTECTION YIELDS RETALIATION AND ECONOMIC INSTABILITY (72)
K. PROTECTIONISM UNDERMINES ENVIRONMENTAL GOALS (73)
L. PROTECTIONIST TRADE POLICIES RESULT IN WORLD-WIDE RETALIATION (74)
M. STRONG TRADE RELATIONS ARE KEY TO STOPPING PROLIFERATION OF WEAPONS OF MASS DESTRUCTION (75)

DISADVANTAGE: COMPETITIVENESS

I. SHELL

   A. EVEN WELL-INTENDED PLANS WILL BE MICROMANAGED AND COERCED TO HARM BUSINESS (76)

   B. INCREASED REQUIREMENTS TREATED AS A COMPLETE CERTAINTY OF DECREASING COMPETITION (77)

II. LINK EXTENSION: SCHEMES TO INCREASE CLEAN ENERGY TECHNOLOGIES DECREASE COMPETITIVENESS

   A. ENVIRONMENTAL PROTECTIONS
1. Business is anti-environment. Any effort to protect it will be feared. (78–80)
2. New protections create fears that destroy investment. (81)
3. Public will perceive the cost to be higher than it actually is. (82)
4. Other nations lack stringent environmental regulations. Increasing regulations guarantees a decline in competitiveness. (93)

   B. INCENTIVE SYSTEMS
1. Government industrial relationship adversarial. Business wants government out of the market. (83)
2. Timing of investment is critical to R&D success. Business must control timing decisions. (88)
3. If government pays for the plan, bureaucratic disaster and government spending results. (120)

   C. EMISSION CONTROLS
1. Compliance costs will be greater than anticipated. (84)
2. Empirically high end estimates win out. (85)
3. Regulations have unintended consequences. (86)
4. Compliance costs are underestimated. (87)

   D. REGULATORY SCHEMES
1. Regulations increase compliance costs. Decreases international competitiveness. (89)
2. Stringent regulations increase other nations' comparative advantage, crushing U.S. competitiveness. (90)
3. Regulations have a large impact on competitiveness. (91)
4. Inefficient regulations cause higher relative compliance costs. Damages our competitive edge. (92, 99)
5. Lack of flexible standards harms competitiveness. (94)
6. Regulations directly decrease international competitiveness. (95)
7. Regulations massively increase management overhead costs. (98)

   E. RESEARCH, DEVELOPMENT, AND INNOVATION
1. Compliance costs discourage innovation and efficiency. (96)
2. Required development allows government to pass on billions of dollars in costs to business. (97)
3. Technology requirements are counterproductive. They increase costs and harm competitiveness. (100)
4. Investment decisions today are critical to future technology. Requirements now have disastrous long-term consequences. (101)
5. Regulations delay investment in technology due to uncertainty. (102)
6. Regulatory costs decrease resources for R&D. (103)
7. Industry is investing in environmentally clean technology R&D now. Changes threaten to disrupt investment. (121)
8. Regulations force diversion of R&D investment funds to compliance costs. (122)
9. Forced technology choice prevents profit realization of past investments. (123)
10. Technology requirements divert money from R&D to lobbying and litigation. (124)
11. Industries are strapped for R&D cash now. Requirements rob them of necessary capital. (125)
12. Innovation requires on the spot capital. Requirements rob industry of critical capital. (126–127)
13. Government targeting of specific technology sector is an assured failure. (128)
14. Technology development takes a long time. Requirements force failure in the interim. (129)
15. Regulatory penalties deter investment. (130)
16. Lack of phase-in kills innovation. (131–133)

   F. TECHNOLOGY
1. Industry will not develop new technologies because they fear that it will be required. (134, 143)
2. New technologies are only possible with a paradigm shift (135, 144–145)
3. Technological innovation will not prevent cost increases. Only industry assumes the risk. (136)

- 4. Technology innovations cause massive unemployment. (137)
- 5. Anti-trust regulations eliminate innovation incentives. (138)
- 6. Investments in new technologies diverts investment monies from other key investment areas. (139)
- 7. Innovation can't offset compliance costs. Business loses in the end. (140)
- 8. Corporate financiers won't invest in clean technologies. Short-term profits are the key. (141, 147–148)
- 9. Shift to sustainable technology development guarantees a stagnant economy. (142)
- 10. Uncertain benefits prevent investment in new technologies. (146)
- 11. New technologies force managerial and organizational change. High cost of transition deters investment. (149)

F. CLIMATE
- 1. Clinton focusing on voluntary CO_2 reductions now. Mandatory actions anger business. (104)
- 2. Carbon tax opposed for fear of decreasing competitiveness. (106)

G. AUTO EMISSIONS
- 1. Initial auto emission reduction causes domestic industry backlash. (107)
- 2. Increased CAFE standards increase car prices, harming auto industry competitiveness. (108,110)
- 3. GM opposed to increased CAFE standards. (109)

H. RECYCLING
- 1. Recycling requirements harm business. (111)
- 2. Recycling increases transportation costs. (112)

I. SPECIFIC INDUSTRIES
- 1. Steel and chemical industries depend on low power prices. Regulations crush both sectors (113)
- 2. Chemical industry struggling. No time for start-up costs. (114)
- 3. Future of electronics industry uncertain. (115)
- 4. Electronics industry critical to U.S. economy. (116)
- 5. Technological development key to electronics industry. (117)
- 6. Investment in R&D is the lifeblood of the semiconductor industry. Specific requirements kill the industry. (118)
- 7. Semiconductors are a strategic industry. Key to economic growth. (119)

III. CLEAN TECHNOLOGY PROGRAMS ENHANCE COMPETITIVENESS

A. MULTIPLE INCENTIVES AND EFFECTS IMPROVE MARKET POSITION (150–151)

B. PRODUCTION CHANGES YIELD MASSIVE GAINS IN EFFICIENCY (152–156)

C. REQUIREMENTS ENCOURAGE INNOVATION CRUCIAL TO WINNING THE GREEN TECHNOLOGY RACE. SUBSTANTIAL TRADE BENEFITS RESULT (157–159)

D. ENVIRONMENTALLY SOUND TECHNOLOGY REQUIREMENTS TRIGGER MOTIVE FOR INCREASING HIGH-TECHNOLOGY R&D (160)

E. REGULATIONS STIMULATE INVESTMENT IN ENVIRONMENTALLY FRIENDLY TECHNOLOGIES (161–163)

F. DEMAND STIMULATION: REGULATIONS ALTER SUPPLY-DEMAND EQUILIBRIUM CREATING BETTER ENVIRONMENT FOR TECHNOLOGICAL DEVELOPMENT (164–165)

G. CLEAN TECHNOLOGY REQUIREMENTS ALLOW FOR A SMOOTH TRANSITION FROM POLLUTION INTENSIVE INDUSTRIES TO EXPANDING ENVIRONMENTALLY SAFE INDUSTRIES (166)

H. REQUIREMENTS SHIFT JOBS FROM DYING INDUSTRIES TO SUSTAINABLE BUSINESSES (167)

I. REGULATIONS HELP COMPETITIVE INDUSTRIES TO STAY ON TOP. ONLY HARM IS TO DECLINING INDUSTRIES (168)

J. FAILURE OF NON-COMPETITIVE INDUSTRIES INCREASES STANDARD OF LIVING (169)

K. COMPLIANCE FREEDOM ALLOWS INCREASING EFFICIENCY AND DECREASES COST (170)

L. FLEXIBILITY IN COMPLIANCE IS KEY TO REALIZING GAINS IN COMPETITIVENESS (171)

M. FLEXIBILITY INCREASES PROFIT OPPORTUNITY SPURRING INNOVATION IN PRODUCT AND PROCESS (172)

N. THE GREATER THE FLEXIBILITY THE GREATER THE ENHANCED COMPETITIVENESS (173)

O. AUTO INDUSTRY EMPIRICALLY RESPONDS TO INCREASED REQUIREMENTS WITH INCREASES IN EFFICIENCY AND COMPETITIVENESS (174-175)

P. CHEMICAL INDUSTRY EMPIRICALLY ATTAINS ENHANCED PROFITABILITY AND COMPETITIVENESS FROM REGULATORY DRIVEN INNOVATION (176–177)

Q. STEEL INDUSTRY EMPIRICALLY BENEFITS (178)

R. CLEAN TECHNOLOGIES ATTRACT OUTSIDE INVESTORS. BUSINESS PROFITS (170–180)

S. MARKET HACKS WRONG. INDUSTRY WON'T OPTIMIZE INNOVATION AND PRODUCTION WITHOUT OUTSIDE INFLUENCE (181–185)

T. ENVIRONMENTAL TECHNOLOGY IS ONE OF THE FASTEST GROWING INDUSTRIES IN THE U.S. (186)

U. LEADERSHIP IN ENVIRONMENTAL TECHNOLOGIES CRITICAL TO GLOBAL COMPETITION (187)

V. CLEAN TECHNOLOGY INDUSTRY WILL BE THE LARGEST MARKET ON EARTH. WE MUST LEAD NOT TO SAVE OUR COMPETITIVENESS (188, 190)

W. ECOLOGICAL EFFICIENCY IS KEY TO OPENING FOREIGN MARKETS AND SAVING OUR COMPETITIVE EDGE (189)

X. ENVIRONMENTAL TECHNOLOGY REVOLUTION IS KEY TO THE TRANSITION TO SUSTAINABLE GROWTH (191–193)

Y. R&D IS KEY TO COMPETITIVENESS AND ECONOMIC GROWTH (194)

Z. INVESTMENT AND INNOVATION CRITICAL TO U.S. LEADERSHIP (195)

AA. SHORT TERM PROFIT HORIZONS STOP R&D (196)

BB. LACK OF INVESTOR PROTECTION CRUSHES R&D IN THE STATUS QUO (197)

CC. MANAGERIAL DECISIONS ARE KEY TO R&D. NOT SOLELY BASED ON ECONOMIC FACTORS (198)

DD. MULTIPLE FACTORS ARE KEY TO TECHNOLOGY DECISIONS (199)

EE. PRIVATE INVESTMENT IRRELEVANT. GOVERNMENT R&D IS CRITICAL (200)

DISADVANTAGE: GERMANY

I. SHELL

   A. GERMAN ECONOMIC STABILITY IS ON THE WAY (201)

   B. LINK: GERMAN CLEAN TECHNOLOGY EXPORTS ARE CRUCIAL TO CONTINUED GERMAN RECOVERY (202)

   C. GERMAN ECONOMIC DIFFICULTIES FOSTER RADICALISM AND DESTABILIZE EUROPE (203)

   D. EUROPEAN INSTABILITY ESCALATES TO NUCLEAR WAR (204)

II. UNIQUENESS EXTENSIONS

   A. GERMANY LEADS THE WORLD IN ENVIRONMENTAL EXPORTS (205)

   B. GERMANY WINNING COMPETITION FOR WORLD ENVIRONMENTAL EXPORTS (206)

   C. GERMANY HAS LEAD IN ENVIRONMENTAL TECHNOLOGY (207)

   D. GERMANY IS HEAVILY INVESTED IN CLEAN TECHNOLOGIES (208)

   E. GERMAN INVESTMENTS IN CLEAN TECHNOLOGY ARE KEY TO PULLING GERMANY OUT OF RECESSION (209)

   F. U.S. LEADERSHIP IN CLEAN TECHNOLOGIES FALTERING. WE REFUSE TO SPEND MORE ON R&D (210)

   G. GERMANY SURPASSING U.S. IN CLEAN TECHNOLOGY (211)

   H. LAX U.S. REQUIREMENTS OFFER AN OPENING FOR FOREIGN BUSINESS'S ENVIRONMENTAL TECHNOLOGIES (212)

IV. LINK EXTENSIONS

   A. CLEAN TECHNOLOGIES IS GERMANY'S FASTEST GROWING INDUSTRY (222-223)

   B. MEDIUM-SIZED GERMAN FIRST NEED A BOOST FROM NEW TECHNOLOGY EXPORTS (224)

   C. GERMAN ENVIRO-TECH INDUSTRY DEPENDING ON 20% EXPORTS (225)

- D. GERMANY AND THE U.S. COMPETE FOR THIRD WORLD CLEAN TECHNOLOGIES MARKET (227)
- E. GERMANY MUST BEAT OUT THE U.S. FOR ENVIRONMENTALLY SOUND TECHNOLOGIES LEADERSHIP (228)
- F. U.S. FOCUS ON NEW TECHNOLOGIES WILL CROWD OUT GERMANY IN EMERGING MARKETS (230)
- G. HIGH ENVIRONMENTAL STANDARDS BREED COMPETITION IN ENVIRO-TECH: GERMANY PROVES (231)
- H. GERMANY CONTROLS CLEAN COAL TECHNOLOGY NOW (232, 235–236)
- I. GERMANY DOMINATING POLLUTION CONTROL TECH NOW (233–234)
- J. GERMANY EXPANDING IN NUCLEAR REACTOR MARKET (237–238)

III. INTERNAL LINK EXTENSIONS
- A. SUBSTANTIAL GERMAN JOBS ARE LINKED TO CLEAN TECHNOLOGIES (213)
- B. EXPORTS OF ENVIRONMENTALLY FRIENDLY TECHNOLOGIES WILL BOOST GERMAN EMPLOYMENT (214)
- C. GERMANS VIEW ENVIRONMENT AS CRITICAL ECONOMIC MARKET (215)
- D. ENVIRONMENTAL TECHNOLOGY IS THE ENGINE OF GERMAN ECONOMIC GROWTH (216–217)
- E. CLEAN TECHNOLOGIES ARE KEY TO EMPLOYING ONE MILLION GERMANS (218)
- F. GERMAN COMPETITIVENESS DEPENDS ON CLEAN TECHNOLOGIES GROWTH (219–221)
- G. TRADE CRITICAL TO GERMAN ECONOMIC GROWTH (226)

IV. IMPACT EXTENSIONS
- A. ECONOMIC STRAINS COULD REIGNITE GERMAN NATIONALISM (239)
- B. ECONOMIC DOWNTURN AND UNEMPLOYMENT SPURS GERMAN RADICALISM (240)
- C. GERMAN PROBLEMS SPILL OVER, DRAGGING EUROPE TO DISASTER (241)
- D. GERMAN STABILITY DECIDES EUROPEAN STABILITY (242–243)
- E. EURO-CONFLICT CAN ESCALATE TO NUCLEAR WAR (244)
- F. FURTHER DECLINE IN GERMAN ECONOMY WILL RESULT IN EMU FAILURE (245)
- G. EMU FAILURE WOULD COLLAPSE GERMAN ECONOMY (246)
- H. GERMANY IS THE CORNERSTONE OF NATO AND THE EU (247–248)
- I. ALLIANCES AND NATO IN JEOPARDY (249)
- J. NATO KEY TO GLOBAL STABILITY (250)

DISADVANTAGE: SMALL BUSINESS
- A. SMALL BUSINESS RESISTS ENVIRONMENTALLY MOTIVATED RESTRICTIONS. THEY FEAR SNOWBALL (251)
- B. SMALL BUSINESS FEARS "SPECIAL TREATMENT" BY GOVERNMENT PROGRAMS (252)
- C. REGULATIONS ARE CURRENTLY ON THE DECLINE (253)
- D. SMALL BUSINESS PRESENTLY EXEMPT FROM ENVIRONMENTAL REGULATION: GASOLINE REFORMULATION PROVES (254)
- E. SMALL BUSINESSES PRESENTLY EXEMPT: CONSERVATION REQUIREMENTS PROVE (255)
- F. SMALL BUSINESSES PRESENTLY EXEMPT: AUTOMOBILE REGULATIONS PROVE (256)
- G. EPA POLICY OF DIMINISHED PENALTIES FOR SMALL BUSINESS IS DISCRETIONARY—NOT NORMAL MEANS (257)
- H. DIMINISHED REGULATION PENALTIES INDUCE QUICKER COMPLIANCE (258)
- I. SMALL BUSINESS HAS BIG TIME POLITICAL CLOUT (259)
- J. SMALL BUSINESS SWING VOTE IS POWERFUL ON THE HILL (260)
- K. REPUBLICANS COURT SMALL BUSINESS (261)
- L. CLINTON COURTS SMALL BUSINESS (262)
- M. EPA IS WORKING HARD TO SATISFY SMALL BUSINESS (263)
- N. SMALL BUSINESS IS A KEY ENGINE OF ECONOMIC GROWTH (264–265)

O. 80% OF SMALL BUSINESS CLASSIFIED AS "SMALL" (266)

P. FAILURE TO GIVE SMALL BUSINESS SPECIAL TREATMENT RISKS JUDICIAL REVIEW AND STRIKE-DOWN (267)

Q. LACK OF SMALL BUSINESS OUTREACH IS GROUNDS FOR LEGAL OVERTURN (268)

R. REGULATIONS MUST BE PRE-REVIEWED IF THEY HURT SMALL BUSINESS (269)

S. ALL AGENCIES FACE THE WRATH OF REGULATORY FAIRNESS REVIEW (270)

DISADVANTAGE: PUBLIC OPINION AND POLITICAL CAPITAL

I. PUBLIC OVERWHELMINGLY SUPPORTS GOVERNMENT POLICIES THAT PROTECT THE ENVIRONMENT

  A. SUPER-MAJORITIES SUPPORT ENVIRONMENTAL REGULATIONS AT THE EXPENSE OF ECONOMIC GROWTH (301–302, 321, 324–326, 333)

  B. PUBLIC MASSIVELY PREFERS CLEAN AIR ENVIRONMENT OVER ECONOMIC GROWTH (303–304)

  C. PUBLIC WILL ACCEPT INFLATION IN EXCHANGE FOR CLEAN ENERGY (305, 308)

  D. SUPPORT FOR ENVIRONMENT RESISTANT TO RECESSION, GOVERNMENT SPENDING, INFLATION, AND ENERGY SHORTAGE (306)

  E. 80% OF PUBLIC SUPPORTS ENVIRONMENTALLY SOUND POLICIES REGARDLESS OF COSTS (307)

  F. "STIFLING ECONOMIC GROWTH" IS AN ACCEPTABLE COST FOR ENVIRONMENTAL PROTECTION (322)

  G. PUBLIC SUPPORTS COMPREHENSIVE APPROACH TO PROVIDING A CLEAN ENVIRONMENT (323, 330)

  H. MAJORITY SUPPORT HIGHER TAXES AND SPENDING DEFICITS FOR MEASURES THAT HELP THE ENVIRONMENT (309)

  I. THIS IS ONE OF THE ONLY AREAS WHERE THE PUBLIC SUPPORTS GOVERNMENT ACTION IN THE MARKET (327, 332)

  J. ENVIRONMENTAL ISSUES ARE AN IMPORTANT EXCEPTION TO THE PUBLIC'S ANTI-REGULATORY MOOD (336)

  K. PUBLIC OVERWHELMINGLY SUPPORTS INCREASED GOVERNMENT SPENDING ON MEASURES THAT IMPROVE ENVIRONMENTAL QUALITY (310–312, 314)

  L. PUBLIC IS STRONGLY PRO-ENVIRONMENTAL (328–329)

  M. PUBLIC SUPPORT IS NOT NEW. MANY YEARS OF SUPPORT FOR ENVIRONMENT FRIENDLY POLICIES (313)

  N. PRO-CLEAN COMMITMENT IS PERSISTENT AND ENTRENCHED (331)

  O. PUBLIC FEARS THAT THE ENVIRONMENT IS WORSENING (342)

II. ENVIRONMENTALLY FRIENDLY POLICIES ARE KEY DETERMINANTS OF POLITICAL CONTROL

  A. SUBSTANTIAL IMPORTANCE AS AN ELECTION ISSUE. CRUCIAL FACTOR IN TIGHT RACES (314)

  B. ENVIRONMENT HOLDS A SWING VOTE OF 11% (315)

  C. ANTI-ENVIRONMENT OFFICIALS ARE EXTREMELY VULNERABLE AT ELECTION TIME (316)

  D. WITHOUT MAJOR ECONOMIC AND FOREIGN POLICY CRISIS, ENVIRONMENTAL ISSUES DETERMINE THE OUTCOME OF ELECTIONS. BUSH ELECTION PROVES (317)

  E. REAGAN'S ENVIRONMENTAL BACKLASH GAVE ENVIRONMENT ELECTORAL POTENCY (318-319)

  F. 80% CONSIDER ENVIRONMENTAL DECISIONS WHEN VOTING (320)

  G. ENVIRONMENTAL POLICY ADOPTION ALWAYS INVOLVES LEGITIMATION, MOBILIZING SUPPORT, POLITICAL SPIN AND EXTENSIVE BARGAINING (343)

III. SALIENCE OF ENVIRONMENTAL ISSUES

  A. SALIENCE IS HIGH FOR ENVIRONMENTAL ISSUES (334-335)

  B. ENVIRONMENTALISTS PUBLICIZE ALL ENVIRONMENTAL POLICIES. PUBLIC IS AWARE (337)

  C. MEDIA GIVES ENVIRONMENTAL ISSUES ENORMOUS IMPORTANCE (338)

D. CONGRESS OVERSTATES THE PUBLIC ATTENTIVENESS. THEY WILL INTERPRET THE POPULARITY OF A PROPOSAL AND VOTE ACCORDINGLY (339)

E. ENVIRONMENT IS A HIGH VISIBILITY ISSUE (340)

F. MEDIA SHAPES PUBLIC OPINION ON ENVIRONMENT (341)

IV. PRESIDENTIAL AND CONGRESSIONAL POPULARITY:

A. PUBLIC SUPPORT IS THE FOREMOST COMPONENT OF PRESIDENTIAL POWER. IT DEFINES THE LIMITS OF RESISTING HIS AGENDA (344)

B. CONGRESSIONAL KNOWLEDGE OF PRESIDENTIAL POPULARITY INFLUENCES THEIR BEHAVIOR. THEIR REELECTION IS DEFINED BY THEIR REACTION TO THE PRESIDENT (345)

C. PRESIDENTIAL POPULARITY SERVES AS THE PREVAILING ATMOSPHERE FOR POLICY-MAKING. SUPPORT IS DIRECTLY CORRELATED TO THE OUTCOME (346)

D. NUMEROUS STUDIES SHOW A POSITIVE RELATIONSHIP BETWEEN SUPPORT AND SUCCESS (347)

E. WASHINGTON COMMUNITY ACTIVELY WATCHES PRESIDENTIAL POPULARITY (348)

F. POPULARITY AUGMENTS ALL ASPECTS OF PRESIDENTIAL POWER. IT'S THE MOST LIKELY AGENT FOR CHANGE (349)

G. PUBLIC SUPPORT IS THE PRIMARY RESOURCE FOR LEADERSHIP (350)

H. HIGH POPULARITY DISCOURAGES CONGRESS FROM VOTING AGAINST THE PRESIDENT. IMPLICIT THREAT TO MAKE IT A CAMPAIGN ISSUE (351)

I. HIGH POPULARITY PROVIDES A POLITICAL COVER FOR SUPPORTERS (352)

J. LOW POPULARITY SEVERELY UNDERCUTS PRESIDENTIAL SUCCESS RATE. NO POLITICAL CONSEQUENCES (353)

K. NEGATIVE CORRELATION BETWEEN POPULARITY AND POLITICAL CAPACITY IS COUNTERINTUITIVE (354)

L. PRESIDENTIAL IMAGE-MAKING DOES NOT WORK. HE IS BLAMED FOR THE SUBSTANCE OF HIS POLICIES (355)

M. THE DROP IN BUSINESS CONFIDENCE IN CLINTON IS PRECISELY WHY THE PUBLIC APPROVES OF THE POLICY. THEY SEE BUSINESS AS RESPONSIBLE FOR THE PROBLEM (356–357)

COUNTERPLAN: ENVIRONMENTAL PROTECTION AGENCY

I. UNILATERAL EPA ACTION BEST

A. EPA NOTICE AND COMMENT PROCEDURES ENHANCE CONSENSUS-BUILDING AND IMPROVE IMPLEMENTATION (271)

B. ADMINISTRATIVE RULEMAKING PROCESS IS SUPERIOR TO CONGRESSIONAL DECISION-MAKING (272)

C. CONGRESSIONAL INTERFERENCE DESTROYS THE PROCESS BENEFITS OF UNILATERAL EPA ACTION (273)

D. EXISTING LAWS GIVE THE EPA BROAD POWER TO MAKE ANY POLICY THAT IMPACTS THE ENVIRONMENT (274–276)

E. EPA HAS COMPREHENSIVE AUTHORITY (277)

F. POLICY-MAKING/IMPLEMENTATION DICHOTOMY IS AN ANACHRONISM. ADMINISTRATIVE AGENTS LIKE THE EPA FREQUENTLY MAKE LEGISLATIVE DECISIONS (278–279)

G. EPA HAS ENORMOUS DISCRETIONARY POWER TO MAKE AND SHAPE POLICIES THAT IMPACT THE ENVIRONMENT (280–281)

H. POLITICAL CAPITAL NET BENEFIT: EPA REGULATORY IMPLEMENTATION HAS NO NOTICEABLE POLITICAL EFFECTS (282)

I. DETAILS OF IMPLEMENTATION AND REGULATORY REQUIREMENTS ESCAPE THE PUBLIC'S ATTENTION (283)

J. UNILATERAL EPA ACTION INSULATES PUBLIC OFFICIALS FROM CRITICISM (284-285)

K. ADMINISTRATIVE ACTION ALLOWS PUBLIC OFFICIALS TO TAKE CREDIT FOR POPULAR PROGRAMS AND DIVERT BLAME FOR UNPOPULAR ONES (286–287)

L. EPA DOES NOT ENGAGE IN OVERREGULATION (388)

M. EPA IS STREAMLINING REGULATIONS, REDUCING BURDENS ON BUSINESS (289-291)

II. CONGRESSIONAL OVERSIGHT OF EPA ACTIONS BEST

A. EPA REGULATIONS ARE ESPECIALLY FEARED BY INDUSTRY. DECADES OF BAD RELATIONS HAVE CREATED A HUGE AMOUNT OF MISTRUST BETWEEN BUSINESS AND THE AGENCY (292)

B. EPA RULES AND GUIDELINES PREVENT TECHNOLOGICAL INNOVATION BY INDUSTRY (293–294)

C. ON BALANCE, CONGRESSIONAL OVERSIGHT IMPROVES ENVIRONMENTAL POLICY-MAKING (295)

D. ONLY BROADENED CONGRESSIONAL SCRUTINY CAN SOLVE EPA DEFICIENCIES (296)

E. INCREASED OVERSIGHT INCREASES EPA ENFORCEMENT EFFORTS (297)

F. CONGRESSIONAL SCRUTINY STIMULATES EPA INNOVATION (298)

G. OVERSIGHT BY AND CONSULTATION WITH CONGRESS DOES NOT CHILL EPA CREATIVITY (299)

H. OVERSIGHT IMPROVES EPA PRIORITIZATION (300)

COUNTERPLAN: ALTERNATIVE AGENTS

A. PRESIDENT AND CONGRESS ARE LOCKED IN A BATTLE FOR CONTROL OVER ENVIRONMENTAL POLICY (358–360)

B. EXECUTIVE AND LEGISLATIVE BRANCHES COMPETE FOR CONTROL OVER EPA POLICY-MAKING (361–365)

C. EXECUTIVE ORDERS HAVE THE FORCE OF LAW (366–367)

D. LEGISLATIVE POLICY-MAKING PROCESS SUPERIOR TO EXECUTIVE FIAT (368)

E. EXECUTIVE ORDERS ARE FREQUENTLY CIRCUMVENTED OR IGNORED BY BUREAUCRATS (369)

F. EXECUTIVE ORDERS THAT RELAX ENVIRONMENTAL STANDARDS WILL RECEIVE MUCH MEDIA ATTENTION (370)

G. CONGRESS WILL EXPLOIT EXECUTIVE EXEMPTIONS FOR POLITICAL GAIN (371)

H. PUBLIC IS CONCERNED WITH THE ENFORCEMENT OF ENVIRONMENTAL LAWS. NOT JUST ENACTMENT (372)

I. DEPARTMENT OF INTERIOR HAS EXTENSIVE RESPONSIBILITIES AND RESOURCES TO REGULATE ENVIRONMENTAL AND LAND MANAGEMENT (373–375)

J. DEPARTMENT OF ENERGY HAS JURISDICTION AND RESOURCES TO ADMINISTER ENVIRONMENTAL PROGRAMS (376–377)

K. NUMEROUS FEDERAL AGENCIES ARE AUTHORIZED TO DESIGN AND IMPLEMENT POLICIES WITH ENVIRONMENTAL IMPLICATIONS (378–379)

L. EPA SHARES JURISDICTION OVER ENVIRONMENTAL POLICIES WITH NUMEROUS OTHER FEDERAL AGENTS (380–381)

M. COUNCIL OF ENVIRONMENTAL QUALITY FAILS (382–388)

N. RELATIONS BETWEEN THE REPUBLICAN CONGRESS AND THE EPA ARE STRAINED (389)

O. EVEN A HINT OF EPA NONCOMPLIANCE WITH STATUTORY MANDATES WILL PROMPT CONGRESSIONAL RETALIATION (390)

P. NON-ADHERENCE TO LEGISLATIVE INTENT WILL SPARK A CONGRESSIONAL CRACKDOWN (391)

Q. EXERCISE OF ADMINISTRATIVE DISCRETION CAUSES CONGRESSIONAL RETALIATION (392–394)

R. FAILURE OF ADMINISTRATIVE AGENCIES TO HONOR LEGISLATIVE WISHES ENSURES CONGRESSIONAL RETALIATION (395–396)

S. CONGRESSIONAL BACKLASH AGAINST THE EPA RUINS ITS EFFECTIVENESS (397)

T. HEIGHTENED OVERSIGHT SKEWS NATIONAL ENVIRONMENTAL PRIORITIES (398)

U. HEIGHTENED OVERSIGHT FORCES THE EPA TO ENGAGE IN UNDERGROUND RULEMAKING (399)

V. INTENSIFIED CONGRESSIONAL OVERSIGHT CHILLS EPA INNOVATION (400)

CHAPTER V

EVIDENCE

1. Renato Ruggiero. FOURTH ANNUAL SYLVIA OSTRY LECTURE, OTTAWA. 1996, (http://www.unicc.org/wto/whats_new/press49.htm) The challenges I have outlined are contributing to a growing awareness that trade is not just a technical question, but a matter of high political importance. In the WTO, the world now has a permanent trade policy forum as well as a more effective means of negotiating commitments and making and enforcing trade rules. Trade and trade policy have been put back in the front row of international concerns, where they were intended to be by the architects of the post-war international institutions. With the establishment of the WTO, and the conclusion—which is expected very soon—of comprehensive agreements for cooperation at every level with the World Bank and the IMF, the matrix of trade and finance and development is not only completed but updated to contribute to global prosperity and stability in the new century. This improved institutional cooperation is a major step towards fulfilling the mandate given to the WTO by governments to work for improved coherence in international economic policy-making.

2. "Overarching Issues and Recommendations," WTO IMPLEMENTATION REPORT. 9 April 1996,(http.//www/ustr.gov/reports/wto/issues.htl) First and foremost, it is critical to remember that the United States has been the driving force behind post-World War II global economic liberalization. We provided the prime impetus and leadership for the creation of the GATT and for seven rounds of multilateral trade negotiations under that organization, including the Uruguay Round. While liberalization cannot succeed without full engagement by all WTO members, U.S. leadership remains a crucial ingredient in the WTO's success.

3. Bernard Hockman and Michael Leidy. "Environmental Policy Formation in a Trading Economy: A Public Choice Perspective," GREENING OF WORLD TRADE ISSUES. 1992, Chapter 11, p. 233. It was noted earlier that quantity regulation will lead to an increase in imports. And it can be shown that corresponding to the increase in imports, import-competing firms in the regulated industry will experience 'injury' in several dimensions. As a result, the likelihood that unfair trade laws or voluntary export restraint agreements (VERs) will be pursued increases following the institution of a quantity-based approach to production-based environmental pollution. Four factors contribute to this.

First, declining domestic sales, market share, production, and employment offer evidence of the injury that is required under current rules, and these occur while the level and share of imports rise. Because satisfying the injury criterion is the principal constraint in gaining protection in this area, the prospect of protection rises, other things equal.

Second, the regulatory regime established to enforce production quotas provides a formal institutional setting for cooperative behaviour that reinforces the ability of firms to pursue other (non-pollution abatement) areas of mutual interest, including efforts to petition for protection. Empirical studies have shown that on average it pays to present a united front in petitions for protection. Internalising industry-wide incentives, including the incentive to petition for protection, is part of the attractiveness of the regulatory approach.

Third, the regulatory regime establishes a precedent for market sharing that may pave the way to cover foreign firms as well. Domestic firms might be able to use the market-sharing arrangement imposed for environmental reasons to argue for its extension to foreign firms. That is, should protection be granted it may be marginally more likely that a negotiated voluntary restraint agreement will be the chosen instrument. Other things equal, such quantity-based protection offers greater scope for the consolidation of market power and thus greater profits.

Fourth, because of all the barriers to domestic entry established under the regulatory pollution-abatement scheme, the prospective profits of protection will not be dissipated by competition over time. This means that the regulatory scheme serves to increase the expected present value of the profits of protection, thereby increasing its appeal.

As a result of these four conditions the pressure for protection and the likelihood that it is granted under the regulatory scheme is enhanced, since the domestic industry experiences injury in several dimensions, this injury coincides with a surge in imports, regulatory barriers to entry increase the expected capitalised value of protection, and the cooperative behaviour enforced by the regulations furthers the industry's ability to speak with one voice.

4. Daniel C. Esty. GREENING THE GATT: TRADE, ENVIRONMENT AND THE FUTURE. 1994, p. 20. As the number of global environmental issues has multiplied and agreements incorporating trade measures as tools to encourage participation and compliance have proliferated, trade officials have looked on nervously. With no guidelines to ensure that trade restrictions are applied in a limited, consistent, and appropriate manner, the trade community sees a danger that the trying task of preserving an open world market will be made more difficult by the indiscriminate (and perhaps even the discriminating) use of trade penalties to promote environmental policies. Moreover, powerful countries such as the United States appear willing to impose trade measures unilaterally to discipline countries that it alone determines to be dam-

aging the "global commons." Such actions threaten not only to unravel the delicate balance of mutual commitments to freer trade but risk breaking down international harmony more broadly.

5. Dale C. Copland. "Economic Interdependence and War," INTERNATIONAL SECURITY. Vol. 20, No. 4, Spring, p. 1. The key to moderating these potential conflicts is to alter leaders' perceptions of the future trading environment in which they operate. As the Far Eastern situation of the late 1930s showed, the instrument of trade sanctions must be used with great care when dealing with states possessing manifest or latent military power. Economic sanctions by the United States against China for human rights violations, for example, if implemented, could push China toward expansion or naval power-projection in order to safeguard supplies and to ensure the penetration of Asian markets. Sanctions against Japan could produce the same effect, if they were made too strong, or if they appeared to reflect domestic hostility to Japan itself, not just a bargaining ploy to free up trade.

The value of maintaining an open trading system through the new World Trade Organization (WTO) is also clear: any significant trend to regionalization may force dependent great powers to use military force to protect their trading realms. In this regard, my analysis tends to support the liberal view that international institutions may help reinforce the chances for peace: insofar as these institutions solidify positive expectations about the future, they reduce the incentive for aggression.

6. Bernard Hockman and Michael Leidy. "Environmental Policy Formation in a Trading Economy: A Public Choice Perspective," GREENING OF WORLD TRADE ISSUES. 1992, Chapter 11, p. 232. The possible relationships between environmental policies and international trade are twofold. First, environmental policies of any kind may affect trade flows. They may do this by directly restricting imports of a product, or indirectly by inducing consumers to either substitute away from or towards imports. Second, environmental policies may lead to an endogenous trade policy response. That is, given that an environmental policy is under consideration, interest groups may be induced to concurrently or subsequently petition the government to limit imports for reasons connected to the environmental policy change. These two possibilities are related, of course, as a necessary condition for the latter is that the environmental policy can be argued to discriminate 'unfairly' against domestically produced goods.

The extent to which trade patterns are altered by environmental policy will depend on the nature of any endogenous trade policy response. The pressure for protection may be direct or may emerge through increased invocation of existing administered forms of protection such as antidumping procedures. In general, because of the anticipated profits associated with restricting imports, the greater prospect of protection is likely to make the regulatory approach to pollution control (as opposed to a market-based approach) additionally appealing to the industry concerned.

7. Robert J. Morris. "A Business Perspective on Trade and the Environment," TRADE AND THE ENVIRONMENT. 1994, p. 128. In addition to these general propositions as the outline of a workable framework within which governments can develop mutually compatible trade and environmental protection policies, business has also set out a series of more specific recommendations to deal with those issues it believes merit the priority attention of governments. The main reason for this priority is that the rules and policies (or lack of them) that governments apply in setting product or process standards, in resolving disputes that arise over the trade effects of national differences, and assisting or not assisting enterprises in meeting new regulations and standards, directly impact how businesses organize their investment, production, and marketing strategies and can become significant factors influencing international trade and investment. Some rules already exist on these issues but greater clarity is needed in all three areas.

8. Robert J. Morris. "A Business Perspective on Trade and the Environment," TRADE AND THE ENVIRONMENT. 1994, p. 128. Turning to the development of environmental policies that minimize trade distortions, business believes that environmental policies should rely to the maximum extent possible on market-oriented measures that encourage innovation by industry to find the most cost-effective ways to achieve environmental goals. The greater the reliance on market policies and instruments, the less the scope for distortion in international trade and investment flows. Harmonization of policies using economic instruments that harness market forces to achieve the goals of regulatory efforts should be encouraged. Among economic instruments, there should be a preference for those involving the least market distortion. Thus, for example, the choice embodied in the U.S. Clean Air Act for the use of tradeable permits is preferable to the proposals under discussion for a carbon or broad-based energy tax, which is burdened with market-distorting effects on competitiveness.

9. Richard H. Snape. "The Environment, International Trade, and Competitiveness," GREENING OF WORLD TRADE ISSUES. 1992, Chapter 4, p. 87. The analysis in the previous two sections looked at the correction of the effects of external diseconomies in a manner designed to raise real world income. It also examined the international income distribution implications of these actions, and the consequential incentive effects for governments which were attempting to increase their own countries' real incomes. Individual firms do not see matters from this perspective. They see

taxes on polluting production and requirements to satisfy environmental regulations as affecting their ability to compete on the world stage, and naturally are likely to attempt to influence policy. Partly under this influence and partly for mercantilist and 'strategic' trade policy reasons (including the moving of the terms of trade favourably for their own countries), governments also are much concerned for the competitiveness of their countries' industries on world markets.

It is clear that taxation or other restrictions on a polluting industry which is engaged in international trade will make it more difficult for that industry to compete. One view of the matter is expressed in Shrybman (1990):

> For a country wanting to maintain stringent environmental standards, while not undermining the competitiveness of its domestic industry, the choices are simple: (a) Establish import tariffs to offset pollution costs so that domestic producers will not be at a disadvantage when competing with imports from jurisdictions without similar environmental regulation, or: (b) Subsidize the cost of environmental protection with general revenue by underwriting pollution control costs.

10. L. Alan Winters. "The Trade and Welfare Effects of Greenhouse Gas Abatement: A Survey of Empirical Estimates," GREENING OF WORLD TRADE ISSUES. 1992, Chapter 5, p. 110. A second potential effect of abatement could be to curtail the growth of trade in manufactures. This might occur as prices of manufactures rose relative to those of services and non-tradeables, and possibly also because protectionist barriers may be used to try to defend local producers from abatement costs. The post-war period has witnessed a rapid expansion in manufactured trade, which, to many commentators, has been one of the principal explanations of the world's unprecedently large increase in prosperity. The causal chain behind this observation entails not merely the fact that trade makes more goods available, but also that it increases competition, allows the exploitation of economies of scale, and stimulates preferences by providing variety. If abatement policy curtailed such growth it could induce a decline in GDP beyond that suggested by the current generation of static or simple growth models. This, in turn, would feed back onto trade patterns in general. This possibility—which has something in common with Hogan and Jorgenson's (1990) views about technical progress—would bear further exploration with regard to both the abatement/manufactures trade link and the consequences of a decline in the growth rate of trade in manufactures.

11. Jagdish Bhagwati. "Trade and Environment: The False Conflict?" TRADE AND THE ENVIRONMENT. 1994, p. 167. But the complaints and political pressures brought by specific industries regarding "unfair trade" by countries with differential and "lower" environmental standards compared to one's own remain a potent force on the political scene. They have now become part of the more general case for what the Americans call "level playing fields"—that is, competition subject to common rules and handicaps.

Demands for harmonization of all kinds of domestic policies are growing apace: in environment, in labor standards, in competition (antitrust) policy, in technology policy. Without such harmonization, it is increasingly asserted that free trade cannot be "fair," and countervailable protection becomes necessary. There are many reasons for this trend which poses a real threat to free trade because, in the end, countries are not likely to become clones of each other in domestic objectives and policies. But the chief reason lies, in my view, in the intensification of competition that the globalization of the world economy has brought about.

12. Jagdish Bhagwati. "Trade and Environment: The False Conflict?" TRADE AND THE ENVIRONMENT. 1994, p. 169. Chief among these arguments is the fear that competition with the imports and exports in third markets from countries with lower standards will put pressure on domestic industries, triggering political action by them to lower standards down to the levels abroad. Having seen former U.S. Vice President Quayle's Competitiveness Council do precisely this, the environmental NGOs in the United States, and their friends in the EC and elsewhere, have come to see this as a real threat to their goals if free trade is embraced and harmonization up is not imposed simultaneously by coercion on foreign countries, especially the poor ones. As Walter Russell Mead put it in a much-cited article in Harper's magazine:

> Either the progressive systems of the advanced industrial countries will spread into the developing world of the Third World will move north. Either Mexican wages will move up or American wages will move down. Environmentalists, labor unions, consumer groups, and human-rights groups must go global—just as corporations have done.

This concern reflects at the global level the debate within the EC: the fear that the Common Market's free trade and free capital flows will lead to harmonization down of standards "from below" and the efforts of many in consequence to impose harmonization at a higher level of standards "from the top."

13. Kym Anderson and Richard Blackhurst. "Trade, the Environment and Public Policy," GREENING OF WORLD TRADE ISSUES. 1992, Chapter 1, pp. 5–6. The lower the costs of protecting and improving the environment, the larger will be both the degree of public support across countries and the amount of environmental improvement accomplished. Hence it is helpful to examine the efficiency of using trade policy instruments to achieve environmental objectives. As with other areas of policy intervention, however, environmental policies also affect the distribution of income and asset values both within and between countries. This in turn creates incentives for inter-

est groups to seek to influence the policy outcome in their favour. Experience has shown that protectionist groups can be skilful at promoting coalitions with environmental groups. There is thus a growing concern that environmental issues are creating indirect as well as direct opportunities to erect new barriers to trade.

For example, if an environmental regulation increases the operating costs of domestic firms sufficiently for them to meet an injury test, this could trigger the imposition of an anti-dumping duty on competing imports. Also, calls for new environmentally-motivated product or production process standards may stimulate domestic producers to advocate measures that impose disproportionate costs on their foreign competitors. More generally, there are many opportunities for domestic producers to attempt to camouflage a reluctance to compete with foreign producers by claiming that increased regulation of imports is an essential part of efforts to improve the quality of the environment.

14. L. Alan Winters. "The Trade and Welfare Effects of Greenhouse Gas Abatement: A Survey of Empirical Estimates," GREENING OF WORLD TRADE ISSUES. 1992, Chapter 5, p. 110. Abatement policy will impinge differently on different fuels and hence is likely to disturb trading patterns *within* the energy sector. Obvious issues include the possible growth of trade in natural gas and uranium, the decline of trade in coal and shale oil, and corresponding price effects. These consequences will be greatly complicated if abatement is compounded by trade liberalization, or at least by the evening out of distortions across countries. For example, if reducing the use of coal encouraged the closing of high-cost production in Europe, the costs of abatement on the major coal users in those countries would be mitigated or possibly even reversed. An important dimension of the political economy of such a situation, however, would be the extent to which high-cost producers would be able to prevent their governments from signing abatement-cum-liberalization packages. There is a danger that spurious arguments about employment and national security would lead countries to seek to place the burden of reduced coal consumption on foreign rather than domestic producers. The resulting import restrictions would clearly be a matter for the GATT.

15. L. Alan Winters. "The Trade and Welfare Effects of Greenhouse Gas Abatement: A Survey of Empirical Estimates," GREENING OF WORLD TRADE ISSUES. 1992, Chapter 5, p. 109. The second dimension of competitiveness is potentially more important. If abatement entails the determination of rigid quantitative emission targets for each country, the implicit carbon tax is almost bound to differ between countries. Indeed, even if it were identical initially, differences would soon emerge as countries grew at different rates. Different tax rates obviously imply competitiveness effects, and one would expect to see energy-intensive production migrating to places where the emission targets were loosest (i.e., where the tax was lower). If demand patters remained unchanged this would affect international trade patterns and would raise the ugly possibility of a carbon tariff.

16. Peter J. Lloyd. "The Problem of Optimal Environmental Policy Choice," GREENING OF WORLD TRADE ISSUES. 1992, Chapter 3, p. 67. In all models of environmental damage with international transmission of the pollutant or emission, the optimal policy affects international trade in some commodities. A good example is the set of optimal taxes on carbon which would substantially lower the price of fuels to producers, especially that of coal, as well as raising the relative price of these goods to users. The volume and terms of trade of coal-exporting countries, for example, would necessarily deteriorate, but such consequences need to be borne as part of the required change in the competitive general equilibrium. (Second-best arguments for using instruments of environmental policy to correct other market failures are ruled out). There will be pressures on governments to prevent some of the changes in real income, including demands for protection from imports when world prices fall. If governments accede to these demands the costs of pollution reduction will be increased. The next section considers the direct use of trade-based instruments.

17. John Piggot, John Whalley, and Randall Wigle. "International Linkages and Carbon Reduction Initiatives," GREENING OF WORLD TRADE ISSUES. 1992, Chapter 6, p. 116. Our results highlight the importance of the additional effects listed above in affecting country incentives to reduce carbon emissions, especially in so far as initiatives for unilateral reduction are concerned. While benefit-related incentives to participate in schemes to reduce carbon can be large, these effects are offset to a surprisingly large degree by production (or consumption) responses in non-participating countries. Large effects on trade volumes can also accompany carbon reduction initiatives. The model results we report thus suggest the likely need for trade or other third-country sanctions, or incentive payments to support global agreements on carbon limitation. This may have important implications for the international trading system over the next few decades.

18. John Piggot, John Whalley, and Randall Wigle. "International Linkages and Carbon Reduction Initiatives," GREENING OF WORLD TRADE ISSUES. 1992, Chapter 6, p. 121. Unilateral consumption of 50 per cent involve losses by the region making such cuts and benefits for all other regions except oil exporters. The positive effects for non-participants come not only from shared benefits in emission reductions but also from terms of trade and other effects. Regions cutting consumption lower consumer prices of energy which allows other energy importing regions to increase consumption.

19. L. Alan Winters. "The Trade and Welfare Effects of Greenhouse Gas Abatement: A Survey of Empirical Estimates," GREENING OF WORLD TRADE ISSUES. 1992, Chapter 5, p. 100. Substitutability helps to determine changes in the share of energy in total costs of production when relative input prices change. It also affects responses to the curtailment of energy-use, although in a somewhat complex fashion. The more highly substitutable is energy for other factors, the greater the extent to which any firm can offset reductions in energy-use without large losses of output by using other factors. This does not necessarily imply, however, that the aggregate output losses are smaller. If there are fixed endowments of the other factors, higher substitutability implies greater pressure on them, and in fact, for a given real increase in the energy price, aggregate output will fall further the greater the degree of factor substitutability. Of course with greater substitutability a smaller rise in the price of energy is required to achieve a given abatement, which helps to offset the previous effect. Indeed if we think of the energy price as being set at whatever level is necessary to achieve a given percentage cut in energy use, the degree of substitutability does not affect the output loss directly, for the latter depends only on the energy share. Under these circumstances substitutability matters only through its influence on the energy cost share.

20. Sandra L. Walker. ENVIRONMENTAL PROTECTION vs. TRADE LIBERALIZATION: FINDING THE BALANCE. 1993, p. 13. Another threatened category may be certain tax measures. A tax on product emissions, if applied domestically, might not raise any GATT problems. However, if because of pressure from industry a government decided to impose equivalent taxes on imports, several GATT incompatibilities could exist. The measure could violate Article III's national requirement if it were applied on the basis of the quantity of emissions in the production of the good. In that event, the measure would not apply to the product but to the production process, and therefore, as the Panel held in *Tuna/Dolphin*, would not fall within Article III and would be considered a violation of Article XI. Such a measure might also run afoul of the MFN requirements of Article I since the level of the tax would vary with the volume of emissions in each country.

21. Sandra L. Walker. ENVIRONMENTAL PROTECTION vs. TRADE LIBERALIZATION: FINDING THE BALANCE. 1993, p. 44. The imposition of a levy or tax on products which cause pollution in their use of disposal would appear to be unassailable under GATT Article III.2 provided that the levy was applied equally to domestic and foreign products. If the tax on imports were applied at the border and equivalent taxes were imposed on domestic products, it would not be considered an import charge because of Article II.2(a), which permits "a charge equivalent to an internal tax imposed consistently with the provisions of paragraph 2 of Article III in respect of the like domestic product." In addition, the policy purpose of the tax would not be considered relevant.

In practice, however, taxes which are applied equally to both foreign and domestic goods may not have equivalent effects on their sale. The tax could apply to goods which are primarily imported and in competition with untaxed, like domestic goods. An example is the 1992 environmental levy on non-refillable alcoholic beverage container in the Canadian province of Ontario. Although this measure has an impact on wine and spirit bottles which are not refilled, the most contentious trade issue has been over beer in metal cans. US beer producers have strongly denounced this levy as discriminatory since US beer is almost exclusively packaged in metal containers (which may be recyclable but not refillable as required for the Ontario levy not to apply) while Ontario beer is largely sold in reusable bottles (for which no levy is charged). In response to the argument that US beer producers could change to bottles, the US industry would have to incur the capital costs of changing over to bottling as well as the shipping costs which are likely higher than for exporting the lighter metal cans.

22. John H. Jackson. "World Trade Rules and Environmental Policies: Congruence or Conflict?" TRADE AND THE ENVIRONMENT. 1994, p. 230. The problem of subsidies in international trade policy is perhaps the single most perplexing issue of the current world trading system. Some of the major controversies and negotiation impasses, such as the question of agriculture, relate to this problem. GATT rules have become increasingly elaborate and contain several different dimensions. Not only are there provisions in GATT Articles VI and XVI, but there is also the Tokyo Round "code" on subsidies and countervailing duties that provides obligations to the signatories of that code. The following hypothetical cases can illustrate some of the problems that could occur:

- Suppose an exporting country establishes a subsidy for certain of its manufacturing companies to allow them grants or tax privileges to assist them in establishing environmental enhancement measures (such as machinery to clean up smoke or water emissions, or other capital goods for environmental or safety/health purposes). When those producers export their goods, the goods could be vulnerable to foreign nations imposing countervailing duties. Is this appropriate or should a special exception for environmental measures be carved out?

- Can an importing country argue that the lack of environmental rules in the exporting country is the equivalent of a "subsidy" and impose a countervailing duty?

- Similarly, suppose a nation lacks environmental rules such that its domestic producers can produce cheaper and thus compete to keep out goods that are imported from other countries that have substantial environmen-

tal rules. Thus the lack of environmental rules becomes an effective protectionist device.

Obviously these hypotheticals are not so "hypothetical." A lot of the discourse about NAFTA expresses the worry that if Mexico lacks environmental rules, this will give Mexico a competitive advantage, vis-à-vis American (or Canadian) producers. Consequentially, these problems illustrate the need for careful examination of the subsidy rules to design appropriate environmental exceptions or rules without destroying the advantages of the subsidy rules.

23. Jennifer Schultz. THE AMERICAN JOURNAL OF INTERNATIONAL LAW. April 1996, Vol. 89, Lexis/Nexis. The Subsidies Agreement, Art. 8.2(c), states that only facilities that have been in existence for two years qualify for support to adapt "to new environmental requirements imposed by law and/or regulations which result in greater constraints and financial burden on firms." Thus, new investments would bear the cost of new technology for environmental protection as part of the financial portfolio, and would not qualify for government support. Moreover, for older facilities to qualify for nonactionable status, the subsidy must be a one-time investment; must be no greater than 20% of the cost of compliance; may not be used for installation or operation of the new technology; must be linked to reducing pollution and not to increasing a firm's savings or profit margin; and must be available to all industries that can use the new technology or process methods, in order to avoid using subsidies to favor one domestic industry over another when both sectors are in equal need of compliance measures.

24. Geraldo Vasconcellos and Richard Kish. COMPETITIVENESS AND AMERICAN SOCIETY. 1993, pp. 90–91. The Organization for Economic Cooperation and Development (OECD), in a recent publication titled *International Direct Investment and the New Economic Environment,* reckons that "investment has conventionally been regarded as a combination of capital, technology, management and commercial expertise. International direct investment can be seen as a well balanced package enabling technology to be implanted. . . ." Yet the control of emerging frontier technologies poses particular economic and political questions. A few pages later, the same OECD report concedes that "if the technology is fundamentally new, the host country has difficulties in negotiating with the small number of firms possessing the technology and creating the conditions to protect it, such as a high level of R&D, product differentiation and *barriers to entry.* Moreover, says the OECD, "in a situation . . . where the new technologies and the know-how concerning them represent a major competitive asset for the enterprises concerned, the latter have no desire to lose any advantage. . . ." [ellipses in original]

25. Marc Levinson. "Kantor's Cant," FOREIGN AFFAIRS. March/April 1996, p. 17. The greatest threat to an open international economy at the end of the twentieth century is not Japan or the European Union. It is public opinion. Voters in many countries are now looking inward, more concerned with conserving what they have than with building the world economy of tomorrow. In the United States, the distrust of things foreign—be they trade, immigration, or investment—is apparent in opinion polls and on the presidential campaign trail. Regrettably, Kantor and Clinton have fostered that trend with their rhetoric. By pledging to pursue a trade policy that would bring good jobs at good wages, the administration promised far more than it was capable of delivering and left voters all the more disenchanted with free trade when it delivered neither. By repeatedly engaging in on-camera brinkmanship over disputes large and small, the Clinton team has strengthened the misimpression that the United States alone favors open markets and that other countries are simply exploiting American naiveté. And by suggesting again and again that other countries' promise cannot be trusted but that the United States always keeps its word, it has reinforced latent public suspicion of trade. If he is to succeed in expanding trade, the U.S. trade representative must focus not on wielding the crowbar abroad but on his most important job: shaping public opinion at home.

26. Bernard Hockman and Michael Leidy. "Environmental Policy Formation in a Trading Economy: A Public Choice Perspective," GREENING OF WORLD TRADE ISSUES. 1992, Chapter 11, pp. 233–234. It should be noted that the endogenous contingent trade policy response leading to increased trade barriers might also appear through a more direct path. Adoption of an environmental policy may cause new pressure for protection in any event, as the domestic industry can now claim that this creates a 'nonlevel playing field'. While injury criteria play no formal role in implementing newly legislated trade barriers, the fact that domestic industries *can* claim to be injured by increased imports subsequent to the imposition of the environmental policy is likely to facilitate the adoption of such measures if they are pursued.

27. Robert Howse and Michael J. Trebilcock. INTERNATIONAL REVIEW OF LAW AND ECONOMICS. Vol 16, No. 1, March 1996, p. 75. Precisely because the implicit benchmark of fairness is so illusory—i.e., a world where governmentally imposed labor and environmental protection costs are completely equalized among producers of like products in all countries—trade measures based upon this kind of fairness claim are likely to be highly manipulable by protectionist interests. Since, of course, protectionists are really interested in obtaining trade protection, not in promoting environmental standards or labor rights, the fact that the competitive fairness claim in question does not generate a viable and prin-

cipled benchmark for alteration of other countries' policies is a strength not a weakness—for it virtually guarantees that justifications for protection will always be available, even if the targeted country improves its environmental or labor standards.

28. Jagdish Bhagwati. "Trade and Environment: The False Conflict?" TRADE AND THE ENVIRONMENT. 1994, p. 168. The notion of unfairness is also attractive to those who seek relief from international competition. If you go to your Congresswoman and ask for protection because the competition is tough, it is going to be difficult to get it. After all, many of them have been sufficiently educated, or perhaps brainwashed (depending on your point of view) into thinking that protection, while not a four-letter word, is not something you want to embrace if you aspire to anything like statesmanship. But if you go to her and say that your successful rival is playing by "unfair" rules, that is just music to protectionist's ears. In the United States, in particular, the "unfairness" notion can take you very far since the economic and social ethos reflects notions of fairness and equality of access (rather than success) more than anywhere else.

The fact that the United States has also been undergoing the "diminished giant" syndrome, vis-à-vis the Pacific nations and her fear of consequent deindustrialization, has also made the American politicians more susceptible to these "unfair trade" arguments from interested lobbies. The continuing dominance of the United States in setting the world's trading agenda powerfully reinforces, in turn, the trend toward "fair trade" and "level playing fields."

While the "unfair trade" argument for rejecting free trade with countries with different environment standards is therefore part of the generic and more general demands for harmonization and level playing fields in world trade, environment (whose protection is legitimately a virtue in itself) brings to this trend additional arguments with perhaps even more powerful appeal.

29. Pietro Nivola. THE BROOKINGS REVIEW. Winter 1996, p. 1. As domestic regulatory costs mount, so can protectionist reflexes. Major industries—steel, chemicals, autos, agriculture, and so on—beset by legal strife over environmental impacts and consumer safety complaints have been especially aggressive in seeking import duties, market quotas, or price supports. During the 1980s steel companies alone were responsible for fully a third of all the antidumping cases submitted by U.S. firms. Likewise, when regulatory burdens, on top of Japanese competition, began to accumulate in the early 1980s, the Big Three automobile manufacturers acquired a multibillion-dollar protective shield against imports.

A connection between self-regulatory hardships and the incidence of trade restraints imposed on international competitors may be more than merely coincidental. When distressed businesses run to the government, they often knock first on the accessible doors of Washington's trade bureaucracy. A beleaguered steel producer or automobile company is far less likely to gain regulatory relief from the Environmental Protection Agency than to find compensatory trade remedies in the Commerce Department. The latter agency's mission, standard operating procedures, and congressional overseers have traditionally served the interests of its business clientele, whereas the EPA and its watchdogs have had exactly the opposite bias, keeping U.S. business on the straight and narrow.

It is also natural to react to one's regulatory handicaps by trying to push them on everyone else, if not by demanding trade protection, then by adjusting the playing field. Efforts to haul additional legal freight into other societies have already complicated international negotiations. Much as the question of "social dumping" came to dominate deliberations about regulatory harmonization in Europe, the North American free trade treaty teetered for a time over the question of environmental and labor standards for Mexico.

30. Bernard Hockman and Michael Leidy. "Environmental Policy Formation in a Trading Economy: A Public Choice Perspective," GREENING OF WORLD TRADE ISSUES. 1992, Chapter 11, p. 235. Product standards provide an example where environmental regulation in itself may have direct effects on international trade. Because they are usually applied to all products, irrespective or origin, they will inhibit imports in all cases where: (a) the product standards differ from those (if any) applying to the exporters' home and other markets; and (b) this difference implies an increase in unit costs of production and/or transportation. To the extent that either occurs one speaks of stands forming a technical barrier to trade. Clearly, import-competing firms have an incentive to adopt unique standards and to ensure that they are strictly enforced at the border. Even if identical standards are applied to domestic and foreign goods, a tariff-like effect may result because the standard differs from those applied in other markets. Any standard, even if applied on a national treatment and nondiscriminatory basis, may inhibit trade if the standard differs in a significant manner from those used in other markets.

31. Sandra L. Walker. ENVIRONMENTAL PROTECTION vs. TRADE LIBERALIZATION: FINDING THE BALANCE. 1993, p. 29. Product- related environmental measures address the harm to the environment which the product may cause in use or at the end of its useful life. Examples of such measures would include those which regulate pesticides, car emissions, discarded refrigerators or air conditioners containing CFCs, as well as product packaging. Regulations of the use and disposal of such products may cause certain restrictions on trade. For instance, a ban on the use of CFC refrigerators by some countries would mean that countries still producing such items would find such markets closed. Regulations aiming to limit the amount of waste from packaging might

require the use of certain kinds of packaging or recycling; foreign manufacturers may bear relatively higher costs to adapt to such requirements particularly where their home countries do not have similar regulations and where they are required to establish recycling or re-use facilities within the export market. The main issues which product regulations raise for international trade law are therefore market access and discrimination between domestic and foreign products.

32. John H. Jackson. "World Trade Rules and Environmental Policies: Congruence or Conflict?" TRADE AND THE ENVIRONMENT. 1994, p. 227. Suppose on the other hand, the government feels that an automobile plant in a foreign country is operated in such a way as to pose substantial hazards to human health, either through dangers of accidents from the machinery, pollutants, or unduly high temperatures in the factory. On an apparently nondiscriminatory basis, the government may wish to impose a prohibition on the sale of domestic or imported automobiles that are produced in factories with certain characteristics. However, note in this case that the imported automobiles themselves are perfectly appropriate and do not have dangerous or polluting characteristics. Thus, the target of the importing country's regulation is the "process" of producing the product. The key question under the GATT/MTO system is whether the importing country is justified in taking a process-related measure either under national treatment rules of nondiscrimination, or exceptions of Article XX, which do not require strict national treatment nondiscrimination treatment. The worry of the trade policy experts is that to allow the process characteristic to be the basis for trade restrictive measures would be to open a pandora's box that could cut a swath through GATT.

The tuna-dolphin case relates to these issues. Although the GATT panel report is not entirely clear on this matter, it seems fair to say that there were two important objections to the U.S. embargo on the importation of tuna because of its objection to the way the tuna were fished, which caused danger to dolphins. First, the question of eco-imperialism arises, where one nation unilaterally imposes its fishing standards (albeit for environmental purposes) on other nations in the world without their consent or participation in the development of the standard. Second, the problem of the inconsistency of the import embargo with GATT rules emerges. Unless there is some GATT exception that would permit the embargo, which relates to the "process-product" interpretation problem and therefore to the national treatment rule and the general exceptions of GATT, the inconsistency remains between Articles III and XX.

The approach in the GATT system so far has given great weight to this slippery slope concern, interpreting both the Article III (including some Article XI questions) and the Article XX exceptions to apply to the product standards and to life and health within the importing country, but not to extend these concepts and exceptions to "processes" outside the territorial limits of jurisdiction. The alternative threatens to create a great GATT loophole and is a serious worry.

33. Jagdish Bhagwati. "Trade and Environment: The False Conflict?" TRADE AND THE ENVIRONMENT. 1994, p. 176. But environmentalists are also worried, with a little more cogency, about the roadblocks that current and prospective GATT rules can pose for environmental regulations standards aimed entirely at domestic production and consumption, matters that are conventionally and properly within domestic jurisdiction.

Now, as long as these rules are applied without discrimination between domestic and foreign suppliers and among different foreign suppliers, there is really little that GATT rules can do to prevent a country from doing anything that it wants to do. For domestic conservation, safety, and health reasons, under Articles XX(b) and XX(g), a contracting party of GATT can even undertake discriminatory, selectively targeted, trade-restraining action, subject to safeguards.

Thus, if you insist on safety belts or airbags in cars, you can impose them on cars as long as both imports from all sources and domestic production are symmetrically treated. The same is true for requiring catalytic converters to reduce environmentally harmful emissions.

The most significant and contentious conceptual question arises when you have a rule that says that consumption (from both domestic and foreign sources) of a product will be restricted if the product is produced using a *process* you disapprove of. Objecting to a process used in a foreign (or, strictly, non-domestic) jurisdiction is, under GATT rulings, not acceptable. There are two types of such process-related problems that we might distinguish:

(1) where the process used is objected to because of "values": for example, purse seine nets or leghold traps; and

(2) where the process used is objected to because it creates cross-border physical spillovers such as acid rain or global warming.

I have already indicated that official trade restrictions under category (1) are properly dismissed as inappropriate unilateralism and the refusal by GATT panels to endorse such actions is good law.

34. Candice Stephens. "The Organization for Economic Cooperation and Development and the Re-emergence of the Trade and Environmental Debate," TRADE AND THE ENVIRONMENT. 1994, p. 87. The opportunities for clashes between environmental and trade policies are increasing with the trend toward life-cycle concepts. Eco-labeling schemes, eco-packaging rules, and recycling programs are no longer fads, but specific examples of the implementation of life-cycle concepts. They are widespread in OECD countries and even in some non-

OECD countries. Many schemes are voluntary, but more and more are government sponsored. Current trade rules, however, do not allow trade discrimination against a product based on how it is produced, and regulations on the disposal or recycling of a product are often seen as potential trade barriers.

35. Candice Stephens. "The Organization for Economic Cooperation and Development and the Re-emergence of the Trade and Environmental Debate," TRADE AND THE ENVIRONMENT. 1994, p. 88. Eco-packaging approaches are also examples of the trend to life-cycle management in environmental policies. New government programs and regulations to promote the return, recycling, or safe disposal of packages and containers are being announced daily. In the effort to cut down on mounting waste and to deal with waste disposal problems, some countries are demanding that all goods, both domestic and imported, be encased in recyclable packaging. Other countries are enacting deposit-refund schemes for bottles and containers or proposing a tax on products depending on the recyclability or environmental friendliness of its packaging. These are innovative and valuable approaches to coping with waste problems, but they can pose problems for imported products. Products that travel long distances may be of necessity be packaged in greater amounts of more durable materials with fewer options for recycling. In addition, imported products may not have access to recycling facilities or deposit-return distribution networks in their countries of destination. There is also the real possibility that the structure and vertical integration of local recycling networks may act to shut imports out of domestic recycling markets.

36. Sandra L. Walker. ENVIRONMENTAL PROTECTION vs. TRADE LIBERALIZATION: FINDING THE BALANCE. 1993, p. 45. Similarly, non-tax measures risk challenge under GATT rules. For example, a requirement that certain products used in Country X meet specified emission standards which are higher than other countries' standards might be incompatible with GATT rules. If the regulation applied equally to like imported and domestic products, there would be no violation of de jure non-discrimination standard under Article III. However, since GATT establishes a non-discrimination standard based on differential impact, it is possible that such a standard would not be legal under the GATT. Foreign companies might be required to produce a special product line for one country, raising the cost of these products, particularly if X's market is not a principal one for country Y exporters.

37. WASHINGTON TIMES. 12 February 1996, Lexis/Nexis. Indeed, on Jan. 17, 1996, in its first major ruling, the WTO decided against the United States and deemed a critical section of the Clean Air Act unfair to foreign oil refineries. The WTO did not determine specific sanctions but instead ordered the United States to change its rules on imported gasoline.

Venezuela and Brazil brought the case against the United States to end a dispute that began before the WTO existed. During the last three years, Venezuela and Brazil have attempted to ship gasoline that appeared to violate U.S. smog standards to the northeast region of the states. The Clean Air Act required domestic producers to meet standards based upon the emissions evaluations of gasoline they produced in 1990.

Many foreign companies, however, did not maintain similar emissions records, so the United States decided that foreign countries could only send America gasoline that met an average U.S. standard. The WTO interpreted this action as discrimination against foreign companies because it required them to meet higher standards than some American producers.

38. Jennifer Schultz. THE AMERICAN JOURNAL OF INTERNATIONAL LAW. April 1996, Vol. 89, Lexis/Nexis. Nations retain the power to select their preferred level of protection. Thus, a government does not have to follow international standards in determining the level of risk its citizens must bear. But this power is circumscribed by the limitation set out in Article 5(5) of the SPS Agreement: each nation "shall avoid arbitrary or unjustifiable distinctions in the levels it considers to be appropriate in different situations, if such distinctions result in discrimination or a disguised restriction on international trade."

Under GATT jurisprudence, the technical meaning given to "discrimination" makes it possible to characterize any regulation or tax that puts foreign products at a de facto disadvantage as discriminatory. Thus, nations that maintain arbitrary or unjustifiable variations in the maximum allowable risk levels (e.g., differences among states) could run into problems under Article 5(5).

39. Joel L. Silverman. THE GEORGIA JOURNAL OF INTERNATIONAL AND COMPARATIVE LAW. Fall 1994, Lexis/Nexis. The serious problem with this provision is that the balance of power in the WTO is held by developing nations, which are generally hostile to unilateral environmental trade actions such as eco-tariffs. Voting in the WTO is by "one nation, one vote," not consensus as under the old GATT. Developing nations will hold eighty-three percent of the votes in the WTO. More than seventy-five percent of WTO members (each a developing nation) voted against the United States on over half of all votes taken in the United Nations in 1993.

40. Jennifer Schultz. THE AMERICAN JOURNAL OF INTERNATIONAL LAW. April 1996, Vol. 89, Lexis/Nexis. Following the logic of the Tuna/Dolphin dispute, many commentators have noted the potential conflict between the GATT and environmental treaties that rely on trade discrimination, such as the Montreal Protocol

on Substances That Deplete the Ozone Layer and the Basel Convention on the Control of Transboundary Movements of Hazardous Wastes and Their Disposal. In an effort to protect these treaties, commentators have suggested that the Montreal Protocol (of 1987) and the Basel Convention (of 1989) can trump the GATT (of 1947) via the later-in-time rule of international law. The GATT/WTO defeats such trumping, however, by resetting GATT's date to 1994, permitting it to leapfrog over all environmental treaties that use trade measures. Henceforth, such treaties can be judged by the GATT/WTO on their merit. The subject of international environmental agreements should be taken up as a priority by the Trade and Environment Committee. The NAFTA's approach, of protecting a list of IEAs with trade provisions, may be a useful precedent.

41. Sandra L. Walker. ENVIRONMENTAL PROTECTION vs. TRADE LIBERALIZATION: FINDING THE BALANCE. 1993, pp. 98–99. The conclusions of the *Tuna/Dolphin* Panel are that: 1) products are alike regardless of how environmentally damaging their production is and 2) that a country cannot require all products sold in it to be produced according to domestic standards. In other words, environmental regulation cannot be applied extraterritorially. This ruling has significant repercussions. First, proposed and existing environmental laws in member countries could run afoul of GATT obligations. Second, international environmental agreements which use trade restrictions to enforce their provisions may be judged as violations of GATT. Third, the GATT may be forced to address unilateral responses to the problem of disparities in environmental regulation given the strong support in environmentally advanced jurisdictions for counteracting trade measures and the importance of disciplining such responses.

42. Hilary French. COSTLY TRADEOFFS: RECONCILING TRADE AND THE ENVIRONMENT. 1993, p. 49. One added policy instrument that could be challenged as a violation of free trade rules is the use of subsidies for environmental goals. Trade agreements generally discourage subsidies, and provide for countervailing duties in some cases to compensate for their continued use. On balance, this could benefit the environment if rules are developed to restrict subsidies that cause environmental harm, such as those which promote unsustainable forestry, energy mega-projects, mines, and intensive farming. However, subsidies that are expressly aimed at financing environmental protection could be threatened. Ironically, while NAFTA frowns on *most* subsidies, including environmentally-oriented ones, it winks and looks the other way when the subsidies are for oil and gas exploration. On the other hand, amendments are being considered in the Uruguay Round would exempt some agricultural land set-aside programs such as the U.S. Conservation Reserve Program from the threat of countervailing duties. Still, a large number of other more general environmental subsidies would be vulnerable under the latest draft of GATT revisions.

43. Paulette L. Stenzel. ALBANY LAW REVIEW. Vol. 59, 1995, Lexis/Nexis. Another example involves a pre-NAFTA challenge under GATT to a state law. In a case known as Beer II, a GATT panel examined a challenge to laws of Texas and other states which gave tax benefits to local breweries and exempted them from certain common-carrier requirements. The GATT panel held that those benefits violated GATT by discriminating against the producers of beer and wine in Canada and other countries.

44. Hilary French. COSTLY TRADEOFFS: RECONCILING TRADE AND THE ENVIRONMENT. 1993, p. 43. Environmentally related export controls, too, can be restricted by trade agreements. When resources are scarce—be they energy, minerals, timber, or water—countries sometimes want to reserve supplies for their own people. This can allow for conservation while still leaving room for domestic consumption. But treating domestic and overseas markets differently runs counter to free trade principles. As a result, such restrictions are discouraged by some trade bodies.

45. Hilary French. COSTLY TRADEOFFS: RECONCILING TRADE AND THE ENVIRONMENT. 1993, p. 43. Another recent skirmish involves European Community complaints that two U.S. automobile taxes intended to promote fuel efficiency—the Corporate Average Fuel Economy Law and the gas-guzzler tax—violate GATT. The EC has said that it plans to call for a GATT dispute resolution panel. The U.S. claims the two taxes would meet GATT tests, as they are applied to domestic producers equally. The E.C. counters that though this appears to be true, they are applied in such a way that European producers are disadvantaged.

46. Robert J. Morris. "A Business Perspective on Trade and the Environment," TRADE AND THE ENVIRONMENT. 1994. Subsidies designed to assist particular sectors or enterprises to meet new environmental protection standards or regulations undercut the Polluter Pays Principle by preventing the market mechanism from incorporating environmental protection costs in product prices. Such subsidies should continue to be actionable under GATT if they cause injury to other parties.

47. Christopher Thomas. "Litigation Process Under the GATT Dispute Settlement System: Lessons for the WTO," JOURNAL OF WORLD TRADE. Vol. 30, No. 2, April 1996, pp. 71–72. In addition to the features of the fact-finding process discussed above, there was an important correlative approach to the application of the governing law. As will be shown below, GATT panels tended to focus on the measure's operation in theory and avoided attempting to evaluate its actual effects. They did so by engaging in an analysis which combined strict treaty interpretation with the use of presumptions to shift the burden of proof. This relieved panels of having to enter into difficult areas of proof.

The potentially challenging area of proof involved considering the trade or economic effects of a trade measure. By the end of the 1970s it seemed relatively clear that if an obligation was found to be breached, it was normally unnecessary to go further to examine any resulting trade effects. The 1979 Understanding generally relieved panels of any duty to estimate the trade effects of the breach of a GATT rule because normally a breach was presumed to result in an adverse impact (although the presumptions theoretically could be rebutted by the respondent). Although contracting parties sometimes urged panels to examine the trade effects of the measures which they sought to defend, panels generally declined to accept the invitation. An important exception was *Spain—Measures concerning Domestic Sale of Soyabean Oil*, where the Panel accepted a trade damage test when construing Article III. The Report was not adopted by the Council.

48. Christopher Thomas. "Litigation Process Under the GATT Dispute Settlement System: Lessons for the WTO," JOURNAL OF WORLD TRADE. Vol. 30, No. 2, April 1996, pp. 72–73. In *Superfund*, the United States sought to rebut what it considered to be a minor breach of the General Agreement by showing that the tax differential in question was so small that it had minimal or no trade effects. This argument raised the question of whether the presumption was absolute or rebuttable and, "if rebuttable, whether a demonstration that a measure inconsistent with Article III:2, first sentence, had no or insignificant effects on trade is a sufficient rebuttal." In analysing GATT practice, the Panel reviewed how the CONTRACTING PARTIES made recommendations or rulings on measures found to be inconsistent with the General Agreement. It noted that, when making recommendations or giving rulings on measures inconsistent with the General Agreement they had not considered the impact of the offending measure. That had been done "only in connection with the authorization of compensatory action." This led the Panel to conclude that the impact of a measure inconsistent with the General Agreement "is not relevant for a determination of nullification or impairment by the CONTRACTING PARTIES."

49. Christopher Thomas. "Litigation Process Under the GATT Dispute Settlement System: Lessons for the WTO," JOURNAL OF WORLD TRADE. Vol. 30, No. 2, April 1996, pp. 74–75. Citing previous panel reports (including one that specifically rejected the relevance of trade flow to a finding of nullification or impairment), the Panel commented "governments can often not predict with precision what the impact of their interventions on import volumes will be." Second, if trade effects had to be proved, contracting parties could not make a claim to prevent a change in politics until adverse trade effects were produced. Third, and most relevant here, was the problem of proof:

"If Article III was considered to be protecting expectations on trade flows, it would be necessary for the CONTRACTING PARTIES to determine what export volumes a contracting party can reasonably expect after having obtained a tariff concession. The Panel is not aware of any criteria or principles that would be applied to make such a determination. The Panel further noted that changes in trade volumes result not only from government policies but also other factors, and that, in most circumstances, it is not possible to determine whether a decline in imports following a change in policies is attributable to that change or to other factors."

Thus, it is evident that the panel's strict interpretation of Article III was partly motivated by problems of proof.

50. WASHINGTON TIMES. 12 February 1996, Lexis/Nexis. Despite these auspicious projections, the WTO experienced a tumultuous year after it officially opened on Jan. 1, 1995. Controversy continues to surround its dispute-settlement panel, composed of international delegates whose decisions cannot be blocked but only appealed by the countries involved.

The domestic debate over the process has been ongoing. During the 1994 GATT talks, a bipartisan group of U.S. legislators, economists and commentators claimed the WTO's dispute-resolution panel would hold too much authority—that its binding decisions would undermine U.S. sovereignty by placing American economic interests at the whims of foreign judges. This election year the stakes are particularly high as protectionists and free traders square off.

51. WASHINGTON TIMES. 12 February 1996, Lexis/Nexis. During the first week of the new year, for example, U.S. Trade Representative Mickey Kantor announced the White House's intention to effect a more aggressive posture with its trading partners. "We will not tolerate a failure to honor agreements or to honor U.S. trade laws," Kantor said. The administration will set up a new unit in the trade representative's office to focus on agreements with Japan, Canada and the European Union, as well as strengthen its effort to open markets in Latin America and Asia.

In December, Dole placed legislation on the Senate calendar that would set up a commission of five federal judges—the WTO Dispute Settlement Review—to interpret WTO decisions. Should the commission find three "inappropriate" rulings within a five-year period, Congress could withdraw U.S. participation in the WTO, a provision Dole refers to as "three strikes—we're out." Says the Republican presidential front-runner, "I believe the legislation goes a long way toward ensuring that America retains full control of her destiny, that no international organization staffed by unelected bureaucrats will dictate what we do here at home."

52. Peter Morici. "Export Our Way to Prosperity," FOREIGN POLICY. Winter 1995–1996, p. 12. Ultimately, real improvements in U.S. market access abroad will strongly depend on foreign governments rewriting and applying laws and regulations to be consistent with the letter and the spirit of the URAs. Disagreements over the interpretation will inevitably arise. For the WTO to be effective, national governments must honor the decisions of dispute-settlement panels.

Some in Congress may lament the loss of sovereignty to the WTO; however, it is clear that American exports and prosperity are, in some measure, connected to its success. In turn, the WTO can succeed only if the United States is perceived abroad to be playing by the rules, and this will prove delicate because of its pursuit of regional trade agreements and continuing trade problems with Japan.

53. WASHINGTON TIMES. 12 February 1996, Lexis/Nexis. WTO proponents warn that if the United States withdraws from the organization, increased global competitiveness and the strength of burgeoning foreign markets ultimately will diminish its economic dominance. "Every day in every way somebody is getting better and better," says Joe Cobb, a research associate at the Heritage Foundation. Still, the WTO must overcome great obstacles if it hopes to accomplish its goal of open markets and equitable dispute settlements. Surviving the onslaught of political criticism during the U.S. election may be the litmus test of the WTO's credibility and resilience.

54. Vincent Cable. "The New Trade Agenda: Universal Rules Amid Cultural Diversity," INTERNATIONAL AFFAIRS. Vol. 72, No. 2, April 1996, p. 234. At worst, the transatlantic dialogue could be a signal of an emerging division within the trading system between countries at different levels of development (or between occidental and oriental). There is already a degree of (positive and negative) discrimination: developing countries already face relatively severe tariff and non-tariff access barriers and have suffered differentially from anti-dumping duties. Hitherto, the threat of a divided, discriminatory trade system has been kept at bay by GATT, by the progress towards widening GATT membership and by a broadly liberal consensus. This tolerance could change. One straw in the wind may be the apprehension expressed in the West about the competitive potential, in many areas of manufacturing (and services), of the major low-wage non-OECD countries, notably China and India. This concern is being given populist expression by the likes of Pat Buchanan (spiced with appeals to religious and racial solidarity) and some intellectual respectability by work from academics, suggesting that real wages of unskilled workers in rich countries may be being depressed by international trade (whatever the broader mutual benefits of trade). The debate is currently on the political periphery but it has incendiary potential. A growing sense that there are fundamental differences rooted in culture and social values which justify exclusion could set the pyre alight.

55. Sandra L. Walker. ENVIRONMENTAL PROTECTION vs. TRADE LIBERALIZATION: FINDING THE BALANCE. 1993, p. 118. An advantage of the implicit subsidy and eco-dumping approaches over trade restrictions such as bans is that it does not cut off trade totally but merely increases the costs of the foreign product. The weakness in these approaches is that the view that another country is subsidizing or eco-dumping is based on a unilateral determination of the « correct » level of environmental regulation. This was illustrated in the case of the Canadian softwood lumber dispute between the US and Canada. The US International Trade Administration found that the Canadian tree-cutting stumpage fees were so low as to constitute a subsidy and threatened to impose countervailing duties against this alleged subsidy. The Canadian government responded to the US. The subjectivity of determining a « correct price » for cutting timber is readily apparent. As well, the ability of an influential country like the US to establish unilaterally such reference standards and to pressure the Canadian government to comply reflects the potential role for strongarm tactics which powerful economies could use against weaker ones.

56. Robert Howse and Michael J. Trebilcock. INTERNATIONAL REVIEW OF LAW AND ECONOMICS. Vol 16, No. 1, March 1996, p. 73. In considering the systemic threat from environmental and labour rights-related trade measures, it is important to distinguish between purely unilateral measures and those that have a multilateral dimension. The former measures are based upon an environmental or labor rights concern or norm that is specific to the sanctioning country or countries. Here, there is a real risk of dissolving a clear distinction between protectionist "cheating" and genuine sanctions to further non-trade values—the sanctioning country may well be able to define or redefine its environmental or labor rights causes so as to serve protectionist interests. Measures with a multilateral dimension, by contrast, will be based upon the targeted country's violation of some multilateral or internationally recognized norm, principle, or agreement—for instance, a provision in an accord to protect endangered species or one of the ILO Labor Conventions. These norms, principles, or agreements are typically not the product of protectionist forces in particular countries, nor are they easily captured by such forces (although the example of the Multi-Fibre Arrangement suggests that this is not invariably the case).

57. Robert Howse and Michael J. Trebilcock. INTERNATIONAL REVIEW OF LAW AND ECONOMICS. Vol 16, No. 1, March 1996, p. 73. Even in the presence of indeterminate welfare effects many free traders may still reject environmental or labor rights-based trade measures on the basis that such measures, if widely permitted or entertained, would significantly erode the coherence and sustainability of rule-based liberal trade. We ourselves, in earlier work, have argued that competitiveness-based or level playing field "fair trade" measures,

such as countervailing and antidumping duties, already pose such a threat. This is based on the notion that the legal order of international trade is best understood as a set of rules and norms aimed at sustaining a long-term cooperative equilibrium, given that the short-term political payoffs from cheating may be quite high (depending, of course, on the character and influence of protectionist interests within a particular country, the availability of alternative policies to deal with adjustment costs, etc.). In the presence of a lack of fundamental normative consensus as to what constitutes "cheating" on the one hand, and the punishment of others' cheating on the other, confidence in the rules themselves could be fundamentally undermined, and the system destabilized.

58. Ambassador Michael Smith. "Afterword," TRADE AND THE ENVIRONMENT. 1994, p. 288. It follows then that if every nation is free to use the trading system to impose its values, environmental or otherwise, willy nilly on other nations, the entire trading system will surely collapse under the weight of these obstacles. Today we will use trade to dictate to the rest of the world how many parts per million of benzene is permissible, tomorrow it will be how many hours in the day a worker can work, next it will be the per capita number of schools a country must have. Surely, these seemingly innocent and laudable social goals will sooner or later be hijacked by protectionist interests. Having served as an international trade diplomat, and knowing the adroitness with which our trading partners use trade measures, I fear that the United States will find itself on the losing end of this protectionist stick more often than not. We will have opened a Pandora's box of protectionism.

59. Jagdish Bhagwati. "Trade and Environment: The False Conflict?" TRADE AND THE ENVIRONMENT. 1994, pp. 171–172. The 1991 GATT report on *Trade and the Environment* drew attention not to the disturbing asymmetry of power to enforce effectively the "values" of the North versus the equally autonomous "values" of the South. Rather, it advanced the "slippery slope" scenario that if any country could suspend another's trading rights in products produced in an "unacceptable" fashion (when no international physical spillovers could be cited as a possible justification and only "values" were at stake), the result was likely to be a proliferation of trade restrictions without any discipline or restraint:

> ... it is difficult to think of a way to effectively contain the cross-border assertion of priorities. If governments suspend the trading rights of other nations because they unilaterally assert that their environmental priorities [in other words, "values"] are superior to those of others, then the same approach can be employed on any number of grounds. Protectionists would welcome such unilateralism. They could exploit it to create embargoes, special import duties and quotas against rivals by enacting national legislation that unilaterally defines environmental agendas that other countries [with different "values"] are likely to find unacceptable.
>
> Changing the world trading rules so as to permit the suspensions of trading rights of others by individual contracting parties, based simply on the unilateral and extraterritorial assertion of their environmental priorities, undoubtedly would be difficult because many countries would consider such a change to be a big step down a slippery slope. [ellipses in original]

60. Vincent Cable. "The New Trade Agenda: Universal Rules Amid Cultural Diversity," INTERNATIONAL AFFAIRS. Vol. 72, No. 2, April 1996, p. 228. Confidence, however, easily flows into hubris, and hubris into complacency about the dangers to the system. There is a danger of regression. Two world wars, economic depression, trade conflict and the intellectual respectability of state control and economic autarky in many countries together did immense damage to the nineteenth-century liberal economic order between 1914 and 1945. Even in the relatively open economies of the OECD it took almost 30 postwar years of rebuilding the global economy to reach the level of integration through trade which existed in 1913. Only those who are hopelessly ahistorical would refuse to countenance something similar happening again.

A specific danger relates to the architecture of a postwar system designed 50 years ago to prevent a repetition of events in the three previous decades. It is questionable whether it is an entirely appropriate model for the twenty-first century. Without structural repairs the old edifice might not hold.

61. Andrea Boltho. "The Return of Free Trade," INTERNATIONAL AFFAIRS. Vol. 72, No. 2, April 1996, p. 259. Reversions to protectionism cannot, of course, be excluded. Negative shocks to the world economy or setbacks in the reform process are always possible. Similarly, regionalism may still impair 'globalism', as may misguided concerns with environmental and labour standards. Most importantly, perhaps US policies remain vulnerable to sudden isolationist lurches and to the influence of powerful protectionist lobbies (as seen in the narrowly averted conflict over trade in cars with Japan in 1995). The failed attempt to restore freer trade in the 1890s by the German Chancellor of the time (Caprivi) may be taken as a reminder that the successes of the Uruguay Round or the creation of the WTO do not guarantee that the present momentum will continue. Two things, however, could both preserve and strengthen it. At home, the presence of a social safety net, quite apart from its other favourable consequences, helps in successfully defusing protectionist pressures, as amply shown by the history of postwar Europe. Internationally, the continuation of cooperation, not only in trade but also in other economic and non-economic areas, strengthens interlinkages and multilateral dependence. Unfortunately, neither of these two features is

as firmly established as one could wish. The present urge to deregulate the state out of existence is threatening both domestic welfare provisions and international cooperation in many areas of macroeconomic policy.

62. Sandra L. Walker. ENVIRONMENTAL PROTECTION vs. TRADE LIBERALIZATION: FINDING THE BALANCE. 1993, p. 57. Article XX(b) relates to the environment to the extent that environmental measures are « necessary to protect human, animal or plant life or health ». The scope of this exception is not clear; the extent to which it will cover environmental protection depends on how broadly a GATT panel will construe it. The reference to human life or health can only be said to be aimed at environmental protection in a broad sense if the concern is to safeguard human health from changes in the environment which threaten the fundamental conditions necessary for human life—clean air, water, and a sufficient resource base—and if the individual is seen as an element of the environment, of the ecosystem and not merely in his capacity as a consumer or worker. Such a broad interpretation of « human health » has not yet been made in the GATT. The inclusion of plant and animal life and health increases the possibility that GATT dispute settlement panels would consider the value of protecting nature in itself where it has no « use » for man. There is, however, no indication from the GATT jurisprudence that degradation of other elements of the environment such as air and water would bring a measure within the « life and health » designation. Where pollution of these elements was serious enough, the link between air and water, and human, plant and animal life or health would not likely be difficult to make. Nevertheless, the exemption is likely too narrow to save all legitimate environmental protection regulations.

63. Vincent Cable. "The New Trade Agenda: Universal Rules Amid Cultural Diversity," INTERNATIONAL AFFAIRS. Vol. 72, No. 2, April 1996, pp. 227–228. Those of us who believe in the value of international economic integration—through freedom of trade and open markets—have a good deal to be optimistic about. Perhaps not for a century have political ideas and economic trends so helpfully coincided. The Uruguay Round agreements is successfully concluded. The World Trade Organization is up and running with the few remaining outsiders, notably China and Russia, waiting to join. Trade disputes between the United States and Japan are apparently being contained without commercial—let alone military—warfare breaking out. Governments in the developing world which once swore by the therapeutic benefits of import restrictions are now unilaterally dismantling them. In Europe and Latin America, and across the Pacific and Atlantic, there are bewildering numbers of 'summit' meetings designed to seek political credit for freer trade. Although occasional voices of seemingly eccentric dissidence are raised—Ross Perot, Sir James Goldsmith and, more ominously, Pat Buchanan—there are few issues around which there is currently a greater degree of intellectual and policy consensus than freedom of trade. Rhetoric is backed by substance. World merchandise trade has grown twice as fast as output over the past decade; trade in services even faster. Global economic integration is not just growing but accelerating.

64. C. Fred Bergsten. "Globalizing Free Trade," FOREIGN AFFAIRS. May/June 1996, p. 109. In addition, there are enormous opportunities for further economic gain in eliminating remaining tariff and nontariff border barriers. The Uruguay Round teed up these remnants of traditional protection for decisive action by converting all agricultural quotas into tariffs, phasing out quota protection for textiles and apparel, and binding most tariffs of developing countries. One last big push could condemn these practices to the dustbin of history.

65. James Miskel. STRATEGIC REVIEW. Summer 1996, p. 74. The public may well view the debate about the economic merits of NAFTA and GATT as confusing and, perhaps, even inconclusive. One thing can be said for this debate: at least it is taking place. The national security implications of NAFTA and GATT termination are, on the other hand, being virtually ignored.

Apart from an occasional reference to the fact that our diplomatic relations with Mexico will sour if NAFTA is terminated, next to nothing has been said about the effects that protectionism might have on defense-related issues. This is unfortunate because a protectionist trade policy will have several insignificant effects on U.S. national security. One is that it could significantly distort defense spending.

The philosophical core of protectionism is the principle of self-sufficiency; i.e., the belief that the nation's greatest good is served when American consumers rely upon American producers. Nowhere does the self-sufficiency credo resonate more forcefully than in the defense industry. If it is good for American citizens to buy domestically manufactured televisions and cars it must be even better for the Pentagon to buy American-made military hardware.

66. James Miskel. STRATEGIC REVIEW. Summer 1996, p. 75. Repatriating such a volume and variety of work to the United States will require substantial subsidies and tight import restrictions, steps similar in design if not in scale to the measures that opponents of NAFTA and GATT envisions for many domestic industries. Unless subsidies and import restrictions are designed with Solomonic wisdom and implemented with surgical precision, little more than higher prices for defense products and larger federal budget deficits will result. Few students of American government can be optimistic that defense industry subsidies and import restrictions will be either wise or well calibrated.

Even if the nation were willing to invest enough to make the entire defense industrial base self-sufficient, or

even to make selected defense industries autarchic, it is by no means clear that the short-term benefits would outweigh the long-term costs. The huge sums of money that would have to be invested in repatriating each and every defense-related supplier of parts, design shop, subcontractor and production facility would inevitably be siphoned away from other accounts in the pool of national treasure that Congress allocates to the Defense Department. One of the accounts most likely to be raided in the name of defense industrial self-sufficiency is operations and maintenance. The other is research and development. Raiding operations and maintenance would jeopardize the readiness of the armed forces. If research and development were to take the hit, the Defense Department would be forced into investing too much in the hardware used in today's reduced-threat environment and investing too little in the hardware that will be essential to maintaining technological superiority over the threats that will emerge tomorrow.

67. James Miskel. STRATEGIC REVIEW. Summer 1996, p. 75. Another negative effect of protectionism on national security is that it will make it more difficult for the United States to orchestrate multi-national cooperation in crises like the Gulf War of 1991 or the near-invasion of Haiti in 1994. If the United States restricts imports, at least some of our trading partners are sure to retaliate by raising their own barriers against American goods and services. Plainly, nations that are engaged in tit-for-tat trade wars with the United States will be less willing to cooperate with us in military crises that do not directly affect them. They may also be less willing to vote our way in the United Nations (UN) and other international bodies like the Organization of American States (OAS). These are not inconsequential matters.

68. James Miskel. STRATEGIC REVIEW. Summer 1996, p. 76. Another area where cooperative trade relations could be important to national security involves oil. One of the measures that the industrialized nations undertook in response to the oil shocks of the 1970s was the establishment of international mechanisms (oil sharing, strategic stockpiles, conservation) to minimize the overall impact of oil shortages. These mechanism have been taken for granted because oil prices have been relatively low since the mid-1980s and the oil market relatively stable. If the oil market were to become less stable, as many think it must since Western reliance upon Persian Gulf oil has begun to approach 1970s levels, international cooperation among the industrialized, oil-consuming nations will be essential to dampen prices and ensure predictable supplies to military and civilian consumers. If industrialized nations are sparring with each other over trade restrictions affecting automobiles, aircraft and TVs, will they agree to cooperate on the trade in oil? Or are they more likely to try to get a jump on the United States by cutting separate deals with oil exporting countries—deals that could conflict with U.S. security interests in moderate Arab states or Israel.

69. Stanley Kober. THE FALLACY OF ECONOMIC SECURITY. Policy Analysis No. 219, 24 January 1995, http://www.cato.org The danger in this situation is not a Japanese-American political-military rivalry duplicating the earlier Anglo-German one (although even that might not be unthinkable if the administration truly believes that "continuity in our strategic and security relationship with Japan requires discontinuity in our economic relations"). Rather, it is that U.S. economic nationalism will give rise to similar policies in other industrialized democracies that will then victimize not only each other but other countries that previously practice economic nationalism but are now seeking to integrate themselves into the global economy.

Nowhere is that truer than in the countries of the former Soviet empire. In striking contrast to his secretary of state, President Clinton has identified the future of democracy in Russia as "the major crisis this world has faced since I've been President." It is important, therefore, to understand what Russians and other East Europeans view as crucial to their development as industrialized democracies. "The Americans are talking about expanding aid to Russia," Sergei Y. Glazyev, Russia's foreign trade minister, complained in 1993. "What we demand is a free access of Russian goods to the markets of developed countries." Ivan Szabo, Hungary's finance minister, similarly warned that "unless the countries of our region are allowed to integrate further into the world economy, things could get worse." Unfortunately, those warnings have gone unheeded. "The EU lets us export everything—except those things in which we have a competitive advantage," Radek Sikorski, a former Polish deputy defense minister, recently complained in the Wall Street Journal. "Such attitudes have helped return former Communist parties to power; if you were put out of business by EU regulations, wouldn't you be tempted to vote for those who never had illusions about the West?"

70. Laura Katz Olson. COMPETITIVENESS AND AMERICAN SOCIETY. 1993, pp. 251–252. The growing integration and interconnectedness of key industrial sectors around the world have produced entirely new modes of competition. Consequently, in lieu of concentrating on the competitiveness of American industry per se, U.S. leaders should be addressing issues related to the global corporation.

The interdependence of the world economic and financial systems in the 1990s compels the United States to alter its concept of competition in additional ways. Leonard Silk aptly remarks, "the line between domestic and international economic policy has been rubbed out. . . . [T]he unit for policy thinking must become the world economy rather than the national economy, although this flies in the face of traditional national politics and economic pressures." Because of the need for a global perspective—along with the realities that the United States is no longer self-sufficient, nor can it any longer mold the world monetary sys-

tem to its own design—national self-interest lies increasingly in international synergism.

It has become more difficult for individual nations to control their own economies, or to solve their domestic problems unilaterally; nations must act within the context of the wider, international system. Furthermore, policy interventions designed to have an impact on specific, circumscribed national problems often activate interdependencies, with unanticipated and significant outcomes for other countries. These consequences can also reverberate, affecting the nation originating the measures.

Accordingly, policies providing for short-term gains at the expense of other countries can threaten international stability overall as well as produce serious long-term domestic crises. Given recent trends, including public policies aimed at protectionism, a new definition of competition incorporating international synergism will not be achieved easily. [ellipses in original]

71. David Dollar. COMPETITIVENESS, CONVERGENCE, AND INTERNATIONAL SPECIALIZATION. 1993, pp. 17–18. Although we have not examined the institutional structures that support (or retard) innovation and investment, nevertheless we feel that our research is very relevant to some of the pubic debates about which institutions are likely to support successful growth. Most important, we find our results a strong endorsement for maintaining an open door to foreign trade and investment. If it turned out that there were a small number of subindustries that were the key to advanced productivity, some might interpret that as a good argument for protecting or subsidizing the development of these crucial sectors (trade as war).

Our finding that advanced countries have leading sectors–but that they differ among nations—supports the notion that open trade is an important institutional support to growth. Without trade this kind of specialization is impossible. And without the large and competitive international market the incentives to innovate would not be so great. In principle, there might be an argument for protecting one's own leading sectors as they develop new products—if, a crucial caveat, there were no threat of retaliation. In reality, the advanced economies cannot afford to play this game. There is too great a danger that what is purported to be selective and temporary protection will result in the permanent erection of trade barriers all around, with detrimental consequences for growth in all of the rich countries.

72. Peter Morici. "Export Our Way to Prosperity," FOREIGN POLICY. Winter 1995–1996, p. 11. Receding into protectionism would increase the demand for workers adversely affected by structural change and lessen somewhat the wage gap between workers with no post-secondary education and others. However, protection that reduces imports would reduce exports as well by driving up the dollar and encouraging retaliation by our trading partners. It would give false importance to fading industries like apparel, low-end auto parts, and furniture, and stifle the growth of technology-intensive industries like aircraft, computer equipment and software, and financial services. Productivity and wage growth would be constrained, and the employment aspirations of increasingly well-educated, younger workers would be frustrated to placate less-educated, older workers. By reducing the availability of truly attractive jobs, protectionism would exacerbate racial and ethnic divisions and jealousies: It could help make America a poorer, balkanized society.

73. Robert J. Morris. "A Business Perspective on Trade and the Environment," TRADE AND THE ENVIRONMENT. 1994, p. 123. Those who argue that trade fosters environmentally unsound production or environmentally unsustainable exploitation of the world's resources often find themselves aligned with, and their arguments co-opted by, those whose main objective is to restrict competition in order to maintain a dominant market position. Environmentalists need to be wary of alliance (even inadvertent) with protectionists, not because it may not be politically advantageous in advancing their views (it often is), but because protectionism and uncompetitive markets will ultimately deprive society of the resources needed to achieve their environmental protection goals.

74. Laura Katz Olson. COMPETITIVENESS AND AMERICAN SOCIETY. 1993, p. 275. In addition, the rise of the multinational corporation and new technological innovations allowing for rapid, inexpensive worldwide communication and transportation, have altered significantly the meaning of competition in the international economy. The emergence of global firms operating outside national boundaries and interests, and the growing integration among transnational corporations, have rendered traditional nation-centered concepts of competition less relevant to current realities.

The greater integration and interdependence of the world's economic and financial systems suggest that national self-interest lies in international synergism. For example, a nation attempting to improve its own short-term competitive position through protectionist trade policies will generate negative ramifications worldwide, resulting in long-term problems for the country originating the policy.

75. James Miskel. STRATEGIC REVIEW. Summer 1996, p. 76. More is involved here than economic sanctions. Among our most serious national security challenges is the proliferation of the weapons of mass destruction. Apart from unilateral, preemptive military action like Israel's 1982 bombing of an Iraqi nuclear reactor, there is little that Washington can do to contain proliferation except restrict trade to rogue states. The only way to prevent nations like Libya from developing ever-more sophisticated chemical and biological warfare capabilities—apart

from bombing Libya's facilities—is by limiting their access to sensitive technologies. In other words, by getting all exporting nations to agree not to sell these items to Libya. Similarly, the only way short of military preemption to contain nuclear proliferation is to prevent nations like Iraq from acquiring items like supercomputers that could be used to develop nuclear weaponry. This requires active cooperation by all of the nations that manufacture supercomputers or that could buy and re-export supercomputers. Unfortunately, it is exactly this kind of active cooperation that will become more difficult if the United States and its major trading partners allow themselves to get caught up in bitter trade disputes.

76. Pietro Nivola. THE BROOKINGS REVIEW. Winter 1996, p. 20. Whether the target is unfair employers, environmental polluters, tobacco companies, or almost anyone else, a distinctive feature of the American regulatory regime is its querulous and punitive style. The objects of social regulation are never just businesses trying to meet a payroll; instead, the selective employers are "racists" or "sexists," the polluters are "criminals," the cigarette manufacturers are "drug" dealers. Firms are presumed to be amoral calculators, whose malfeasance has to be fiercely deterred and compliance sternly coerced. The distrustful mentality makes for considerable complexity.

A British environmental statute or a Japanese labor contract is more likely to be short and broadly worded, allowing decisionmakers on the spot considerable leeway to resolve problems by mutual consent. By contrast, as Robert A. Kagan, of the University of California at Berkeley, has aptly observed, the U.S. documents tend to resemble arms control treaties by mutually suspicious nations. The presumption is that the parties are surely antagonists to be closely monitored and controlled. Thus, long covenants attempt to micromanage every eventuality. Administrative discretion is narrowed; "invisible handshakes" are dreaded; each detail must be inspected and reinspected through formal proceedings, oversight, and judicial review.

77. Robert Repetto. JOBS, COMPETITIVENESS, AND ENVIRONMENTAL REGULATION. 1995, p. 3. The proposition that differential environmental standards lead to loss of competitiveness and employment is so obvious to many businessmen, labor leaders, and politicians that it is regarded as axiomatic. Its validity needs no demonstration: if U.S. firms are forced to incur costs that their international rivals are not and these costs are not matched by market benefits, then profitability or market share will suffer, so output and employment will be reduced.

78. Carl Frankel. TOMORROW. March/April 1996, pp. 14–15. For the extremists, 'regulatory reform' is essentially a pretext for rolling back environmental regulation. Tom Delay, who as the majority whip in the House of Representatives is one of Newt Gingrich's chief deputies, is not one to control his rhetoric: he has called the Environmental Protection Agency (EPA) "the Gestapo of government" and labeled the Nobel Prize, which was awarded in 1995 to two researchers for their work on CFCs, "the Nobel appeasement prize." Many Republican proposals reflect this unrepentant anti-environmentalism.

Many business and trade associations have been only too eager to jump aboard the anti-environmental gravy train. To take one example, big business had a big hand in writing the House revision of the Clean Water Act, which would substantially reduce protection for wetlands and controls on polluted runoff from farms and city streets. The paper industry successfully inserted a provision calling for the EPA to give greater weight to cost-benefit analysis and risk assessment in determining what technologies should be used to meet water quality goals. The Chemical Manufacturers' Association (CMA) won a clause softening pollution discharge requirements.

79. Carl Frankel. TOMORROW. March/April 1996, p. 14. In 1994 the Republican party gained control of Congress for the first time in forty years, setting in motion a no-holds-barred assault on business-as-usual in the federal government. Among the targets: substantial portions of the system of environmental protection that had been painstakingly built up over the previous 20 years.

So where have the corporations that fervidly profess their environmental commitment been doing this onslaught? Good question—for the most part, the silence has been deafening. Says Greg Wetstone, legislative director for the Natural Resources Defense Council, "We have not seen any of the big *Fortune 500* companies out front in terms of seeking to restrain the Republicans attacks on the environment."

Not only that, but according to Wetstone "many trade associations have supported the proposals." So perhaps it is not entirely accurate to say that the Republican campaign has been met with silence: there has also been applause.

80. Carl Frankel. TOMORROW. March/April 1996, pp. 10–11. U.S. corporate environmental performance is similarly contradictory. In our feature story ("Where Have All the Green Corporations Gone?"), we examine how companies have responded to the conservative assault on the current environmental regulatory system. Given the fervor with which many *Fortune 500* corporations regularly proclaim their environmental commitment, one might expect at least a few of them to have objected to the Republican campaign in Washington but with rare exceptions companies have stood by silently—or clapped like seals. That's disheartening news if you want corporate environmentalism to be more than mere greenwashing.

81. Claas van der Linde. ENVIRONMENTAL POLICIES AND INDUSTRIAL COMPETITIVENESS, 1993, p. 76. Regulation should be efficiently and

rapidly enacted. This will reduce uncertainty, shorten costly waiting and negotiation periods and give the affected industry as much time as possible to react with new, innovative technologies and develop early mover advantages over foreign competitors, or, if competing countries already have similar regulations, assist it in not falling too much behind. Once enacted, standards must be administered in a consistent manner. If there is room for interpretation and negotiation, companies will only waste time and resources and be distracted from their real mission, innovation. Standards must finally be stable over time. Frequent changes cause uncertainty and prevent companies from committing resources for research and development of new innovative technologies.

82. James Medhurst. ENVIRONMENTAL POLICIES AND INDUSTRIAL COMPETITIVENESS. 1993, p. 39. The perceptions held as to the potential scale of these adjustment costs, by countries, is emerging as a real obstacle to the implementation of environmental policies. Thus, while the costs are arguably of a relatively modest proportion, their expected increase combined with uncertainty as to the final level and economic impact causes concern. An emerging issue is whether the avoidance of these costs and the continuation of lower environmental controls is consistent with the longer-term values of society and hence whether the resultant long-term loss of competitive advantage, by failing to reach the required environmental performance, is worth short-term savings in expenditure and adjustment.

83. Eric Marshall Green. ECONOMIC SECURITY AND HIGH TECHNOLOGY COMPETITION IN AN AGE OF TRANSITION: THE CASE OF THE SEMICONDUCTOR INDUSTRY. 1996, p. 165. Government and industry in the United States have generally had what may be considered an adversarial relationship. Fundamental to the American perspective is the notion that government intrusion in the free-market economy is not only inefficient but counterproductive. Industrial policy in the United States conjures up images of fat bureaucrats meddling in the affairs best left to the market. In principle, government intervention in specific industrial sectors is justified in two respects: on the grounds of protecting an important economic asset in the interest of national security, and in response to unfair trade practices. For the most part, government policy has the objective of creating favorable macroeconomic conditions for industry to convert innovation into commercial profit.

84. Candice Stephens. ENVIRONMENTAL POLICIES AND INDUSTRIAL COMPETITIVENESS. 1993, p. 9. The second panel addressed the effects of environmental policies on the competitiveness of industries at the micro-economic or sector and firm level. Participants agreed that environmental control costs can have more significant impacts on particular sectors or firms which are not revealed in the macro-level studies. Negative impacts will be greater for certain pollution-intensive and resource sectors (chemical, mining, oil refining, pulp and paper) where environmental compliance costs are an above-average share of total costs (over 2 per cent) and investment (environmental costs are an estimated 18 to 20 per cent of total investment in some sectors). Environmental regulations may have negative impacts at the margin in the case of sectors or firms which have competitive weaknesses in other areas relating to labour, capital or technology.

85. Carole Gorney. COMPETITIVENESS AND AMERICAN SOCIETY. 1993, p. 161. It is reasonable to accept a higher rather than a more conservative price tag after comparing pre-amendment costs. In 1987, for example $31 billion was spent on air pollution control alone in the United States. The figure is for both operating and capital costs. It compares with $81 billion spent the same year for pollution abatement of all types. A year later, combined operating and capital costs for air pollution rose to more than $42 billion; expenditures for all pollution jumped to nearly $93 billion.

86. Bruce Stram. SHAPING NATIONAL RESPONSES TO CLIMATE CHANGE: A POST RIO GUIDE. Ed. Henry Lee, 1995, p. 232. Command-and-control policy options look more attractive because they appear to provide relatively detailed plans (for example, what new equipment will be installed and by whom). Thus, political compromise is often fostered. Of course, these specific plans will inevitably have significant unintended consequences and are not nearly so certain as their precision leads proponents to believe.

87. Candice Stephens. ENVIRONMENTAL POLICIES AND INDUSTRIAL COMPETITIVENESS. 1993, p. 8.
Data and Methodological Limitations—Limitations in the availability and quality of data on pollution control costs in OECD countries constrain empirical testing of the macro-economic effects of environmental policies. The absolute costs of compliance with environmental regulations may be underestimated by difficulties in measuring the costs of cleaner production technologies and plant modifications (as opposed to investments in end-of-pipe technologies). Most macro-level empirical studies have been based on data from the 1970s and early 1980s and may need to be updated to include more recent cost data as well as to factor in environmental costs and benefits.

88. Roger Nagel. COMPETITIVENESS AND AMERICAN SOCIETY. 1993, p. 179. Since the rise of the modern industrial corporation little more than a century ago, technological innovation has been a determinant of successful manufacturing in open, competitive markets. The impact of a firm's innovation policies on its

competitiveness is, however, far from straightforward. Decisions to innovate or not to innovate are not a matter of "picking winners." The success of a firm is not guaranteed by committing it to what becomes the next technology success story; nor does deciding not to adopt a successful innovation necessarily imply competitive decline. the timing of innovation decisions is crucial, as are the particular form in which an innovation is introduced and the scale on which it is introduced. Technology pioneers often pay a heavy price for the privilege, while firms can sometimes find niche markets that allow them to bypass an innovation and remain profitable.

89. Carole Gorney. COMPETITIVENESS AND AMERICAN SOCIETY. 1993, p. 162. How these direct and indirect costs affect the nation's competitiveness depends on the specific industry, its product and location, and the degree and nature of its competition. Industries in highly competitive markets are unable to add environmental control costs on to the price of the product. Instead, they may decrease output to reduce their average total cost (unit cost). Some industries may actually go out of business, thus reducing productivity. Still others, in less competitive markets, may choose to pass through the costs to the consumer. Whatever the case, "if domestic producers face higher pollution control costs than competitive foreign suppliers, they will eventually lose in international competitiveness.

90. James Tobey. ENVIRONMENTAL POLICIES AND INDUSTRIAL COMPETITIVENESS. 1993, p. 1. Trade effects of environmental policies have been explored in some detail using standard models of international trade (Asako 1979; Baumol and Oates 1988; McGuire 1982; Pethig 1976; Siebert 1974). It has been shown that environmental control costs encourage reduced specialization in the production of polluting outputs in countries with stringent environmental regulations. In contrast, countries that fail to undertake an environmental protection program should increase their comparative advantage in the production of items that damage the environment. These potential trade effects have a strong element of a priori plausibility and proposed environmental regulations are, in fact, often opposed vigorously on the grounds that they will impair the international competitiveness of domestic industries.

91. Pietro Nivola. THE BROOKINGS REVIEW. Winter 1996, p. 18. Amid this global economic integration, peculiarities in regulatory environments can magnify national variations in the fortunes of key industries or even in the margins of national living standards. Just as the strategic industrial policies of trading partners might contribute to their comparative commercial advantages (and disadvantages), so eccentricities in native legal cultures or regulatory styles may no longer be of little consequence in the global marketplace. In short, how we regulate ourselves now makes a bigger difference, for us and for the rest of the international community.

92. Pietro Nivola. THE BROOKINGS REVIEW. Winter 1996, p. 19. Frequently the costs of environmental policies run high in the United States not because U.S. standards are better, but because the chosen mode of intervention is less efficient. The principal U.S. energy conservation program, for example, consists of ungainly command-and-control regulations (the so-called Corporate Average Fuel Economy, or CAFE, mandates) imposed on the automobile industry. Every other advanced country relies, instead, on a simpler technique: higher fuel taxes, which have helped keep consumption per capita far below the U.S. level. Similarly, some U.S. mandates regulating water quality, however harshly enforced, may well be less cost-effective than the Dutch and German experiments with effluent fees.

93. Carole Gorney. COMPETITIVENESS AND AMERICAN SOCIETY. 1993, pp. 163–164. The point was made earlier that if U.S. producers face more stringent regulations than their foreign counterparts, they will lose competitiveness. In fact, the United Kingdom, France, and Germany have very few legally enforceable codes in the areas of health, safety, and the environment. They rely instead on limits or guidelines that can be modified quickly and easily in response to current technical and economic constraints. For example, the British Nuclear Installations Inspectorate uses a system of consultative regulation and inspection, with elements of public participation. The British Alkali and Clean Air Inspectorate works according to the following official description:

> The Chief Inspector, with the help of his deputies, lays down the broad national policies and provided they keep within their broad lines, inspectors in the field have plenty of flexibility to take into account local circumstances and make suitable decisions.

94. Carole Gorney. COMPETITIVENESS AND AMERICAN SOCIETY. 1993, p. 161. Related to this point, and another reason why regulatory critics accuse the U.S. system of being cost inefficient, is the policy of requiring uniform emissions reductions from all sources. Continuing with our previous example, suppose that three plants in an area each emit sulfur dioxide, but that it costs each $100, $600, and $1,000 per ton, respectively, to reduce emissions. Let us further suppose that the goal of applicable regulations is to reduce emissions by thirty tons. Cost-wise, it would make more sense to reach the thirty ton goal by requiring more reductions from sources that spend less to achieve those reductions. When we look later at comparisons with other countries, many of whose industries are major marketplace competitors of U.S. industries, we will see that they have developed much more flexible systems that allow for adaptation to specific industry costs and needs on a localized basis.

95. Claas van der Linde. ENVIRONMENTAL POLICIES AND INDUSTRIAL COMPETITIVENESS, 1993, p. 69. Environmental regulation affects two broad categories of industries. It directly affects those industries which have to comply with the regulation. It also indirectly affects pollution control industries which supply their goods and services to the directly affected industries. In looking at the economic effects of environmental regulation, there are disadvantages in the directly affected industries, which may experience a decline in domestic demand as a result of cost increases and may also experience a weakening of their relative international competitiveness.

96. Richard B. C. Stewart. ENVIRONMENTAL AFFAIRS LAW REVIEW. Review 547, Lexis/Nexis. Over the last twenty years in the United States, the main model for the design of regulation has been the technology-based command-and-control standard. In some cases the central government specifies the precise technologies or designs that industry must install. In other cases, it expresses central standards in performance terms, for example, requiring all polluters in a given category to limit emissions to a specific percentage or amount. In theory, polluters have flexibility to meet such standards through a variety of techniques. In practice, however, regulators typically base such standards on the use of a specific technology, and polluters face strong administrative incentives to use that technology in order to demonstrate.

Many United States laws regulating pollution and hazardous waste, including the Clean Water Act (CWA) and the Clean Air Act (CAA), embody standards that result in high compliance costs, restrict innovation, discourage efficient use of resources, and require detailed central planning of economic activity. While some of these regulatory controls have been initially effective in limiting environmental degradation, they also have proved to be costly and less effective over the long run.

97. THE ECONOMIST. 27 July 1996, p. 13. Suppose that a government wants to reduce pollution. It could do so by buying, say, $1 billion-worth of anti-pollution devices and giving one to every power plant, raising the money through taxes or borrowing. Or it could simply decree that all power plants must be equipped with anti-pollution devices, forcing power companies to fork out the $1 billion of its resources in order to cut pollution.

Nothing wrong with that, you might think: pollution is, after all, a bad thing. But there are differences between these two forms of spending. A government that spends directly must go through the reasonably transparent exercise of taxing or borrowing. A government that merely regulates passes these costs on to customers, employees and owners of power plants, making the costs hard to track and measure. Is there an even better use for that $1 billion? Since this spending is invisible to it, a government need not care; you can call any sort of tune if someone else is paying the piper.

98. Carole Gorney. COMPETITIVENESS AND AMERICAN SOCIETY. 1993, pp. 161–162. Another cost of regulation is management overhead. With the expansion of regulation since 1970 has come confusing overlaps in agency jurisdiction and staggering amounts of written codes and paperwork. Frequently a company must receive permits from myriad federal, state, and local regulators before beginning, or even continuing specified operations. A graphic example is Dow Chemical Company's facility in Pittsburg, California, which must file 563 separate permit applications each year to cover its direct emissions into the air. The company also must obtain 370 additional permits for the source that generates the materials that escape through the emissions point.

In another example, Dow canceled plans to build a $300 million petrochemical complex in California because of regulatory red tape. After two years, and expenses of more than $4 million to secure environmental approval, the project had obtained only four of the required 65 permits from federal, state, local, and regional regulatory agencies.

99. Bernard S. Black and Richard J. Pierce, Jr. COLUMBIA LAW REVIEW 1341. October 1993, p. 1341. Most environmental regulation in the U.S. is "command and control" (C&C) in nature. Regulators specify how much of each pollutant each pollution source can emit. Often, they prescribe specific technology designed to meet those emission limits. A large factory or power plant can have dozens, perhaps hundreds of point sources within it; each point source needs a permit for each regulated pollutant.

Command and control regulation has succeeded in reducing air and water pollution in the U.S. in the last 25 years, despite growth in population and industrial output. But these reductions have cost much more than they needed to. Some sources are strictly regulated; others are regulated loosely or not at all. Under EPA's new source performance standards (NSPS), new coal-fired power plants must install expensive SO_2 control technology, technically called flue gas desulfurization and colloquially called "scrubbing." Most old plants, in contrast, emit SO_2 with no limits, except for tall smokestacks that reduce local concentrations of SO_2 and, perversely, increase acid rain SO_4 downwind. Power plants are often required to install control technology that costs far more than smaller businesses pay per unit of pollution prevented. A power plant cannot, in general, increase emissions of pollutant A at one smokestack by 10 units if it decreases emissions of the same pollutant by 10 units elsewhere. Moreover, NSPS rules require all new sources to install the best available control technology even if they are located in clean air regions where pollution harm is minimal.

100. Robert Hahn and Robert Stavins. SHAPING NATIONAL RESPONSES TO CLIMATE CHANGE: A POST RIO GUIDE. Ed. Henry Lee, 1995, p. 180. In contrast, by requiring firms to meet a specific

standard, command-and-control regulations may result in unduly expensive means of a nation's achieving an environmental target. The reason is simple: the costs of controlling pollutant emissions can vary greatly among and even within firms. Technology appropriate in one situation may not be appropriate in another. Indeed, the cost of controlling a given pollutant may vary by a factor of one hundred or more among sources, depending on the age and location of plants and the available technologies.

101.   Roger Nagel. COMPETITIVENESS AND AMERICAN SOCIETY. 1993, pp. 186–187. The very developments that today are driving the transformation of the mass production system of manufacturing can serve as a springboard for the United States to regain manufacturing competitiveness by the year 2006. The technologies required by the new system are well defined and will become available, at a price, to whoever has the ability to use them. This ability, however, will depend upon innovations in the organization of the management of manufacturing enterprises. American industry must acquire the necessary technologies in a timely manner, but the technologies alone are not enough. The technologies will have to be integrated into organizational frameworks that fully utilize the knowledge, creativity, and, above all, the initiative of the human resources available to industry. Even this is not enough. Industrial enterprises must have access to generic social resources—an appropriately educated workforce, adequate communication and information networks, a supportive political, legal, and economic climate. Together, technologies, their organization within enterprises, and the linkages of these enterprises to their social contexts, constitute a system.

102.   Arnold W. Reitze. 21 ENVIRONMENTAL LAW 1549. Summer 1991. The federal command-and-control approach has had successes but has run out of steam, and has little chance of dealing effectively with the major air pollution problems that threaten our atmosphere on a global basis. We cannot save the environment just by creating more regulations. Most people working in the field cannot find the time to read, let alone understand, the regulations EPA has promulgated, and if they are significant, they will probably be embroiled in litigation for several more years. When we begin to implement the 1990 Amendments we are likely to find the governmental costs are far greater than the resources given to EPA to do the job. Delay in implementation is almost certain to occur. We can not expect to protect our environment to the degree necessary merely by using more stringent controls that stress the limits of our technology and that have high marginal costs as well. More stringent laws will also be politically costly because effective controls will impinge more directly on both the wallets and the freedoms of individual citizens and voters. Pollution controls also threaten the ability of our industries to compete internationally—our industries are already carrying heavier tax burdens than foreign competitors and paying higher costs for capital due to our three trillion dollar national debt. The 1990 Amendments will add to both the cost and complexity of our air pollution and control program.

103.   Carole Gorney. COMPETITIVENESS AND AMERICAN SOCIETY. 1993, p. 163. Nevertheless, other economists argue that composition, rather than level of GNP, is more likely to be affected by regulation. By this is meant that while certain sectors of the economy might decline, there would be spending and price tradeoffs that would keep total GNP fairly constant. Regulatory costs also are believed to be linked to a reduction in available resources need to modernize facilities and support research and development. At the 1990 competitiveness hearings in Washington, the Office of Technology Assessment reported that investment in machinery and equipment in the United States during the past fifteen years has remained static at about 7 to 8% of GNP, while Japan's comparable investments have continued to rise to a present high in excess of 20% of GNP. Furthermore, nondefense expenditures for research and development have been stagnant at about 1.8% of U.S. GNP, while West Germany and Japan not only have outstripped us for the past twenty years, but also have continued to widen the gap significantly each year.

The president of the Council on Competitiveness summarized the situation for hearing members: "Last year, with an economy that is about sixty percent of ours in terms of size, Japan actually invested (gross) more dollars than the United States. That puts the latest technology to work in Japan's factories and the best tools in the hands of the Japanese worker."

104.   Ben Bonifant, Matthew Arnold, and Frederick Long. BUSINESS HORIZONS. July 1995, Lexis/Nexis. Voluntary programs have also been used to stem the buildup of greenhouse gases by encouraging companies to reduce their energy use. The EPA has identified a variety of areas where energy use can be significantly reduced while providing attractive economic returns. However, many cost-effective energy saving projects are not undertaken because firms are unaware of their benefits or because methods of capital allocation do not encourage these types of investments. In one example, the EPA has attempted to overcome these obstacles with the development of the Green Lights program, wherein the EPA provides technical advice, assistance in financial analysis, and recognition for program accomplishments. In return, companies that have become "Green Lights Partners" provide reports on lighting up-grade programs.

The Clinton administration's "Climate Change Action Plan" makes extensive use of voluntary programs. Relying on partnerships between government and the private sector, the plan anticipates returning greenhouse gas emissions to 1990 levels. For example, a reduction of 8.8 million metric tons of (carbon equivalent) releases are

anticipated through the greater use of energy efficient motors. In a separate program called the Climate Challenge, electric utilities pledge to voluntarily lower their greenhouse gas releases. The administration reports that more than 50 initiatives of this type are being undertaken as part of the effort to stem global warming.

105. TOMORROW. January/February 1996, p. 50. In the United States, where insurance is a singularly unpopular business, the industry has kept a low profile on the issue of climate change, despite the gargantuan losses it has suffered from catastrophes since 1992, when Hurricane Andrew caused over $15 billion in losses. Nonetheless, in 1994 the industry organized the Insurance Institute on Property Loss Reduction to promote safer building codes, sponsor research on disaster loss problems and educate the industry and the public. Frank Nutter, president of the Reinsurance Association of America, has forged industry ties with U.S. Vice President Albert Gore to work on climate change initiatives. Progress is slow. Of his efforts to mobilize action in the U.S., Nutter says: "It's been lonely."

106. Daniel C. Esty. GREENING THE GATT: TRADE, ENVIRONMENT AND THE FUTURE. 1994, p. 161. Finally, the suggestion that environmental factors do not affect competitiveness (and should be ignored as a policy variable) is demonstrably untrue if one looks beyond the narrow category of pollution control spending. Specifically, broader environmental policies such as energy pricing unequivocally have competitiveness effects. The European Community's decision to postpone proposed energy taxes based on the carbon content and energy content (BTUs) of fuels until the competitiveness implications for European industry could be addressed— by comparable tax increases in the United States and Japan—offers a very concrete example of this phenomenon (*Financial Times*, 25 February 1992, 7). Similar fears about competitiveness impacts led to the demise of President Clinton's 1993 proposal for a new BTU tax in the United States.

107. Pamela Cohn. EMORY INTERNATIONAL LAW REVIEW. Spring 1995, Lexis/Nexis. The international regulations established by the Montreal Protocol are effective. Toyota has planned for a complete elimination of CFCs and is close to reaching its goal. The Japanese automobile industry set 1994 as the deadline for the installation of alternative refrigerants to CFCs in automobile air conditioners. Japan is ahead of the Protocol's deadline of 1996 for completely eliminating the use of CFCs in air conditioners. The Montreal Protocol exemplifies effective global cooperation in the protection of the environment. Perhaps, a similar type of treaty could help the problems associated with automobile emissions. In order for an agreement to be effective, it probably will have to set strict dates for compliance and regulations. The United States, however, will run into major adversity from automobile companies that fear a loss of competition to Japan.

108. John Shanahan. THE CANDIDATE'S BRIEFING BOOK. 1996, Chapter 4, http://www.townhall.com/heritage/issues96/chpt4.html Energy conservation mandates also cause other unintended consequences. Increasing CAFE standards would raise the average price of cars by as much as $1,500. And with higher priced autos, many people would keep their autos on the road longer, thus minimizing the very decreases in fuel consumption regulators are hoping to achieve. On top of that, older autos kept on the road longer would mean increased pollution, not lower pollution. Saving some oil hardly justifies the death and suffering imposed by ill-considered command-and-control federal regulations such as CAFE.

109. Robert Repetto. JOBS, COMPETITIVENESS, AND ENVIRONMENTAL REGULATION. 1995, p. 31. This regulatory approach is so much more costly that even some industries have come out in favor of having environmental charges instead. For example, General Motors has argued for promoting fuel efficiency with a gasoline tax, which applies to all vehicles, rather than with CAFE standards, which only hits new car sales. However, what's good for General Motors is apparently not good for the politicians.

110. John Shanahan. THE CANDIDATE'S BRIEFING BOOK. 1996, Chapter 4, http://www.townhall.com/heritage/issues96/chpt4.html Conservatives are not against conserving energy when it makes sense to do so, but responsible adults who must weigh the pros and cons of their actions and who must pay for the energy are the best judges of when it makes sense to conserve or expend energy in their daily lives. The fact is that federally mandated energy conservation is costly in terms of money, human suffering, and increased deaths, and sometimes is even counterproductive. For instance, Corporate Average Fuel Economy (CAFE) standards are now supported primarily by the Clinton Administration and environmentalists as a way to preserve fuel. The most serious problem with CAFE standards is that the weight reductions necessary to meet them make cars less safe because lighter cars are less crashworthy. In fact, about 2,500 additional deaths and 25,000 additional serious injuries occur per model year. Although raising the standard would raise the death toll even more, that is exactly what Clinton wants to do.

111. John Hood. POLICY REVIEW. No. 74, Fall 1995, http://www.heritage.org An article of faith among environmental advocates is that recycling will help American business conserve natural resources and demonstrate its environmental responsibility. This assumption doesn't account for the potential costs, including environmental costs, of pursuing recycling regardless of whether it is truly profitable.

For example, curbside recycling programs usually require more collection trucks. That means more fuel consumption and engine emissions. Some recycling programs produce high volumes of wastewater and use large amounts of energy.

112. John Hood. POLICY REVIEW. No. 74, Fall 1995, http://www.heritage.org Filling disposable cardboard boxes takes half as much energy as filling recyclable glass bottles. For a given beverage volume, transporting the empty glass bottles requires 15 times as many trucks as does transporting disposable boxes. Transporting the containers once they are filled also costs less when using disposable boxes. Juice boxes don't break like glass bottles can, they are easily packed or frozen, and they seem to encourage juice consumption among the young, with whom they have proven to be popular.

113. Carole Gorney. COMPETITIVENESS AND AMERICAN SOCIETY. 1993, p. 167. Both the steel and chemical industries are subject to a double whammy when it comes to environmental legislation. Both are heavy users of electricity—electric power production being another industry reeling under environmental and safety regulatory costs. According to Douglas Biden, economist and secretary-treasurer of the Pennsylvania Electric Association, the United States used to have the second lowest electricity costs in the world. "It is now fifth or sixth," he said. "Canada has always had the lowest costs because it has a wealth of hydro (water) resources and it is run by the government and subsidized by it. France has passed us as it is 75% nuclear, and it is also subsidized. Sweden is 40 to 45% nuclear."

114. Carole Gorney. COMPETITIVENESS AND AMERICAN SOCIETY. 1993, pp. 166–167. Less publicly known is the competitive struggle of the U.S. chemical industry. While this industry is the third largest in the world and a mainstay of America's international competitiveness, there has been some worrisome erosion in its position. Michael T. Kelley, Deputy Assistant Secretary in the Department of Commerce, testified at a 1988 Senate hearing on the Competitiveness of the U.S. Chemical Industry, that in the period 1983–88, chemical imports grew an average of 11% while exports lagged at only 6% annual growth. He concluded that heavy environmental and safety regulation, particularly in the pharmaceutical, biotechnology, pesticides, and organic chemical producing industries, should be recognized as adversely affecting the American chemical industry's position in the world marketplace."

115. John Kenly Smith. COMPETITIVENESS AND AMERICAN SOCIETY. 1993, p. 272. At this juncture the future course of events is not clear. The Japanese manufacturers have built their dominating position in electronic products on the base of process technology, which represents innumerable man-years of learning by doing. Although it is always quicker and cheaper to copy the leader rather than to pioneer, it is unclear whether American firms will be able to catch up with the Japanese. But, if the entire technological structure of electronic product manufacture changes dramatically, say, to one in which standardized electronic components are assembled into a wide variety of products, then the cost of entering and competing in this industry might become quite low. This is exactly what happened to the chemical industry in the postwar era.

116. Robert Pfahl. THE GREENING OF INDUSTRIAL ECOSYSTEMS. 1994, p. 208. With more than 2.4 million employees, electronics is the largest industrial sector in the United States. Some firms produce thousands of different products, each with hundreds of components, most of which are procured from other firms in the industry. Approximately 70 percent of all firms in the industry are small companies with fewer than 200 employees. Thus, the infrastructure is extremely complex, with many firms functioning as both suppliers and consumers.

117. Robert Pfahl. THE GREENING OF INDUSTRIAL ECOSYSTEMS. 1994, p. 209. The heart of electronics design is the electrical and mechanical interconnection of individual components and subassemblies by fastening them to a substrate that contains electrically conductive wiring. These designs are manufactured by using mass-production processes that accommodate the assembly of a broad range of components.

Because of the rapid advances occurring continuously in electronic components—advances that reduce size and cost while increasing speed and functionality—periodically the performance of the component packaging or the interconnection substrate is exceeded and new technology (including design, materials, and manufacturing technology) must be introduced (Pfahl, 1992). Thus, unlike many other manufacturing industries in the secondary sector, the American electronics industry periodically goes through fundamental shifts in manufacturing technology.

118. Eric Marshall Green. ECONOMIC SECURITY AND HIGH TECHNOLOGY COMPETITION IN AN AGE OF TRANSITION: THE CASE OF THE SEMICONDUCTOR INDUSTRY. 1996, pp. 173–174. Another major weakness in the American development of technology concerns capital formation. The development of advanced microelectronics is a task requiring enormous commitments of capital, equipment, and expertise. It is true that most enterprises at one time or another experience problems of insufficient financing. Capital investment flows are directed in venues that maximize private return with little concern for any inherent strategic value. The result is that risk technologies tend to be underfunded. While European and Japanese firms operate in an institu-

tional framework that lessens the problems of capital formation, the United States has not developed an institutional apparatus to augment private investment in specific R&D. Moreover, in an era of constrained resources, lack of strategic direction makes problems of capital shortfall more acute.

Capital investment has been described as the lifeblood of the semicondutor industry. The advantages held by the Japanese producers in capital formation are enhanced by vertical integration and diversified revenue streams. This advantage enables Japanese producers to maintain investment in technology and manufacturing in periods of slack semiconductor demand. Escalating development costs in technology generation and production facilities are a particularly heavy burden for American merchant firms, which do not have the same diversified revenue streams or capacity to develop capital reserves. Although the merchants have a record of being more innovative in response to changing market demands, they cannot match Japanese manufacturing advantages and staying power. Whatever dynamic advantage that American merchants have in advanced chip design is jeopardized by capital requirements that are more easily met by larger electronic producers. This may threaten America's leading position in microprocessors, which are entering an era of commodity pricing. Since investment in R&D and new manufacturing systems is integral to the success of American semiconductor producers, the question is how to make the capital available for requisite investments.

119.   Eric Marshall Green. ECONOMIC SECURITY AND HIGH TECHNOLOGY COMPETITION IN AN AGE OF TRANSITION: THE CASE OF THE SEMICONDUCTOR INDUSTRY. 1996, p. 42. Semiconductors are the essential framework of the information society and meet our criteria of a strategic industry. Semiconductors generate technological spillovers that impact upon the competitiveness of downstream products and production processes and are indispensable components in the exercise of power in military systems and commercial technology. The National Advisory Committee on Semiconductors stated in 1989 that "today's $50 billion world chip industry leverages a $750 billion global market in electronics products and 2.6 million jobs in the United States." In 1993 the world semiconductor market had grown to $77 billion, and according to the Semiconductor Industry Association, it was over $120 billion by 1995. Semiconductors are at the beginning of a decisive industrial chain forming the basis of innovation in all electronic applications. Establishing a position in this industry can give a firm or a nation a dominant position in a stream of product and process innovation.

120.   THE ECONOMIST. 27 July 1996, p. 10. In theory, the simple solution is for governments to avoid all forms of indirect spending, and pay for everything directly. They would reimburse companies and individuals for the full costs of implementing their policies, paying for all of these activities with debt and taxes. But this sort of radicalism is impossible, for two reasons.

First, many government policies are much simpler, and cheaper, to implement through regulation: it is easier to require firms to use good safety standards, for example, than to have government agencies implement them in every office and factory. And trying to reimburse each firm for its share of the burden would create a bureaucratic nightmare. Second, even if the government could compensate everyone for complying with regulations, this would create a huge incentive problem. Just think what would happen to wages if the government agreed to pay every employer the difference between the minimum wage and the wage that it would have offered to its workers in a free market.

121.   Mina Mohammadioun. TEXAS BUSINESS REVIEW. June 1995, Lexis/Nexis. Continued economic expansion and population growth will increasingly lead to exhaustion of nonrenewable natural resources, increases in air and water pollution, and an overload of landfills from household and industrial solid waste. As this happens, global demand for environmentally friendly products will likely increase. Consumers valuing environmentally benign products and green manufacturing will be willing to pay a premium for these products and services. Competitive advantage and market share will increasingly belong to companies and nations that develop and use green technologies. Many firms recognize that it is in the interest of long-term competitiveness to invest in improved technology now and eliminate the need for investment in waste management in the future.

122.   Robert Repetto. JOBS, COMPETITIVENESS, AND ENVIRONMENTAL REGULATION. March 1995, p. 7. Regulations limit industry's choice of technologies, product design and mix, plant location, and other important production decisions. Firms must allocate investment and operating funds to reduce environmental impacts, with scant hope of recovering all these expenditures through materials or energy savings or higher product prices. In addition to direct compliance costs, industries face delays and uncertainties in dealing with regulatory requirements.

123.   François Leveque. ENVIRONMENTAL POLICIES AND INDUSTRIAL COMPETITIVENESS. 1993, p. 82. The flexibility of industries in coping in the new environmental context is strongly correlated with this factor. It takes nearly ten years before launching a new active ingredient in the agro-chemical industry and fifteen years of sales to pay back the research and development expenditure. A large part of the installed production capacity in pulp and paper, oil refining, chemical or nuclear power station is not yet amortized. In these cases, very stringent regulation could lead to plant closures or

reduced product lines and thus induce significant capital reductions and worker lay-offs.

124. Henry Lee. SHAPING NATIONAL RESPONSES TO CLIMATE CHANGE: A POST RIO GUIDE. Ed. Henry Lee, 1995, p. 28. Theoretically, while command-and-control regulation can push some technologies into the marketplace and can require the purchase of others, it can also lock in the existing array of technologies by removing the incentive for a company to exceed its emission targets. Furthermore, regulators rarely require the use of a technology that is not already available and proven. As Hahn and Stavins point out, under command-and-control regulation, money that could be invested in technology development too often gets diverted to lobbying and litigation.

125. Eric Marshall Green. ECONOMIC SECURITY AND HIGH TECHNOLOGY COMPETITION IN AN AGE OF TRANSITION: THE CASE OF THE SEMICONDUCTOR INDUSTRY. 1996, p. 41. The results of strategic competition may be heavily influenced, positively or negatively, by government policy. Competition in these industries does not conform to models of perfect competition. These firms typically operate in imperfectly competitive markets with economies of scale, learning by doing, and R&D capabilities defining comparative advantage. It is a man-made comparative advantage. Maintaining that advantage can be costly, however, because the cost of innovation has risen sharply over the past several years owing to the increasing complexity of technology and the efforts required to extend the present technological frontier. It is becoming increasingly difficult for a single firm to marshal sufficient resources and expertise to promote innovation from research to development. If government promotion can augment critical resources, the possibility exists for government to tilt the terms of competition in favor of a domestic industry.

126. John Kenly Smith. COMPETITIVENESS AND AMERICAN SOCIETY. 1993, p. 270. The other key to the success of Silicon Valley was the fact that the military took much of the risk out of innovation, a risk that for many firms threatens survival. In any innovation, the development phase is the critical one because large sums of money are needed to develop prototypes, manufacturing processes, and markets. Generally, the amount of money spent during this phase is much greater than in the earlier research phase. Using modern accounting techniques, such as discounted cash flow analysis, money invested in development is treated as a loan that accumulates interest charges as long as it is not paid back. Therefore, if the initial investment is large and sales are slow to develop, large "interest" payments, when combined with the initial capital input, may keep the product permanently in the red. For a small company the problem is even greater, because it might actually have had to borrow money to develop a new product. Failure in this instance can lead to bankruptcy. The military helped the Silicon Valley companies over the development hum by providing capital and by buying the initial small batches of product at relatively high prices.

127. Eric Marshall Green. ECONOMIC SECURITY AND HIGH TECHNOLOGY COMPETITION IN AN AGE OF TRANSITION: THE CASE OF THE SEMICONDUCTOR INDUSTRY. 1996, p. 39. Since many firms are unable to generate internally the capital required to finance new ventures, capital markets assume an important role in the development of new R&D activities. To a greater or lesser degree, all business investment has some element of risk. However, additional risks are evident in R&D because a maximum return may be generated only if the firm is the first to apply it. Moreover, the evaluation of that risk can be more difficult, because in many cases it is an endeavor that establishes new directions with uncertain precedents. Although borrowers will have a better sense of the inherent risk of the project, conveying that knowledge to lenders can be problematic, and the consequent uncertainty of R&D investment promotes a trend to overestimate the risk of new activities, thereby raising the cost of capital formation. Higher capital costs make investments more expensive and therefore reduce the overall level of investment.

128. David Dollar. COMPETITIVENESS, CONVERGENCE, AND INTERNATIONAL SPECIALIZATION. 1993, p. 18. The fact that different nations have their advanced production concentrated in different product lines also suggests something about what kind of government intervention is desirable. Again, if there were a few key technologies that were the secret to success, then a strong case could be made for ex ante promotion of these activities. In reality, the progress of nations is focused on highly specialized subindustries that are different in each country. It is difficult for anyone, in government or not, to predict in which areas new innovations are going to occur. Furthermore, once a particular country has pioneered some new technology, the best strategy for other nations is not necessarily to try to catch up in that exact product line. Rather, it may be preferable for other countries' firms to take off in related—or even in totally different—directions. What all this means is that it is generally not a good idea for a government to target the development of particular technologies. There is ample evidence that this kind of targeting has not worked well in a wide range of countries.

129. François Leveque. ENVIRONMENTAL POLICIES AND INDUSTRIAL COMPETITIVENESS. 1993, p. 82. The investment cycle factor emphasizes industrial diversity according to the end-of-pipe vs. clean technology issue. The adoption of more environmentally-friendly technology could be a short-term undertaking in the food packaging sector and a very long one in the

minerals sector. Environmental policy should take into account this key factor, especially in proposing stable objectives and policies based on long-term planning horizons to reduce the uncertainties which act as obstacles to major investments.

130.   John Hanson. CANADA-U.S. LAW JOURNAL 299. 1995, Lexis/Nexis. There are other impediments in this market, testing and demonstration impediments. The risk of noncompliance penalties dampens industrial willingness to use unproven technologies. That is almost selfexplanatory. If you have a choice between two technologies to handle a problem in this command and control environment, you are almost always going to prefer the one that is proven versus the one that is not quite yet proven for fear that you would fall on the other side of the line in terms of performance, and then you end up with noncompliance penalties.

131.   Robert Repetto. JOBS, COMPETITIVENESS, AND ENVIRONMENTAL REGULATION. March 1995, pp. 26–27. One-size-fits-all regulations that necessitate uniform technological solutions to pollution problems within an industry drastically raise the costs of compliance. They're usually end-of-pipe solutions that discourage innovative product redesign and process changes that could save money in the long run. They ignore important differences among firms in the scale, remaining lifetime, location, and design of facilities, though such differences often make other pollution-control options more cost-effective. As a result, the same amount of expenditure buys drastically varying reductions in emissions in different industries, among emissions sources within industries, and even within the same facilities. Instead of starting with the least expensive ways of reducing emissions and moving gradually up the cost curve until the emission reduction target is achieved, this regulatory approach sometimes requires a great deal of expenditure for relatively meager results. A recent Resources for the Future report summarized numerous studies of the cost-effectiveness of regulation under the Clean Air Act: most studies show that actual expenditures are several times those that would be required to achieve the same goals under a least-cost approach.

132.   Ben Bonifant, Matthew Arnold, and Frederick Long. BUSINESS HORIZONS. July 1995, Lexis/Nexis. Manufacturers are often assisted by suppliers as they search for other means of production. Because suppliers are anxious to maintain their markets, they dedicate substantial resources to the development of substitute materials that achieve similar outcomes without damaging the environment. Further, the competition among suppliers accelerates the pace at which new technologies are put into place. This can be critically important because in many instances manufacturers are compelled to adopt end-of-pipe controls if alternative means of production have not been demonstrated within the implementation schedules of the regulation.

133.   James Medhurst. ENVIRONMENTAL POLICIES AND INDUSTRIAL COMPETITIVENESS. 1993, p. 47.
Mode of Enforcement: long vs. short timescales. The speed with which policy measures are introduced can be critical in determining the response adopted. Experience in the West suggests that too short a timescale will favour the adoption of end-of-pipe techniques, since alternative clean technologies may not be available immediately. On the other hand, too long a timescale may create the impression that improvement in environmental performance need not be a high priority for firms, and so send the wrong signal to industry. Hence, there is a need to strike the right balance in terms of a timescale which will enable industry to respond in the desired manner.

134.   Bernard S. Black and Richard J. Pierce, Jr. COLUMBIA LAW REVIEW 1341. October 1993, p. 1341. A further problem with C&C regulation is its tendency to freeze compliance at current technology levels. Polluters have no incentive to clean up any more than required by their C&C permit, even if they can do so cheaply. They have only limited incentives to develop new, more effective methods of pollution control, since those methods, once EPA learns about them, can lead to stricter emission limits. Moreover, they have no incentive to develop cheaper methods of pollution control unless those methods permit full compliance with the current standards. There is no opportunity, in short, for the cost/quality tradeoffs that are routine in other areas of business.

135.   Frank Popoff. HARVARD BUSINESS REVIEW. November/December 1995, Lexis/Nexis. Industry has a big responsibility here, too. Companies must invest in innovation today to build trust. If we act now, change will become more palatable to representatives of government and the environmental community. Over the long term, changes will make traditional environmental regulation less necessary and less costly for businesses and taxpayers.

I think successful companies will embrace environmental initiatives as a value and a responsibility, not as a burden. This requires a separate paradigm shift for industry—from compliance to voluntary action.

136.   Frederick Anderson. THE GREENING OF INDUSTRIAL ECOSYSTEMS. 1994, p. 101. As industry becomes more frugal and waste-free, opportunities for trimming waste inevitably rise in cost, and the market incentive to reduce pollution diminishes. "Technological innovation" is not a complete solution to rising costs. While current innovations in industrial production tend to be inherently less polluting, there is no technologi-

cal reason why "modernization" should always produce cleaner production processes. Innovation also pollutes, although perhaps not in the same manner as an outdated technology. New technology may also produce "disbenefits" such as flimsier, less attractive, less reliable, or more dangerous products. Economists and engineers point out that optimizing for environmental protection *alone* means some loss in safety, efficiency, durability, convenience, attractiveness, or price. The proliferation of lightweight, energy-efficient vehicles has meant some loss in safety as well as comfort. Reducing CO2 emissions by increasing the use of nuclear power raises questions relating to long-term natural resources use, safety, and waste generation.

137. Willis Harmon. THE FUTURIST. July/August 1996, p. 14. In an economy-dominated value system there is a pervasive belief that quality of life is to be measured in technological terms, that technological advance equals societal progress.

Some technological advance does indeed add to quality of life. But new technology is often developed and applied without being guided by a strong moral sense and social vision. Technological advance often replaces high-paid, low-tech workers with low-cost automated processes and low-wage foreign labor. It pushes those formerly high-paid production workers into low-paying service jobs and shifts wealth from displaced workers to those who own or control the technology.

138. Deanna Richards et al. THE GREENING OF INDUSTRIAL ECOSYSTEMS. 1994, p. 5. Impediments to implementing environmentally preferable alternatives, however, are not restricted to environmental laws and regulations. For instance, anti-trust laws, especially in the United States, can hinder industry cooperation that would be critical to developing comprehensive product and material recycling systems (Anderson, in this volume); inappropriate standards and specifications by large customers, such as the federal government, can also stifle the diffusion of environmentally preferable technologies (Morehouse, in this volume); and consumer protection laws that classify "remanufactured" products as "used" (second-hand) discourage product-life extension activities such as refurbishment and component reuse.

139. Theodore Moran. INTERNATIONAL ORGANIZATION. Vol. 50, Winter 1996, p. 178. For more than a decade, private consultants, led (according to Krugman) by the likes of the Boston Consulting Group, have advised corporations to shift resources internally from lagging activities to high-growth, high value-added areas to enable them to compete more successfully in international markets. The faulty logic of using industrial policy to strengthen national competitiveness, he suggests, can be traced to this legacy.

But a nation trying to follow the same path will discover, however, that it can devote extra resources to such preferred sectors only by withdrawing them from other activities. Unless there are externalities that generate extra benefits for the nation greater than what private actors would receive anyway to compensate for the penalty imposed on the rest of the economy, governmental intervention will leave the country in worse condition than would have been the case with no intervention. This is particularly true when the targeted sectors are high-wage, high value-added sectors, which require much greater use of other resource inputs (such as capital) per unit output than the activities from which they are drawn, imposing an especially harsh burden on more labor-intensive sectors.

140. John Hood. POLICY REVIEW. No. 74, Fall 1995, (http://www.heritage.org) In Gore's case, he makes no conceptual distinction between profiting from innovation and profiting from regulation. Through much of his book, he argues that governments should make environmental regulations more strict, repeating many of the doomsday scenarios of which Hawken is fond, but then concludes that American firms can make money designing technologies to meet the higher standards. This is no doubt true, but it ignores the firms—actually shareholders, workers, and consumers—who clearly lose when regulatory standards change.

Innovative compliance might reduce the cost of regulation, but it doesn't eliminate it. For example, a state might decide to require that only cars fueled by electric batteries will be sold within its borders. This might well lead to the development of better, cheaper electric cars, as firms struggle to capture as much as they can of the new automobile market, but that's not the only consideration. Under the mandate, consumers would lose access to the products they prefer. Some individuals and businesses would leave the state entirely. The resulting costs and dislocations would far exceed the environmental benefits of lower auto emissions, which are questionable in any event. (The generation of electricity for the cars, for example, would still produce some air or water pollution.)

141. TOMORROW. May/June 1996, p. 28. Such hurdles create confusion, and increase risk for financiers. One of the major problems faced by developers is convincing financiers that the technology can generate sufficient returns for investors. "It's always difficult to convince the finance people when you are dealing with technical issues," says Neil McLeod, a U.K developer of a new method to treat contaminated land, called Envirotreat.

His company received seed funding from the U.K. government and is now using the London-based WHEB Partnership to help find financial and licensing partners. WHEB was set up in 1995 by former partner, Kim Heyworth, and former consultant, Rob Wylie, at consultants KPMG.

"Much of our work is in helping developers commercialize their technology. Our clients range from entrepreneurs and university scientists to multinationals and governments," says Heyworth.

He says potential investors are wary of environmental technologies because of two main risks. First, the potential legal liabilities if the technology fails to work. And second, the lack of a track record in most new environmental technologies.

142. Ernst Ulrich von Weizsacker. EARTH POLITICS. 1994, p. 204. There are ways and means of encouraging this developments, for example through special state programmes for employment and the environment. The basic idea is for the state to create jobs in the environmental sector, raising the necessary funds to do so either through higher taxation on polluters or the Keynesian approach of deficit spending which it is hoped will be paid off later through a higher national income. A third, more conventional, strategy would be further to tighten environmental laws to such an extent that private industry is forced to employ more people in the area of environmental protection.

According to current economic theory, all the three categories—if carried out by a single nation in isolation—lead to economic losses due to the competitive situation in international trade. The problem can be looked at in a different way, however. If there are other strategies available such as ecological tax reform, which appear to achieve the same degree of environmental protection but without jeopardizing international competitivity at the same time, one should not resort to artificial job creation at the expense of the economy. What might be justifiable, at least for a transitional period, would be conversion from military programmes (which are, of course, state-run) towards environmental protection by the state.

In the long term, there is not much prospect in either the public or the private sector for jobs in remedial 'end-of-the-pipe' environmental protection. If the strategies discussed in this book become a reality, the number of jobs even in environmental protection will actually decrease because the technological and cultural transformation envisaged is meant to lead to the prevention of environmental damage through process and attitudinal changes rather than through generating additional activity. Thus a sustainable economy may indeed be an economy with a shrinking formal economy (Hueting et al., 1991), but not one with less well-being.

143. Bruce Stram. SHAPING NATIONAL RESPONSES TO CLIMATE CHANGE: A POST RIO GUIDE. Ed. Henry Lee, 1995, p. 224. In contrast, a general technology standard is likely to establish levels of abatement that are inappropriate for a wide range of circumstances, likely requiring either too little or too much abatement. The polluting industries have a perverse incentive to discourage innovation in abatement technology, since evolution of better technology might lead to regulatory standardization of more expensive technology.

144. Frank Popoff. HARVARD BUSINESS REVIEW. November/December 1995, Lexis/Nexis. In "Green and Competitive: Ending the Stalemate," Michael Porter and Claas van der Linde thoughtfully assert that environmental regulation, at its theoretical best, can catalyze innovation. I think this notion is right on target. The question that must be answered is, How can their vision of our regulatory framework be made a reality?

I, too, can cite cases in which regulations hastened the development of cleaner technologies or accelerated performance improvements. Strangely enough, a fundamental business principle holds true here. Significant improvements in technology leads to greater efficiency and reduced waste, thereby enhancing competitiveness.

The authors' argument is sound and worth discussion among all sectors. Environmental regulation is necessary. I certainly don't advocate a rollback in regulations; rather, constructive change is needed in the regulatory framework. We need to embrace a new paradigm for regulatory supervision, one that fosters a spirit of innovation and responsibility rather than merely an obligation to comply.

If we can change the paradigm, regulations tomorrow will be more inspiring and less prescriptive than those on the books today. They will set overarching performance-based environmental goals, then introduce flexibility and market incentives to stimulate innovation in industry.

145. Robert Pfahl. THE GREENING OF INDUSTRIAL ECOSYSTEMS. 1994, p. 213. As one considers developing policies that will stimulate shifts to environmentally preferable industrial ecosystems, one must be careful to create a structure that accommodates—indeed, takes advantage of–the major paradigm changes in product and process technology that occur in industries such as the electronics industry. Policymakers must recognize, as they evaluate alternative policies that encourage DFE practices, that industries experiencing periodic changes in their product and manufacturing technology provide a unique opportunity for implementing major changes with positive environmental impact. These changes in secondary sector industries such as the electronics, automotive, and aircraft industries can have a leveraged impact on primary-sector industries (those concerned with the extraction of raw materials [Ayres, 1992]) by reducing, changing, or eliminating the demand for raw materials (Allenby, 1992a). However, the risks from an unknown and fluctuating regulatory environment at present discourage corporations from considering major environmentally preferable changes in manufacturing processes and materials.

146. Deanna Richards et al. THE GREENING OF INDUSTRIAL ECOSYSTEMS. 1994, p. 5. If the environmental preferability of products cannot be determined unambiguously—and it is doubtful that the methodologies and sufficient data to identify environmentally preferable options do generally exist—regulators, design

teams, and product and process engineers face a serious conundrum. they are being asked to promote and develop environmentally desirable materials, technologies, processes, and products with little valid guidance on what "environmentally preferable" means in practice, how these choices can be identified, or how their choices may affect other parts of industrial ecosystems, including raw material suppliers or component manufacturers, delivery, maintenance and collections systems, waste handlers, recyclers, and consumers.

The comprehensive life cycle assessment is data intensive and can vary depending on the quality of data available, the biases of the assessor, and the assumptions made. The need, however, is for simple methodologies that are inexpensive to use. As a corollary, no one has yet developed a methodology that easily identifies first-order environmental effects and separates them from the innumerable second- and third-order effects that any design choice entails.

147. Laura Katz Olson. COMPETITIVENESS AND AMERICAN SOCIETY. 1993, p. 279. In the face of their declining economic position, American corporations have attempted since the 1970s to maintain high short-term profits through strategies that have had increasingly deleterious effects on the national interest, as well as on workers, consumers, and local communities. Efforts to cut labor costs by relocating firms nationally and internationally in order to retain or secure special economic protections and concessions, and by investing in mergers and acquisitions rather than in new plants and equipment, have not served community and social needs.

148. Neil Gunningham. LAW AND POLICY. January 1995, p. 65. There are a number of reasons why this is so. Because corporations are judged by markets, investors, and others principally on short-term performance, they have difficulty justifying investment in environmentally benign technologies that may make good economic sense in the long-term, but rarely have an immediate or medium-term payoff. Most areas of reform, including stopping harmful emissions to land, water, and air, replacing harmful chemicals with safer, more expensive ones, and cleaning up contaminated land, are vulnerable to these short-term market pressures.

Individual managers face a similar dilemma. They, too, will be judged essentially on short-term performance, and if they cannot demonstrate tangible economic success in the here and now, there may be no longer-term to look forward to. Significantly, business unit mangers in particular tend to exhibit scepticism about recovering environmental costs in the market. This is a serious obstacle to improved corporate environmental performance, because, if middle managers who must implement company-wide changes perceive "environment" as a threat, then senior managers probably cannot achieve a top-down vision.

149. Roger Nagel. COMPETITIVENESS AND AMERICAN SOCIETY. 1993, pp. 179–180. The impact of innovation commitments by the management of a firm is affected by the coordination between management practices and distinctive organizational requirements of new technologies. The optimal utilization of a technology may require significant changes in a firm's managerial values, in its internal "culture," even in its organizational structure. Conversely, attempting to assimilate a new technology into existing managerial practices may well vitiate, and will certainly limit, the competitive impact of an innovation, as we will illustrate below and in our "vision" of an emerging new era in manufacturing.

150. Robert Repetto. JOBS, COMPETITIVENESS, AND ENVIRONMENTAL REGULATION. March 1995, p. 2. Is all this necessary? Must we pay a heavy price to international trade and investment for environmental protection, or can we have our cake and eat it too? The counter-argument, articulated first by business school professor Michael Porter, asserts that stringent environmental regulations may lead firms to develop new, less-polluting and more efficient products and manufacturing processes. Such innovations give firms that have responded creatively to regulation a competitive advantage over sluggardly rivals as environmental standards tighten worldwide.

> "Ultimately, nations succeed in particular industries because their home environment is the most forward-looking, dynamic, and challenging . . . Strict government regulation can promote competitive advantage by stimulating and upgrading domestic demand. Stringent standards for product performance, product safety, and environmental impact pressure companies to improve quality, upgrade technology, and provide features that respond to consumer and social demands. Easing standards, however tempting, is counterproductive."

A somewhat different counterargument hinges on the rapid growth in markets for goods and services that "solve" environmental problems. According to recent surveys, these "green" industries, which sell pollution monitoring and abatement equipment, engineering and construction services, and a variety of products with environmentally superior characteristics, have already reached almost 200 billion dollars in sales annually in the industrialized countries alone, and are expected to expand even more rapidly in the newly industrialized countries where environmental conditions demand increased attention. Countries with more stringent environmental standards in their home markets will allegedly develop a competitive advantage in these "green" industries, offsetting whatever disadvantage those standards impose on the "dirty" industries. Both counterarguments suggest that our relatively strict environmental standards are likely to confer benefits on American industry in the long-run. [ellipses in original]

151. Candice Stephens. ENVIRONMENTAL POLICIES AND INDUSTRIAL COMPETITIVENESS. 1993, p. 9. Environmental regulations may also have positive competitiveness implications for specific sectors. In general, pollution and resource degradation measures can spur firms to develop more resource-efficient methods of production and reduce costs. Environmental regulations can be good for competitiveness in yielding: 1) innovation advantages, 2) efficiency advantages, 3) front-runner advantages, and 4) spin-off activity advantages. In Japan, for example, pressure to increase energy efficiency and lower pollution levels has reduced energy and raw material inputs to Japanese-produced goods and yielded competitive cost advantages. Large multinational firms are best placed to realise these benefits and also have strong incentives to improve their environmental management capabilities, particularly to avoid accidents, interruptions to production and liabilities, and to promote a green corporate image.

152. Claas van der Linde. ENVIRONMENTAL POLICIES AND INDUSTRIAL COMPETITIVENESS, 1993, p. 72. Properly designed environmental regulation may trigger a number of different advantages. It may induce firms to develop products with lower production costs, improved attributes, lower operating costs, or, in a more general form, to develop products with an early mover advantage over competing foreign products.

Standards can pressure firms into developing new, more *resource efficient ways to produce their goods* and thus lower production costs. A standard, for instance, which aims at reducing usage of a production input that is harmful to the environment, can pressure firms into finding new ways which use less of this input, thereby not only reducing pollution, but also the cost of inputs. Regulation which aims at increasing energy efficiency can also pressure firms into designing more resource efficient production methods and again bring about benefits which, more or less, offset the costs of compliance. In Japan, for instance, pressure to increase energy efficiency has considerably reduced energy and raw material usage of Japanese-produced goods and given them a cost advantage over competing foreign goods.

153. Claas van der Linde. ENVIRONMENTAL POLICIES AND INDUSTRIAL COMPETITIVENESS, 1993, p. 72. Standards that originally aim at reducing pollution in the production process or making it safer may also result in better products, for they will often force the firm to better control and optimize its production process which can then lead to *products with higher or more consistent quality*. A paper plant, for instance, which is required to very carefully control its emissions and reuse any by-products to the largest extent possible will have to develop sensing and optimization capabilities which may also allow it to make better paper.

154. Ben Bonifant, Matthew Arnold, and Frederick Long. BUSINESS HORIZONS. July 1995, Lexis/Nexis. Second, and perhaps more important, new regulations are focusing heavily on substances with essential functions in production, creating opportunities for manufacturers to reduce materials use and for suppliers to develop effective substitutes. In a recent study of industries affected by regulations (conducted under the guidance and within the conceptual framework of Professor Michael E. Porter of the Harvard Business School, with support from the U.S. Environmental Protection Agency), we found that this previously unrecognized factor may be the most important reason environmental concerns are becoming a more important characteristic of competition.

155. Claas van der Linde. ENVIRONMENTAL POLICIES AND INDUSTRIAL COMPETITIVENESS, 1993, p. 72. Not only can a standard trigger innovation to reduce the usage of a primary production input, it may also lead to the reuse of an intermediate or by-product, or a method of more fully utilizing or recycling a by-product. Standards can also induce firms into developing better, safer, or more *environmentally friendly products* for which customers are willing to pay more money. Thus, it may cost a great deal to develop a product that complies to a new noise standard, but customers may sufficiently value the low-noise product that they are willing to pay extra for it. Or, in the process of meeting the standard, the company may end up with a better designed product which can also fetch a better price in the market.

156. Ben Bonifant, Matthew Arnold, and Frederick Long. BUSINESS HORIZONS. July 1995, Lexis/Nexis. The markets developing for less environmentally damaging raw materials involve substitution. New formulations are being offered that achieve the outcomes of previously used materials. In many cases these new formulations are initially either more expensive or of somewhat lower quality than their predecessors. However, the imposition of environmental requirements changes the economics for the users. The costs of using traditional inputs are increased substantially by the addition of capital expenditures for control technologies. Further, the initial costs associated with implementing new methods often decrease over time as suppliers gather scale and learning benefits. Such factors force manufacturers to take a long-term view when considering their environmental strategies.

157. Claas van der Linde. ENVIRONMENTAL POLICIES AND INDUSTRIAL COMPETITIVENESS, 1993, p. 73. Assuming that concern with the environment and energy conservation spreads all over the world, environmental standards may also pressure firms into creating innovation offsets earlier or more effectively than their foreign competitors who do not yet have to comply with this regulation, but will have to do so in the future. A properly designed standard may thus lead to an

early mover advantage or a relative competitive advantage and cause a positive trade effect. In Germany, for instance, recycling standards which have been enacted earlier than in most other countries, have given German firms early mover advantages in developing products and packages which can be recycled. Likewise, zero-pollution standards in California may provide early mover advantages to Californian makers of electro-mobiles.

These early mover advantages are a way of turning the previously described static competitive disadvantages relative to foreign competitors who are not yet subject to similar regulation into dynamic advantages of international competition. In a world of international competition, it is thus possible to gain competitive advantages and enhance the innovation offset if environmental regulation triggers the innovation offset more effectively or earlier than in foreign nations.

Early mover advantages due to environmental regulation are most likely to be developed with respect to products and product attributes, because it is safe to assume that a large part of international demand is moving in the direction of valuing safe, energy-efficient, or low-pollution products. However, environmental standards can also trigger early mover advantages with respect to processes. Here it won't be as much the low-pollution characteristics of the production process, but improvements in the cost or quality of the process which can result in advantages relative to foreign competitors.

158. Candice Stephens. ENVIRONMENTAL POLICIES AND INDUSTRIAL COMPETITIVENESS. 1993, p. 8. Some analysts maintain that industrial competitiveness can be enhanced by the investment and production changes which are stimulated by environmental policies. Environmental legislation can be a driver spurring technological changes which lead to efficiency and competitive advantages. Environmental policies can provide an incentive for the use of cleaner and fewer inputs, cleaner and more efficient technologies, and waste minimisation and recycling. The result can be the production of clean products with environment-related marketing advantages, the development of supporting industries producing environmental goods and services, and front-runner advantages for industries who invest early in environmental technologies.

159. Ian Hodge. ENVIRONMENTAL ECONOMICS: INDIVIDUAL INCENTIVES AND PUBLIC CHOICES. 1995, p. 26. It is sometimes argued that environmental regulations *stimulate* economic growth; an argument often illustrated by reference to the successes of the German or Scandinavian economies. Within this it is sometimes observed that manufacturers with high environmental standards (selling products or using processes which have less environmental impact than their competitors) are apparently more profitable than their less environmentally aware competitors. There are a number of reasons why this may be the case.

As environmental problems are identified and become more critical, public pressure develops for higher environmental standards and new regulations are introduced by government. This creates market opportunities for pollution control equipment, for more environmentally sensitive products and for the equipment with which these products can be manufactured. Countries or firms which are early to develop the new technology for meeting this environmental concern will not only be ahead in marketing their product, they may also sell the technology to those firms behind in the development of new technology. This is sometimes referred to as 'first mover advantage.' It might also be anticipated that, in such circumstances, consumers may well be responsive to the promotion of 'green' products, creating new marketing opportunities.

160. Claas van der Linde. ENVIRONMENTAL POLICIES AND INDUSTRIAL COMPETITIVENESS, 1993, p. 69. Much of the recent debate about the economic implications of environmental regulation has assumed a dichotomy between the environment and industry competitiveness. Regulation to protect the environment, while socially beneficial, was assumed to be costly and could only be established at the expense of a healthy economy. Discussions, then, have centered around industry's responsibility towards the environment, the costs that regulations have inflicted on the economy or particular industries, on how to share in a fair way among companies the burden of maintaining a clean environment, and on ways to internalize the costs and benefits of environmental regulation.

However, it can be argued that there is no necessary dichotomy between having a healthy and competitive economy and a clean environment and strict laws to protect it. It is even likely that private industry and individual firms can actually benefit from strict, but well-designed regulations designed to create or maintain a clean environment. Properly designed environmental standards may be able to trigger innovative activities that at least partially offset the costs of complying with them. The inherent nature of pollution as some type of waste and the role of change and innovation in creating competitive advantage give rise to this expectation.

161. James Medhurst. ENVIRONMENTAL POLICIES AND INDUSTRIAL COMPETITIVENESS. 1993, p. 42. In summary, the absence of a "green backlash" scenario is likely to mean that there is a direct relationship between environmental expenditure and competitive advantage. Moreover, to the extent that expenditure is used to create factor endowments for sustainable development such as the development of environmental technologies, the advantage is more likely to be created, and in a shorter timescale. To the extent that improved levels of environmental performance are required in future, those countries with limited environmental expenditure are likely to be in the process of losing advantage.

162. John Atcheson. THE ENVIRONMENTAL FORUM. August 1996, p. 21. More recently, Porter and Class van der Linde published research on the effects of environmental regulation. They support Romm's thesis, concluding that the best corporations have responded to stringent regulations by innovating, and ultimately enhancing, their competitiveness. In "Green and Competitive," published in the *Harvard Business Review*, Porter and van der Linde suggest that properly constructed regulations can actually stimulate this process and be good for industry.

163. Scott Barrett. ENVIRONMENTAL POLICIES AND INDUSTRIAL COMPETITIVENESS. 1993, p. 164. One final point is that it is conceivable that governments may wish to impose "strong" standards in order to encourage domestic industry to innovate and develop less expensive technologies for reducing pollution. Once the new technologies have been developed, R&D costs are sunk, and the domestic manufacturers can sell the technologies abroad for a lower price than would be required to induce foreign manufacturers who hadn't already incurred the costs of R&D to enter the market.

164. James Medhurst. ENVIRONMENTAL POLICIES AND INDUSTRIAL COMPETITIVENESS. 1993, p. 43. Demand conditions reflecting environmental legislation and preferences for environmentally sustainable products will stimulate firms to innovate ad compete on the basis of environmental performance. The availability of factor conditions, especially technical resources, will facilitate the opportunity for rivals to pursue competition on the basis of cleaner products/processes. The existence of a strong environmental protection industry feeding off factor and demand conditions also provide the means by which domestic rivalry on the basis of environmental performance can be pursued. The uncertainty is whether the pursuit of environmental performance is viewed by firms as conferring any long-lasting commercial benefits. Thus demand conditions remain the most important factor in deciding whether competitive advantage results from environmental expenditure. The choice of scenario is therefore fundamental to the views taken towards the relationship between environmental costs and industry competitiveness.

165. James Medhurst. ENVIRONMENTAL POLICIES AND INDUSTRIAL COMPETITIVENESS. 1993, p. 42. Where environmental performance is seen as providing a competitive edge, domestic rivalry will stimulate demand for products which are viewed as environmentally sustainable. Increases in market share for those products will stimulate other firms to integrate environmental performance into marketing and production activities. Vigorous domestic rivalry can raise foreign demand by raising the profile of the firms and reducing the perceived risk of foreign buyers of sourcing internationally. In the case of this determinant, environmental expenditure associated with investments in new product/processes leads to competitive advantage if consumer preference/values are sufficiently developed in favour of sustainability.

166. Robert Repetto. JOBS, COMPETITIVENESS, AND ENVIRONMENTAL REGULATION. March 1995, pp. 22–23. The real issue is not the environment vs. jobs. The issue is what we want our economy to produce. If we want it to produce a clean environment along with other goods and services, the industries that contribute to a clean environment will have higher output and employment; those that do damage to the environment will have less. While jobs in particular industries may rise or fall, total employment will not be systematically affected.

What we want the economy to produce is continually changing, and industries expand and contract as a result. As sales of personal computers have risen, sales of portable typewriters have declined. Jobs for typewriter repairmen have disappeared while opportunities for computer programmers have multiplied. No doubt this shift has created hardships for some households, but no politician or lobbyist has said, "The U.S. economy can't afford to have personal computers because it will destroy jobs in the typewriter industry." Yet, they routinely claim that we can't afford clean air because it will destroy jobs in coalmining or some other industry.

Those who complain about the effect of environmental controls on employment are usually thinking about *particular* jobs, in their own firms, industries or communities—and well they should. Losing a job hurts the individuals affected and their families and, if clustered in particular localities, those communities as well. The role of public policy, however, is not to guarantee particular jobs but to ease the transition from declining to expanding industries—through unemployment compensation and retraining programs, by giving people opportunities to acquire skills and resources they need for greater occupational mobility, by macroeconomic policies that maintain high aggregate employment, and through measures that moderate abrupt economic shocks.

167. Robert Repetto. JOBS, COMPETITIVENESS, AND ENVIRONMENTAL REGULATION. March 1995, pp. 2. The usual riposte is that environmental protection actually creates more jobs than are lost: limits put on the extraction of natural resources may threaten jobs in extractive industries, but will save or create jobs in recreation industries and in footloose high-tech industries attracted to a high-quality environment. Environmental regulations that require pollution abatement or raise energy prices create jobs in industries supplying pollution-control or energy-conservation equipment and services. Since these industries are more labor-intensive than the heavily polluting industries (e.g., energy supply, basic metals, and

chemicals) it is argued that greater expenditures on environmental protection will create jobs on balance, even if it's at the expense of employment in the polluting sectors.

168. Claas van der Linde. ENVIRONMENTAL POLICIES AND INDUSTRIAL COMPETITIVENESS, 1993, p. 73. Whether there will be an offset as well as the extent of it is not only a function of the regulation. It partly depends on the affected industry's current state of competitiveness—its diamond. An industry which is already competitive and has a favorable home diamond is much more likely to take up a new standard as a challenge and respond to it with beneficial innovation than an uncompetitive industry which may have lost its ability to change and innovate. The more favorable the home diamond, then, the higher the expected innovation offset. It is no coincidence that competitive Japanese and German automobile makers innovated and developed lighter and more fuel-efficient cars in response to new fuel consumption standards while the less competitive American car industry fought similar standards and hoped they would go away.

169. Robert Thornton. COMPETITIVENESS AND AMERICAN SOCIETY. 1993, pp. 93–94. The general presumption among economists is that companies that possess the necessary technology, management skills marketing skills, and so forth to enable them to earn at least a normal rate of return on their invested capital will thrive and prosper, while those that cannot earn a normal rate of return will go out of business. The failure of companies that are unable to sell their products for a normal profit has always been considered a good thing. The resources (workers, machinery, and so forth) that such companies were employing became available to other companies more capable of selling their products at a profit. This is not to say that bankruptcies are an unalloyed blessing. Those persons formerly employed at a bankrupt firm must find jobs elsewhere; investors in the company have lost some part of their wealth; firms that had been supplying product to the bankrupt company see their sales decline. From the point of view of the economy as a whole, however, bankruptcies are the market's way of ensuring that scarce resources are employed in the best way. Donald Dull cannot make a profit running a diner at 4th and Main, but Sally Sharp realizes that the business that corner really needs is a pharmacy. She buys Dull's building at a foreclosure sale, and her venture becomes successful. Economists believe that the most important factor in explaining the economic prosperity of the United States is the willingness of the American government to refrain from impeding the free flow of resources to their most productive uses. By and large the American government has been willing to step aside and allow those incapable of earning a normal profit to go into bankruptcy. The economy is better off for these inefficient firms having disappeared.

170. Ben Bonifant, Matthew Arnold, and Frederick Long. BUSINESS HORIZONS. July 1995, Lexis/Nexis. The two factors discussed above are increasing the importance of environmental issues in strategic competition. Figure 3 displays the changing characteristics of competition in achieving environmental goals. For regulated firms, increased freedom in compliance methods allows those firms that are discovering low-cost means of compliance to gain advantage over their competitors. Similarly, suppliers of environmental technologies are in an increasingly dynamic competitive environment. Each must consider how its technology compares on cost and performance measures against alternatives in raw materials and process technologies that yield comparable environmental performance.

171. Ben Bonifant, Matthew Arnold, and Frederick Long. BUSINESS HORIZONS. July 1995, Lexis/Nexis. Advocates for encouraging innovation are building strong voices in policy, academic, and business forums. The ensuing competition both within and among industries is bringing down the cost of responding to emerging environmental issues. Increased flexibility in regulations is only partially responsible for this as industry attitudes about the role of business in environmental matters change and the characteristics of the materials being addressed lend themselves to a wider range of alternatives. These positive outcomes should encourage regulatory designers to further substitute technology-based standards with alternative approaches. It is encouraging that many of the short-term challenges for the EPA offer similar opportunities for flexibility and competitive response.

Understandably, the business community has been among the loudest advocates for greater flexibility. The opportunities now being afforded, however, put new demands on both the regulated firms and their suppliers. The most skillful competitors are finding that their efforts to lower environmental costs through innovative approaches are being rewarded. Similarly, supplier firms identifying emerging opportunities are taking market share away from competitors that continue to supply products that regulations are making increasingly expensive to use. Less innovative companies are now losing and will continue to lose pace.

172. Ben Bonifant, Matthew Arnold, and Frederick Long. BUSINESS HORIZONS. July 1995, Lexis/Nexis. Figure 1 maps out some recent environmental initiative factors that have led to the flexibility allowing some firms to gain advantage over others. The remainder of this article discusses how this flexibility has led to profit opportunities for suppliers and reduced compliance costs for regulated manufacturers. Our conclusion is that creative approaches to environmental regulation are only partially responsible for the environmental issues in many firms' planning. Competition among alternative raw materials and processes has emerged even under traditional

methods of regulation as the regulatory targets have switched to materials used in manufacturing and away from those generated as byproducts.

173. Ben Bonifant, Matthew Arnold, and Frederick Long. BUSINESS HORIZONS. July 1995, Lexis/Nexis. The greater flexibility resulting from the nature of the substances being regulated and the design of those regulations has dramatically increased the level of competition among suppliers. Growing competition among entirely different methods of compliance contrasts sharply with the strict adherence to a prescribed method of control resulting from earlier regulations. For example, when the 1977 Clean Air Act Amendments effectively required all coal-fired power plants to install scrubbers, the ensuing supplier competition was limited to producers of the prescribed technology. However, when more flexible requirements were established in the 1990 Clean Air Act Amendments, scrubber producers began to compete with the costs of transporting low-sulfur coal. Not only were the utilities able to choose the lowest-cost means of compliance for their facilities, but the increased pressure on suppliers drove down the cost of competing technologies.

174. Bob Ferrone. TOTAL QUALITY ENVIRONMENTAL MANAGEMENT. Summer 1996, p. 41. The complex web of worldwide markets and technological interactions taking place today have been driving the auto industry to take a system-wide perspective on vehicle material selection, production and use, material recovery and reuse, and waste disposal. As the global market continues to become increasingly important to the auto manufacturers, new approaches like the Resource Productivity model and life-cycle management are being employed. Faced with such issues as product-take-back requirements in Europe, the auto industry has been developing new design tools and assembly techniques for ease of disassembly.

The auto industries have responded to regulatory and market pressure by developing more efficient, lighter-weight vehicles that are cost-effective and competitive. The designers employed a number of techniques to integrate environmental knowledge and date into vehicle design. They have also become extremely efficient in material use and recovery (Exhibit 2).

175. Pamela Cohn. EMORY INTERNATIONAL LAW REVIEW. Spring 1995, Lexis/Nexis. In recent years, preservation of the environment has become a prevalent issue for countries throughout the world. Countries are particularly concerned with acid rain, global warming, and the depletion of the ozone layer. Motor vehicle emissions comprise a major source of environmental pollution, Therefore, improving emissions, developing clean, alternative fuels, and creating environmentally safe automobiles are essential in order to prevent and control pollution. Globally, politicians and environmentalists are working to regulate automobile pollution. Automakers must comply with these laws and policies through development of automobiles that are compatible with these regulations. The regulations on motor vehicles emissions also generate competition among automakers to develop technology in order to meet standards as quickly as possible, "the chief player in this drama is mankind. We must ask how humanity will be using fossil fuel in the decades ahead. We must ask whether alternative energy resources will be either proven economicly sic competitive to fossil fuels or mandated by governments." The automobile manufacturers must immediately anticipate the future in order to develop the technology necessary to compete in the automobile industry.

176. Deanna Richards et al. THE GREENING OF INDUSTRIAL ECOSYSTEMS. 1994, p. 16. There is also evidence of some companies experimenting with providing functionality of products in their marketing efforts. For example, a new business group at Dow Chemical, Advanced Cleaning Systems, intends to maintain and increase the sales and profitability of the company's chlorinated solvents business, which is threatened by increased regulation of ozone-depleting chemicals and federal and state programs to control toxic air emissions (Dillon, in this volume). The group is developing not only alternative chemicals and processes but a range of customer services as well—improved process controls and the recycling of spent chemical solvents. It offers delivery of new chemicals to its customers in conjunction with "take-back" of the used chemical in reusable containers. (Dillon, in this volume). The spent chemical can then be cleaned or reprocessed and be either returned to service or disposed of appropriately. This practice is an example of asserting control over the product life cycle while providing the customer with the functions of the product.

177. Ben Bonifant, Matthew Arnold, and Frederick Long. BUSINESS HORIZONS. July 1995, Lexis/Nexis. Meanwhile, Union Carbide has developed an entirely different approach: a system that uses carbon dioxide in a supercritical state as a solvent and is even said to enhance the quality of the finish in many operations. Clearly, the royalty Union Carbide will be able to charge by licensing the use of this technology will be determined by the competitive costs of the capture-and-control devices as well as the quality of the water-borne systems. The pressure of these competing technologies is compelling incinerator firms to search for means of bringing down costs, driving coating firms to rapidly explore the means of improving finish quality, and forcing Union Carbide to competitively set its licensing fees. Overall, this lowers the costs of environmental compliance for wood furniture manufacturers and reduces the trade-offs that might be required to achieve compliance if fewer options were available.

178. Heather Timmons. CHEMICAL MARKETING REPORTER. 14 November 1994, Lexis/Nexis. Mr. Cable also sees fuel costs affecting oxygen use. "Part of this growth is environmentally driven, but there are fuel efficiency issues. As the price of fuel goes up, we expect demand for oxygen in glass manufacture to increase," he says.

Additionally, environmental pressures are pushing oxygen demand in the steel industry, while an overall increase in US steel production has spurred total consumption of bulk gases. Currently, according to Mr. Campbell, mills are running at 90 percent of capacity. "We're seeing many steel mills operate at redline," adds Mr. Rowland.

Praxair is reporting substantial double-digit growth for argon, thanks to the steel industry. And BOC is working on developing low purity oxygen at better cost.

Most producers are looking toward regulatory issues as one of the major forces behind long-term growth for the entire industrial gas industry. "Environmental issues have been fueling opportunity in glass, petrochemicals and iron and steel manufacturing, where oxygen is used to reduce emissions and/or enhance combustion," says Mr. Fisher. As a result, he adds, BOC has been concentrating on improving both its cryogenic and noncryogenic technology to become more efficient in the on-site production of oxygen.

The desire for greener systems that emit less and recover more is also fueling growth for gases in new applications. Air Products is using oxygen to recover sulfur in SO_2 streams, and nitrogen to strip volatile organic compounds (VOCs) from waste streams.

179. TOMORROW. May/June 1996, p. 31. The report says that investors are now recognizing that environmental aspects can strongly influence both the present and future value of a company's assets.

The key decision factor for investing is the expected growth of earnings and asset values. "As a consequence, all the factors influencing the shareholder value have to be assessed, including environmental considerations."

180. TOMORROW. May/June 1996, p. 30. Environmental Reporting (ER) is increasing as more and more companies realize the value of communicating their environmental performance—especially achievements—publicly. Not only does the information collected for reporting purposes encourage continuous improvement, but the reporting process demonstrates environmental commitment to key audiences. EA provides much of the information needed for effective environmental communication. EA is also strongly linked to the financial markets—in particular to the way banks, insurers and investors are now examining and measuring companies. These markets will increasingly expect businesses to provide information on how their products and processes are impacting the environment, and what steps they are taking to reduce this.

181. Robert Repetto. JOBS, COMPETITIVENESS, AND ENVIRONMENTAL REGULATION. March 1995, pp. 11–12. Economists view with enormous skepticism the hypothesis that firms typically overlook opportunities to reduce costs or improve product quality. One of the most important insights in economics is that market competition continually pressures firms to maximize profits by reducing their operating costs and improving their products. This explains why firms in a market economy are most efficient in providing goods and services than organizations not subject to market competition—the U.S. Congress, for example. However, in their formal analyses, economists take this insight a step further and stipulate that most firms throughout the economy have already optimized their operations, an assumption that absolutely dumfounds anybody who has actually worked inside a corporation for more than a week. This extraordinary assumption is analytically convenient: economists can say much more about some observed behavior if it reflects the maximum attainable value of some objective—such as profitability—than if it is just part of a general muddling along. However, if it were true that companies typically operate at maximum efficiency, it would be hard to understand exactly what the hordes of management consultants swarming around them are being paid to do. To take a specific example, it would be hard to understand how the Ford Motor Company, after watching their Japanese rivals at work, could achieve radical cost savings in producing new models—after almost a century in the business—by *starting* to have their designers talk with their manufacturing engineers and marketing experts while the designs are being worked out.

182. John Atcheson. THE ENVIRONMENTAL FORUM. August 1996, p. 21. The market, as Joan Robinson and others have pointed out, is filled with ways to fall far short of optimal efficiency. Access to information, competing demands that limit a manager's focus, rate of technological change, demand for short term profits, foreign and domestic fiscal and regulatory policies can all conspire to make the market "fail." an increasing number of economists, led by Brian Arthur, point to much more fundamental market imperfections such as technological "lock-in" and "increasing returns" that conspire to make the market select suboptimal technologies even when exogenous factors such as unequal access to information are not present. A growing number suggest that the market does not optimize—it "suffices." Indeed, the empirical evidence from research by Romm and Porter, and from facilities operating in states with facility planning laws, shows clearly that huge untapped efficiencies (the proverbial $10 bills that many economists say cannot be there) are there to be had for the taking–but only when companies are encouraged by good public policies and private practices to look for them.

Given the right policies and practices, these environmental challenges can translate into economic opportuni-

ties in the form of more efficient and more competitive domestic industries, and huge emerging markets for cleaner methods of production, and more efficient infrastructure.

183.   Claas van der Linde. ENVIRONMENTAL POLICIES AND INDUSTRIAL COMPETITIVENESS, 1993, p. 74. If companies can create innovation offsets that directly result in competitive advantages, we have to ask why government standards are necessary in the first place and why companies won't innovate on their own when environmental issues are concerned. Among the answers are: 1) incomplete offsets, 2) information deficits and 3) the need for pressure.

Environmental regulation can not always lead to benefits that exceeds all its costs. Only in some subsets of all cases will innovation offsets and indirect benefits actually offset all costs of compliance. Regulation is needed, because without regulation a nation would neither get the social environmental benefits, nor would it get any economic offsets. Conceptually that would be a worse social equilibrium than a case where regulation results at least in some incomplete offsets.

Innovation is difficult to achieve if information is hard or too costly to obtain. If government standards or regulation can facilitate the procurement of information they can be an advantage. Government-backed standards can also be signals to industry that certain product aspects will take on greater significance in the future. Change and innovation, finally, are always unsettling and likely to be avoided by firms, unless there is some pressure to change, be it from competitors, customers or some other source. Environmental standards can create the needed pressure to prod firms into initiating necessary changes.

184.   Robert Repetto. JOBS, COMPETITIVENESS, AND ENVIRONMENTAL REGULATION. March 1995, p. 29. Although such market trends as these demonstrate that people value a clean, safe environment, their willingness to pay is more often masked by the need to decide collectively on the quality of the environment they share. Not everybody can afford a family estate on the coast of Maine as an escape from urban summer smog. Most of us have to breathe the air in the cities and suburbs where we live, and it's more or less the same quality air for the entire urban population. It's virtually useless for one person to spend the extra money for a cleaner car unless others do the same. Without that assurance, people tend to underspend on environmental quality. That's the main reason why governments have to get involved in the collective decision-making process.

185.   John Atcheson. THE ENVIRONMENTAL FORUM. August 1996, p. 21. The government *does* have to get out of the way if we are to have a technological revolution. But it cannot get off the field and rely on the market to deliver that revolution. It must continue to be a partner in research and development of new, efficient technologies. It must encourage and stimulate efficiency and prevention at facility, community, and ecosystem levels through an enlightened environmental policy. Finally, it must retain a strong and credible enforcement capability for those who fail to seize the opportunities presented by lean and clean corporate practices.

As Alan Kay of Apple computers said, "The best way to predict the future is to invent it." We can invent a future in which innovation and risk are rewarded, in which economic and environmental objectives are mutually achievable, and in which industry has greater flexibility in how it achieves environmental objectives, with loss of accountability. We can have a highly competitive U.S. industry that increasingly approaches environmental solutions through efficiency of resources, energy, and capital and through prevention of waste. We can move toward a society with all sectors of our economy operating in a framework of cooperation which encourages systemic, facility-wide, community and regional approaches contributing to a sustainable economy and ecology. But we will not get there if we wait for parachutes to appear or if we attempt to parachute cats onto our environmental problems.

186.   John Hanson. CANADA-U.S. LAW JOURNAL 229. 1995, Lexis/Nexis. Let us talk about the nature of this industry today. The environmental technologies and services industry is one of the fastest growing in the United States in this decade. The industry could exceed over 170 billion dollars annually this year. It employs more than one million people. This industry is larger than many of our other key industries. It is larger than the computer industry. It is larger than the pharmaceutical industry. And it is even larger than the plastics industry. So any of you who remember the movie, "The Graduate," think of yourselves as Dustin Hoffman or someone going to talk to a young person like Dustin Hoffman. The answer today is not "plastics." It may be environmental technologies.

This market could be three percent of our GNP by the year 2000. Capital investment in environmental technologies has increased thirty-four percent in just the time period between 1987 to 1990. The forty-seven billion dollars invested in 1990 was about three percent of all U.S. capital investments that year.

187.   Sheldon Friedlander. THE GREENING OF INDUSTRIAL ECOSYSTEMS. 1994, p. 226. The design of clean technologies offers the United States an opportunity to obtain a competitive advantage on world markets. The U.S. market is so large that the demands placed on processes and products to satisfy our own environmental regulations have a major impact on technology worldwide. It is in our national interest to remain the leaders in identifying existing or potential undesirable environmental effects and to develop regulations based on special U.S. geographic and demographic requirements and the

political factors peculiar to our federal system and its relationship to industry. Surely U.S. industry is in a better position to respond to such regulations than foreign industry. To this end, the nation will need an engineering profession highly skilled in advanced methods for designing clean technologies, through education and basic research.

188. Bob Ferrone. TOTAL QUALITY ENVIRONMENTAL MANAGEMENT. Summer 1996, pp. 41–42. As the 21st century approaches, a period of dramatic and profound change will affect the way we do business and design products in the global marketplace. We are in the early stages of transformation from nonintegrated systems to integrated *eco-efficient* systems that control material flow in product creation for the benefit of both the environment and the bottom line.

From fossil fuels to reusable power, from mechanics to organics, from product support to life-cycle management of resources, the new realities of continued global economic and financial competition are upon us. With the next millennium, we must change our basic approach to business with respect to the environment. Environmental skepticism must give way to an age of environmental optimism and market drivers that continue to push for enhanced eco-responsible business performance.

Industries and governments are aware of the increased interest in the development of international environmental standards (ISO 14001) taking place today. This awareness creates a need to pursue a multidisciplinary approach that links technology, business, environment, and economic systems in an "industrial metabolism" that can be sustained in the future. The era of energy/material-intensive design and fixed manufacturing processes is beginning to give way to an era in which economic growth will be dominated by low energy, low material content, and process that are responsive to rapid changes in technology. The computer and electronics world, where products change on the average of every 3 to 12 months, is a perfect example.

Secretary of State Warren Christopher, in a speech at Stanford University, discussed the relationship between ecological understanding, competitiveness, and national security that is beginning to permeate American foreign policy. He said the future market for environmentally sound products and technology may be bigger than the current market for defense products. Countries such as China, he said, are under enormous environmental pressure because of rapid economic growth and projected increasing demands for energy, water, and materials. If we are to stay competitive in the coming era of eco-efficient business, we must begin to consider new approaches to design and operations management.

189. Joseph Fiksel. TOTAL QUALITY ENVIRONMENTAL MANAGEMENT. Summer 1996, p. 47. "Eco-efficiency" is a term that does not yet appear in dictionaries but has already gained considerable force in shaping the environmental policies and practices of leading corporations. The Business Council on Sustainable Development (BCSD) sounded a trumpet call with their 1992 manifesto, "Changing Course." Due to the credibility of the companies that constitute BCSD's membership—including Dow Chemical, 3M, Northern Telecom, Ciba-Geigy, Volkswagen, Nissan, Mitsubishi, and many others—their message has had a substantial influence on the strategic thinking of company executives around the world, BCSD's concept of eco-efficiency suggests an important link between resource efficiency (which leads to productivity and profitability) and environmental responsibility.

Eco-efficiency makes business sense. By eliminating waste and using resources wisely, eco-efficient companies reduce costs and become more competitive. As environmental performance standards become commonplace, eco-efficient companies will be at an advantage for penetrating new markets and increasing their share of existing markets. This article describes the business practices companies are adopting to increase their eco-efficiency and improve their competitive advantage.

190. TOMORROW. May/June 1996, p. 28. Many developers believe that untold riches await them if they can get their environmental technology from their garage to the global marketplace.

The OECD says the market for environmental technology and related services will be worth US$ 300 billion by the end of the decade. The U.S. Export-Import Bank estimates its worth to be already $400 billion world wide.

191. John Atcheson. THE ENVIRONMENTAL FORUM. August 1996, p. 19. There is far more at stake here than inside-the-beltway bragging rights. It is important that the United States reinvent environmental policy to stimulate a technological revolution for both environmental and economic reasons. The environmental facts of life over the course of the next quarter century are distinctly different than they were for the first quarter century. The first environmental revolution was fueled by a set of acute and obvious problems. This next one will be based on a subtler but no less pernicious set of problems. It must focus on chronic stresses resulting from humanity's exponentially increasing appropriation and use of land, energy, materials, and ecosystems.

To use a medical analogy, the first decades of environmental policy found the planet wounded, and applied bandages, casts, and slings. Having healed the wounds, we are now discovering that the patient is suffering from hear disease, a systemic disorder. In the United States, there is a nascent but growing recognition that our current environmental programs are no more relevant to these new problems than band-aids are to a heart attack victim.

192. Lester Brown. THE FUTURIST. July/August 1996, p. 12. Satisfying the conditions of sustainability—whether it be reversing the deforestation of the

planet, converting a throwaway economy into a reuse-recycle one, or stabilizing climate—will require new investment. Probably the single most useful instrument for converting an unsustainable world economy into one that is sustainable is fiscal policy.

193. Eric Marshall Green. ECONOMIC SECURITY AND HIGH TECHNOLOGY COMPETITION IN AN AGE OF TRANSITION: THE CASE OF THE SEMICONDUCTOR INDUSTRY. 1996, pp. 34–35. The intersectoral linkages and positive externalities associated with the high technology enterprise are fundamental reasons why this economic activity is of special relevance to the economic power and preeminence of a nation. It is not so much a question of nurturing national champions as it is of sustaining a vital industrial network where certain industries become instrumental for long-term productivity growth and economic competitiveness. It should be noted that direct evidence that high technology industries generate returns to society above their direct returns is elusive. Technological spillovers are difficult to measure, and since externalities have no price, the level of benefit to the national economy is uncertain. The difficulty in establishing a direct causal relationship should not be surprising, however. Externalities and spillovers, as their definitions would suggest, are beyond the effective predictive ability of the economics profession. Yet, it is reasonable to assume that external benefits can be generated from technological innovation and dispersed through intersectoral linkages.

194. David Dollar. COMPETITIVENESS, CONVERGENCE, AND INTERNATIONAL SPECIALIZATION. 1993, p. 186. We believe that R & D is particularly important to promote. There are three problems associated with firm-level R & D. First, it is a high fixed-cost activity, which can dissuade the individual firm from undertaking it. Second, there is duplication of effort if many firms undertake the same or similar projects. Third, there are difficulties in "appropriating" the returns to R & D (see Nelson and Wolff 1991, for example) and there is thus a strong incentive to be a follower in an industry.

Economists have long recognized that these externalities involved in the production of commercially valuable knowledge are inevitable—and even to some extent desirable. For this reason, direct and indirect support for R & D promotes our national welfare and may greatly enhance our competitive position in the future. Such subsidies can take the form of tax breaks for R & D or direct government assistance to the research effort. Governments of all the major economies—including the United States—provide these kinds of subsidies. In Japan, government support has tended to favor the development and use of new technologies for consumer products, and has met with notable successes in automobiles, consumer electronics, and medium-range mainframe, high-quality personal, and laptop computers.

195. David Dollar. COMPETITIVENESS, CONVERGENCE, AND INTERNATIONAL SPECIALIZATION. 1993, p. 17. Nevertheless, we would not want the results of our research to be interpreted as a call to complacency. Developments to date are highly consistent with the notion of convergence: the large U.S. productivity lead after World War II was something of an aberration, and the world has now returned to a more normal state, in which there are a number of advanced economies whose productivity levels are close together. As Japan and Germany have approached the U.S. level, their growth rates have slowed down, which is further evidence in support of the convergence hypothesis.

Today, industrial production is distributed among OECD countries more or less in proportion to population. Recent history gives us good reason to expect that the advanced economies will grow at similar rates, so that aggregate productivity differences among them will remain small. However, there is nothing automatic about this process. The source of productivity growth at the industry and subindustry level is innovation and investment. For a nation to remain a member of the convergence club, its firms will have to continue to create new technologies and make new investments at a rapid rate.

196. Laura Katz Olson. COMPETITIVENESS AND AMERICAN SOCIETY. 1993, pp. 283–284. Importantly, the performance of American companies as well as that of its mangers, is measured almost exclusively by the firm's current stock price and by its ability to reach high targeted rates of return on investment, usually on a quarterly basis. This is an extremely short time frame, especially since major product innovations in most firms take from seven to fifteen years before they become profitable. More recently, if quarterly profits are not sufficiently high, the company may become susceptible to hostile takeover.

Pressure for a short managerial time frame is further buttressed by the American tendency to reward senior executives for immediate results. These executives, who tend to stay with a firm only five or six years, are too often motivated by their profit-related bonus and stock option plans. Moreover, middle-managers usually are promoted and rewarded based on the quarterly earnings of their particular profit centers. Consequently, corporate managers tend to pursue projects with short-term profits and to avoid long-term commitments, including new technology and new products. As Thurow sums it up, "short time horizons are produced by an economic environment where everyone responds rationally to individual incentives, but the sum total of those individually rational choices is social stupidity."

197. Eric Marshall Green. ECONOMIC SECURITY AND HIGH TECHNOLOGY COMPETITION IN AN AGE OF TRANSITION: THE CASE OF THE SEMICONDUCTOR INDUSTRY. 1996, p. 34. Insuffi-

cient investment in R&D may be attributed to the two problems of appropriation and asymmetries of information in capital markets. Several studies have indicated that public returns on innovation are greater than private returns, and one estimates that society's overall return on R&D is at least twice the private return. Private enterprises that do invest heavily in R&D rarely capture the full value of the investment. Despite patent and copyright protection much of the reward from private innovation is enjoyed by imitators and consumers. More fundamentally, much of the reward for innovation is captured as improvements to the industrial framework which many other enterprises are linked to and depend on. Questions of appropriation may therefore diminish the incentive of a firm to invest in R&D, despite indications that society may benefit by an amount that more than justifies its expenditure. It is this incongruity of risk to reward that may create a condition of underinvestment.

198.   Roger Nagel. COMPETITIVENESS AND AMERICAN SOCIETY. 1993, p. 180. At the same time, innovation commitments by management cannot be derived objectively either from technical knowledge itself or from an examination of new technologies. The reverse is closer to the truth, namely, that the technical resources (the knowledge and the technologies) available to the management of a firm at a given time are themselves the products of value-laden, hence subjective, managerial decision-making processes. It is the management of a firm that decides, on managerial grounds, the kinds of technical expertise and the directions of research and development that will be supported internally, or acquired from other firms in the form of personnel, patents, licensing agreements, or reverse engineering. It is, again, management that decides, for each firm, what technical knowledge and which technologies are to be used, how they are to be used, what they are to be used for, and on what scale. Technological innovation is thus driven by extra-technical value judgments. These value judgments are embedded, in part explicitly but more often tacitly, in enterprise-specific decisions that are invariably the prerogative of management. They are the expression of a firm's managerial agenda together with the management processes organized for the express purpose of executing that agenda.

Such a view of innovation flies in the face of the popular view that innovation is a universal and objective process, driven by innate characteristics of theoretical knowledge, and of artifacts that somehow dictate the terms of their own unfolding. This view was supported by traditional notions of material progress that saw as inevitable the introduction, application, improvement, and dissemination of steam power, electricity, and internal combustion engines, for example, in just the forms that those technologies actually took. Many critics of technology, however, especially in the 1960s and 1970s, warned of an "imperative" in technical knowledge and in artifacts that threatened to overwhelm humanity's ability to manage their humane implementation. Both of these groups ignored the fact that the domain of the technical, in spite of having been shaped by value judgments, cannot include the terms of its own application. The application of technical knowledge can only be determined by the addition of extra-technical value judgments, such as those mentioned above.

199.   Roger Nagel. COMPETITIVENESS AND AMERICAN SOCIETY. 1993, p. 181. The task that management confronts in formulating innovation policies for a firm (or for a government agency, or a laboratory) is very much like the task that applied scientists and engineers confront in solving a design problem. There is no one correct solution to a design problem, nor can the reductive analytical techniques familiar from mathematics and physics be applied to design problems. Their solution requires assigning values on extra-technical grounds, as relevant both to the definition of the problem and to what will be considered an acceptable solution to it. Typical factors in industry, for example, would include manufacturability, cost, compatibility with existing products, facilities and inventory, competitive pressures, legal restrictions, marketing projections, aesthetics, corporate image requirements, and managerial experience and prejudices. The very same sorts of considerations apply, as well, to the solution of the managerial problem of designing a role for technological innovation within a corporation's strategic plan.

200.   Murray Weidenbaum. SOCIETY. May/June 1996, pp. 97–98. What about the future? Recall that the first of these eight points began with an illustration of the awesome power of technology. Nobody can forecast which specific technologies will succeed in the coming decade. But the prospects for U.S. companies being in the lead are very bright. There is a special reason for optimism.

In the 1990s, the United States will be benefiting from the upsurge of industrial R&D during the 1980s. A key but undramatic crossover occurred in the early 1980s. For the first time in over a half century, the magnitude of company-sponsored R&D exceeded the total of government-financed R&D. That primary reliance on private R&D continues to this day.

201.   FINANCIAL TIMES. 23 August 1996, Lexis/Nexis. Its decision to lower the securities repurchase (repo) rate from 3.30 per cent to 3 per cent, rather than to around 3.20 per cent, appeared to have been made at least as much for international as for domestic reasons. Concerns about France's economic outlook, and its chances of meeting the criteria for European monetary union, have put pressure on the franc recently.

"You can make a case for saying the Bundesbank acted as much for Europe as for themselves," said Mr. John Butler, bond analyst at WestLB Research. "This could be considered the first European act of the German central bank."

Certainly, the repo cut was greeted enthusiastically in financial markets, with an immediate rise in the dollar against the D-Mark, higher German bond and share prices and an easing of pressure on the French franc. Helping to bolster the French currency, the Banque de France also cut its intervention rate from 3.55 to 3.35 per cent.

Mr. Klaus Friedrich, chief economist at Dresdner bank, said the repo move showed the Bundesbank's "European awareness." However, apart from paying closer attention to European economic needs ahead of Emu, it was also looking to the needs of German exporters.

Thus, while the repo cut would not have a big direct impact on the German economy, it would lower the D-Mark against the dollar and other currencies and make it easier to see German goods abroad. "The Bundesbank knows the external view of the D-Mark is a very critical factor for the economy," Mr. Friedrich added.

With inflation low, economic recovery hesitant and growth in the money supply easing, the scope for a cut in the repo rate has been increasing for some time. Bundesbank directors have been saying the M3 trend would be watched closely to see if the room for manoeuvre was finally large enough for action.

202. GREENWIRE. 21 July 1995, Lexis/Nexis. In a speech on 7/19 at the German Environmental Foundation, German Pres. Roman Herzog urged industry leaders to integrate economic and enviro concerns. Herzog, referring to the dispute over Shell Oil's plan to dump the Brent Spar oil rig at sea: "Shell Oil has just seen what sort of consequences environmentally unsound practices can have even in the short term."

Herzog urged business to realize that "sustainable production and consumption patterns" are economically profitable, claiming that Germany is the world's largest exporter of environmental technologies. With an annual growth rate of 6%, more jobs are created in the environmental sector than any other, he said (FRANKFURTER RUNDSCHAU, 7/20).

203. David M. Keithly. ORBIS. Spring 1994, p. 220. Economic uncertainty provides a fertile feeding ground for political radicalism. Germany's unification travails are having an impact throughout Europe and could still have unanticipated consequences. The United States should show its understanding of these problems and assist Germany in the face of current challenges. U.S. Assistance should be part of an overall Western economic strategy to rebuild East Central Europe. Problems in the five new states of Germany are severe, ranging from ecological degradation to unemployment to health care to political extremism, and it should be borne in mind that difficulties there are less serious than in other former East bloc countries.

The most palpable threat to Germany and to Europe stems from economic difficulties and dislocations. Political extremism might even threaten democratic developments in Germany, and ethnic rivalry already menaces it in former satellite countries—both are the handmaiden of economic decline.

204. David Calleo. THE FUTURE OF AMERICAN FOREIGN POLICY. 1994, p. 191. We should not forget that Europe remains probably the one place in the world where destabilization could still easily trigger a general nuclear war. Europe's postwar progress toward inner stability has been an incomparable advance toward world stability. Trying seriously to reverse that progress would be a crime against history and a frivolous disregard of our own vital interests.

205. REUTER EUROPEAN BUSINESS REPORT. 19 June 1995, Lexis/Nexis. "Germany has the strictest environmental legislation in the world," he said.

Angela Merkel, environment minister in Chancellor Helmut Kohl's cabinet, said these tough laws fostered a German "green" technology sector that leads the world market with a 21 percent share of the 55 billion mark ($39 billion) total in 1993.

"This demand, and the products offered, have not fallen like manna from heaven," Merkel noted in her opening address. "Production often began to develop only after state requirements, norms and restrictions gave the orders and guaranteed a certain minimum need."

The sector employs 635,000 Germans and should grow to 1.1 million by 2000, surpassing the electronics industry, she said.

206. MANAGEMENT INTERNATIONAL REVIEW. January 1995, Lexis/Nexis. Germany has set considerable store in its environmental technology in international competition. However, there is some effort necessary to save this industry from the fate of other high-tech branches where Germany slipped into second place. Japan has already announced its claim to take over the world leadership in environmental technology by the end of the century (Servatius 1993).

The position of German products and companies on the world market can be assessed by different criteria. It can, for example, be derived from export shares (OECD, 1992), from the country's foreign trade specialisation (UBA 1993, Letler et al. 1992) and from the opinion of experts (BMFT 1993). Summarising corresponding studies, it can be stated that both in trade and R&D, Germany is playing a major role world wide. The main competitors come from the U.S.A. and Japan (Schweinbenz 1994).

207. GERMAN BUSINESS SCOPE. February 1995, Lexis/Nexis. In worldwide comparisons, German environmental technology is a leader. This was the finding of a study conducted by the Rhine-Westphalian Institute for Economic Research and the Institute for Economic Research Halle at the request of the Federal Ministry of Economic Affairs. In addition, the number of patents

related to environmental technology has increased significantly, from 215 in 1983 to 876 in 1993. According to the study, in 1992 the German environmental technology industry share of the market worldwide is 21 percent, double the average of German industry in general. By comparison, the U.S. share is just 17 percent, the Japanese is 13 percent.

With the industry, the priorities in western and eastern Germany are quite different. Total sales in the western part in 1993 were DM 51.4 billion, in the eastern part DM 4.1 billion. Priorities in the western part are water pollution and sewage treatment, air pollution and climate protection. The new "Bundeslaender" are still catching up in the area of wastewater treatment and prevention of water pollution.

In the next few years, according to the two institutes, recycling and solid water will be the major areas of interest throughout Germany.

208. HAZNEWS. 1 August 1996, Lexis/Nexis. The environment industry of the south German state of Bavaria has set up an environment fund of DM340 million ($222.6 million). "Enormous investments ... in innovative environmental protection technology will secure the markets of the future," said Thomas Goppel, the Bavarian environment minister. The ministry says that by 2000, the German environmental technology market will be worth some DM100,000 million, and for the rest of the European Union, DM217,000 million. Demand for environmental technology will "massively increase" in Asia and Latin America in the 21st century, said Dr. Goppel (see HAZNEWS, May 1996, pp. 10–11). The Bavarian Government is also investing some DM145 million in a new State Office for Environmental Protection in Augsburg. Dr. Goppel hopes the Office will stimulate private investment in environmental technologies through its administrative and consultancy support. [ellipses in original]

209. SLUDGE. 15 August 1995, Lexis/Nexis. The environmental technology field soon will explode in Germany, with huge increases in the number of employees and money spent, reports a new market study. With a worth of 71 billion deutsche marks (DM) expected to grow to DM 110 billion by 2005, the German environmental technology (ET) market leads all of Europe.

The German market for wastewater treatment and biosolids management technologies grew by 7 percent in both 1993 and 1994, Kaiser said. In all of Western Europe, however, the combined water purification, wastewater treatment and biosolids marks is slowing from DM 119 billion in 1993 to an estimated DM 88 billion this year. The market will shrink to DM 58 billion in 2005, Kaiser said.

Despite a worldwide recession, Western Europe has become an ET center. Tougher rules forcing greater environmental investments have fueled this development, according to the study. The 1994 Market for Environment Technology, by the Helmut Kaiser Consultancy, ET will account for more than 2 million jobs in Germany by 2000, the report, which polled more that 1,450 companies, said.

210. SIERRA. January 1995, Lexis/Nexis. In the United States, however, governments and businesses alike remain so focused on short-term profits and quarterly earnings that they overlook the true source of long-term wealth: innovation. Necessity breeds invention, and during the 1970s, when protecting the environment and saving energy were seen as essential elements of national policy, the United States brought hundreds of new products and processes to the verge of commercial reality.

These ranged from systems to generate electricity from wind and sunlight with zero pollution to little-known devices such as fuel cells that can power everything from homes to locomotives with zero or near-zero pollution and noise, while requiring minimal space. Yet these and thousands of other born-in-the U.S.A. environmental products were abandoned during the 1980s as the Reagan and Bush administrations, the Congress, and many state officials turned their backs on environmental protection, orphaning technologies that now stand to generate billions, perhaps trillions, of dollars, for their new proponents.

211. WASHINGTON POST. 18 December 1994, Lexis/Nexis. Had the country capitalized on the lessons of the '60s and '70s, conclude Curtis Moore and Alan Miller in Green Gold, by the late '80s it could have been positioned to become "the unrivaled industrial power of the 21st century." Instead, its dominant position in environmental technology had been allowed to slip away. The Reagan administration had bought into the cheap-energy doctrine of budget chief David Stockman. The quest for energy efficiency and independence from imported oil dramatically faded at the same time that regulatory and environmental programs were slashed.

The notion that the environment has suffered because of cheap oil and cheap cola is hardly new. Nor is the idea that one day the future will belong to "green" technologies. But in Green Gold, we have a lucid and compelling, if a bit repetitious, account of how the United States squandered its technological advantage in the '80s and how Germany and Japan have seized the opportunities.

212. BUSINESS WIRE. 21 May 1996, Lexis/Nexis. Commenting on first quarter 1996 results, Frans Luttmer, Chairman of the Executive Board, said: "The increase we achieved in revenue through autonomous growth as well as in net income are important accomplishments. In the markets in which Heidemij is active, conditions were diverse. The Dutch market was solid, however, in Germany, business conditions continued to weaken and are not expected to improve until 1997. In the United States, the environmental market is difficult at best as a result of political disagreement which has prevented important environmental legislation and the attached budgets and appropriations from being passed. The changing

U.S. market, however, also offers opportunities which Heidemij is actively pursuing."

213. EUROPE ENVIRONMENT. 8 March 1996, Lexis/Nexis. Speaking in Berlin on February 23, the German Secretary of State for the Environment revealed that some 680,000 jobs in Germany are, to a greater or lesser extent, linked to the environment, a figure that is expected to rise above a million by the year 2000. Opening the UTECH 96 Forum on environmental technologies, Ulrich Klinkert, claimed that Germany remains the world's second exporter of such technology behind the United States. Notwithstanding, Ernst Ulrich von Weizsecker, Chairman of the Wuppertal Institute for climate, the environment and energy, regretted that in spite of its leading rank "Germany still dallies far too long in weighing the pros and cons of new ecological ideas 'before deciding to invest,' unlike Japan, where similar decisions are taken in just 20 minutes."

214. HAZNEWS. 1 May 1996, Lexis/Nexis.
Environmental technologies will be of major significance to the German economy and employment, says HKU. One of the most significant "export weapons" is the German discovery of the closed loop concept, which handles products and production processes from "cradle to grave" and attempts to re-integrate them back into the production process. "Here we have the know-how like no other nation in the world," says HKU. However, it is questionable whether many will actually take these undisputed and good opportunities on board, says the firm.

According to HKU, exports accounted for 20% of German environmental industry activity in 1995; and will account for 19% in 1996. However, only 9% of German environmental technology firms are exporting companies. "A lack of competence in marketing as well as a lack of knowledge about local conditions in the target import countries are the main reasons for this," says HKU. However, HKU estimates that if the German environmental technology firms are able to intensify their export activities, employment in the environmental technologies market could double by the year 2000.

215. SIERRA. January 1995, Lexis/Nexis. Little wonder that Edda Muller, former chief aide to Germany's minister for the environment, declares emphatically that "what we are doing here is economic policy, not environmental policy."

She is not alone in this view of the future, nor is Germany. For example, Takefumi Fukumiza, U.S. representative of Japan's powerful Ministry of International Trade and Industry, says that industrialists in his country see "an inescapable economic necessity to improve energy efficiency and environmental technologies, which they believe would reduce costs and create a profitable world market.

With virtually no coal, oil, or natural gas, and limited mineral resources, Japan has historically been forced to do more with less than its principal industrial competitors, the United States and Germany. As a result, it makes steel, automobiles, and a wide range of other goods with greater efficiency and less pollution than any other nation. That national thrift and the technologies it has spawned are now global commodities as other nations increasingly search for cleaner, more efficient manufacturing methods and energy use. "The potential profit in such a market," explains Fukumizu, "is limitless."

216. SIERRA. January 1995, Lexis/Nexis. As ethically committed as Germany's citizens and government are to protecting the earth, they also perceive the process of eliminating pollution as an opportunity to further strengthen their nation's economy.

Already running a close race with the United States as the world's leading exporter of merchandise, Germany is convinced that its environmental regulations, easily the world's most stringent, will stimulate the development of a wide range of new "green" technologies that can be market globally just as demand for them is beginning to increase sharply.

217. FINANCIAL TIMES. 29 June 1995, Lexis/Nexis.
The state, Germany's most populous, owes its wealth to the steel mill and coal mines which flourished there at the beginning of the century and well into the 1960s.

Since then, such industries have largely disappeared and a concerted attempt has been made to find an alternative for the ageing steel mills. Media-related business has been successfully transplanted into parts of the state but in areas in and around Mulheim one much talked-about new sector is environmental technology, an area in which Germany leads the world and which has raised local people's awareness of environmental issues.

218. WORLD ENVIRONMENT REPORT. 14 September 1994, Lexis/Nexis. Developing technologies and practices that conserve energy and resources not only promotes sustainable development but will boost Germany's economy, public and private-sector leaders told a conference in Cologne, Germany.

Germany is widely recognized as one of the world's leaders in environmental technologies. Environment Minister Klaus Toepfer said in a statement that environmentally friendly products will drive much of the economy in the future.

That sentiment was echoed by Tyll Necker, president of the BDF Federation of German Industry, who said German industry has a 21 percent share of the world environmental industry and intends to maintain and grow that slice.

Germany registered environment product sales of $132.8 billion in Europe in 1993. Worldwide sales only hit $112.6 billion two years later, government records show.

Toepfer estimated that more than 1 million people will be employed by Germany's environment industry in 2000.

The country is expected to actively market its environmental industry goods and services in the Third and former Second World.

219. REUTER EUROPEAN COMMUNITY REPORT. 31 AUGUST 1994, Lexis/Nexis. German government and business leaders are hoping to convince manufacturers that environmental protection is more a business opportunity than a burden.

Investment in energy- and resource-saving products and methods will promote innovation and create jobs, they said at a conference on the environment on Tuesday.

The ruling coalition, with its eye on the October 16 national elections, may also be trying to convince voters it is as concerned about the environment as the opposition parties.

"The future, world-wide, belongs to environmentally friendly products and production processes," Environment Minister Klaus Toepfer told the conference sponsored by his ministry and the BDI Federation of German Industry.

220. THE ECONOMIST. 28 October 1995, Lexis/Nexis. Governments have a poor record in fostering high-technology industries. Germany's Jürgen Rüttgers reckons he can change that.

Mr. Rüttgers' guiding vision is probably correct, if not particularly original. Mobile information and capital are "dissolving space and time," he says, citing as an example a company that writes software around the clock, using programmers in different time zones. Germany's task is not to resist this, he says, but to add as much value as it can in industries such as information technology, genetic engineering and environmental technology. Among these, Germany stands out only in environmental technology, and that is mainly because fierce regulations has turned it green.

221. CONTROL AND INSTRUMENTATION. June 1995, Lexis/Nexis. The environmental industry will create profits, trade surpluses, employment, technological advantage, as well as environmental protection. It will be a key industry in the 21st century.

The industry covers a heterogeneous set of goods and services, from remote sensors to high-temperature incinerators, designed to prevent or to minimise pollution.

A few far-sighted Governments, such as Germany, Japan and the U.S., now perceive the environment industry as being of strategic importance. They already dominate the market, and their Governments have developed supportive policies on R&D funding, export promotion, tax incentives and regulation, to help them with the lion's share of the fast-growing world markets.

222. CONTROL AND INSTRUMENTATION. June 1995, Lexis/Nexis. Furthermore these markets are likely to significantly increase in the future, with estimated average growth rates of between 5.5% and 8%. There are only a few other business sectors (notably information technology) that are expanding at such a rate.

The environmental business is bigger than the aerospace industry, and is (at least) as large as the pharmaceutical chemical industry. The OECD calculate, for example, that 1,780,000 people will be employed in the environmental work in the.

223. HAZNEWS. October 1994, Lexis/Nexis. The global market for environmental goods and services is already the same size as that of the aerospace industry: well over $200,000 million a year, said John Gummer, UK Secretary of State for the Environment. Opening a 2-day conference, European Environmental Initiative 1994, held in London to examine the opportunities for US firms in the European market, he said, "by the turn of the century we expect it to reach twice that sum . . . and by 2010, nearly three times as large, which would make it as big, or bigger, than the global chemical industry." [ellipses in original]

224. HAZNEWS. April 1995, Lexis/Nexis. The International Transfer Centre for Environmental Technology (ITUT), which opens in September 1995 in Leipzig, East Germany, will create "an international platform for the transfer of modern environmental technologies" into central and eastern Europe, South-east Asia, India, and China. According to the German environment ministry, this would include the experience that Germany has gained in environmental clean-up and restructuring of industry in the former East German states. The ITUT would also help establish links between technology centres, companies, embassies and chambers of commerce. A recent study has suggested that medium-sized German environmental technologies companies need more support in their entry into international markets.

225. HAZNEWS. December 1994, Lexis/Nexis. Sales of environmental technologies in Germany totalled DM55,000 million ($33,250 million) in 1993, according to a recent study by the economic research institutes of the Rhineland-Westphalia region (RWI) and Halle. Estimates by different institutes often vary, as there is no general definition of the environmental technologies industry, said RWI, whose figure does not include the environmental services and integrated environmental protection. According to the study, the environmental technologies industry is dominated by medium-sized companies: over 60% of firms have less than 100 employees. The German environmental technologies industry is competitive, but only exported 20% of its production in 1993, mostly to western industrialized countries, with only 2% to Central and Eastern Europe. The study said that the strong dependence of demand for technologies on environmental policies, should not be overlooked.

226. David Keithly. ORBIS. Spring 1994, p. 220.

Finally, and perhaps most important, the United States must strive to regain faith in its postwar trade policies. Free trade and open markets helped tremendously in rebuilding Europe after the Second World War. Nothing could more quickly choke off the second European reconstruction than trade barriers and beggar-thy-neighbor policies. At least one in every three jobs in Germany is export-dependent, and in coming years, this ratio will be even higher in the eastern German states as companies endeavor to secure cost advantages in highly competitive international markets by availing themselves of lower wage levels and, in some cases, of newly refurbished plants and equipment. In short, trade is a crucial element in any second economic miracle in Germany.

227. FEDERAL DEPARTMENT AND AGENCY DOCUMENTS. 5 July 1996, Lexis/Nexis. U.S., Japanese, and German firms largely supply the global market for air pollution control equipment and services (APC), reports the U.S. International Trade Commission (ITC) in its publication Global Competitiveness of U.S. Environmental Technology Industries: Air Pollution Prevention and Control.

The report indicates that the U.S. APF market is significantly larger than the markets in Japan and Germany, and while firms in each of these countries dominate their own domestic markets, they compete with each other and others in third country markets as well as in the United States. Large multinational companies dominate the industry in all three countries, though the U.S. industry has a greater number of small specialized firms.

The ITC, an independent, nonpartisan, factfinding federal agency, recently completed the report as the second of a two-part study for the U.S. Senate's Committee on Finance. The report examines the global competitiveness of U.S. industries that supply goods and services for air pollution control and prevention for stationary sources (such as electric power producers and industrial manufacturers) and for mobile sources (such as cars, buses, and trucks). The report also examines the APC industries of Japan and Germany.

228. MANAGEMENT INTERNATIONAL REVIEW. January 1995, Lexis/Nexis. Germany has set considerable store in its environmental technology in international competition. However, there is some effort necessary to save this industry from the fate of other high-tech branches where Germany slipped into second place. Japan has already announced its claim to take over the world leadership in environmental technology by the end of the century (Servatius 1993).

The position of German products and companies on the world market can be assessed by different criteria. It can, for example be derived from export shares (OECD 1992), from the country's foreign trade specialisation (UBA 1993, Legler et al., 1992) and from the opinion of experts (BMFT 1993). Summarising corresponding studies, it can be stated that both in trade and R&D, Germany is playing a major role world wide. The main competitors come from the U.S.A. and Japan (Schweinbenz 1994).

229. THE REUTER BUSINESS REPORT. 9 August 1996, Lexis/Nexis. But the high standards seen at both the state and federal levels can be turned to environmental technology company's advantage, Newell said.

"Our commitment to high environmental standards should become a competitive tool," Newell said, especially in small countries that do not have the resources of an Environmental Protection Agency to review competing technologies.

The domestic environmental technology market generates about $170 billion in annual revenues per year, of which $10 billion is derived from export, Newell said. There are more than 100,000 U.S. companies working in the environmental field, employing 1.2 million people.

230. AIR CONDITIONING, HEATING AND REFRIGERATION NEWS. 19 February 1996, Lexis/Nexis. "In his book *Green Gold: Japan, Germany, the United States, and the Race for Environmental Technology*, University of Maryland Professor Allen Miller states that if American companies are going to effectively compete in the global marketplace, we must focus on providing technologies that are both business-wise and earth-wise," said Schultz.

"According to Professor Miller, some countries are pursuing environmental technologies because their people want strong laws and strong environmental protection.

"Companies are entering the market because they believe it will provide competitive advantage. At Trane, we believe that we can have both strong environmental technologies can play a key role in achieving these goals.

"And if the United States are going to continue to be world-class leaders, we must continue to provide solutions that are good for the environment and good for business."

231. REUTERS. 19 June 1996, Lexis/Nexis. Dobson said strict environmental standards in countries such as Germany, Japan, and the United States had spurred companies there to develop new processes and technologies.

As a result, Germany had a trade surplus in environmental technology of about seven billion stg in 1990 whereas Britain's surplus was just 300 million stg, he said.

232. COAL AND SYNFUELS TECHNOLOGY. 19 September 1994, Lexis/Nexis. Like so many U.S.-developed technologies, clean coal technologies that got their start here are coming to fruition at the hands of German and Japanese companies, say two industry observers.

And this could spell trouble for U.S. companies aiming to play a part in the burgeoning global market for envi-

ronmental technologies, say Curtis Moor and Alan Miller in a recent book on that market.

"The United States has the most to gain and the most to lose" in the race among industrialized nations to gain a niche in the so-called "green technology" market, said Moore, who, along with Miller, spoke with reporters in Washington last week.

Estimates of the size of the global market for clean technology over the next 20 years range from $270 billion to $750 billion (C&ST 3/28).

In particular, the Moore-Miller book—entitled *Green Gold*—points to integrated gasification combined cycle and pressurized fluidized bed combustion as examples of U.S.-started technologies that are being developed overseas. IGCC and PFBC are viewed by many industry experts as the cleanest, most efficient and most cost-effective commercial-ready clean coal technologies to date.

233. CNN. 1 October 1994, Lexis/Nexis.
TUCHMAN: Although coal is still the source of electricity for most Americans, the negative environmental impact has for many decades limited the enthusiasm about this most abundant of American energy sources. But this cleaner technology, which is also being used in Germany and Japan, is seen as a way to increase usage of coal and lessen dependence on oil. In addition, dozens of research projects are now being conducted throughout the U.S. in pursuit of other clean coal-burning concepts.

234. WASHINGTON POST. 18 December 1994, Lexis/Nexis. In the environmental technology field that the United States dominated a generation ago, the overall leader in manufacturing pollution-abatement is now Germany and Japan dominates in the export of air-pollution-control devices.

235. COAL AND SYNFUELS TECHNOLOGY. 31 October 1994, Lexis/Nexis. A German electric company and regional government are targeting clean coal technologies to help reduce CO_2 in North Rhine-Westphalia.

The government of North Rhine-Westphalia and the German electric company RWE AG recently made a pact to cut CO_2 emissions in the region by 27%, according to reports from Reuters news service.

According to Reuters, RWE and its Rheinbraum AG subsidiary plan to spend $13.42 billion during 1995–2030 to replace their brown coal-fired power plants with cleaner, more efficient units and to build an integrated gasification combined cycle power plant. A spokesperson for the European Community in Washington confirmed that such a plan is in place, but at press time was unable to provide details about the clean coal projects.

236. GLOBAL WARMING NETWORK ONLINE TODAY. 23 September 1994, Lexis/Nexis. According to Curtis Moore and Alan Miller, who have recently published the book GREEN GOLD, U.S. companies are missing the potential on clean coal technologies, and once again, German and Japanese companies are reaping the benefits of U.S.-developed technologies.

"The United States has the most to gain and the most to lose" among the industrialized countries in the green technology market, according to Moore. The global clean technology market is valued at least $270 billion, but could potentially reach $750 billion over the next 20 years.

In their book, Moore and Miller cite integrated gasification combined cycle (IGCC) and pressurized fluidized bed combustion (PFBC) as the cleanest, most efficient and most cost-effective industries that are also commercial-ready. Yet, these two U.S. technologies, developed with funding from the Department of Energy and Environmental Protection Agency are seeing their potential utilized outside the U.S. For example, while the IGCC Cool Water plant in California demonstrated this technology, this plant is idle, while Texaco Corporation is refining the technology with its own gasifier in China. Additionally, Asea Brown Boveri, a Swedish-German-Swiss conglomerate, is utilizing PFBC in Sweden, Poland, and Spain at commercial-scale plants.

237. AP WORLDSTREAM. 12 July 1995, Lexis/Nexis. Germany and France will spend 220 million marks (dlrs 158 million) in the next several years to create a nuclear reactor secure enough to prevent the release of radiation in case of a core meltdown, they announced Wednesday.

A consortium of France's Framatome and Germany's Siemens AG is in the design phase of a new generation of pressurized water nuclear reactors.

Under tough regulations passed last year, any new nuclear reactor must be built so that in case of core meltdown dangerous isotopes are contained in the reactor building, said Gertrud Sahler, spokeswoman for the environment ministry.

Germany and France hope the Siemens/Framatome reactor will be a model for the international nuclear industry, she said. The German and French environment and nuclear security agencies also are involved in the research project announced Wednesday.

238. WASHINGTON POST. 18 December 1994, Lexis/Nexis. Although American companies nearly monopolized the market for nuclear power plants outside the old Soviet bloc, the most promising of the new, far more safe designs belong to German, Japanese and Swedish manufacturers. And a quarter-century after being forced to make stupendous improvements in efficiency and emission controls, American automobile manufacturers have yet to design the drastically cleaner and more efficient vehicles of the future.

239. Alan Tonelson. CURRENT HISTORY. November 1995, p. 358. In Germany, Helmut Kohl's

long years in office indicate a fundamental stability in that country's politics. Yet because of reunification, the potential for tectonic change among the triad countries may be the greatest here. Although the cost of integrating eastern and western Germany has been huge and not yet fully paid, the likely long-term economic benefits are equally impressive.

To date, reunification has not generated any unusual upswing in German nationalism. Moreover, 45 years under sharply different political and economic systems have created a stubborn social and cultural gap between "Wessies" and "Ossies." Nonetheless, a reunified Germany is simply too big, too wealthy, and too powerful to remain a diplomatic and military pygmy; its increased weight has already been felt in European Union economic policy and diplomacy. As memories of the Nazi era fade, so will Germany's inhibitions. Moreover, resentment of its defeated-power status (as exemplified by its lack of a permanent United Nations Security Council seat and the presence on its territory of Western military forces obviously intended for its own containment as well as Russia's) is likely to grow.

Economic strains could also fuel German nationalism. Germany still faces a challenge that America has in many respects already confronted squarely–paying the social price for enhancing national competitiveness. For all its strengths, German industry is still weighed down by heavy regulation and an enormous welfare state. Moreover, the sheer magnitude of this welfare state and German social history suggest that Germans will not accept a new austerity as meekly as Americans have.

240. David M. Keithly. ORBIS. Spring 1994, p. 219.
Economic uncertainty breeds frustration, unemployment promotes radicalization, and dispirited people are inclined to demand hard answers. In most of the former GDR, the official unemployment rate is over 15 percent, with the actual rate considerably higher, since many of those who are employed are working part-time. With virtual certainty, economic dislocation will be a drawing factor for neo-Nazi groups and other political extremists in the foreseeable future. As jobs disappeared following the collapse of the GDR, and in the wake of recession following unification, radical trumpeting became more shrill and political violence spiraled upward.

241. David M. Keithly. ORBIS. Spring 1994, p. 208.
And what of the international consequences of Germany's growing political extremism? One must bear in mind that Germany is the lodestar of East-Central Europe—the leading economic power, the largest trader, the copious supplier of aid.

242. Gary Geipel. CURRENT HISTORY. November 1995, p. 375. If Europe has a middle, then it is Germany. This is much more than an observation about geography. Economically, politically, and even culturally, the Federal Republic of Germany occupies a central place in any map of Europe's future. Britain, France, and especially the United States retain considerable influence through the powers of initiative or veto in key organizations, while Russia's potential for belligerent or cooperative behavior is significant as well. Even small nations can give rise to diplomatic and military nightmares, as has been recently demonstrated once again in the Balkans. By its acquiescence, misbehavior, or leadership, however, only Germany now can shape virtually every major decision and trend that is of consequence for Europe as a whole.

243. David M. Keithly. ORBIS. Spring 1994, p. 208.
Thus, Germany is the pivotal country on the Continent when, as the British weekly *The Economist* maintains, Europe's future is in the balance. What happens in Germany will have profound repercussions elsewhere, and the way Germany goes others will go as well. Given this fact, the United States, for its part, should demonstrate a firm commitment to assist Germany in the face of current challenges. But what form should this assistance take?

244. Charles Glaser. CHICAGO. Summer 1993, pp. 8–9. However, although the lack of an imminent Soviet threat eliminates the most obvious danger, U.S. security has not been entirely separated from the future of Western Europe. The ending of the Cold War has brought many benefits, but has not eliminated the possibility of major power war, especially since such a war could grow out of a smaller conflict in the East. And, although nuclear weapons have greatly reduced the threat that a European hegemon would pose to U.S. security, a sound case nevertheless remains that a major European war could threaten U.S. security. The United States could be drawn into such a war, even if strict security considerations suggested it should stay out. A major power war could escalate to a nuclear war that, especially if the United States joins, could include attacks against the American homeland. Thus, the United States should not be unconcerned abut Europe's future.

245. THE BANKER. March 1996, Lexis/Nexis. But in a curious juxtaposition, the latest German economic data shows that even Germany may be struggling to meet the convergence criteria for union that it has stressed should be strictly applied (*The Banker* 12/95, p. 22).

The slowdown in the European economy and the prospect that key countries such as France (and even Germany) may fail to get their budget deficits down to the 3% of gross domestic product level and meet other criteria, have not only raised the spectre of postponement, but questioned the fundamentals of monetary union.

The linchpin of the project is Germany, the biggest economy in the EU and the country with the strongest currency, the Deutschmark.

But Chancellor Kohl sees EMU as more than just currency union and views it as an important step towards political union. His motivation, notes one German banker, is not to repeat the suffering of the past and 'to balance out the fears' of those in Europe. This may be a noble goal given Germany's past but is it the only alternative, and must all of Europe be subject to Helmut Kohl's view of history?

Views differ, and inside Germany opinions do not necessarily mirror those of the Chancellor. for many the prospect of EMU postponement is not a nightmare and, as German unemployment rises beyond 4 million, there is plenty of scepticism.

246. THE BANKER. March 1996, Lexis/Nexis. So what would happen if EMU fails? The consensus among many German bankers is that the franc/D-mark relationship could not be maintained, with the franc coming under strong pressure to devalue, and that massive waves of speculation and capital outflows would force abandonment of the project.

The problem for Germany of other countries devaluing their currencies is already a reality (note the impact of the UK and Italy following expulsion from the exchange rate mechanism). In fact, Deutsche Bank suggested recently linking the franc and D-mark prior to EMU, a suggestion not taken up by the Bundesbank.

The failure of EMU would have adverse repercussions for Germany. Bankers suggest the existing European Monetary System (EMS) would collapse, with adverse effects on the D-mark and the German economy. The vacuum caused by the failure of EMU could be devastating at worst and destabilising at best.

With no ready policy alternative, Europe could be in for considerable turbulence.

247. Gary Geipel. CURRENT HISTORY. November 1995, p. 377. The pursuit of balance also has a newer, second axis in German foreign policy that is no less challenging than the East-West axis. Germany resides at the pivot point between designs for European security that emphasize European institutions and those that emphasize NATO and the transatlantic connection. Thanks in large part to German mediation, few alliance leaders now perceive a fundamental contradiction between the development of a "European security and defense identity" and the maintenance of a strong connection to the United States. It remains far from clear, however, where the European identity should be built—in NATO, in the still-independent Western European Union (WEU), or inside the EU itself. That decision, in turn, has major implications for the membership policies of each organization. For example, if the EU is to take on collective responsibility for defense, then new Union members must be screened for their potential impact on a common military policy, regardless of their economic qualifications. And if the EU is to speak for "Europe" on defense matters inside the transatlantic alliance, then NATO may be pressured not to add new members who are not also EU members.

248. Masashi Nishihara. ADELPHI PAPER 285. January 1994, p. 248. If Western Europe should become an unstable area, for example, by failing to contain a strong, self-assertive Germany, European political cooperation or the WEU, not to mention NATO, may collapse. An unstable Western Europe would be a serious source of concern because it would probably fail to offer economic support or fail to help settle regional conflicts in the rest of Europe, and because many countries depend upon Western Europe for economic interactions. Such a Europe would also have a negative impact on the international economic system of which it is a key pillar.

249. Walter Hahn. GLOBAL AFFAIRS. Fall 1991, pp. 56–57. In short, the political prospects in Germany are uncertain at best, and with them not only Germany's future role in NATO, but also the future of the U.S. military presence. Even if the Kohl government can cope in the meantime with domestic problems, while keeping an internally preoccupied Germany on an Alliance track, an electoral victory by the Social Democratic Party, with its built-in antipathies to NATO and quasi-pacifist tendencies, could substantially alter not only the German picture but the broader European and Alliance scenarios as well.

250. Jacques Delors. ADELPHI PAPER 284. January 1994, p. 8. There is no doubt that NATO is the most effective machine for international cooperation on defence and security in Europe. It is the ultimate insurance policy against a nuclear war or a conventional attack on the territory of its member-states. At the same time it ensures an American presence in Europe and provides a forum for transatlantic dialogue on all issues affecting security in the broad sense of the term. It has an effective crisis-management capability, an integrated command structure and even its own military resources. It is therefore an important element of stability which its members are reluctant to abandon. Further, the interests of NATO members broadly coincide.

251. MODERN PAINT AND COATINGS. September 1996, Lexis/Nexis. "If the EPA and Mary Nichols are willing to stick their necks out in an election year and defy the White House, what are they and the environmentalists willing to do once the election's over? Mitchell asked a recent industry meeting in Los Angeles.

Pointing out that Nichols was once head of the Natural Resource Defense Council before joining EPA in Washington, Mitchell questioned whether Nichols may have a conflict of interest. "She has no scientific background and no background to make these kinds of decisions," Mitchell charged.

"There's a dirty little secret in Washington. They formed a Clean Air Act without learning the science." Not-

ing that EPA had indicated it will review the progress of the coatings industry to comply with its rule and VOC reductions in the future, and then determine if further regulation is needed. Mitchell noted: "We have no assurance of what that (further regulation) will be. Once you are regulated, you have no chance to get out."

Needless to say, Mitchell and more than 100 other smaller local and regional paint makers are not yet ready to heel or play dead. A number of western manufacturers are prepared to seek legal remedies if necessary.

Clearly, despite their size, they are willing to prove there's as much to their bite as their bark.

252. MODERN PAINT AND COATINGS. September 1996, Lexis/Nexis. And now, on top of everything else happening in Washington, word comes down that the Senate Small Business Committee has had cause to question recent activities of the EPA. The committee held an oversight hearing in Washington in July to question, among other actions, how EPA happened to rush through the publishing of its proposed national architectural coatings rule. EPA, among other federal agencies, was being questioned on its compliance with the Small Business Regulatory Enforcement Fairness Act (SBREFA), a new law designed to guarantee that federal agencies minimize the burdens of regulations on small businesses. President Clinton signed the law in March and it went into effect on July 1. The most important aspect of the law to small businesses is its requirement that agencies consider carefully the impacts of new regulations on small business. The law provides for judicial review of federal agencies' flexibility analyses and sets up review boards to evaluate regulations.

253. GEORGIA ENVIRONMENTAL LAW LETTER. June 1996, Lexis/Nexis. On April 11, 1996, EPA published a direct final rule that will delete dozens of rules contained in the federal regulations implementing the Clean Air Act (CAA). 61 Fed. Reg. 16050. EPA is taking the action in an effort to purge its regulations of rules found to be obsolete, ineffective, or duplicative. The rule became effective June 10, 1996.

254. CLEAN AIR NETWORK ONLINE TODAY. 4 September 1996, Lexis/Nexis. In an effort to make refiners and additives manufacturers pay for the bulk of costly, complex health tests, rather than hit small business blenders with duplicative tests, the Environmental Protection Agency (EPA) has proposed exempting fuel marketers with less than $50 million annual sales from paying for fuel or fuel additives health testing. However, some say the proposed rule does not go far enough.

"EPA is to be congratulated for recognizing that the requirement for certain terminal blenders to participate in the fuel and fuel additive test program is duplicative and onerous," said the Reformulated Fuels Association (RFA) in written comments to EPA. "But the proposal creates new problems that need to be addressed. The proposed rule erects substantial and unnecessary marketplace barriers to the use of ethanol."

255. ENERGY USER NEWS. August 1996, Lexis/Nexis. PHILADELPHIA—The U.S. Environmental Protection Agency (EPA) and the Small Business Administration (SBA) jointly announced the national kick-off of the Energy Star Small Business Program at a meeting here in June.

Focusing on the needs of businesses occupying less than 100,000 square feet of space, the voluntary program approaches the small business sector separately from EPA's existing Green Lights and Energy Star Buildings programs.

"Unlike when we were reaching the Wal Marts, Coca-Colas, and McDonalds of the business world, we are going to have to take a very different approach in reaching small business," explained Jerry Lawson, director of the new program. "We think that we will be recruiting lots of small businesses because of the dollar savings and our networks and partnerships. We believe we can save a lot of money together and reduce pollution."

256. CLEAN AIR NETWORK ONLINE TODAY. 29 October 1996, Lexis/Nexis. The Automotive Warehouse Distribution Association (AWDA) has asked the Small Business Administration to take action concerning an Environmental Protection Agency (EPA) proposal to revise national ambient air quality standards for ozone and particulate emissions.

EPA's proposal would make existing Clean Air Act national air quality standards more stringent and increase the number of national nonattainment areas. According to AWDA, the proposal would place substantial constraints on the national transportation industry, and the many small businesses that make up the industry.

AWDA pointed out that up to 70 percent of all interstate motor carriers operate only six or fewer trucks, with small businesses representing the majority of the automotive and heavy-duty truck service industry.

EPA is required under the Small Business Regulatory Enforcement Act to convene review panels to gauge the effect of the agency's regulations on small business entities. On September 6, EPA announced that it would assemble review panels to determine the potential impact of the revised standards.

The agency agreed to form the review panels in order to be prepared in the event it is determined that the proposed changes have a significant economic effect, said Thomas Kelly with EPA. The proposed changes could be modified based on the findings of the review panels to minimize the negative effects on small business.

257. VIRGINIA ENVIRONMENTAL COMPLIANCE UPDATE, SEPTEMBER 1996. 3 June 1996, Lexis/Nexis. Readers are cautioned that the policy is intended only as guidance; it is not binding on the agency.

Additionally, it applies only to penalties assessed by EPA—it has no effect on enforcement actions initiated by DEQ. To the extent that the policy may differ from the terms of other EPA enforcement policies (including penalty policies) under media-specific programs, this new policy supersedes those policies.

258. VIRGINIA ENVIRONMENTAL COMPLIANCE UPDATE, SEPTEMBER 1996. 3 June 1996, Lexis/Nexis. EPA has issued a final policy on compliance incentives for small businesses that provides incentives for these companies to correct environmental violations promptly. The policy, which was effective on June 10, 1996, sets forth guidelines for the agency to use in handling violations, and it offers guidance to state and local authorities for enacting similar programs.

The policy states that EPA either will not initiate enforcement action for civil penalties or will mitigate civil penalties whenever a small business makes a good-faith effort to comply with environmental requirements. A good-faith effort may be shown by receiving on-site compliance assistance or by promptly disclosing the findings of a voluntarily conducted environmental audit. In addition, the violation (1) must be the small business's first violation of the particular requirement; (2) must not involve criminal conduct; (3) must not cause a significant health and safety or environmental threat or harm; and (4) must be remedied within a set amount of time.

In all cases, the business must report the violations promptly to EPA or the appropriate regulatory agency within 10 days. If the business has conducted an audit, the report must be in writing. An environmental audit is defined by EPA in its audit policy published at 60 Fed. Reg. 66706 (December 22, 1995).

259. LOS ANGELES TIMES. 6 July 1996, Lexis/Nexis. The last time the minimum wage was raised, in 1989, the voice of small business was barely a whisper over the roar of labor unions and big corporations.

But this time, small business has dominated the debate.

Although the half-dozen Washington-based small-business lobbies have not beat back the pay raise—a potent election year issue—they nonetheless managed to wring out as compensation a list of long-sought-after measures that are likely to pass Monday.

"The small-business lobby hasn't always been so successful, but after the White House Conference on Small Business in 1995, things began to change," said Judith Meador, publisher and editor of the *St. Louis Small Business Monthly*.

Congress and President Clinton "have begun to realize that big business is laying people off and that future jobs are going to come from small business," Meador said.

Small-business associations now are considered key economic players in debates on the Hill, said Bennie Thayer, head of the National Assn. for the Self-Employed. Small-business leaders get quoted in the media more frequently than many Congressional members, and top-ranking Capitol politicians ask them to help write pro-small-business legislation, said Jamie Wickett, a lobbyist for the National Federation of Independent Business.

260. LOS ANGELES TIMES. 6 July 1996, Lexis/Nexis. "If you hurt small business . . . they can make things happen," Kotkin said. "It doesn't take that much of small business to switch over to a political party to make a big difference."

Key to the change in the political climate has been the trend of downsizing. Many of the employees shed by Fortune 500 companies and federal, state and local governments have gone on to start their own companies.

With 54% of the nation's employees currently working in small businesses and 98% of all jobs since 1989 being created by businesses with fewer than four employees, Congressional leaders have come to recognize that small-business issues are the nation's economic issues, small-business leaders say. [ellipses in original]

261. LOS ANGELES TIMES. 6 July 1996, Lexis/Nexis. Small business demonstrated its newfound clout in 1994 when, displeased with Clinton's 1993 tax increase and his proposal for employer-mandated health care plan, they went to the polls and helped vote in business-friendly Republicans to take control of the House of Representatives, said John Galles, NSBU president.

Now those same Republicans are working diligently to retain small-business votes by pushing for small-business legislation, Galles said.

262. LOS ANGELES TIMES. 6 July 1996, Lexis/Nexis. Similarly, Clinton has become more small-business-friendly in the past two years, small-business leaders said. He convened the 1995 White House Conference on Small Business, bringing together nearly 2,000 small-business owners from across the country and has pushed for many of their recommendations.

263. "Air/Water Pollution Report," ENVIRONMENT WEEK. 28 October 1996, Lexis/Nexis. EPA has been working "diligently" to develop procedures for full implementation, writes assistant administrator David Gardiner. Steps include preparing a guide for plain-English explanations of rules affecting small entities: small business compliance assistance centers for four industrial sectors with heavy concentrations of small businesses; a policy on small business rights in enforcement actions; and studying new guidelines for regulatory flexibility.

The agency has been submitting all final rules promulgated since March 29, to the House, Senate and congressional General Accounting Office. It has appointed Tom Kelly, director of the office of regulatory management & information, as its small business advocacy chair and contact for EPA actions under SBREFA.

264. Robert Easton. SMALL BUSINESS INCENTIVES HEARINGS. Y4.En2/3:103-148, p. 182. Small businesses are essential to the U.S. economy and are the breeding ground for entrepreneurial imagination and creativity, which has served to make the U.S. an economic leader. Most of the major companies in the U.S. today began in their infancy as one of these small businesses which would not have survived absent access to financing at critical times in their development. Today, these small businesses face the reality of an increasing tightening of credit and credit requirements. It is essential that we seek every opportunity to facilitate the flow of capital to these entities, especially in circumstances in which customer protection is not jeopardized.

265. BUSINESS WIRE. 26 June 1996, Lexis/Nexis. The approved proposals span a wide range of technologies, including software development, robotics, microelectronics production, computer memories, solar energy plastics recycling, and wireless communication. The companies receiving the awards are located in Alameda, San Francisco, San Mateo and Santa Clara counties. All but one of the recipients is a small business (less than 500 employees). "Small, high-tech businesses provide the engine of growth in our economy," Dr. Glass noted. "They not only create new jobs, but through innovation, they launch entirely new products and industries. We're trying to apply our limited funds at the point of maximum leverage."

266. LOS ANGELES TIMES. 4 March 1995, Lexis/Nexis. Economists and analysts say that small businesses such as Space Applications are the engine of growth that will drive California's economy into the next century. And, taken together, they make a significant contribution to the county's economy.

According to estimates by the Orange County Chamber of Commerce & Industry, more than 80% of the businesses in the county are small, with fewer than 100 employees.

267. PESTICIDE AND TOXIC CHEMICAL NEWS. 21 August 1996, Lexis/Nexis. Another aspect of that law could pose an additional test for EPA. SBREFA contains a section making regulations that have a significant effect on small businesses subject to judicial review, if an agency has failed to conduct sufficient outreach to those businesses.

268. PESTICIDE AND TOXIC CHEMICAL NEWS. 21 August 1996, Lexis/Nexis. Kevin Bromberg, a lawyer in the Small Business Administration's Office of Advocacy, and a frequent thorn in EPA's side on TRI matters, said the rule could be subject to judicial review if EPA did not do "adequate outreach" to small businesses.

Existing regulations already require agencies to conduct outreach to small businesses. But Bromberg said that SBREFA now makes the lack of adequate outreach judicially reviewable. In proposing the addition of chemical distributors, EPA has "made a major error, if not a legal error," Bromberg said. "We're advising EPA that they should be extremely careful here."

269. MISSOURI ENVIRONMENT COMPLIANCE UPDATE, May 1996, Lexis/Nexis. Parts of the law, directed specifically at OSHA and EPA, requires those agencies to conduct special pre-promulgation reviews of the impacts of proposed rules on small businesses. The Act also directs the Small Business Administration to designate an ombudsman and regional Small Business Regulatory Fairness Boards to oversee the enforcement of federal regulations against small businesses.

270. PESTICIDE AND TOXIC CHEMICAL NEWS. 21 August 1996, Lexis/Nexis. Since it became law June 28, SBREFA requires all federal agencies, including EPA, to systematically consider the impact major new regulations will have on small businesses (See June 5, Page 3). EPA and OSHA are also required to convene panels of agency officials to solicit input from the small business community.

271 Yvette M. Barksdale. 42 AMERICAN UNIVERSITY LAW REVIEW 273. Winter 1993, pp. 334–335. It is both unconstitutional and inappropriate for the President to direct administrative value selection. Presidential value selection exceeds the limits on executive power delineated in Article II of the Constitution by impairing structural protections within the Constitution for basic legislative, democratic decisionmaking values of consensus building, public participation, deliberation, and diffusion of power. Even if these legislative values are not deemed to be constitutionally mandated for administrative decisionmaking processes, they are important values for governmental decisionmaking generally because they provide both stability and democratic legitimacy to administrative value selection. These democratic values extend beyond the mere question of who makes social value choices to broader process issues and additionally are more significant than electoral accountability in legitimizing and justifying administrative value selection.

Agency processes for value selection, as currently constituted better protect these decisionmaking values than presidential processes because agency processes better provide for the inclusion of the public. This openness of administrative processes to public input supports the values of consensus, participation, deliberation, and diffusion. The insularity of presidential decisionmaking does not. Accordingly, the President should not direct administrative value selection.

272. Robert B. Reich. 94 YALE LAW JOURNAL 1617. June 1985, p. 1617. Public administrators

may be in a better position than legislators to foster a national debate over certain value-laden issues simply because the administrators deal with specific applications of general principles. Legislators, on the other hand, often have an incentive to keep their discussions to a fairly high level of generality; delving into knotty details will likely be seen as an invitation to controversy. Yet it is through concrete examples of difficulties in applying regulation—real-life stories about community struggles over civil rights for the handicapped, sex discrimination in schools, advertising directed at children, and other intersections of regulation and people's lives—that social learning takes place, even at a national level. Ultimately, such public deliberations may encourage legislators to alter general standards, or cause administrators to design rules that better reflect the citizenry's concerns.

273. Louis Fisher. CONSTITUTIONAL CONFLICTS BETWEEN CONGRESS AND THE PRESIDENT, 3d Ed. 1991, p. 103. To carry out the laws, administrators issue rules and regulations of their own. The courts long ago appreciated this need. Rules and regulations "must be received as the acts of the executive, and as such, be binding upon all within the sphere of his legal and constitutional authority." Current law authorizes the head of an executive department or military department to prescribe regulations "for the government of his department, the conduct of its employees, the distribution and performance of its business, and the custody, use, and preservation of its records, papers, and property."

These duties, primarily of a "housekeeping" nature, relate only distantly to the citizenry. Many regulations, however, bear directly on the public. It is here that administrative legislation must be restricted in its scope and application. Regulations are not supposed to be a substitute for the general policymaking that Congress enacts in the form of a public law. Although administrative regulations are entitled to respect, the authority to prescribe rules and regulations is not an independent source of power to make laws. Agency rulemaking must rest on authority granted directly or indirectly by Congress.

274. Patricia M. Wald. B. C. ENVIRONMENTAL AFFAIRS LAW REVIEW 519. Spring 1992, p. 519. Congress has recognized both the rapidly changing state of our knowledge about environmental dangers and its own short attention span and inevitably political orientation in dealing with any one subject, even the environment. As a result, it has delegated authority and discretion under those laws to the executive, and primarily to the EPA, to create and operate the apparatus for carrying out the laws' goals. For instance, over the past twenty years Congress has revisited the CAA for major overhauls only twice, in 1977 and in 1990. In the interim, it has relied on the EPA and the courts to keep the United States's environmental protection programs on track.

275. J. William Futrell. ENVIRONMENTAL LAW: FROM RESOURCES TO RECOVERY. Ed. C. Campbell-Mohn, 1993, pp. 76-77. Congress is where organized industry, environmentalists, and state and local representatives hammer out their various compromises. These compromises result in complicated environmental statutes that, nevertheless, delegate great power to the agencies. Because Congress sometimes seeks to escape interest group displeasure by avoiding hard decisions, important policy decisions are left by default for the agency to decide. This process leads to the expansion of federal environmental programs in a piecemeal uncoordinated fashion.

276. Harold Krent and James Rossi. LEGAL TIMES. 3 April 1995, p. 22. In considering legislation, Congress is often unable to reach its own decisions regarding regulations for a variety of reasons, such as technical complexity or political deadlock. Congress has punted on hard policy choices, empowering bureaucrats in administrative agencies to make decisions regarding the content, application, and enforcement of regulations. The EPA determines how best to meet pollution goals, the Occupational Health and Safety Administration determines which workplace hazards to prohibit, which to limit, and which to ignore and the U.S. Sentencing Commission determines the appropriate trade-offs in establishing sentence lengths.

277. Marc K. Landy, Marc J. Roberts, and Stephen R. Thomas. THE ENVIRONMENTAL PROTECTION AGENCY: ASKING THE WRONG QUESTIONS. 1994, pp. 5–6. First of all, an agency like EPA has substantial power, because of its expertise and its formal authority. Even when disputing its findings, all parties to a controversy often find themselves focusing on the information and analysis that EPA provides.

Moreover, agencies like EPA exercise much discretion in choosing which master to serve, and for what purposes. Constitutional law leaves their relationship to Congress and the President ambiguous. Statutes do not and cannot fully guide their behavior. Thus they possess substantial power that needs to be understood, managed, and used to advance democratic purposes.

The Environmental Protection Agency is unique among environmental regulatory agencies in that it deals with both public health and resource management issues. Its comprehensive authority is reflected in its position in the Executive Branch. It is the only regulatory agency whose administrator reports directly to the President.

278. Glen O. Robinson. AMERICAN BUREAUCRACY: PUBLIC CHOICE AND PUBLIC LAW. 1991, p. 69. Public choices are increasingly made not by the legislature, but by bureaucratic surrogates pursuant to a delegation of power from the legislature. I refer here not to the fact that legislative decisions require executive implementation that entails a degree of executive judgment and discretion. The fact is unremarkable. What is

remarkable is the extent to which political responsibility for policy choices is transferred to bureaucratic organizations. The line between policy-making judgments and implementive decisions has always perhaps been an indefinite one; in the modern administrative state it has become increasingly irrelevant as well.

279. Walter A. Rosenbaum. ENVIRONMENTAL POLITICS AND POLICY, 3d Ed. 1995, p. 105.
Federal agencies concerned with environmental affairs and closely related matters such as energy, consumer protection, and worker health have grown explosively in the past three decades. More than 150 major new federal laws, most concerned with broad regulation of business and the economy in the interest of public health and safety, have been enacted since 1970. More than twenty new regulatory agencies have been created to implement these programs, including the EPA, the Occupational Safety and Health Administration (OSHA), and the Department of the Interior's Office of Surface Mining Reclamation and Enforcement.

The Power of Discretion. The significance of the environmental agencies mentioned in the previous section rests less on their size and budget than on the political realities obscured by a constitutional illusion. The Constitution *appears* to vest the power to formulate policy primarily in Congress, while leaving to the president and the executive branch the task of seeing that the laws are "faithfully executed." Although implemented and enforced principally in the bureaucracy, public policy actually develops in both branches of the government.

280. Walter A. Rosenbaum. ENVIRONMENTAL POLITICS AND POLICY, 3d Ed. 1995, p. 113.
In the political conflicts of environmentalism, the federal bureaucracy is rapidly displacing Congress as the strategic heartland. A great diversity of administrative agencies share increasing responsibility for environmental administration and, with it, the enormous delegated and discretionary power inherent in implementing the law. By far the most important, most visible, and largest of the environmental regulators is the Environmental Protection Agency.

281. Mark Seidenfeld. 80 IOWA LAW REVIEW 1. October 1994, pp. 8–9. In other words, the size and complexity of the government's regulatory role make it impossible for Congress to set forth sufficiently detailed criteria in statutes that would both dictate regulatory decisions in particular contexts and still be sufficiently flexible to allow wise and efficient regulation. Pragmatically, Congress is forced to grant agencies much policy-setting discretion. Unless there is some constraint other than detailed statutory prescription to keep agency regulations consistent with the values of the polity, such broad delegation of policy-setting functions runs counter to the principle that the government should rely on democratic process to define fundamental policy.

282. Robert V. Percival. 54 LAW AND CONTEMPORARY PROBLEMS 172. Autumn 1991, p. 195.
Because environmental protection has been consistently popular throughout the last two decades, the enactment of environmental legislation can produce immediate political benefits for Congress and the president. Politicians can proudly point to new laws as demonstrating that they have acted to protect the environment, even though the impact of the laws on environmental quality may depend largely on the issuance and enforcement of implementing regulations by administrative agencies.

Yet the process for developing regulations to implement the environmental statues does not promise great political rewards for several reasons. First, the regulatory process is far less visible and understandable to the environmentally concerned public than is the legislative process, despite the participatory model of regulatory decisionmaking embodied in the Administrative Procedure Act. Second, industries subject to regulatory action have greater incentives to participate in the regulatory process than does the environmentally concerned public. While national environmental groups will participate in rulemaking proceedings of national importance, industries subject to environmental regulation will have a great incentive to participate in any rulemaking that may impose significant compliance costs on them.

Third, the benefits of environmental regulation, though often substantial, typically accrue over long periods of time in ways that are not nearly as visible as the impacts of compliance costs. Thus, by postponing compliance costs, a delay in implementing an environmental regulation may generate benefits that are more immediate and tangible to consumers and the regulated industry than are the long-term costs to the environment.

Fourth, because environmental regulation inevitably creates winners and losers, environmental legislation creates new opportunities for gains by those who can influence the course of the implementation process. The regulated community has a tremendous incentive to seek to delay or weaken implementing regulations. The president, in the short run, can reap political rewards from the regulated community by exercising Executive Office oversight that moderates the impact of environmental regulations on industry. If this oversight is largely hidden from public view, the political costs are likely to be minimal and are greatly offset by the benefits accruing to the president from the grateful regulated industry.

283. Zachary A. Smith. THE ENVIRONMENTAL POLICY PARADOX. 1992, p. 50. Most of what elected officials do escapes public attention. Although major issues may generate headlines, the details related to these issues are largely ignored by the public. Furthermore, although we may know how elected officials vote from roll call and recorded vote figures, we know much less about what motivates them to vote one way or the other. Many decisions are made outside the public spotlight. Much new

legislation and many refinements in existing legislation receive very little, if any, attention from anyone other than the parties directly involved in the legislation.

284. Harold Krent and James Rossi. LEGAL TIMES. 3 April 1995, p. 22. By delegating decisionmaking authorities to agencies, Congress can reap substantial benefit from constituents by appearing to combat the pressing social issues facing the nation. Who can argue with the goal of a better environment, a safer workplace, or a more consistent sentencing system?

But while members of Congress claim credit for attempting to solve the problems of the environment and crime at the same time they distance themselves from agency regulations regarded unfavorably by their constituents. Such delegation permits members of Congress to maximize their credit while minimizing their blame for unpopular regulations. The result is a loss of democratic accountability.

285. Walter A. Rosenbaum. ENVIRONMENTAL POLITICS AND POLICY, 3d Ed. 1995, p. 132. Environmental legislation is often vague and contradictory because Congress cannot or will not resolve major political conflicts entailed in the law. Instead, Congress often papers over the conflict with silence, confusion, or deliberate obscurity in the statutory language. This results in a steady flow of political hot potatoes to the bureaucracy, which must untangle and clarify this legal language—often to the accompaniment of political conflict and legislative criticism—or leave the job to the courts.

286. Glen O. Robinson. AMERICAN BUREAUCRACY: PUBLIC CHOICE AND PUBLIC LAW. 1991, pp. 76–77. What needs to be explained is why a congressman would prefer the uncertainty attendant on delegation of legislative power to an agency, rather than press for immediate legislative resolution of the substantive policy content desired. One obvious answer is that legislators may not know what the desired content is, either because of the present complexity of the problem or because changes in circumstances over the life of the legislative program will require changes in solutions devised today. I think a more important answer is that legislators may be ambivalent because of conflicts among their constituents. If, of course, there is an obvious net advantage to favoring one set of constituents over another, a legislator will presumably take a decisive position where he or she can, accepting both credit and blame on the assumption that the former will outweigh the latter. But it may not be obvious whether, on net, credit will exceed blame. Such a situation calls for a strategy of ambiguity in which the legislator can seek credit for doing something while shifting blame for outcomes that cause unexpected political harm.

The delegation of broad power to an agency, the actions of which are indeterminate, can be seen as such a strategy. One might suppose that, given an alert, intelligent constituency, a legislator would not be able to do both. But the credit claiming and the blame shifting do not occur simultaneously or with respect to the same action. Credit is claimed for the creation of a "responsible" agency to study the problem and act on it. Granted, constituents are not so stupid as to give full credit for taking action, even if it is incomplete. No one likes a do-nothing legislator, but everyone understands the need to seek the "assistance" of expert agents in resolving "tough" problems, If, when the agent takes action years later, the results prove to be injurious to constituents, the legislator may receive some blame for her earlier delegation (if constituents have good memories), but this can be offset by the legislator's claiming that the agent abused its discretion and by offering to intervene on behalf of the constituent. Anyone who thinks this scenario does not describe real life is invited to examine the action of legislators in oversight committees.

287. Cathy Marie Johnson. THE DYNAMICS OF CONFLICT BETWEEN BUREAUCRATS AND LEGISLATORS. 1992, pp. 6–7. Position taking and credit claiming involve considerable posturing and symbolic action yet little direct involvement in the administration of policies. Position taking and credit claiming also imply that legislators write vague, general laws, allowing them to avoid conflict and making it appear as though they are tackling the tough, difficult issues, while all along they are leaving the specific decisions to the bureaucracy. Bureaus are given programs to implement yet few detailed instructions on just how they should do that.

Of course, when Congress delegates authority, bureaus make mistakes and constituents complain. Legislators, then, have the opportunity to chide the bureaus and tell them just what they did wrong. But casework rather than oversight is the mechanism most used. Through casework a legislator handles the specific problems or questions of an individual constituent, such as difficulty with a Social Security check or a complaint about the safety of children's toys. Casework allows a legislator to respond directly to the demands of a constituent, demonstrating his or her concern and responsiveness to the voter's needs. Oversight, a committee investigation into agency actions, provides fewer opportunities to advance legislators' reelection aims because there is no guarantee that constituents will pay any attention to the proceedings. When it does occur, it is often political and symbolic, allowing the legislators to present a picture of dedicated, hard-working representatives protecting the public interest while still avoiding the nitty-gritty, often conflictual, and sometimes boring issues involved in the administration of the laws.

288. Robert V. Percival. 54 LAW AND CONTEMPORARY PROBLEMS 172. Autumn 1991, p. 192. Another variant on the agency failure theory suggests that EPA has some kind of institutional bias in favor of "too much" regulation. This idea may stem from the perception that some of the statutes EPA administers contain unrealis-

tically ambitious regulatory standards. Yet EPA's track record in implementing such legislation suggests that sufficient deterrents already exist to prevent it from regulating too extensively. The example usually used to illustrate this point is that EPA has regulated only seven hazardous air pollutants during the twenty years it has had to implement the provisions of Section 112 of the Clean Air Act. A similar pattern appears to prevail at OSHA. OSHA established new workplace exposure limits for only ten toxic chemicals during the same period that the private standard-setting organizations upon whose recommendations OSHA had relied in establishing its original standards lowered their recommended exposure limits to hundreds of chemicals.

289. Daniel W. Gottlieb. PURCHASING. 11 April, p. 41. At the same time, the EPA is trying to reform itself through the Clinton administration's "Reinventing Government" campaign which includes moves to "simplify, streamline, and reduce the cost of" regulations to industry and bring affected industry into the fashioning of standards earlier in the regulatory process.

To do this, the administration proposes deleting more than 1,000 pages or 10% of the total volume of regulations on the books; developing a one-stop form for reporting various types of emissions from plants; proposing to make trading of emission credits more flexible, and conducting common sense initiative (CSI) discussions with different industries on how to rationalize and simplify regulations and enforcement for their members.

290. Barbara Mikulski. ROLL CALL. 24 June 1996, Lexis/Nexis. For instance, the EPA is moving ahead to implement recommendations by the National Academy of Public Administration. These will significantly reduce the regulatory burden on well-intentioned companies and continue the Clinton Administration's efforts to reform regulatory action throughout the executive branch.

And I believe that continuing to invest in programs like the Environmental Technologies Initiative, which is being threatened with elimination, would help businesses take advantage of these new technologies and new marketing opportunities.

291. Carol M. Browner. ENVIRONMENTAL PROTECTION AGENCY'S FISCAL YEAR 1996 BUDGET REQUEST. Senate Committee on Environment and Public Works, Y4.P96/10:S.Hrg. 104–48, p. 16. The President's 1996 budget for the U.S. Environmental Protection Agency reflects the President's New Covenant between Government and the American people. Within an era of limited Federal resources, we must continue to protect our health, the health of our families, the health of our communities, and the health of our economy.

This budget shows modest growth. It will allow us to continue to expand the efforts to which I committed myself when I came to the Agency 2 years ago. and to achieve the very best environmental results at the least cost—industry by industry, community by community. Two years ago, I began directing the Agency away from the pollutant-by-pollutant, crisis-by-crisis, one-size-fits-all regulatory system of the past.

In place of that approach, we are moving toward a new, comprehensive approach to achieve results that are cleaner for the environment, cheaper for business and taxpayers, and smarter for America's future. This new approach is reflected in our Common Sense Initiative, in our community environmental protection program, and in our participation in the President's Environmental Technology Initiative, and National Performance Review. Through NPR we are continuing to examine all that we do at EPA, to ensure that we do everything necessary to protect our health, our air, our land, and our water in a common-sense, cost-effective, and consensus-based manner.

This approach is reflected throughout the budget that we present to you today. This is a budget in which we target the highest risk environmental problems. It is a budget in which we will strengthen our partnerships with States, tribes, communities, business and the public; a budget in which we strengthen our science, and a budget which will allow us to firmly enforce the law, but also work with businesses to comply with those laws in the very first instance.

292. Peter Fairley. CHEMICAL WEEK. 1 November 1995, p. 22. EPA and industry sources also acknowledge that they are fighting cultural barriers to change. "The projects we've laid out are really near-term activities to get people pointed in the right direction," says Jay Benforado, deputy director/research and science integration in EPPC's office of R&D, who is on loan to Hansen's office as reinvention coordinator. Benforado says the projects are opportunities to experiment with new ways of thinking after 25 years of command and control.

The cultural barriers go beyond EPA: They permeate EPA's relationship with industry. "It's almost like a religious war or something," Benforado says. "People are just fighting because their forefathers fought." Industry sources generally admit that their behavior has contributed to the mistrust. "Redemption is possible, but it's not going to happen overnight," says Delcambre.

293. ENVIRONMENTAL REMEDIATION TECHNOLOGY 1996. Vol. 13, No. 4, 28 June 1996, Lexis/Nexis. Gardiner's testimony went into detail about barriers to innovation and how EPA is attempting to alleviate inflexibility. "American businesses spend over $130 billion a year to comply with environmental mandates," he said. "But EPA's rules, guidelines and administrative procedures sometimes hinder technology innovation by making it difficult or undesirable to use new monitoring assessment, pollution prevention, remediation of control techniques.

"For example, many environmental standards that are currently in place tend to 'lock in' the use of existing technologies because they are based on reference technologies that are already well-demonstrated when standards are promulgated (see chart p. 124). Even when companies are legally permitted to use alternative methods to meet a standard, they are usually unwilling to risk non-compliance by implementing a relatively unknown or unproven technology. Enforcement personnel do not normally grant exceptions for businesses that make bona fide attempts to comply using an innovative approach but need extra time or fall short of the regulatory mark. Since companies are given no reward for trying a new approach and no protection against failure, conventional technologies tend to be used over and over again, freezing out newer and more effective options."

294. ENVIRONMENTAL REMEDIATION TECHNOLOGY 1996. Vol. 13, No. 4, 28 June 1996, Lexis/Nexis. Amid tough questioning by Republicans concerning the Environmental Protection Agency's ability to make good on reforms, Assistant Administrator for Policy, Planning and Evaluation David Gardiner recently told a joint House Subcommittee on Technology and Energy and Environment, "The U.S. EPA agrees ... that the deployment of cleaner, safer and cheaper environmental technologies ... is lagging and that we are not able to maximize the benefits of high technology for the environment or industry."

Rep. Bill Baker (R-Calif.) led the hard line of questioning June 20, at the Environmental Regulation: A Barrier to the use of Environmental Technology? hearing, focusing on what he feels is the agency's inability to match rhetoric with actual performance. Partisan rankling aside, Baker told real-life horror stories of private sector initiatives mired in agency bureaucracy. In prepared testimony, Gardiner outlined internal and external impediments that limit private-sector investment in innovative environmental technologies:

Current statutes, regulations, policies, and procedures at the federal, state and local levels favor the use of conventional, often less efficient or cost-effective technologies;

Private industry and the financial community are reluctant to fund development of new technologies because of changes in the tax code and uncertainties as to whether innovative methods will be approved for use by regulatory agencies;

Technology users, financiers, consulting engineers and regulators are unable to obtain credible, independently verified data on the performance and cost of promising new technologies; and

Lack of established information networks that provide users with awareness of (and easy access to) better, cleaner, safer and lower cost technologies. [ellipses in original]

295. Richard J. Lazarus. 54 LAW AND CONTEMPORARY PROBLEMS 205. Autumn 1991, pp. 237–238. The adverse effects of congressional oversight appear to be substantial. But do they mean that congressional oversight should be eliminated? Absolutely not. The practical advantages of oversight are too numerous, and its importance to our system of government is too central. Congressional oversight is, after all, Congress's most effective way to curb abuses by another branch of government and is also necessary for Congress's development of legislation. The adverse effects on EPA and on the fashioning and implementation of federal environmental law also do not support singling out the environmental law for drastic reductions in congressional oversight. Though the adverse effects are substantial, they do not appear to be greater than the associated benefits. The federal environmental statutes and EPA have made significant achievements during the last twenty years. Substantial improvements in environmental quality have been made in some areas and at least a resistance to more environmental degradation has emerged in many other areas, despite a growing level of industrial activity. Congressional oversight has played an instrumental role in many of those accomplishments.

296. Joel A. Mintz. ENFORCEMENT AT THE EPA: HIGH STAKES AND HARD CHOICES. 1995, p. 128. In many respects congressional oversight of the agency's enforcement work has been a success. While not free of difficulties, this oversight has, on a number of occasions, effectively identified significant inadequacies in EPA's enforcement approaches and then prodded the agency to remedy those shortcomings. At the same time, certain chronic problems–difficulties that have plagued EPA enforcement programs through several administrations—have received relatively scant congressional attention. Thus, although Congress has focused on inaction, timidity, and inconsistency in the agency's enforcement of particular statutes (often in the context of particular enforcement cases), it has largely ignored such persistent questions as turnover among professional enforcement employees, the nature and quality of EPA's management of its workforce, the adequacy of working conditions for the agency's enforcement staff, and the extent to which civil service personnel rules pose impediments to the recruitment of skilled EPA employees. While these matters may be dismissed as mundane or insignificant, they go to the very heart of the agency's long-term ability to sustain an effective enforcement effort. EPA's congressional overseers must broaden their focus so that issues of this sort are periodically and systematically reviewed.

297. Joel A. Mintz. ENFORCEMENT AT THE EPA: HIGH STAKES AND HARD CHOICES. 1995, p. 108. Beyond these direct effects on EPA enforcement policies, congressional oversight has also influenced EPA decision making indirectly in other critical respects. Thus,

for example, Congress's sustained insistence on active and vigorous enforcement activity has encouraged the agency to emphasize the *volume* of its formal enforcement actions (including its annual count of civil judicial referrals, administrative orders, criminal prosecutions, civil penalties, etc.) in measuring and evaluating its enforcement successes. Congressional interest has motivated EPA officials to expand the scope of EPA's criminal enforcement work, and congressional inquiries have periodically placed demands on the time and energy of the agency's top managers, as well as its professional enforcement staff.

298. Joel A. Mintz. ENFORCEMENT AT THE EPA: HIGH STAKES AND HARD CHOICES. 1995, p. 110. In his examination of congressional scrutiny of EPA, Lazarus takes the position that Congress's actions have "chilled decision making and innovation within the EPA" and denied the agency needed flexibility to respond to "the uncertain contours of environmental problems." Whatever validity this conclusion may have with respect to the EPA's rule-making activities, it seems of little relevance in the enforcement arena. EPA's management and staff have devised some innovative approaches in several areas of the agency's enforcement work, as illustrated by EPA's multimedia enforcement efforts and its thematic enforcement initiatives. In addition, as described earlier, EPA's RCRA LOIS enforcement initiative and Superfund enforcement first policy were bold changes in agency practices that were stimulated rather than inhibited, by congressional interest.

299. Steven Shimberg. 54 LAW REVIEW AND CONTEMPORARY PROBLEMS 241. 1991, p. 244.
Disadvantage 3. The threat of oversight chills decisionmaking and innovation at EPA.
Response. EPA is always free to propose new approaches to Congress. This can be done in the context of a proposed rulemaking developed in consultation with the "feared" congressional overseers or as a formal bid or informal legislative proposal.

A good example of this consultative approach is the recently enacted acid rain program. It is innovative. It was proposed by EPA and the president, and was approved by Congress. In fact, the Agency's experience with pollution prevention programs shows that whatever chilling does occur results not from congressional oversight but from internal executive branch disputes. For example, EPA's legislative proposal on pollution prevention was held up at OMB for several months. When it became clear that OMB was not going to approve the EPA proposal, it was abandoned as a legislative initiative.

300. Joel A. Mintz. ENFORCEMENT AT THE EPA: HIGH STAKES AND HARD CHOICES. 1995, p. 111. Another objection that Lazarus raises regarding congressional oversight of EPA relates to EPA's policy priorities. He contends that extensive congressional investigations "skew" EPA's agenda as it responds to the multiple requests and complaints of individual committee and subcommittee chairs with oversight leverage. At least as it relates to EPA's enforcement policies, this concern seems exaggerated. In responding to Lazarus, Steven Shimberg aptly noted: "While oversight does affect EPA's priorities, this is not necessarily a bad thing. EPA must be responsive to public concerns. If the agency cannot garner public support for its priorities, those priorities *should* be 'skewed' by Congressional oversight."

Shimberg's point is borne out by the experience of Congress' oversight of EPA in enforcement matters. Although Congress has indeed had considerable influence, both directly and indirectly, over the shape and thrust of the agency's enforcement programs, that influence has, for the most part, been constructive. Administrative agency autonomy is, in many instances, efficient and worthwhile. However, it should not be seen as an end in itself. To the extent that EPA's congressional overseers have succeeded in guiding the agency in the direction of improved effectiveness and societal benefit, it seems inappropriate to condemn their efforts.

301. Riley Dunlap. PUBLIC OPINION QUARTERLY. Vol. 55, 1991, pp. 655-656. The public's support of environmental regulations on business and industry is also compatible with their increasing preference for environmental quality over economic growth. This trend has grown so markedly over the past decade that environmental protection is now endorsed by large majorities and economic growth by only small minorities (table 28). A similar trend is apparent in support of environmental protection "regardless of the cost" (table 29).

302. CAPITAL TIMES. 8 June 1992, p. 6D. Others with approving majorities were Uruguay, Chile, Germany, Denmark, Norway, Switzerland, Great Britain, Canada, Ireland, Finland, the Netherlands, South Korea and the United States. In the United States, 65 percent of those surveyed said they would trade higher prices for a safer environment.

Comparable answers resulted when participants in the 22 countries were asked if they would give priority to protecting the environment "even at the risk of slowing down economic growth."

Only in India and Turkey did majorities give economic growth the priority, even if the environment would suffer as a result.

In the United States, 59 percent said they would accept slower economic growth to protect the environment.

303. Riley Dunlap. ENVIRONMENT. October 1991, p.13. The fifth item, used annually by Cambridge Reports since 1981, poses a direct tradeoff between environmental quality and economic growth by asking respondents which one ought to be sacrificed (see Figure 5 on page 33). Again, the pro-environment position grew

steadily over the last decade, and in 1990 the percentage opting for environmental quality exceeded that preferring economic growth by more than 4 to 1 (64 percent versus 15 percent). It appears that the increase in support for environmental quality was drawn about equally from the ranks of those who earlier favored economic growth and those who were unsure of their position (note the decline in the percentage of "don't know' responses to this item).

304. Craig Humphrey. SOCIAL SCIENCE QUARTERLY. December 1994, p. 805. Comparative public opinion research suggests that a new worldview with distinctive beliefs about social institutions may be emerging in advanced industrial countries. Using survey research from European countries, for example, Inglehart (1977, 1990) found substantial numbers of respondents expressing interest in participatory democracy, protecting nature, reducing the size of the military, nuclear disarmament, and attaining satisfying work, even if it means a lower income. Inglehart found these interests to be intercorrelated, and the results led him to infer that a "postmaterialist" worldview was emerging in Western countries (but see Easterlin and Crimmins [1991]).

Concern about the environment is an important part of postmaterialism. Residents in the United States and Germany favor environmental protection over national economic growth by a ratio of better than two to one.

305. Riley Dunlap. ENVIRONMENT. October 1991, p. 13. Although the fourth item made only a general reference to the costs associated with an environmental protection, several sets of shorter-term trend data dealing explicitly with personal willingness to pay for environmental protection have yielded similar results. For example, between 1986 and 1990, Cambridge Reports used the following item: "Increased efforts by business and industry to improve environmental quality could lead to higher consumer prices. Would you be willing to pay higher consumer prices so that industry could better preserve and protect the environment, or not?" The percent responding yes rose from 56 to 72 percent over the five years.

306. Robert Shapiro. PUBLIC OPINION QUARTERLY. Vol. 50, 1986, p. 14. The survey data during the 1980s reveal substantial and in some cases clearly increasing support for environmental regulation. The year 1983 was pivotal: the Reagan administration had wanted to overhaul the major environmental laws, but it misread or failed to anticipate public opinion (Lake, 1983; Business Week/Harris Poll, 1983; Shabecoff, 1983). In response to survey questions about specific environmental problems, large segments of the public viewed nearly all of them as important *national* issues in need of significant attention, although they may rarely be perceived as *local* problems.

Support for environmental protection has been resistant thus far to energy shortages and price increases, recession, Reagan's landslides, and the high costs of environmental programs and regulation. There was little, if any, appreciable backlash in opinion. The proportion of the public wanting less regulation or less spending has remained small, with relatively little change, even as support for deregulation or "less government" rose at the beginning of the Reagan administration (cf. Shapiro and Gilroy, 1984). The public has been reluctant to support environmentally harmful policies which might improve the economy and energy resources, with possible exceptions such as increases in strip-mining and offshore oil exploration, although citizens may waver if faced with hard choices close to home (Rosenbaum, 1985:68–70). It is evident, however that people do not—perhaps rationally—see as mutually exclusive a healthy environment and other important needs, or they do not want to have to choose between them (Ladd, 1982; Mitchell, 1980; Keefer, 1984).

307. Walter A. Rosenbaum. ENVIRONMENTAL POLITICS AND POLICY, 3d Ed. 1995, p. 25. The strength of public support for environmental protection in the late 1980s, measured by public opinion polls, appears robust, widespread, and sometimes unprecedented. In mid-1989, for instance, a *New York Times*/CBS News poll, repeating a question often posed since 1970, asked a representative sample of the American public whether it agreed with the statement: "Protecting the environment is so important that requirements and standards cannot be too high, and continuing environmental improvements must be made regardless of cost." The results of that poll are found in Figure 1–1. Almost 80 percent of the respondents agreed with this assertion while 14 percent disagreed. By comparison, only 45 percent of the respondents had agreed to the same assertion in 1981 and about 65 percent agreed in 1986. Numerous studies suggest that substantial majorities in almost all major socioeconomic groups support the environmental movement and governmental programs to protect the environment—and have done so since the outset of Environmental Era I. The Reagan administration contributed handsomely to its own vexations by failing to appreciate the breadth and tenacity of public support for environmental protection in the face of regulatory reform aimed directly at major environmental programs. Environmental interests effectively mobilized the public against much of the Reagan reform agenda, thereby greatly limiting its scope and effectiveness.

308. Riley Dunlap. PUBLIC OPINION QUARTERLY. Vol. 55, 1991, p. 656. One might argue that it is easy to express support for environmental protection over economic growth in the abstract, where the personal implications are not apparent. However, a variety of evidence shows an increase in the public's expressed willingness to pay higher prices for goods and services (including grocery items), to the point that willingness to absorb

the cost of environmental protection has clearly become the majority position (tables 32–36). This increased willingness to pay higher prices for environmental quality is dramatically illustrated in table 37.

309. Scott Segal. TEXAS ENVIRONMENTAL LAW JOURNAL. 1996. Moreover, on Earth Day 1995, Vice President Gore gave a speech in which he referred to a recent *Time Magazine* poll indicating "that 54% of Americans supported tax increases to help the environment." Even when the environment is compared with the issue of budgetary restraint, for example, the public still tends to side with environmental protection. Two-thirds of persons polled in a January 1995 *Los Angeles Times* survey said they "opposed budget cuts on the environment to reduce the federal deficit."

310. Riley Dunlap. ENVIRONMENT. October 1991, p. 10. Three items focus explicitly on public support for environmental protection by the government. The first, used continuously by the National Opinion Research Center since 1973, asks respondents whether they think the government is "spending too much, too little, or about the right amount" on "the environment" (see Figure 1 on page 12). After leveling off in the mid 1980s, the percentage of people responding that "too little" is being spent has climbed steadily and substantially over the past four years. By the end of the decade, the ratio of respondents wanting more rather than less spending on the environment rose to nearly 18 to 1, or an astounding 71 percent versus only 4 percent—the former percentage exceeding not only the 1980 level of 48 percent but also the 1973 level of 61 percent.

311. Riley Dunlap. ENVIRONMENT. October 1991, p. 12. It could be argued that the foregoing items are weak indicators of support for environmental protection because they make no reference to the costs associated with such protection. This is not true, however, of the next item. Used intermittently in *New York Times*/CBS polls since 1981, the fourth item asks respondents to agree or disagree with a very strongly worded statement: "Protecting the environment is so important that the requirements and standards cannot be too high, and continuing environmental improvements must be made regardless of cost." Agreement with this statement rose steadily over the past decade, with the result that in 1990 three-fourths of the public was expressing agreement and only one-fifth was expressing disagreement (see Figure 4 on page 32).

312. Kent Portney. CONTROVERSIAL ISSUES IN ENVIRONMENTAL POLICY. 1990, pp. 23–24. This apparent concern for environment quality also seems to carry over to opinion about governmental policies. For example, in a 1982 Gallup survey, people were asked whether they favor or oppose a number of different proposals including "strict enforcement of air and water pollution controls as now required by Clean Air and Clean Water Acts." Some 83% of the public favored strict enforcement, more than favored not cutting Social Security, or even prohibiting the Justice Department and federal courts from ordering busing to achieve social balance (Mitchell, 1984, p. 65). A 1986 American National Election Studies survey conducted by the Institute for Social Research at the University of Michigan asked people about federal spending on environmental protection. This survey showed that about 50.6% of the people surveyed felt that federal spending on improving and protecting the environment should be increased. Only about 4.6% felt that spending should be reduced. More recently, a Gallup survey focused on the 1988 presidential election revealed that having a president interested in "proposing new laws to increase protection of the environment" was a very high priority for a significant majority of people, second only to reducing the federal deficit.

313. Riley Dunlap. PUBLIC OPINION QUARTERLY. Vol. 55, 1991, pp. 654–655. Given the foregoing it comes as no surprise that support for government action on behalf of environmental quality has risen substantially, particularly in the last few years. Thus, a large majority believes that government is "spending too little" on the environment (table 16), and majorities say that government regulations have "not gone far enough" (table 17) and that there is "too little" government regulation in the area of environmental protection (table 18). These strong multi-year trends are bolstered by two-point-in-time data as well (tables 19 and 20). Especially striking is the proportion of "pro-" to "anti-environmental" opinion found with these items in 1990.

Public support for government action on specific types of environmental problems is also strong. Majorities believe that too little is being done to solve waste problems (table 21) and large majorities want "urgent" or "prompt" government action on a wide range of problems (table 22). Even when attention is focused on specific restrictions on individual behavior, there is generally strong support (table 23).

The strong endorsement of government action to protect environmental quality is understandable given that the public sees government as having primary responsibility for environmental protection and is somewhat skeptical of the efficacy of efforts by individuals in the absence of government regulations (see, e.g., Roper Organization 1990). Such strong support of government action makes the public's negative reaction to the Reagan administration's environmental agenda understandable (Dunlap 199a; Mitchell 1990).

314. HOTLINE. 19 April 1990, Lexis/Nexis. *The New York Times*/CBS polls (offer) a statement that's worded very strongly: "Protecting the environment is so important that requirements and standards cannot be too high, and continuing environmental improvements must be

made regardless of cost." The first time they asked it was back in '81; 45% agreed. In an early '89 survey it was up to 74%, and in a more recent one it was up to 80%. Again, an incredible, consensual percentage . . . (for) an extremely strong pro-environmental position. [ellipses in original]

315. Riley Dunlap. ENVIRONMENT. July/August 1987, p. 34. At the time Anthony was writing, Louis Harris was beginning to attract considerable attention for his inferences (from his survey results) about the political impact of environmental quality. As Anthony noted, in a February 1982 survey, Harris found 15 percent of those polled indicating that they probably would not vote for a congressional candidate whose views on "controlling air and water pollution" disagreed with their own, even if they mostly agreed with the candidate on other issues. Further, since 13 percent held pro-environment view and only 2 percent held anti-environment views, Harris concluded that environmental issues represented a potential swing vote of 11 percent—ranking it third among nine issues examined.

316. Sheldon Dunlap. ENVIRONMENTAL POLITICS AND POLICY. Ed. James Lester. 1995, p. 108. Finally, it should be kept in mind that public support for environmental protection is crucial not only at election time, but when lobbying for new legislation or more effective implementation and enforcement of existing legislation. Thus, the true test of the strength of public concern for environmental quality lies not just in voting, but in signing petitions, writing and phoning public officials, attending meetings, testifying at hearings and so forth. Such political behaviors, along with providing support to environmental organizations (and their staffs of lobbyists, lawyers, and scientists), are all ways in which the American public can support environmental protection. The results reported in this chapter suggest that the public is amenable to providing such support; the key lies in how well the environmental movement and political leaders can mobilize it. At a minimum, the existence of broad public support in favor of environmental protection provides legitimacy for those working on behalf of this goal and a major obstacle to political candidates and others who suggest that protecting the environment is less important than stimulating the economy. This represents an enormous change from a quarter century ago.

317. Norman Vig. ENVIRONMENTAL POLICY IN THE 1980s. 1990, pp. 45–46. Bush had been searching for issues on which to differentiate himself from Reagan during the year before the 1988 election. One issue that surfaced repeatedly among his campaign advisers was the environment. During the spring of 1988 his policy advisory staff, headed by pollster Robert Teeter, identified this as one of several key issues (along with education, drugs, and crime) that could swing independent voters to Bush, and the campaign conducted extensive research on the environmental record of Bush's likely opponent, Gov. Michael Dukakis. During June and July polls indicated rapidly rising public concern over the drought and other environmental issues. Meanwhile, the Bush campaign staff had identified six key "swing" states that were likely to determine the outcome of the election: California, New Jersey, Ohio, Illinois, Michigan, and Missouri. These were all states with a large number of suburban voters interested in "quality of life" issues. In the absence of major economic and foreign policy crises, these issues took on increasing significance among Bush's campaign strategists. In mid-August, the campaign staff, headed by James Baker and Teeter, decided to schedule Bush for a week of environmental speeches at the end of August and in early September. Robert Grady, a former aide to New Jersey governor Thomas Kean, who had emerged as the staff's environmental adviser, drafted the speeches for Bush.

318. Norman Vig. ENVIRONMENTAL POLICY IN THE 1980s. 1990, p. 64. The administration's third misperception about public opinion was to assume that the environmental issue and environmentalists were a negligible political threat to the president's supporters in Congress and to the president himself. If the administration had abided by the implicit rules followed by previous administrations instead of challenging the environmental consensus, this assumption would have been valid. But by making so little effort to placate environmental sentiment in appointments, in legislative initiative, and in rhetoric during its first two years, the Reagan administration created the potential for the issue to have some political potency in the 1984 elections. In the longer run the administration appears to have handed a previously nonpartisan issue to the Democrats.

319. Riley Dunlap. ENVIRONMENT. July/August 1987, p. 38. While the foregoing helps to account for the negligible political damage caused by Reagan's poor environmental record, readers should not conclude that environmental issues have *no* political significance. First, environmental activists can provide significant resources such as labor and money to political campaigns. Through such support, environmentalists frequently claim to have affected the outcome of targeted campaigns. Second, in some campaigns environmental issues are so important they clearly affect the outcome. For example, Brock Adams's (D-Wash.) startling upset of Slade Gorton (R.–Wash.) in the 1986 election for the U.S. Senate is commonly attributed to Adams's staunch opposition to having the nation's first permanent nuclear waste repository located in Washington—a position shared by an overwhelming majority of state residents.

320. Steven Adler. THE WALL STREET JOURNAL. 29 July 1996, www.wsj.com A majority of Americans consider a political candidate's position on environmental issues when deciding how to vote: 81% rate

environmental positions as greater than five on a 10-point scale of importance.

321. Scott Segal. TEXAS ENVIRONMENTAL LAW JOURNAL. 1996. Representative Sherwood Bochlert (R-NY), a key figure in rallying disaffected GOP members against the appropriations riders dealing with the environment, noted that "Republicans can read—every poll members saw showed that the public said we shouldn't be going this way." One frequently cited poll, conducted by NBC and the *Wall Street Journal*, indicated that seventy-nine percent of Americans (both Republicans and Democrats) believe that environmental regulations should be "retained or strengthened." When presented with a choice between strengthening regulations or weakening environmental regulations, seventy percent of Americans chose strengthening in a recent *New York Times*/ABC News poll.

322. Jeremy Rifkin. VOTING GREEN. 1992, pp. 32–33. A majority of Americans, regardless of their party affiliation or past ideological orientation, are in favor of tough new legislative initiatives to clean up the environment. According to a Harris Poll conducted in 1990, three out of every five Americans are now deeply concerned about the environment. In a survey conducted by the Associated Press, 79 percent of all the respondents said that pollution now threatens their quality of life. Interestingly, an ABC/*Wall Street Journal* Poll in 1990 "found no difference in the degree of support for environmental protection between Democrats and Republicans." A majority of Americans from both political parties said they are prepared to make significant economic sacrifices, if necessary, to address the growing environmental threats. In a Gallup Poll conducted in 1990, 71 percent of those surveyed believed that protecting the environment should be given priority, even if it meant "stifling economic growth." In a Harris Poll, 95 percent of those surveyed said they would even be willing to trade a loss of jobs and the closing of polluting plants, if necessary, to clean up the air. Clearly, a majority of the public, Republicans and Democrats alike, are beginning to shift political perspective, aligning themselves with the new Green vision of progress with its emphasis on the quality of life and the protection of the biosphere.

323. Riley Dunlap. ENVIRONMENT. October 1991, p. 12. Taken together, the results from these three items show that a majority—sometimes a large majority—of the U.S. public favors increased efforts by government to protect the environment and that only small minorities (ranging from 4 to 16 percent) favor reducing these efforts. Results such as these have led several public opinion analysts to argue that environmental protection has become a "consensual" issue in U.S. society, "a basic value that no major bloc of voters opposes." This consensus was perhaps most dramatically reflected in a 1989 Harris poll in which respondents were asked, "Do you think this country should be doing more or less than it does now to protect the environment and curb pollution?" An incredible 97 percent felt that the country should be "doing more" and only 1 percent that it should be "doing less." (The remaining 2 percent volunteered, "about what we do now" or were "not sure.") The implications of such lopsided support for environmental protection were captured by another analyst in describing the frustration faced by the business community in opposing environmental regulations: "The problem is that anti-environmental values have no legitimacy in the public debate."

324. Kent Portney. CONTROVERSIAL ISSUES IN ENVIRONMENTAL POLICY. 1990, p. 23. In general, there is much evidence that the American public is very concerned about environmental issues and highly supportive of efforts to protect the environment. Almost all of the major public opinion surveys indicate that usually more than half of the American people want a cleaner and safer environment. Different surveys ask people different questions, but the general pattern seems consistent. For example, a 1988 Gallup Survey revealed that some 66% of the public feels "extremely concerned" about pollution of oceans and beaches, 66% is extremely concerned about pollution of drinking water, and 50% is concerned about air pollution.

325. SAN FRANCISCO EXAMINER. 14 August 1995, p. A15. But, a recent *New York Times*/ABC News poll found that a full 70 percent of Americans believe environmental regulations should be strengthened, not weakened. A separate Luntz poll, conducted for House Speaker Newt Gingrich, found a 2–1 majority favor "doing more to protect the environment' over "cutting regulations,' including a majority of Republicans and Perot voters.

326. THE ATLANTA JOURNAL. 25 September 1995, Lexis/Nexis. Republicans should take note of the recent NBC-*Wall Street Journal* poll finding that 79 percent of Americans, Republicans and Democrats, believe environmental regulation should be retained or strengthened. That's hardly mandate for extreme measures, especially for those that are sneaked through in budget bills.

327. Riley Dunlap. ENVIRONMENT. October 1991, pp. 10–11. The second item, used by Cambridge Reports since 1982, shifts from government spending to respondents' views on the appropriate level of "government regulation and involvement" in environmental protection (see Figure 2 on page 13). After dropping in 1987, there has been a substantial rise in the percentage responding that there is "too little" government involvement in environmental protection. This surge has yielded an increase of 27 percentage points, from 35 to 62 percent, between 1982 and 1990.

328. Sheldon Dunlap. ENVIRONMENTAL POLITICS AND POLICY. Ed. James Lester. 1995, pp. 105–106. Having already summarized the long-term trends in public opinion toward environmental issues, let me end by offering some general conclusions and drawing some implications from the results reviewed in this chapter. It is clear that the direction of public opinion is strongly proenvironmental, as large majorities express support for protecting the environment. Thus, it seems appropriate to conclude that environmental protection has become a highly consensual issue. Large majorities typically take proenvironmental positions on issues such as supporting government regulations and spending, and only small minorities endorse antienvironmental positions (often smaller than those taking the neutral position). This conclusion is bolstered by the fact that huge majorities of Americans belong to proenvironmental orbits, identifying themselves as environmentalists or as sympathetic toward the environmental movement and often engaging in various degrees of environmental activism. Indeed, very few policy issues achieve a level of consensus equal to that of current support for environmental protection.

329. Norman Vig. ENVIRONMENTAL POLICY IN THE 1980s. 1990, p. 85. A particularly striking example of the increase in support for environmental protection during the 1980s is found in a series of seven polls conducted by the *New York Times*/CBS News. In each survey interviewers asked a national sample of Americans whether they agreed or disagreed with the following statement: "Protecting the environment is so important that requirements and standards cannot be too high, and continuing environmental improvements must be made regardless of cost." Given the strongly pro-environmental wording of the statement, the fact that in 1981 a plurality of 45 percent said they agreed with it was regarded as strong evidence of public support for environmental protection. Thus, it is striking that subsequent askings during the 1980s traced an ever-upward level of support for this position, as shown in Figure 4–1. By the end of Reagan's first term, the percentage of respondents saying that environmental improvements must be made regardless of cost had increased to 58 percent. In 1988, shortly before he left office, the percentage had climbed to 65 percent. Two successive surveys in 1989 recorded an increase of an additional 15 percentage points for those who took this uncompromisingly pro-environmental position—another effect of the *Exxon Valdez* oil spill.

330. Norman Vig. ENVIRONMENTAL POLICY IN THE 1980s. 1990, p. 57.
Causes of Environmental Concern. Why has the environment become such an enduring social concern that the environmental consensus persisted in the face of the Reagan attack? First, environmental issues have an inherently broad appeal. In contrast to many social problems that affect a particular group such as blacks or the handicapped, environmental problems potentially affect all citizens. Environmental quality improvements are true public goods in the sense that they are available to everyone. Moreover, and this is the so-called motherhood component, environmental improvements involve non-controversial goals, such as health, cleanliness, natural beauty's preservation, future generations' needs, and wildlife protection. Because numerous issues fall into the environmental category, diverse constituencies have an interest in supporting environmental regulation. Hunters as well as animal protection advocates favor wildlife habitat protection. Urban residents are affected particularly by air pollution, rural dwellers by strip mines and well contamination caused by toxic wastes. Citizens with little interest in nature and wildlife may be sensitive to the public health threat posed by toxic wastes or drinking water contamination.

331. Riley Dunlap. PUBLIC OPINION QUARTERLY. Vol. 55, 1991, p. 651. The last *Public Opinion Quarterly* report on the environment noted that "the issue of environmental protection has had its ups and downs [but] has been a persistent concern" (Gilroy and Shapiro 1986, p. 27). Five years later it is clear that public support for environmental protection not only has persisted but also has risen substantially in recent years. This report documents that upward trend by reviewing a wide range of national trend data. But first, the history of public opinion on environmental issues is briefly reviewed in order to put the recent trends into context.

332. Riley Dunlap. ENVIRONMENT. October 1991, p. 10. The widespread celebration of the 20th anniversary of Earth Day in April 1990 illustrated the prominent position that environmental issues have maintained on the U.S. agenda. This prominence—somewhat surprising given that few social causes manage to persist for two decades—is strongly supported by recent public opinion polls, which show that the upturn in support for environmental protection that occurred during the Reagan administration has definitely not subsided. Indeed, a summary of public opinion trends through 1986 that appeared in the July/August 1987 issue of *Environment* holds true today: "After declining moderately from the early to late 1970s, public support for environmental protection began to rise shortly after Reagan took office and has continued to do so." In fact, this support has increased at an accelerating pace in the last two or three years. A careful analysis of the evolution of public support for environmental protection from 1980 to 1990 should help elucidate some of the implications of the U.S. public's current high level of environmental awareness and concern.

333. THE NEW YORK TIMES. 5 July 1995, p. A11. The compromise, passed by the House, is still awaiting final action in the Senate. Environmentalists had called for a veto of it because of the logging provision; in agreeing to that provision, the Administration promised to

protect the environment in any logging operations that it allows.

What the Administration's political strategists find especially significant is the wide margin of public support for environmental regulations across a variety of issues.

"The voters reject less regulation of toxic and hazardous wastes by 17–76, of water pollution by 26–71, of drinking water by 23–71, of wetlands and coastal areas by 24–63, of air pollution by 35–61 and of historic sites by 35–59," said Mr. Morris's report on his Virginia poll, a telephone survey of 1,000 likely voters taken May 18–21. The sample provided a 90 percent likelihood of accuracy within a margin of plus or minus 2.4 percentage points.

334. Riley Dunlap. PUBLIC OPINION QUARTERLY. Vol. 55, 1991, pp. 652–653. Public support for environmental protection is sometimes questioned because environmental problems are seen as not being very important to the public (see Mitchell 1990, p. 84). The primary evidence is that such problems have low "salience," indicated by the near disappearance of "environment" from volunteered "most important problem" (MIP) responses after the early seventies (Dunlap 1989, pp. 101–4). This situation has changed, at least temporarily, the past couple of years. Results in tables 1 and 2 and from a number of cross-sectional studies (e.g., Opinion Dynamics Corporation 1990, p. 14) reveal that environmental problems have once again become relatively salient, receiving significant MIP mentions (but lagging behind top-rated issues such as drugs).

Most important problem responses are a stringent measure of salience, and it has been argued that public concern for the environment should be measured relative to concern over other problems (Dunlap 1989, pp. 124–30). Results from two different approaches to measuring "relative importance," shown in tables 3 and 4, reveal that environmental problems fare better than in MIP responses (not unexpected given Shuman and Scott's [1987] comparison of question formats), but still fail to reach the very top of the list. Thus, environmental problems have become more salient in recent years but are still seldom seen as *the* most important problem facing our nation.

335. Norman Vig. ENVIRONMENTAL POLICY IN THE 1980S. 1990, p. 84. It now appears that environmental issues are gaining in salience as well as in strength. Evidence of a possible salience shift comes from a "most important problem" question asked four times a year of a large national sample by Cambridge Reports. At the end of 1987 only 2 percent of respondents told the Cambridge interviewers that the environment was "one of the two most important problems facing the United States today." This level was typical of many earlier askings of this type of question in other national surveys. By the end of 1988, however, mention of the environment recorded a modest increase to 6 percent. Then, in Cambridge Reports' second quarter (April-June) 1989 survey, the percent citing the environment as a most important problem jumped to 16 percent, an increase that may confidently be attributed to the *Exxon Valdez* oil spill, which occurred in late March. Significantly, the level of salience did not fall in the third quarter (July-September) 1989 survey, when 16 percent again named the environment.

336. THE CHRISTIAN SCIENCE MONITOR. 27 November 1995, Lexis/Nexis. WHAT happened to public concern about the environment? Proposals to defund environmental agencies and undo decades-old environmental protections are sailing through Congress. These are not minor adjustments; if passed, they will bring substantial degradation of our air, water, forests, and natural species. Are our elected representatives responding to new preferences of the voters?

There is a perception by the new Republican majority that voter weariness of intrusive government includes "burdensome" environmental regulations.

What data support this idea? The primary evidence cited is that antiregulatory candidates swept last year's elections. Also, opponents of environmental regulations have been outspoken, while support is heard mostly from environmental groups and liberal pundits.

A more thorough analysis, based on national polls and in-depth interviews with voters, suggests that the perception on Capitol Hill is faulty. In pushing anti-environmental proposals, congressional Republicans are taking risks that could threaten other goals more central to their agenda.

The opinion data show that solid majorities now oppose government regulations in general. Environmental regulations, however, are a notable exception.

337. Sheldon Dunlap. ENVIRONMENTAL POLITICS AND POLICY. Ed. James Lester. 1995, pp. 107–108. In short, while there may not be a large "green bloc" of single-issue voters, comparable in commitment to the antiabortion or antigun control lobbies, the results presented in this chapter indicate that environmental protection has the support of a large majority of Americans. For this reason, very few candidates wish to be identified as "antienvironmental," especially at a time when environmentalists are increasing their efforts to publicize candidates' records and positions on environmental issues (League of Conservation Voters, 1992; Rifkin and Rifkin, 1992). Furthermore, it should be kept in mind that voting is only one of many political actions that people take. Individual environmentalists and environmental organizations such as the League of Conservation Voters often play crucial roles in political campaigns, either in support of proenvironmental candidates or in opposition to those with poor environmental records. Contributing money, distributing campaign literature, making phone calls, and so on, can be of immense value to a candidate, and if only a tiny fraction of the huge proenvironment majority of the public takes

such actions it could make the difference in many elections.

338.   Sheldon Dunlap. ENVIRONMENTAL POLITICS AND POLICY. Ed. James Lester. 1995, p. 96. Secondly, all kinds of environmental problems have received enormous media attention in recent years (Mitchell, 1990: 88-89). For example, during 1988-89 alone, *Newsweek* carried numerous cover stories on environmental problems, *U.S. News and World Report* (1988) had one on "Planet Earth: How It Works, How to Fix It," and *Time* (1989) captured the most attention by naming the "Endangered Earth" as "Planet of the Year" in lieu of its famous "Man of the Year" for 1988. Ungar (1992) has shown how scientists' warnings and abnormally hot weather generated media (and societal) attention to global warming in the late eighties, while Mazur and Lee (1993) have documented the huge increase in media attention given to the ensemble of global problems—ozone depletion, the greenhouse effect, and rainforest destruction—during this same time period.

339.   George C. Edwards. AT THE MARGINS. 1989, p. 105. A primary task of members of Congress is to preempt an effective challenge to their reelection by demonstrating broad support and a lack of vulnerability in their constituencies. Thus they are in most cases highly attentive to their constituents and eager to avoid votes and stands on issues that might be a catalyst for strong opponents to emerge.

In such an environment, it should come as no surprise that senators and representatives are hypersensitive to the anticipated reaction of their constituents to their actions. They overestimate their visibility to their constituents and the extent to which the electorate is concerned about issues. Whether or not voters follow the issues of the day or know how members of Congress vote on them, the members certainly act as if their votes make a difference in their chances for reelection or elevation from the House to the Senate.

Reinforcing this concern for public opinion are the role perceptions of members of Congress. Although good time series data on congressional role perceptions are lacking, and the way in which these perceptions affect behavior is unclear, there is some reason to believe that many senators and representatives feel they ought to reflect public opinion in their voting in Congress. To the extent that this holds at any given time, it increases the sensitivity of members of Congress to the electorate.

Members of Congress thus face an environment of uncertainty and vulnerability. They want to please their constituents, yet often do not know how the voters feel about matters of public policy or politics. Looming over this arena is the public official who is the most visible and whom senators and representatives must continuously take into account: the president.

As the central figure in American politics, the president is the object of a constant stream of commentary and evaluation by all segments of society, including those that are unlikely to articulate specific policy preferences. Indeed, press coverage of the president and his policies exceeds that of all other political figures combined. The president's standing in the public provides members of Congress with a guide to the public's view.

340.   Norman Vig. ENVIRONMENTAL POLICY IN THE 1980s. 1990, pp. 59–60. A final reason for the continued support for environmental protection is its high visibility in the news media. This stems from the persistence, multiplicity, and local character of environmental threats. Nearly each time Congress passes legislation that addresses a current environmental problem, new ones command the attention of the press and the public, ultimately requiring congressional action. In 1983, for example, toxic wastes and acid rain were two issues prominent in Congress, yet, as recently as 1975, few people had heard of either. By 1980, however, 64 percent of the public expressed "a great deal" of concern about hazardous chemical waste disposal, a level of concern even higher than had been expressed about air and water pollution in earlier surveys. Site-specific toxic waste problems also contributed to this issue's visibility, as the Love Canal and Times Beach stories made for dramatic television and print media news coverage. It is not surprising, therefore, that of all the controversies surrounding Gorusch's tenure at EPA, her administration of the $1.6 billion toxic-waste cleanup program would be the one to ultimately cause her downfall.

341.   Fred Starheim. ENERGY, THE ENVIRONMENT, AND PUBLIC POLICY. 1988, p. 91. Repeated messages by the popular media have molded public opinion on environmental issues. The president of the renowned Air and Waste Management Association (Thielke 1990) has noted that members of the public with whom he has been in contact are generally content with the condition of the environment to which they are exposed day-to-day. However, these same people express real concern over such environmental issues as acid rain, toxic waste, hospital waste on beaches, and destruction of the ozone layer—of which they have become apprised through the media—although reportedly none of these people have experienced these issues or their effects firsthand.

These observations by Thielke and others underscore two important concerns: the strong sentiment people develop in response to publicized environmental issues and the popular media's role in influencing and informing public opinion.

342.   Sheldon Dunlap. ENVIRONMENTAL POLITICS AND POLICY. Ed. James Lester. 1995, p. 96. That such worsening environmental conditions were viewed as growing threats to humans is reflected in results from Roper national surveys that found that the percentage

agreeing that "environmental pollution" is a "very serious threat these days to a citizen like yourself" increased from 44 in 1984 to 62 in 1989 (Dunlap and Scarce, 1991:659–661). More generally, a 1990 Hart/Teeter national survey found 76 percent saying that they "think the environment in the United States" had become "worse" and only 16 percent that it had become "better" over the last twenty years. Similarly, a 1990 Gallup national survey found only 14 percent saying that our country had made "a great deal" of progress in dealing with environmental problems over the past twenty years, and only 18 percent had a similar level of optimism that we would make great progress over the next twenty years (Dunlap, 1991:14–15). In short, the evidence indicates that over time the public has come to see both local and global environmental conditions as worsening, views these problems as threats to human welfare, and is pessimistic about future conditions and our ability to deal with them.

343. Kent Portney. CONTROVERSIAL ISSUES IN ENVIRONMENTAL POLICY. 1990, p. 40. After legislative proposals have emerged, policy-making turns on the process of legitimating one of these. This process consists of the actions taken by any party to line up support for, or opposition to, any given legislative proposal. In this process, various actors may try to mobilize and stimulate support for one piece of legislation. Not uncommonly, interest groups will mobilize opposition to legislative proposals that they consider harmful to their interests. Legitimation also involves activities to formulate political strategies to make one alternative sound much better than others.

Legitimation activities feed directly into the process of adopting environmental policy. Environmental policy adoption processes involve the actions of members of legislatures leading up to and including voting for or against enactment. This process is usually started with the formal introduction of a bill in Congress or a state legislature. It may include the activities associated with legislative committee hearings, the process actually used by the legislature to dispose of the legislation, end even the interpersonal politicking—bargaining, logrolling, and the like—surrounding a legislative proposal.

344. Charles Ostrom. PRESIDENTIAL STUDIES QUARTERLY. Fall 1988, p. 745. For Neustadt, reputation and prestige are the needed ingredients. Reputation pertains to the judgments of other decision makers about the president's skill as a politician. Reputation is essentially the evaluation of the "Washington elite" based upon a president's willingness and ability to play the "insider" game, to protect himself and to win allies, and to maximize the political discomfort of adversaries.

Prestige refers to the president's standing *outside* the Washington community, primarily his standing with the American public. Herein lies the importance of the opinion polls and public support. The president's approval score, as the foremost indicator of prestige, is a vital piece of information which operates to shape the perceptions of Washington decision makers and make them susceptible to persuasion. As Neustadt argues,

> Most members of the Washington community depend upon outsiders to support them or their interests. The dependency may be as direct as votes, or it may be as indirect as passive toleration. Dependent men must take account of popular reaction to their actions. What their publics may think of them becomes a factor, therefore, in deciding how to deal with the desires of a president. His prestige enters into that decision; their publics are part of his. Their view from inside Washington on how outsider view him thus affects his influence with them.

Public support serves as an integral component of power because it provides a president with leverage over other decision makers. A substantial level of public support increases the political risks and uncertainty of opposing the president and thus can be expected to condition the actions of other decision makers. As such, the most recent opinion poll provides an indication of whether "it is wise or foolish to resist presidential persuasion."

345. George C. Edwards. AT THE MARGINS. 1989, p. 106. Because of the high visibility and frequency of presidential approval polls, it is safe to assume that members of Congress are aware of the president's standing with the public. In addition, senators and representatives learn of the public's opinion of the president from other political elites, political activists, leaders of interest groups, the press, attentive publics, and constituents. Some of what they hear may be echoes of their own actions in government.

The public's evaluations of the president are important not only as surrogates for broader opinions on politics and policy, but also as influences on congressional behavior. Members of Congress must anticipate the public's reaction to their decisions to support or oppose the president and his policies. Depending on the president's public standing, they may choose to be close to him or independent from him to increase their chances of reelection. Polls find that a significant percentage of voters see their votes for candidates for Congress as support for the president or opposition to him.

346. Charles Ostrom. PRESIDENTIAL STUDIES QUARTERLY. Fall 1988, p. 744. Simply stated, public support is closely linked to presidential power and as a result, influences the character and outcomes of politics in the nation's capital. As an element of public opinion, the level of public support is, first of all, an ingredient of the Washington political context. Whether interpreted in terms of the president's image, credibility, or competence, public support is an intrinsic component of the power situation in which the president must operate. The public

standing of the president serves to define a prevailing atmosphere within which competition among the president and other decision makers take place. As Neustadt observes, "the prevalent impression of a president's pubic standing tends to set a tone and to define the limits of what Washingtonians do for him or do to him."

If the character and style of this competition are considered, public support assumes an even more significant role. The primary feature of Washington politics is that formal decision-making authority is fragmented among multiple "pockets of power"—the president, the Cabinet, congressional committees and subcommittees, executive bureaus and agencies—each pursuing their own objectives, each responsive to different constituencies, and each exercising checks upon the authority of other decision makers. In this setting, shared partisanship and formal lines of organizational authority offer little guarantee of cooperation in the formulation, passage, or implementation of policies. Initiatives in different substantive domains typically require different supporting coalitions and success on one policy front ensures neither future successes nor its extension to other areas. Policy coalitions are thus fragile and transient. The ability to construct and maintain such alliances is regarded as an acid test of successful political leadership.

347. Jon Bond. PRESIDENT IN THE LEGISLATIVE ARENA. 1990, pp. 179–180. As we observed above, there is some confusion concerning the nature of the relationship between public approval of the president and support for his preferences in Congress. The Moffett quotation at the beginning of this chapter suggests that presidential popularity has a strong, direct effect on congressional behavior—that is, as the president's popularity increases, so does his success in Congress. And political scientists present systematic quantitative evidence in support of this perspective. For example, Edwards (1980) finds strong correlations between presidential popularity and support in Congress. He argues that his "research provides evidence that members of Congress *do* respond to . . . the president's current popularity among their supporters" (110, emphasis in original), and that presidential popularity "must be added to the list of variables that have been found useful in explaining roll call behavior" (109) Rivers and Rose (1985, 195) say that their "analyses show that presidential influence in Congress rises and falls with the president's public prestige." Similarly, Ostrom and Simon (1985, 350) conclude "that the 'public' and 'legislative' presidencies are linked via the reciprocal relationship between approval and success." [ellipses in original]

348. Charles Ostrom. PRESIDENTIAL STUDIES QUARTERLY. Fall 1988, pp. 745-746. Public support is a political signal to which journalists and pundits, particularly those in the Washington community, pay attention. The measurements of public support are treated like readings from a political barometer and provide justification for a variety of judgments offered about the qualities of a given administration. The release of an approval rating is often accompanied by a host of commentaries which attempt to decipher the "meaning" within the latest reading of public sentiment. Thus, the ratings have been employed to compare a president to his predecessors, to assess the image and credibility of the incumbent, and to judge the electoral prospects of the president and his party. Together, the measurement of public support and the attention which this relationship receives constitute an important characteristic of the modern presidency. As de Sola Pool has observed, presidential performance polls have become "a fact of American political life" and "a part of America's unwritten Constitution." Presidents since Truman have pursued the objectives of their administration in a context of constant scrutiny by the media and continuous evaluation via the public opinion poll. It is a context that Hodgson labels the "perpetual election."

349. George C. Edwards. AT THE MARGINS. 1989, p. 125. The strategic position of presidential approval is mixed. It accords the president useful leverage in dealing with Congress. As the most volatile resource for leadership, public approval is the factor most likely to determine whether or not an opportunity for change exists. Public approval makes other resources more efficacious. The president's party is more likely to be responsive if the president is held in high public esteem, the public is more easily moved, and legislative skills become more effective. Public approval is therefore the resource with the greatest potential to turn a typical situation into one favorable for change.

350. George C. Edwards. AT THE MARGINS. 1989, p. 101. In his first debate with Stephen Douglas, Abraham Lincoln voiced what has become a central tenet in the contemporary study of the presidency: "Public sentiment is everything. With public sentiment nothing can fail; without it nothing can succeed." In a more current analysis, Richard Neustadt argues, "While national party organization falls away, while congressional party discipline relaxes, while interest groups proliferate and issue networks rise, a President who wishes to compete for leadership in framing policy and shaping coalitions has to make the most he can out of his popular connection." Without question, public support is a primary resource for presidential leadership of Congress.

351. George C. Edwards. AT THE MARGINS. 1989, p. 107. The White House encourages members of Congress to infer from the president's approval levels the public's support for his policies. Ultimately, the effectiveness of this strategy is tied to the potential for making the support of a senator or representative a campaign issue. Presidents high in the polls are in a position to make such threats. According to an aide to Reagan, for example, the president's contacts with members of Congress before the

tax vote of 1981 were "merely a device to keep the congressmen thinking about what could happen next year. I'm sure Mr. Reagan is charming as hell but that isn't what is important. It's his reminding these people that they could lose their jobs next year.

Members of Congress may also use the president's standing in the polls as an indicator of his ability to mobilize public opinion against his opponents. Senators and representatives are especially likely to be sensitive to this possibility after a successful demonstration of the president's ability to mobilize the public, as appears to have occurred in response to the efforts of Reagan's White House in 1981. As Richard Neustadt put it, "Washingtonians . . . are vulnerable to any breeze from home that the presidential words and sighs can stir. If he is deemed effective on the tube, they will anticipate. [ellipses in original]

352. George C. Edwards. AT THE MARGINS. 1989, p. 113. A president with strong public support provides a cover for members of Congress to cast votes to which their constituents might otherwise object. They can defend their votes as having been made in support of the president rather that on substantive policy grounds alone. Of course, a president without public support loses this advantage and may find himself avoided by members of Congress who will certainly not articulate their decisions as having been made in support of the president if the president is caught in the depths of the polls.

In addition, low ratings in the polls may create incentives to attack the president, further eroding his already weakened position. For example, after the arms sales to Iran and the diversion of funds to the contras became a cause célèbre in late 1986, it became more acceptable in Congress and in the press to raise questions about Ronald Reagan's capacities as president. Disillusionment is a dangerous force for the White House.

353. George C. Edwards. AT THE MARGINS. 1989, p. 107. Looking at the matter from another perspective, low presidential approval ratings free members of Congress from supporting the president if they are otherwise inclined to oppose him. A senior political aide to President Carter noted:

> When the President is low in public opinion polls, the Members of Congress see little hazard in bucking him. . . . After all, very few Congressmen examine issue solely on its merits; they are politicians and they think politically. I'm not saying they make only politically expedient choices. But they read the polls and from that they feel secure in turning their back on the President with political impunity. Unquestionably, the success of the President's policies bear a tremendous relationship to his popularity in the poll. [ellipses in original]

354. Jon Bond. PRESIDENT IN THE LEGISLATIVE ARENA. 1990, pp. 185–186. The negative relationships between presidential popularity and congressional support are counterintuitive. Previous research suggested that members of the opposition party might be less supportive of the president when he is popular for two reasons. First, a popular president might be less compromising with the opposition party, which would lead to more partisan voting. Second, members of the opposition party are not likely to receive credit at election time for helping the president. Because their opponents will benefit from the president's popularity, members of the opposition might be inclined to follow their basic partisan predisposition and try to make the president look bad (Bond and Fleisher 1980). Although this reasoning might explain the negative correlation for members of the opposition in general, it does not explain why members of the opposition party base respond positively while members of the cross-pressured opposition respond negatively. In fact, the proposed explanations for a negative relationship would seem to apply more to members of the opposition party base than to members of the cross-pressured opposition.

355. Charles O. Jones. THE PRESIDENCY IN A SEPARATED SYSTEM. 1994, p. 118. The extent to which White House staff can improve public standing is uncertain, in spite of the efforts made to do so. According to Ron Nessen, former press secretary to President Ford,

> The ability of the White House to manage the image of a president, to manipulate the press, is wildly exaggerated. . . . A President who is doing the right things in a substantive way and is popular because of what he is doing or for what he stands for is not going to have to worry about his image in the press. A president who is doing things that are unpopular, or the economy is bad, or his views are not fully accepted—he's going to have problems with the press. There's only a marginal, small impact that any sort of media management or image making can have on this relationship.

Richard A. Brody and Benjamin I. Page provide evidence in support of Nessen's conclusions. In studying the effect of events on presidential popularity, they conclude: "In general, people seem to cast a broad net of responsibility, blaming the President (or giving him credit) for bad or good news even in matters we might consider beyond his control. . . . If the news was good, the President's popularity rose; if bad, it fell. . . . A popularity-maximizing President, then, would do well to produce good results. [ellipses in original]

356. Riley Dunlap. ENVIRONMENT. October 1991, p. 34. It is tempting and easy to criticize the U.S. public for not behaving in ways that are more consistent with its proenvironment opinions. However, several factors ought to be considered. Rightly or wrongly, the public tends to see business and industry—rather than individuals—as the major cause of environmental problems and,

therefore, as having primary responsibility for solving them. As Roper puts it, "In the mind of the public, business causes most environmental problems, so the perception is that business should bear the brunt of the responsibility for addressing them. And the only way business will do so, in the public's eye, is if it is required to by the government." The result, Roper concludes, is that "the search for solutions . . . is above all an institutional affair. One institution—government—should increasingly intervene with another—business—to ensure that environmental improvements are made." [ellipses in original]

357. NATIONAL JOURNAL. 28 April 1990, p. 1062.
Pro-growth conservatives like to believe environmentalism is an "elitist" issue. As Earth Day chairman Denis Hayes recently pointed out, the membership of environmental organizations has for a long time consisted largely of "upper-middle-class, well-educated, politically active, 35–60-year-old white folks." In other words, liberals.

Minorities and working people were not supposed to be moved by environmental concerns. They were too busy worrying about how to make ends meet. The conservative strategy was to build a solid populist base on the issue of economic growth.

That was a serious miscalculation. The pro-growth conservatives had it exactly backward: environmentalism is the populist position, not the elitist position. More people refuse to believe there is any trade-off between environmental protection and economic growth. If forced to choose, however, the public opts decisively for slower growth and a clean environment. That holds true for rich and poor, young and old, Republicans and Democrats.

Business is frustrated by the environmental consensus. To business, the environmental movement is a greater threat to free enterprise than Communism. The problem is that anti-environmental values have no legitimacy in the public debate. Therefore, business leaders find it hard to defend their interests.

358. Robert V. Percival. 54 LAW AND CONTEMPORARY PROBLEMS 172. Autumn 1991, p. 127.
The rise of the modern administrative state has been fueled by the explosive growth of federal regulatory legislation. By shifting policymaking authority from Congress to administrative agencies, regulatory legislation has helped to transform "the system of shared powers created by the Constitution" into "a system of shared influence over bureaucratic decisionmaking." Nowhere is this transformation more apparent than in the competition between Congress and the president for control of federal environmental policy. This competition has stimulated the development of a new tool for executive influence on regulatory decisions: review of agency action by the Executive Office of the President. Presidential use of regulatory review and the congressional responses it has provoked pose new challenges to theories of the impact of separation of powers on federal policymaking.

359. Patricia M. Wald. B. C. ENVIRONMENTAL AFFAIRS LAW REVIEW 519. 1992, pp. 520–521. How do the courts enter the picture? In each of its environmental laws Congress has specified—with varying degrees of clarity—which actions of the EPA or other agencies a party can challenge in court and what the challenger must prove. Because Congress often has been at odds with the executive branch over the implementation of environmental legislation, the provisions for court review of the EPA's actions are unusually detailed. In many statutes, Congress has included provisions regarding judicial reviews to preempt the less specific Administrative Procedure Act (APA) review provisions that typically govern challenges to regulatory statutes.

In general, the prominent place accorded judicial review in federal environmental statutes must be viewed in the context of our overall concepts of limited government and the separation of powers. There must be a neutral forum in which to decide disputes over whether the executive branch is carrying out the will of Congress as set out in the laws, whether it is exercising authority it was never given, whether it is declining to follow mandates it was given, and whether it is making unreasonable decisions when it has rulemaking discretion. When Congress originally passed the major environmental laws in the early 1970s, there was widespread suspicion among its members that the executive branch would not move fast enough on its own—that it had become a "captive" of private industry. Again in the mid-1980s, there was profound disenchantment with the way the EPA was doing (or not doing) its business. Congress therefore gave the courts the important role of watchdog in order to insure fidelity to the laws' strictures. The fact that, during much of the past two decades, the presidency and Congress have been in the hands of different political parties has intensified the constitutionally embedded tensions between the branches.

360. Robert V. Percival. 54 LAW AND CONTEMPORARY PROBLEMS 172. Autumn 1991, p. 173.
One of the most ignored, but significant, consequences of regulatory review has been its contribution to intensifying institutional competition between the presidency and Congress for control of federal policymaking. The concept of regulatory review by the Executive Office was developed during President Nixon's unprecedented efforts to expand presidential policymaking authority through administrative action. As the White House staff grew in size and influence, Congress expanded the size of its own bureaucracy and developed new devices for asserting its influence. With Congress and the presidency long under control of opposite political parties, partisan rivalry has exacerbated the institutional rivalry for control over environmental policy.

361. Steven Shimberg. 54 LAW REVIEW AND CONTEMPORARY PROBLEMS 241. 1991, p. 241.
Once these complex, controversial laws are enacted, EPA

is left with the unenviable task of implementing them, frequently without adequate staff or financial resources and on the basis of incomplete scientific understanding of the problems. EPA has to contend not only with Congress and the laws it passes, but with powerful, well-funded, and often reluctant regulated industries. In addition, EPA is part of the executive branch and is accountable, first and foremost, to the president and the overseers who work for the president at the Office of Management and Budget ("OMB") and other agencies. This leads to the second point that warrants emphasis: that there has been "a marked lack of consensus" between the executive and legislative branches concerning the proper direction of federal environmental policy. In fact, much of the congressional oversight directed at EPA is really directed at OMB and the president.

362. Richard J. Lazarus. 54 LAW AND CONTEMPORARY PROBLEMS 511. Autumn 1991, pp. 210–211. Whatever the accuracy of the view that Congress has historically shirked most of its responsibilities to oversee agencies, no such neglect is evident in the case of congressional oversight of EPA. As described above, EPA came into existence just as Congress sought to use its oversight authority to strike a new balance of power between the executive and legislative branches. The result has been a persistent struggle between the two branches for control of EPA.

363. Richard J. Lazarus. 54 LAW AND CONTEMPORARY PROBLEMS 511. Autumn 1991, pp. 347–348. EPA is plainly in a dilemma. The agency strives to be responsive to both the environmentalists' vision and the regulated community's pragmatism, but ultimately satisfies neither. EPA is also a pawn in an ongoing struggle between the executive and legislative branches for control over national policy. Finally, EPA is pushed in one direction by public aspirations and pulled in the other direction by the absence of public willingness to change and by the public's proven incapacity for self-sacrifice.

364. Mark Jaffe. THE ATLANTA JOURNAL AND CONSTITUTION. 10 December 1995, p. 1H. One of the EPA's problems has been the bureaucratic equivalent of a child caught in a bitter custody battle between warring parents, in this case Congress and the presidency. "The trust of Congress in EPA has never been high," said William Ruckelshaus, who directed the agency in its infancy, from 1970 to 1973, and then again during the Reagan administration, from 1983 to 1985. At the outset, Sen. Ed Muskie, the Maine Democrat who wrote many of the early environmental laws, "simply didn't trust Nixon," Ruckelshaus said.

The result was that 15 congressional committees claimed oversight of the EPA and many of the laws offered no "wiggle room" for the agency. "They were highly prescriptive and had very specific timetables," Ruckelshaus said.

When the Reagan administration took office, the problem got worse. The administration's openly hostile behavior toward the agency sparked hearings and investigations by the Democrat-controlled Congress. Before the dust had cleared, Anne Gorsuch Burford, then EPA administrator, and 20 other top officials were forced to resign, and one, Rita Lavelle, head of the Superfund toxic waste cleanup program, was convicted of perjury for testimony she gave to Congress. She served a six-month jail term.

365. Walter A. Rosenbaum. ENVIRONMENTAL POLICY IN THE 1990s: TOWARD A NEW AGENDA. 1994, p. 130. Defying a biblical admonition, the EPA serves two masters. The agency is responsible to Congress for implementing environmental legislation and managing its budget as Congress intends. Congress and its committees also exercise continuing legislative oversight of EPA's activities within their jurisdiction. As an executive agency, it must also implement the president's management directives and his policy agenda insofar as these coincide with its legislatively mandated activities. EPA's relations with both Congress and the White House are typically adversarial, with occasional lapses into uneasy cooperation.

366. Robert A. Shanley. PRESIDENTIAL INFLUENCE AND ENVIRONMENTAL POLICY. 1992, pp. 49–50. The Constitution contains no specific provisions for presidential authority to issue executive orders. They are generally issued to direct actions of governmental officials but may indirectly affect private individuals and groups. Orders have the force of law and the presumption of validity when issued under a valid claim of authority. But increasingly, orders have been promulgated under unclear claims of authority. Even when there may be a specific citation of statutory or constitutional authority, it is exceptional for an administration to provide prior notice and solicit widespread public comment or input. There are some constraints in case law, but traditionally the courts have accorded presidents a wide berth and deference in issuing executive orders. Furthermore, there are significant problems in obtaining judicial review of executive orders.

367. Robert A. Shanley. PRESIDENTIAL INFLUENCE AND ENVIRONMENTAL POLICY. 1992, pp. 71–72. The courts have focused on narrower issues of statutory interpretation concerning OMB's exercise of power in regulatory oversight. While the Supreme Court placed some constraints on executive power in several past cases, more recent decisions in *Buckley* v. *Valeo, Immigration and Naturalization Service* v. *Chadha,* and *Bowsher* v. *Synar* have been cited as an acknowledgment of rule making as a legitimate aspect of presidential authority. Historically the Supreme Court and federal courts have given the president considerable deference in the issuance of executive orders and presidential proclamations. Thus, a long-term remedy for purported presidential incursions in

regulatory oversight is perceived to be a bill passed by Congress, subject to a presidential veto, rather than a judicial remedy. However, more short-term, effective accommodations can also be negotiated without legislation between the respective claimants of the White House and in the Congress.

368. Yvette M. Barksdale. 42 AMERICAN UNIVERSITY LAW REVIEW 273. Winter 1993, pp. 304–305. Second, although presidential oversight may increase the electoral accountability of administrative decisionmaking, accountability alone is not sufficient to countenance overriding foundational constitutional restrictions on unilateral presidential policymaking. The Constitution's choice of legislative processes as opposed to executive processes for governmental value selection is not accidental. The preference ensures not merely that basic societal value choices will be made by persons accountable to the public—a grant of full legislative authority to the President would have accomplished this—but also protects important legislative process values such as consensus building, citizen participation, deliberation, and diffusion of power. Legislative value selection processes ensure that the Government makes value choices only through a deliberative and participatory political struggle, and accordingly, legislative selection requires considerable national consensus with respect to policy and value choices before any bill becomes law. Additionally, legislative value selection processes ensure that the power to define societal ends is not concentrated in the hands of one individual, the President, but instead is broadly diffused among the different political branches of government and the many persons who comprise them. These legislative process values are more important than electoral accountability for protecting the soundness and legitimacy of governmental value selection; they are an important part of the structural protections for governmental policymaking and value selection within the Constitution. Accordingly, Article II should be interpreted to preserve rather than to override these important legislative process values. Because presidential value selection does not protect these necessary legislative values, it should not be permitted under Article II.

369. Robert A. Shanley. PRESIDENTIAL INFLUENCE AND ENVIRONMENTAL POLICY. 1992, p. 51. Adequate and prompt implementation of executive orders can also present problems. Although the federal bureaucracy has been cooperative in implementing some executive orders, noncompliance does occur. An exploratory study of Nixon's directives, including executive orders, requests, and commands issued during 1969 and 1970 suggests there may have been widespread noncompliance. The judicial and administrative enforcement of some executive orders can be protracted and burdensome, as demonstrated in the Carter Administration; and the preferred enforcement tool is persuasion. However, the executive order can be an integral part of an administrative presidency strategy, as illustrated in the Reagan Administration's use of executive orders dealing with regulation policy and environmental and natural resources policies.

370. Marc K. Landy, Marc J. Roberts, and Stephen R. Thomas. THE ENVIRONMENTAL PROTECTION AGENCY: ASKING THE WRONG QUESTIONS. 1994, p. 314. The Republicans too have had to struggle with their own extremist demons. As the Gorsuch episode showed, environmental sentiment makes it foolhardy for the executive to try to subvert pollution control laws. The vigilance of congressional staffs and environmental advocates insures that such efforts are likely to glow in the hot glare of media exposure. Moreover, as George Bush recognized, there is a strain of Republican conservationism that dates back to Theodore Roosevelt. It continues to appeal to suburban Republican and independent voters, many of whom have succumbed to the blandishments of Ross Perot. Reilly's emphasis on ecology and the global environment thus made quite good sense in narrow electoral terms.

371. Robert V. Percival. 54 LAW AND CONTEMPORARY PROBLEMS 172. Autumn 1991, p. 196. Having garnered the political benefits from enacting legislation, legislators can reap additional benefits by attacking lax implementation of the laws by administrative agencies and by amending the laws to force more aggressive agency action. Because the president is responsible for the activities of executive agencies, legislators also can score points against a president from an opposing party by exposing efforts to weaken the implementation of environmental statutes. These factors help explain why the exercise of executive oversight has been, and is likely to continue to be, highly controversial.

372. Joel A. Mintz. ENFORCEMENT AT THE EPA: HIGH STAKES AND HARD CHOICES. 1995, p. 2. Environmental enforcement is clearly a crucial component of EPA's implementation work. Enforcement and compliance monitoring efforts now represent approximately 25 percent of the agency's total budget, and they engage the efforts of approximately 3,200 full-time employees. Furthermore, as Peter Yeager has noted,

> To the public mind, enforcement is the centerpiece of regulation, the visible hand of the state reaching into society to correct wrongs. . . . Both symbolically and practically, enforcement is a capstone, a final indicator of the state's seriousness of purpose and a key determinant of the barrier between compliance and lawlessness. [ellipses in original]

373. Zachary A. Smith. THE ENVIRONMENTAL POLICY PARADOX. 1992, p. 60. The Department of the Interior was established in 1849 by combining the Treasury Department's General Land Office, the War Department's Office of Indian Affairs, which later became

the Bureau of Indian Affairs, the Patent Office, the Pension Office, and the Census Office. In its first year of operation, the department had a permanent work force of ten people and a budget of $14,200. Today the Department of the Interior has extensive responsibility over environmental and land management. The Department of the Interior houses the National Park Service, the U.S. Fish and Wildlife Service, the Bureau of Land Management, the Minerals Management Service, the Office of Surface Mining, the U.S. Geological Survey, the Bureau of Reclamation, and the Bureau of Mines. The department is organized with a secretary and five assistant secretaries covering each of the major substantive areas with the department's jurisdiction.

374. Walter A. Rosenbaum. ENVIRONMENTAL POLITICS AND POLICY, 3d Ed. 1995, pp. 119–120. Established as a cabinet-level department in 1845, the Department of the Interior has acquired responsibilities over more than a century that leave few national environmental issues untouched. The Department's important environmental responsibilities include (1) protection and management of more than 549 million acres of public land—roughly 28 percent of the total U.S. land area—set aside by Congress for national parks, wilderness areas, forests, and other restricted uses; (2) administration of Native American lands and federal Native American programs, including authority over western tribal lands containing a very large proportion of the coal, petroleum, uranium, and other largely unexploited energy resources in the western United States; (3) enforcement of federal surface mining regulations through its Office of Surface Mining, Reclamation and Enforcement; (4) conservation and management of wetlands and estuarine areas; and (5) protection and preservation of wildlife, including endangered species. Headed by a cabinet secretary appointed by the president, the Department's programs historically have been a primary concern to environmentalists; the interior secretary, although not always identified with the environmental or conservationist movements, in recent decades has been compatible with their interests.

375. CONGRESSIONAL DIGEST. February 1994, p. 36. **Department of the Interior.** The Department of the Interior was created by an act of Congress in 1849, which transferred to it the General Land Office, the Office of Indian Affairs, the Pension Office, and the Patent Office. Over the many years of its existence, its role has changed from that of general housekeeper for the Federal Government to custodian of the Nation's natural resources.

As the Nation's principal conservation agency, the Department of the Interior has responsibility for most of our nationally owned public lands and natural resources. This includes fostering sound use of our land and water resources; protecting our fish, wildlife, and biological diversity; preserving the environmental and cultural values of our national parks and historical places; and providing for the enjoyment of life through outdoor recreation. The Department assesses our mineral resources and works to ensure that their development is in the best interests of all our people by encouraging stewardship and citizen participation in their care. The Department also has a major responsibility for American Indian reservation communities and for people who live in island territories under United States administration.

376. Walter A. Rosenbaum. ENVIRONMENTAL POLITICS AND POLICY, 3d Ed. 1995, pp. 123–124. By a perverse bureaucratic destiny, the DOE enters its third decade still mired in high-visibility controversy, struggling against the stigma of flagrant mismanagement and saddled with responsibility for unpopular policies. In many respects, the DOE's problems seem almost inevitable in light of its history. The Department was created in 1976 when Congress combined a number of independent agencies with programs already operating in other departments in order to bring the federal government's sprawling energy activities within a single bureaucratic structure. Under the DOE's jurisdiction are regulatory activities and energy programs strongly affecting the environment. The more important of these include (1) promotion of civilian nuclear power activities; (2) regulation of military nuclear facilities and radioactive wastes; (3) administration of the federal government's research and development programs in energy production and conservation; (4) regulation of price controls for domestic petroleum and natural gas; and (5) administration of federal research and development grants for commercial synthetic fuels production in the United States. With approximately 18,700 employees and a budget of $16.8 billion in 1983, the DOE is the principal executive agency involved in the regulation and production of many different energy technologies with significant environmental impacts. Also by design it is expected to undertake a volatile agenda of frequently contradictory and inconsistent missions destined to set it at odds with itself and with the environmental community: to promote environmental risky energy technologies and to minimize the environmental risks; to promote energy use and energy conservation; to stimulate research and development of new energy consuming technologies and energy saving ones; to control energy prices in emergencies; to avert energy shortages and stimulate long-range energy planning.

377. CONGRESSIONAL DIGEST. February 1994, p. 35. **Department of Energy.** The Department of Energy was established by the Department of Energy Organization Act, effective October 1, 1977. The Act consolidated the major Federal energy functions into one Cabinet-level department, transferring to the Department of Energy all the responsibilities of the Energy Research and Development Administration, the Federal Energy Administration, the Federal Power Commission, and the Alaska, Bonneville, Southeastern, and Southwestern Power

Administrations, as well as the power-marketing functions of the Department of the Interior's Bureau of Reclamation. Also transferred to the Energy Department were certain functions of the Interstate Commerce Commission and the Departments of Commerce, Housing and Urban Development, the Navy, and the Interior.

The Department of Energy provides the framework for a comprehensive and balanced national energy plan through the coordination and administration of the energy functions of the Federal Government. The Department is responsible for long-term, high-risk research and development of energy technology; the marketing of Federal power; energy conservation; the nuclear weapons program; energy regulatory program; and a central energy date collection and analysis program.

378. Renu Khator. HANDBOOK OF BUREAUCRACY. Ed. A. Farazmand, 1994, p. 205. Today, the United States has one of the world's most complex and comprehensive environmental bureaucracies. The genesis of the consolidated environmental bureaucracy dates back to the early 1970s, when President Nixon reorganized it by bringing several federal programs under the umbrella of the Environmental Protection Agency (EPA). Since then, more than 20 new agencies dealing with environmental issues have been created to implement more than 150 new laws passed during the last 20 years alone. The EPA's network now includes 10 regional offices, with a staff of over 15,000 and a budget of $4.9 billion. This reflects the doubling of staff and a tenfold increase in budget since 1972 (a large amount of this increase amounts to waste treatment grant money committed for Superfund and local governments). Along with the EPA, several federal agencies—including the Department of Interior, Bureau of Land Management, National Parks Service, Forest Service, OSHA, Council on Environment Quality, Nuclear Regulatory Commission, and Department of Energy—work in the environmental cause.

379. J. William Futrell. ENVIRONMENTAL LAW: FROM RESOURCES TO RECOVERY. Ed. C. Campbell-Mohn, 1993, p. 79. Although the DOI and EPA are the two major environmental agencies, responsibilities to protect the environment are also vested elsewhere. For example, the Department of Agriculture, The Corps, and the Federal Energy Regulatory Commission have significant responsibilities to implement environmental law.

380. Walter A. Rosenbaum. ENVIRONMENTAL POLITICS AND POLICY, 3d Ed. 1995, pp. 133–134. There are other important influences in the agency's working life. The EPA's mission, its regulatory effectiveness and its resources are also affected by the fragmentation of authority for environmental management and regulation among a multitude of federal agencies—a reminder that EPA lacks a unitary environmental charter. At least twenty-seven other federal agencies and bureaus share regulatory authority over the environment and occupational health (fig. 1–1, p. 8).

This fragmentation of authority denies EPA the resources needed to fulfill its regulatory responsibilities. For example, the agency is required by the Clean Water Act to control the discharge of harmful pollutants in navigable waterways of the United States. But the most common source of surface-water contamination is "nonpoint" pollution (sedimentation, pesticides, and other chemicals from agriculture, forestry, and other activities), whose control requires land use management. Federal authority for this purpose is vested in the departments of Agriculture and Interior and several other smaller bureaucracies. More than twenty-five different federal laws concern some aspect of toxic and hazardous waste management, involving not only EPA but the Food and Drug Administration, the Department of Transportation, the Consumer Product Safety Commission, and many others. Although EPA is made responsible for some aspect of groundwater protection by six of the ten major laws it administers, it must share regulatory authority over groundwater with seven other departments and agencies, including seven agencies within the Department of the Interior and five in Agriculture.

381. J. William Futrell. ENVIRONMENTAL LAW: FROM RESOURCES TO RECOVERY. Ed. C. Campbell-Mohn, 1993, p. 74. On the negative side, separation of powers leads to fragmentation and Balkanization. Environmental programs are scattered throughout the government. Water pollution from strip mines is regulated by the Department of the Interior (DOI), water pollution from chemical plants by the Environmental Protection Agency (EPA), water pollution caused by soil erosion by the Department of Agriculture, and water pollution caused by road salts not at all. This is partially because the location of an environmental program in a specific agency results from the congressional committee that created it. For example, the Coastal Zone Management Act drafted by the Senate Commerce Committee is lodged in the Department of Commerce, and the National Estuary Program drafted by the Senate Environment and Public Works Committee is lodged at EPA.

382. Jacqueline Vaughn Switzer. ENVIRONMENTAL POLITICS: DOMESTIC AND GLOBAL DIMENSIONS. 1994, pp. 53–54. Prior to the Clinton administration, the president also received policy advice on environmental matters from the Council on Environmental Quality (CEQ), created as part of the National Environmental Policy Act of 1969. Its members recommended policy to the president and to some degree evaluated environmental protection programs within the executive branch. The CEQ did not have any regulatory authority, and its recommendations were purely advisory. Clinton replaced the CEQ with the White House Office of Environmental Policy, elevating the status of the body within the administration's hierarchy.

383. Walter A. Rosenbaum. ENVIRONMENTAL POLITICS AND POLICY, 3d Ed. 1995, pp. 118–119. Since its first year of operation in 1971, the CEQ has published a widely distributed and densely documented annual report, a periodic appraisal of major environmental trends, issues, and new developments. The Council also administers the process for writing and reviewing EISs within the federal government. The CEQ is a small agency with no regulatory responsibilities or major environmental programs beyond modest research activities, but it has assumed symbolic importance and political value to environmental interests. Its presence within the White House implies a high national priority to environmental programs, and the Council's opportunities to influence the president directly means that it might act, in the words of environmental leader Russell Peterson, as "the environmental conscience of the executive branch" Nonetheless, as with all other presidential advisory bodies, the CEQ can exercise no more influence in White House decisions than the president cares to give it; it may carry on its NEPA-mandated activities, but the president is free to ignore any of its recommendations or other initiatives.

384. Walter A. Rosenbaum. ENVIRONMENTAL POLITICS AND POLICY, 3d Ed. 1995, p. 118. The National Environmental Policy Act included a provision for the establishment of a commission to advise the president on environmental matters. To be headed by three members appointed by the president, the Council on Environmental Quality (CEQ) was to be part of the president's staff. Among the major responsibilities prescribed for the Council in section 203 of NEPA were (1) to gather for the president's consideration "timely and authoritative information concerning the conditions and trends in the quality of the environment both current and prospective"; (2) "to develop and recommend to the President national policies to foster and promote the improvement of environmental quality"; and (3) "to review and appraise the various programs and activities of the Federal Government" to determine the extent to which they comply, among other things, with the requirement for writing EISs. The CEQ was created, then, as were other major presidential advisory commissions, to provide policy advice and evaluation from within the White House directly to the president.

385. CONGRESSIONAL DIGEST. February 1994, p. 44. The Council on Environmental Quality was established within the Executive Office of the President by the National Policy Act of 1969, to formulate and recommend national policies to promote the improvement of the quality of the environment. Additional responsibilities were provided by the Environmental Quality Improvement Act of 1980.

The Council consists of three members appointed by the President with the advice and consent of the Senate; one of the members of designated by the President as chairman.

The Council develops and recommends to the President national policies that further environmental quality; performs a continuing analysis of changes and trends in the national environment; reviews and appraises programs of the Federal Government to determine their contributions to sound environmental policy; conducts studies, research, and analyses relating to ecological systems and environmental quality; assists the President in the preparation of the annual environmental quality report to Congress; and oversees implementation of the National Environmental Policy Act. The chairman of the Council also serves as chair of the President's Commission on Environmental Quality.

386. Zachary A. Smith. THE ENVIRONMENTAL POLICY PARADOX. 1992, pp. 52–53. To back up these lofty goals, "action forcing" procedures were developed in Section 102 of the act. The most important requirement of NEPA was that an *environmental impact statement*, or EIS, accompany "major federal actions significantly affecting the human environment." NEPA directed that an EIS contain the environmental impact of a proposed federal action, any adverse environmental effects that cannot be avoided should the federal action proposed be implemented, any alternative to the proposed action, and any irreversible commitments of resources that an action would involve should it be implemented.

NEPA also established the Council on Environmental Quality (CEQ) within the executive branch. Although the primary purpose of the CEQ was to advise the president about environmental matters, the CEQ became influential by developing regulations governing the EIS process and its implementation by federal agencies. Also, the Supreme Court has recognized the authority of the CEQ in developing regulations for the implementation of NEPA.

387. Walter A. Rosenbaum. ENVIRONMENTAL POLITICS AND POLICY, 3d Ed. 1995, p. 119. The CEQ's rapid decline in status since the Carter administration illustrates how much its effectiveness depends on presidential favor. The Council enjoyed considerable influence under President Carter, a strong environmentalist, but its influence plummeted rapidly during the Reagan administration. One of Reagan's earliest acts after his inauguration was to reduce the CEQ's staff from forty-nine to fifteen and to reduce its budget by 50 percent; the intent was obvious to environmentalists. The CEQ's publications and other research activities immediately declined in number and quality. Throughout the remainder of the Reagan administration, the CEQ seemed to inhabit a White House nether region from which it was seldom seen or heard. President Bush's pledges of environmental concern did not, to the disappointment of environmentalists, portend better days for the CEQ, nor did the arrival of the Clinton administration in 1992. President Clinton's proposal to replace the CEQ with a White House environmental advisor appointed by the president was a shock to the environ-

mental community and seemed virtually an obituary for the Council and evidence of its rapidly evaporating influence in White House policy making. Even should the CEQ remain, it has clearly ceased to be a major player in White House politics.

388. Richard N. L. Andrews. 20 ECOLOGY LAW QUARTERLY 515. 1993, pp. 575–576. The weakness of the CEQ model is its fundamental dependence on the personal support of the President. Through the 1970s the CEQ's professional staff grew to about fifty and included substantial expertise in the ecological and environmental sciences as well as economics, policy, and law. Its budget, while modest, allowed it to initiate the studies discussed above. The Reagan administration, however, led by OMB Director David Stockman, deliberately destroyed it in all but name. Reagan fired its entire professional staff and drastically reduced its budget. Under the Bush administration the CEQ once again was led by an experienced professional, but its critical mass of budget and staff expertise was not restored. Despite a few modest initiatives over the decade, the CEQ has not yet demonstrated its ability to play once again the strong anticipatory, agenda-setting, or interagency coordinating roles it played in the 1970s.

389. Peter Fairley. CHEMICAL WEEK. 1 November 1995, p. 22. Meanwhile, the spirit of cooperation EPA is forging with industry has not been extended to most Republicans in Congress. EPA's relationship with the people who pay its bills has deteriorated as the GOP and Browner parted ways over regulatory reform, rewriting of environmental laws such as the Clean Air Act, and EPS's budget. In each case, Browner—with more than a little help from the environmental movement—has inflicted political damage on House and Senate Republicans and fractured their majorities by tagging their efforts "anti-environmental."

390. Paul R. Portney, Katherine N. Probst, and Adam M. Finkel. 21 ENVIRONMENTAL LAW 1461. Summer 1991, p. 1461. What is an administrator to do? On the one hand, the administrator would like to concentrate on those problems that experts believe pose the greatest risks to ecosystems and human health. This is crucial given the limited national resources available to deal with environmental problems, not to mention the limited resources of EPA itself. For both reasons, an administrator would like to be able to pursue a "worst first" approach. Yet no EPA administrator can be effective if seen as ignoring heartfelt public concerns and legislative imperatives. The period between 1981 and 1983 at EPA illustrates that Congress will strip EPA of its traditionally granted discretion at the first whiff of mis- or malfeasance.

391. J. William Futrell. ENVIRONMENTAL LAW: FROM RESOURCES TO RECOVERY. Ed. C. Campbell-Mohn, 1993, p. 87. Congress has reacted to executive branch efforts to soften the law's bite with softer regulations by following Hayek's advice, Indeed, the environmental legislation passed during the Reagan and Bush administrations shows Congress' distrust of the executive. Congress has enacted complex and detailed statutes designed to circumscribe agency discretion and compel executive action along the lines dictated by the legislature.

The 1984 amendments to the Resource Conservation and Recovery Act (RCRA) are lengthy and unparalleled in their level of detail and, until the 1990 CAA amendments, set a record for intrusion into an agency's management practices. Reliance on *Chevron* by executive agencies to undermine legislative intent will lead Congress to become even more rigid and confrontational, to enact even more statutory deadlines for still more precisely detailed mandatory duties, which in turn are backed up by self-enforcing hammers and by provisions allowing citizen suits if an agency misses its deadlines.

392. Robert A.. Shanley. PRESIDENTIAL INFLUENCE AND ENVIRONMENTAL POLICY. 1992, p. 11. Furthermore, all appointed agency officials must undertake actions required by law, which may not be in accord with the preferences of an administration. When Congress is concerned that agency heads have strained legislative intent in their discretionary power to implement a law, it can retaliate in various ways, including tightening restrictions on an agency head's discretion, imposing "hammer" time deadlines on the reauthorization of programs, and cutting an agency's appropriations.

393. Renu Khator. HANDBOOK OF BUREAUCRACY. Ed. A. Farazmand, 1994, p. 208. Overall, the political system in the United States makes it largely impossible for the environmental bureaucracy to be anything other than regulatory. Moreover, the political arrangements require that its regulatory powers be closely scrutinized. A fear of too much administrative discretion is shared by all parties concerned—the executive branch, the legislature, the judiciary, and the public. They also dread bureaucratic professionalism, for such is again a source of independence and authority. Binding the environmental bureaucracy with hammer clauses and scientific norms is a sign of mistrust as well as apprehension.

394. Cathy Marie Johnson. THE DYNAMICS OF CONFLICT BETWEEN BUREAUCRATS AND LEGISLATORS. 1992, p. 21. In the policy model disputes do not center only around the scope of an agency's authority; they also arise over questions about the way the authority is exercised. Legislators with a strong interest in a particular agency's program will decide to watch the agency's actions closely. Policy decisions may be investigated by a committee to determine if the agency made the correct decision. Committee members may become disgruntled with an agency's policies and tighten controls through legislative changes. Authorization for new pro-

grams may include very detailed procedures that an agency must follow when implementing the program.

395. Cathy Marie Johnson. THE DYNAMICS OF CONFLICT BETWEEN BUREAUCRATS AND LEGISLATORS. 1992, pp. 13–14. Finally, legislators want to exert control over the bureaucracy in order to protect the power of their institution. One cannot deny that members are concerned about the power of Congress and that they will adopt institutional practices and policies in order to bolster and protect it. Reforms of the 1970s, such as the War Powers Act and the Budget and Impoundment Control Act, were designed to enhance the power of Congress vis-à-vis the president. But it is not only the president that Congress worries about. Bureaucrats running rampant over legislative wishes arouse concern that Congress as an institution is being weakened by the bureaucrats' abuse of power. Legislators develop devices, such as the legislative veto, to ensure the primacy of congressional views. In fact, Congress continued to pass legislation with veto provisions, even after the Supreme Court declared the device unconstitutional.

396. William D. Ruckelshaus. FEDERAL NEWS SERVICE. 29 February 1996, Lexis/Nexis. From my experience at the Agency, as well as those of other Administrators with whom I have spoken, I can attest to the truth of this statement. Over the years, Congressional reaction to EPA has been strikingly similar, regardless of the political persuasion of the member. Frustrated with EPA for different reasons, members of Congress have attempted to direct and determine the methods and the outcomes of agency actions. The impact of this on the agency is ultimately debilitating.

397. Richard J. Lazarus. 54 LAW AND CONTEMPORARY PROBLEMS 205. Autumn 1991, p. 229. Second, the oversight process may also have contributed to retarding the evolution of federal environmental law. Much of the myth of EPA's "regulatory failure" has been perpetuated by the oversight process as those in Congress have derided agency officials for failing to meet statutory mandates that were unattainable anyway. In response to these perceived instances of agency failure, Congress usually has returned to the manner in which it historically oversaw the federal bureaucracy prior to its explosive growth this century: detailed prescription of the terms of the agency's implementation of the law.

While legislative prescription had some advantages (for example, it increased congressional accountability), the came at the expense of the kind of flexibility EPA needed to respond to the uncertain contours of environmental problems. Congress and EPA have rarely known the best way to respond to an environmental pollution problem at the time a statute was passed. The implementation of environmental standards has necessarily required substantial groping in the dark because policymakers have chosen not to risk environmental quality and human health by waiting for the elusive notion of scientific certainty. Statutory prescription therefore is an especially risky endeavor. It can lead to wasteful expenditures for pollution control and by way of missed opportunity, to more, rather than less environmental degradation.

398. Richard J. Lazarus. 54 LAW AND CONTEMPORARY PROBLEMS 205. Autumn 1991, pp. 356–357. Excessive congressional, OMB, and judicial oversight also has resulted in poor allocation of agency resources and skewed national environmental priorities. Each overseer can use his or her leverage (that is, power to delay or reduce appropriations, hold up confirmation of agency appointments, create bad publicity, eliminate agency discretion, or impose appropriations riders o redefine agency priorities), but the end result is unlikely to reflect any broad or thoughtful determination of environmental priorities. In fact, quite the opposite is true.

Members of the House and Senate routinely respond to narrow parochial concerns and to their own need to receive maximum favorable publicity. There is thus no reason to assume that the views of a particular subcommittee chair are consonant with those of Congress as a whole, or even consistent with the views simultaneously expressed by a different subcommittee. Oversight is likely to empower a few isolated interest groups that are able to persuade the chair to express their concerns to the agency. Indeed, a congressional subcommittee may be as likely as an agency, if not more so, to be "captured" by a special interest group.

The competence of congressional staff to draft environmental legislation containing increasingly detailed prescription can also be questioned. Because authority among congressional committees is so fragmented, it is extremely difficult, if not impossible, for any one committee to undertake a broad, coordinated look at a complex problem. Rather, each committee tends to "view "the bill through the narrow lens of its own particular mandate. None tr[ies] to critically examine the structure of the program as a whole.

399. Richard J. Lazarus. 54 LAW AND CONTEMPORARY PROBLEMS 205. Autumn 1991, p. 356. Another adverse effect of excessive oversight of EPA is that it has caused the agency to go "underground" in its lawmaking. To avoid overseers, EPA has increasingly resorted to less formal means of announcing agency policy determinations. Instead of promulgating rules pursuant to the Administrative Procedure Act, EPA now frequently issues guidance memoranda and directives. Also, many important agency rulings are not reflected in generic rulemaking, but in individual permit decisions. OMB oversight is thereby avoided, and judicial review of agency action is limited.

400. Richard J. Lazarus. 54 LAW AND CONTEMPORARY PROBLEMS 205. Autumn 1991, p. 229.

The worst result of the current administrative scheme is that it has undermined environmental protection by chilling agency and congressional innovation. Increased statutory prescription comes at the expense of agency discretion and flexibility. Intense agency oversight, repeated regulatory failure, and frequent controversy likewise discourage agency initiative.

EPA officials have long recognized the need for administrative experimentation and reorganization. Congress, however, as increasingly denied the agency the option of exercising administrative innovation. Moreover, even when the opportunity remains, EPA officials have often shied away from innovation because of actual and anticipated objections from those elsewhere in the executive branch, the regulated community. Congress, and environmental organizations who are suspicious of the agency's needs.

For example, Congress's sharp restriction of agency discretion in its regulation of hazardous air pollutants under section 112 of the Clean Air Act has apparently prompted the agency to take less action than it might have under a more flexible statutory scheme. By mandating what is often infeasible, such as disallowing any significant consideration of economic costs, Congress prompted EPA to do very little. EPA chose not to list a pollutant as "hazardous" in order to avoid triggering the statute's rigid requirements. The agency has consequently acted on only seven out of hundreds of toxic air pollutants over the last twenty years, leaving the others unregulated.

CHAPTER VI
ALTERNATIVE ENERGY SOURCES
OUTLINE

I. DEFINITIONS OF RENEWABLE ENERGY

   A. DEFINITION OF RENEWABLE ENERGY (127)

   B. NUCLEAR POWER IS A RENEWABLE ENERGY (38, 43)

II. RENEWABLES GROWING

   A. RENEWABLE USE INCREASING NOW (93, 121)

   B. TRANSITION TO RENEWABLES INEVITABLE. OIL INDUSTRY ANALYSIS PROVES (59)

   C. LOBBIES PUSHING FOR RENEWABLE ENERGY NOW (1)

   D. INCREASED USE OF RENEWABLE ENERGY SNOWBALLS (56)

   E. BEST ANALYSIS PROVES TRANSITION TO RENEWABLE ENERGY IS INEVITABLE (89, 91, 101)

   F. DOE RESEARCH REDUCES COSTS OF RENEWABLE TECHNOLOGIES (92)

   G. NOW IS A CRITICAL PERIOD IN THE DEVELOPMENT OF RENEWABLES. INVESTMENT DECISIONS TODAY PLAY OUT OVER THE LONG HAUL (96, 120)

   H. FEDERAL INVESTMENTS KEY TO DEVELOPMENT OF RENEWABLES (97, 120)

   I. DOE INVESTMENT SOUND (114)

   J. INVESTMENT IN RENEWABLES SNOWBALLS (124)

   K. HIGH OIL PRICES CAUSE RESEARCH INTO RENEWABLES (129)

   L. MUST REGULATE TO STIMULATE GROWTH OF RENEWABLES (130)

III. NOW RENEWABLES IN THE STATUS QUO

   A. RENEWABLE ENERGY USE MINIMAL NOW (122)

   B. INVESTMENT IN RENEWABLES DECLINING NOW (131)

   C. AMERICAN RENEWABLES INDUSTRY HURTING NOW (32)

   D. POWERFUL LOBBIES BLOCK RENEWABLE ENERGY (123)

   E. DEREGULATION OF POWER GENERATING UTILITIES KILLS RENEWABLES (133)

   F. RENEWABLE ENERGY TECHNOLOGY GROWTH SHAKY NOW (125)

   G. STABILITY KEY TO SUCCESS OF RENEWABLE ENERGY (126)

   H. OIL DEPENDENCE INCREASING. U.S. WILL BE COMPLETELY DEPENDENT SOON. NO RENEWABLES IN SIGHT (58)

   I. OIL DEPENDENCE INEVITABLE (88, 90)

   J. PRIVATE SECTOR INVESTMENT SMALL. CURRENT FEDERAL BUDGET CUTS KILL RENEWABLES (98)

IV. BENEFITS OF RENEWABLES

   A. ECONOMIC BENEFITS

     1. Renewables key to solving the trade deficit. (28, 30, 103)

     2. Renewable energy reduces U.S. foreign investment. (31)

     3. Renewables research key to future economic growth. (60, 102, 105, 107)

     4. Must invest in renewables to enhance job creation. (66)

     5. Research in renewables key to maintaining U.S. competitiveness. (86–87, 95, 110, 113, 119)

     6. Huge global market for renewable exports. (109)

     7. Pollution benefits have a huge impact on economic growth. (106, 108)

     8. Research benefits industry. (116–118)

   B. FOREIGN POLICY BENEFITS

     1. Oil independence key to an independent foreign policy. (29)

     2. Commitment needed to fulfill international treaty commitments. (4)

     3. Present international climate negotiations will increase use of renewable energies. (36)

     4. Strong American renewable policy needed to insure U.S. leadership on international climate agenda. (41)

C. ENVIRONMENTAL BENEFITS
 1. Renewables reduce greenhouse gases and global warming. (45, 48)
 2. U.S. must take action to reduce warming via renewables. International efforts fail. (49–51)
 3. Renewables key to reducing pollution. (63–65)

D. OIL PRICE AND SUPPLY BENEFITS
 1. Must fund renewables to prevent next oil crisis. (62, 80, 84, 115, 128)
 2. Oil shocks inevitable without renewables. (67–69)
 3. Oil shortages inevitable. Must begin transition now. (71–72)
 4. No domestic oil supply. (77–78)
 5. Middle East control of oil resources makes need to facilitate transition essential. (73)
 6. Oil crisis impacts. (70, 74–76)

V. LIMITATIONS OF RENEWABLES

A. SPENDING DISADVANTAGE LINKS
 1. Even limited research programs are expensive. (2–3)
 2. Full commitment to renewable transitions extremely expensive. (32)
 3. Expense undermines economic growth. (56)

B. POLITICAL CAPITAL DISADVANTAGE LINKS
 1. Renewables are a source of political conflict. Republicans and Democrats disagree. (5–7)
 2. Environmental issues are the key to Republican unity in Congress. (16–17)
 3. Clean energy issues are essential to Republican control of the agenda. (27)
 4. Environmental issues are key to public support for a Republican Congress. (21, 23–26)
 5. Environmental issues are critical to electoral politics. Disagreements generated over renewables create fertile ground for gridlock. (8–11, 18)
 6. Court involvement insures political conflict. (12)
 7. Energy and environment issues are key to congressional elections. Hit lists are formed early. (13–15)
 8. Public commitment to clean technologies is high. Government involvement is key to satisfying the public. (19–20)
 9. Environmental protection is key to Gore's popularity. (26)
 10. Reducing fossil fuel subsidy politically contentious. (55)
 11. Bipartisan support for renewables funding. (57)

C. FREE MARKET/ VOLUNTARISM COUNTERPLAN
 1. Many voluntary ways to increase use of renewables. (33)
 2. Subsidies discourge growth of renewable energy. (34, 53)
 3. Market competition key to low renewable prices. (35)
 4. Reducing subsidies key to saving international climate negotiations. (54)
 5. Intense political opposition to energy taxes. (79)

D. TRADE CONFLICT DISADVANTAGES LINKS
 1. Japan controls the nuclear power market now. Competition threatens a trade conflict. (40, 42)
 2. Japan and Europe control the fuel cell market. U.S. competition could ignite a trade conflict. (83)
 3. Japan significantly invested in renewable energy now. (111)
 4. Germany pushing renewable energy now. (112)
 5. International competition in renewables is intense. Foreign competitors ahead. (85, 99)

E. POLLUTION
 1. Renewables won't solve the pollution crisis. (104)

VI. SPECIFIC RENEWABLES

A. BIOMASS
 1. Biomass burning increase NOx gases. (160)
 2. Intense public opposition to biomass burning. (161)
 3. Too much waste for biomass industry. (162)
 4. Biomass solves municipal waste crisis. (163)
 5. Biomass industry will grow now. (164)

B. ELECTRIC CARS/LOW EMISSION AUTOS
 1. Technology feasible. (81–82)
 2. Everyone supports "green" gasoline. (179)
 3. Ethanol = broad bipartisan support. (180)
 4. Zero emission car mandates = across the board price hikes. (181)
 5. EV mandate = $20 billion in regulatory cost. (182)

6. Infrastructure = billions in new emissions. (183)
7. Cost shifts = widespread price hikes. (184)
8. EV mandate = billions in cost shifts and lost revenue. (185)
9. National ZEV mandate = automotive disaster. (186)
10. Automakers adamantly opposed to California model. Forced industry introduction = industry collapse. (187)
11. Automakers hate ZEV mandates. (188)
12. Cannot sell EVs outside California. (189)
13. Even the rosiest scenarios concede demand shortfalls. (190)
14. Tech-neutral vehicle policy fragments U.S. innovation efforts. Mandated markets will be captured by focused competition. (191)
15. Divided focus undermines U.S. competitiveness. America will lose all market share in alternative vehicle development. (192)
16. EV decreases energy consumption by 35%. (193)
17. EVs = energy independence. (194)
18. Atlanta proves: Car emissions irrelevant. Most smog is natural. (195)
19. No impact cars emit small particulates. (196)
20. Smog is decreasing to all time lows. (197)
21. Newer car models are significantly cleaner. (198)
22. Consumers reject electric cars. (199)
23. Technological uncertainties doom demand. Environmental benefits irrelevant. (200)
24. Any initial setbacks torpedo future sales. (201)
25. ZEV mandates lead to slower turnover of older models, increasing overall pollution. (202)
26. No alternatives to lead-acid powered electric cars are available and market ready. Electric cars will increase toxic lead exposure. (203)
27. Increased lead exposure = $100 billion economic shock. (204)
28. EVs increase SO_2 emissions and acid rain. (205)
29. Electric car mandates stop development of the smog-eating radiator. (206)
30. Smog-eating radiators = most effective and cost-efficient clean air solution. (207)
31. State's counterplan. (208–220)

C. GEOTHERMAL
1. Geothermal industry isolated geographically now. (143)
2. Geothermal profits high now. (144)
3. Growth of American geothermal industry declining now. (145)
4. Geothermal supply vast. (146)
5. Geothermal energy best used for utilities. (147)
6. Geothermal industry risk. (148)
7. Government cuts in research in hot dry rocks cripple U.S. geothermal competitiveness. (149)
8. Geothermal not a flexible energy source. (150)
9. Geothermal industry will grow now, but long-term outlook shaky. (151)
10. Current cutoff of R&D will doom geothermal energy development. (152)

D. NUCLEAR ENERGY
1. Massive public opposition to nuclear energy. (39)
2. Nuclear power reduces CO_2 and warming. (41, 44, 47)
3. Nuclear power use increasing globally. (46)

E. PHOTOVOLTAIC CELLS AND SOLAR ENERGY
1. DOE research reduces price of photovoltaics. (94)
2. Photovoltaic industry large and global. (153)
3. Technology and innovation key to solar industry. (154)
4. Price key to solar power development. (155)
5. Solar power diverse. (156)
6. Solar power has many applications. (157)
7. U.S. government support key to growth of solar industry. (158)
8. Solar power industry growing and secure now. (159)

F. SOLAR THERMAL POWER
1. Solar thermal power is limited now. (165)
2. State tax laws cripple solar industry. (166)
3. Solar R&D increasing. (167)
4. American solar thermal industry dead now. (168)
5. Solar thermal power losing out to other renewables now. (169)

G. WIND POWER
1. Wind power is a global industry. (134)
2. Windpower industry growing in U.S. now. (135)

3. Wind not a stable energy source. (136)
4. Geographic barriers block effective adoption of wind power. (137)
5. Wind power investment increasing. (138)
6. U.S. losing competitive edge in wind power now. (139)
7. American wind industry blowing strong now. (140–141)
8. Wind power industry vulnerable now. (142)

CHAPTER VI

EVIDENCE

1. REUTERS FINANCIAL SERVICE. 9 January 1997, On-line. Renewable energy advocates Thursday urged the Clinton administration not to cut back spending on energy efficiency and renewable energy programs.

    The administration was reportedly set to ask Congress for about $1 billion for the programs in the 1998 fiscal year, down from a request of $1.1 billion for 1997. Congress provided funding of $840 million for the 1997 year.

2. REUTERS FINANCIAL SERVICE. 9 January 1997, On-line. The $100 million cut in the request would likely eliminate funding for geothermal heat pumps, photovoltaics for utilities, global solar efforts, renewable energy assessment, and climate challenge, the Sustainable Energy Coalition said.

3. REUTERS FINANCIAL SERVICE. 9 January 1997, On-line. The $100 million cut in the request would likely eliminate funding for geothermal heat pumps, photovoltaics for utilities, global solar efforts, renewable energy assessment, and climate challenge, the Sustainable Energy Coalition said.

4. REUTERS FINANCIAL SERVICE. 9 January 1997, On-line. "Accordingly, it is inconsistent with the Administrations' climate change and other environmental commitments and will undermine its ability to meet your economic, technological, and international trade goals," the coalition said in a Jan. 9 letter to President Clinton.

5. INTERNATIONAL SOLAR ENERGY INTELLIGENCE REPORT. 15 November, 1996, On-line. Renewable energy advocates can expect increased influence in Congress by moderate Republicans in the wake of the Nov. 5 general elections, say solar energy advocates.

    The elections warrant cautious optimism for future federal funding of renewable energy programs and electricity restructuring legislation, said the liberal Sustainable Energy Coalition (SEC), based in the Washington suburb of Takoma Park, Md. However, the re-election of President Clinton coupled with increased Democratic membership in the House and tighter Republican reins in the Senate suggest clashes between the White House and Congress, the group warned.

6. INTERNATIONAL SOLAR ENERGY INTELLIGENCE REPORT. 15 November, 1996, On-line. Rep Dan Schaefer (R-Co.), who introduced electricity restructuring legislation in the last Congress, kept his seat. Schaefer's legislation, which never got out of the House Commerce Committee, would have mandated utilities to produce 4 percent of power from new renewable energy sources.

    Schaefer is likely to reintroduce the same bill early in the next Congress, said Larry Sherwood, executive director of the American Solar Energy Society (ASES) in Boulder, Colo. This session will see more hearings and active consideration of the bill than the 104th Congress, he said.

7. INTERNATIONAL SOLAR ENERGY INTELLIGENCE REPORT. 15 November, 1996, On-line. Rep. Sherwood Boehlert (R-N.Y.) a moderating force in the GOP, will "clearly be a leader on the moderate wing," of the Republican party, Sherwood said. Boehlert is an important leader on any environmental legislation in the House, Sherwood added.

    Continued Republican control of the house may work in favor of renewable, SEC said. A Republican-controlled House will pass electric utility restructuring legislation through faster than a Democratic-controlled chamber would, the environmental group speculated.

8. THE CHRISTIAN SCIENCE MONITOR. 14 November 1996, On-line. The dust settling out of last week's elections has a decidedly green tint to it. The environment played a key role in many wins and losses, and this could have an effect on how the reannointed Clinton administration and the 105th Congress address major issues involving pollution cleanup and natural resource protection.

    As with other national issues of major concern, moderation is likely to be the byword.

9. THE CHRISTIAN SCIENCE MONITOR. 14 November 1996, On-line.
'Green' voter power
In general, activists are pleased with the results of the voting. "The environment had unprecedented power in this election," says Carl Pope, executive director of the Sierra Club. "It was a key factor, and in some races the primary factor, voters considered in making their decisions.

10. THE CHRISTIAN SCIENCE MONITOR. 14 November 1996, On-line. The Sierra Club, which pumped $7.5 million into voter education and direct electoral activities this year, claimed victory in two-thirds of the 62 House and Senate races it considered high-priority.

    The League of Conservation Voters (the only other major environmental group that overtly backs candidates) contributed to the campaigns of candidates run against the so-called "Dirty Dozen" in Congress. Of those, half won. And in several other races, incumbents whom activists consider to be anti-environment won by very slim margins.

11. THE CHRISTIAN SCIENCE MONITOR.
14 November 1996, On-line.
Key Clinton issue
In the campaign, President Clinton repeatedly mentioned environmental protection as one of his key issues. And to the extent it plays well to the general electorate, he seems certain to follow through. On Tuesday, he signed a bill that adds to the national park system in 41 states.

12. THE CHRISTIAN SCIENCE MONITOR.
14 November 1996, On-line. One thing for sure: Federal courts will play a key role on the environment. The United States Supreme Court this week heard arguments in a $75 million case in which farmers here in southern Oregon are challenging the Endangered Species Act. How the high court rules here could have as much impact as anything Congress does.

13. INTERNATIONAL HERALD TRIBUNE.
12 November 1996, On-line. The most dangerous place to be in last week's American elections was on an environmental hit list. Sixteen of 19 defeated Republicans incumbents—85 percent—were in this category. That compares with an overall incumbent loss rate of only 6 percent.

Half to two-thirds of the League of Conservation Voters so-called Dirty Dozen and of the Sierra Club's priority opponents lost.

The vote did not amount to a national referendum on the environment, but in a substantial number of congressional races the environment was, for the first time, a decisive factor. "It was an issue that elected," said pollster Stan Greenberg, "and even more, one that defeated."

In eight of nine races that Mr. Greenberg surveyed, the environment was one of the two most important reasons voters cited for opposing a candidate.

14. INTERNATIONAL HERALD TRIBUNE.
12 November 1996, On-line. This is new. In previous elections voters have described themselves as "environmentalist" (including a staggering 83 percent of those who voted in 1994) but the issue did not concern them in the voting booth.

Most likely what has changed is that environmental improvement is no longer taken for granted. People's worries are reserved for what is not working, and until this past Congress environmental progress never seemed in question. This time it had been threatened, and voters cared.

15. INTERNATIONAL HERALD TRIBUNE.
12 November 1996, On-line. Riders to spending bills would have dismantled much of the health, safety and public-lands safeguards erected during the past 30 years. Majority Whip Tom Delay, a Texas Republican, straightforwardly introduced legislation to simply repeal the Clean Air Act.

By late summer and early fall of last year, the public was thoroughly alarmed. The shift in sentiment can be precisely timed by the moment when the environment suddenly appeared in the president's mantra—Medicare, education and the environment.

16. INTERNATIONAL HERALD TRIBUNE.
12 November 1996, On-line. Environment was the first issue to crack Republican unity in the House, eventually provoking a rebellion by some 60 moderates led by Representative Sherwood Boehlert of New York. Early this year the tide began to move in the opposite direction, eventually leading to the defeat of nearly all the anti-environmental measures.

17. INTERNATIONAL HERALD TRIBUNE.
12 November 1996, On-line. In the Senate, Majority Leader Trent Lott (Mississippi), campaign committee chair Christopher Bond (Missouri) and Appropriations chairman Ted Stevens (Alaska) have urged a quick return to various pieces of the most contentious agenda. Influential Senator John McCain (Arizona), on the other hand, has singled out the environment as the key issue on which the Republican Party has to find a new approach.

If Republicans can find their way back to the environmental views that produced the bipartisan legislation of the Nixon era, the party will gain—and so will the country.

18. THE PLAIN DEALER. 9 November 1996, On-line. The election postmortem produces hope that a new bi-partisan congressional coalition will form that is large enough to defend and strengthen the nation's besieged environmental laws.

Basis for this hope rests with a modest change in the composition of Congress. Some of the worst environmental offenders have been sent packing by the electorate. Furthermore, lawmakers are clearly more sensitive to environmental concerns. Lawmakers with poor environmental records who managed to win another term can't help but be chastened by their like-minded colleagues' fate as well as—in most instances—their own narrow margin of victory. This could quite readily translate into enough lawmakers answering the call of moderates of both parties to create a formidable environmentally friendly voting bloc—a bloc unreceptive to waging any further ideological warfare on pollution abatement regulation.

As Sierra Club executive director Carl Pope points out, the voters didn't re-elect President Bill Clinton and a Republican Congress to resurrect gridlock. The White House and Capitol Hill will be expected to cooperate, and no issue, in Pope's view, is more conducive to attracting a consensus than the environment.

"The environment will be the acid test of whether the two sides can truly work in concert, because the environment is a non-partisan, common sense issue," he said.

19. THE PLAIN DEALER. 9 November 1996, On-line. Helping lawmakers to see things Pope's way are the pre-election public opinion polls, which show that

the environmental issue has come of age. In many jurisdiction, it turned out to be of sufficient magnitude to determine victory or defeat for candidates.

This trend seems destined to continue in future elections. A national poll on the eve of the 1996 election found that 85 percent of the respondents considered a candidate's stand on environmental issues reason enough to warrant (or lose) their vote.

20. THE PLAIN DEALER. 9 November 1996, On-line. Surviving members of the congressional wrecking crew who participated in the assault on the nation's environmental laws in the first year of the 104th Congress must be aware by now that their constituents, no matter how antagonistic to big government's intrusiveness, make an exception in regard to environmental regulations. It appears, that most of the public, regardless of political persuasion, wants the government to be active and not scrimp on environmental protection (though of course, they expect the resources to be spent wisely).

21. THE PLAIN DEALER. 9 November 1996, On-line. This leaves the Republican congressional majority in a quandary. They can listen to their constituents and abandon their assault on environmental regulation. Or they can revive the crusade promoted by the anti-federal government ideology of their leaders, and by the ambition of corporate polluters who just happen to have made extremely generous campaign contributions.

One would hope these lawmakers will place the obligation to their constituents ahead of party loyalties. Such a decision in the case of environmental concerns would be not only the most ethical but most prudent, political course of action—assuming the legislators were interested in re-election.

22. THE PLAIN DEALER. 9 November 1996, On-line. Much depends on President Clinton as well. He has proved he can thwart attempted congressional rollbacks of environmental law. But can his administration become proactive and upgrade environmental protection in his second term? With Vice President Al Gore eyeing the presidential race in the year 2000 and sure to be given added responsibilities in the environmental arena, the answer to the previous question is likely to be in the affirmative.

23. John McCain. THE NEW YORK TIMES. 22 November 1996, On-line. As Republicans prepare to begin a second term in control of Congress, a deep skepticism exists in the electorate about the party's commitment to protecting the environment. Polls indicate that the environment is the voters' number-one concern about continued Republican leadership of Congress.

Public skepticism that Republicans share Americans' environmental values raises an important question. Have Republicans abandoned their roots as the party of Theodore Roosevelt, who maintained that government's most important task, with the exception of national security, is to leave posterity a land in better condition than they received it?

The answer must be no. But if we are to restore the people's trust and retain the privilege of serving as the majority party, we better start proving it. We need to assure the public that in the 105th Congress the Republican environmental agenda will consist of more than coining new epithets for environmental extremists or offering banal symbolic gestures.

24. John McCain. THE NEW YORK TIMES. 22 November 1996, On-line. Some Republicans dismiss the public's misgivings as the product of hopelessly partisan environmentalists who have abandoned reason in favor of sky-is-falling hyperbole and who eagerly denounce even the most justifiable reforms as evil conspiracies to "gut" environmental law. But blaming public skepticism exclusively on the influence of interest groups is escapism. We Republicans are responsible for much of the negative perception of our environmental record.

25. John McCain. THE NEW YORK TIMES. 22 November 1996, On-line. Too often the public views Republicans as favoring big business at the expense of the environment, and as too eager to swing the meat ax of repeal when the scalpel of reform is what's needed. Last year, a Republican bill to repeal the Clean Air Act and Republican attempts to pass riders on appropriations bills blocking the enforcement of environmental laws contributed significantly to mistrust and skepticism.

26. John McCain. THE NEW YORK TIMES. 22 November 1996, On-line. Americans know that many environmental laws—from the Superfund to the Clean Water Act—desperately need repair to make them more cost-effective. In fact, these two laws should be top priorities in the new Congress. But killing is a lousy way to treat the disease and squanders our credentials as reformers while adding substance to our critics' accusations of extremism.

Most important, we must learn that protecting the environment requires the bipartisan cooperation necessary for progress on all the great issues of our day. Partisanship breeds mistrust, demagoguery and gridlock.

27. John McCain. THE NEW YORK TIMES. 22 November 1996, On-line. In the 104th Congress, bipartisan consensus enabled Congress to reform and reauthorize the Safe Drinking Water Act and pesticide legislation, salvaging important accomplishments before Election Day.

Republicans should not allow the fringes of the party to set a radical agenda that no more represents the mainstream of Republicans than environmental extremists represent the mainstream of the Democratic Party. Only by

faithfully fulfilling our stewardship responsibilities can we expect to remain the majority party.

28. THE HOUSTON CHRONICLE. 27 October 1996, On-line. Since 1990, oil imports have added over $280 billion to America's trade deficit, accounting for 41 percent of the total. Oil import dependence has required billions of dollars in military expenditures to protect the oil installations of the Middle East.

29. THE HOUSTON CHRONICLE. 27 October 1996, On-line. Recently, retired four-star Air Force Gen. Lee Butler, who was director of strategic planning and policy for U.S. armed forces during the 1991 Persian Gulf War, told a Senate committee that "we are paying an onerous price for this dependency, a price which goes well beyond the billions of dollars that deepen our trade deficit. The greater penalty is that it shapes, conditions, constrains and inhibits our foreign and security policy."

30. THE HOUSTON CHRONICLE. 27 October 1996, On-line. To reduce oil imports, it is vital to eliminate taxpayer subsidies to foreign oil investment and to aggressively promote and develop environmentally friendly domestic energy resources and energy technologies.

31. THE HOUSTON CHRONICLE. 27 October 1996, On-line. The combination of domestic natural gas, conversion of biomass, the use of solar and other renewable energy technologies along with a significant commitment to improve energy efficiency and public transportation would sharply curtail America's need for imported oil, increase jobs and keep more dollars at home.

32. OECD OBSERVER. August/September 1996, On-line. Renewable energy is often more expensive than more traditional sources, and the higher costs of harnessing it have prompted government intervention to promote it, often through government expenditure on R&D and/or by introducing favourable economic or fiscal measures such as capital or output subsidies, tax breaks, and so on. Technical improvements and larger markets for renewable energy have indeed succeed in lowering costs, sometimes dramatically over the last decade. But low energy prices mean there is still a price differential between most forms of renewable energy and other energy sources, especially the low cost and flexibility of electricity production from combined-cycle gasturbines.

33. OECD OBSERVER. August/September 1996, On-line. Fifth, voluntary actions can range from formalised and binding negotiated agreements between government and industry or utilities to more informal approaches aimed at encouraging renewable energy; and are increasingly being used as part of the mix of measures in place to promote renewable energy. The range of VAs is extremely diverse: some emphasise increased dissemination of information; others take the form of more formal or binding agreements (such as the requirement of the Dutch electricity companies to produce 3% of electricity from renewable sources by 2000).

34. OECD OBSERVER. August/September 1996, On-line. The largest barrier to increased use of renewable energy is its cost, despite the reductions achieved over recent years." But other obstacles, particularly for renewable electricity, includes subsidies and other support for competing conventional fuels (especially coal and nuclear power). Lack of full-cost pricing when determining the cost of competing energy supplies also hinders the development of renewable energy since the cost of environmental impacts are usually not included in energy prices. High discount rates disadvantage projects with high capital costs but low running costs (such as renewable electricity schemes). And the increasing deregulation of the electricity industry in many of the IEA countries is stimulating competition between electricity suppliers, sometimes on a short-term price basis. Since renewable electricity is often more expensive than electricity from other fuels, that could hinder its development (unless, of course, governments set up other schemes, such as protected market shares). And there are costs from system integration and capacity backup. Other, non-cost barriers, especially the imperfect flow of information and the lack of integrated planning procedures and guidelines, also inhibit the uptake of renewables.

35. OECD OBSERVER. August/September 1996, On-line. Some policies, particularly premium prices guaranteed over several years, have succeeded in increasing renewable-electricity production and capacity rapidly, showing that investors will invest in such projects under favourable economic conditions, despite regulatory barriers. Recent policies have also demonstrated that competition can bring down the costs of renewable electricity substantially over a short period of time. But sharp increases in electricity production from renewables have, to date, been achieved only at a cost and often with distortions to energy markets.

36. OECD OBSERVER. August/September 1996, On-line. Governments have to improve the environmental performance of the energy sector while maintaining a low-cost and reliable energy supply. Whether the newer policy approaches to promote renewables can help achieve this aim in an era of increasing pressure on national energy budgets has yet to be determined. But the impetus for more widespread use of renewable energy may increase if new commitments to limit or reduce greenhouse-gas emissions beyond the year 2000 are introduced. In that event, governments may choose to accelerate or strengthen policies that promote renewable energy.

37. Lynn Martin. THE CHRISTIAN SCIENCE MONITOR. 31 March 1997, On-line.
'Energy independence' outdated
We can no longer think of energy in terms of "energy independence" for solely the US. With population growing around the world, exploding sales of automobiles, and dramatic increases in electricity consumption, it will be critical to develop new nonpolluting, secure sources of energy in the coming millennium. For America to be a leader in developing a global strategy on climate change it must not only own up to its responsibilities to reduce emissions, but it must also address a full range of renewable resources.

38. Lynn Martin. THE CHRISTIAN SCIENCE MONITOR. 31 March 1997, On-line. The issues raised by environmental threats (not to mention economic and national security considerations) are leading many to recognize the need for a transition away from oil- and coal-based power sources. Industrialized countries are pursuing ways to reduce emissions—ranging from retrofitting plants with energy-efficient lighting to engaging the power of the free market through emissions trading schemes. Throughout the world calls are heard for a massive shift to renewable sources of energy like solar, wind, and nuclear power. The former options are relatively noncontroversial, but nuclear power, the most advanced renewable source of power, remains a lightning rod for many, especially in the US.

39. Lynn Martin. THE CHRISTIAN SCIENCE MONITOR. 31 March 1997, On-line.
The glare of Three Mile Island
American attitudes towards nuclear power are much like the proverbial deer, caught this time in the headlights of Three Mile Island. Public discussions on nuclear power typically emphasize incidents of the past rather than focus on the present. Perhaps it is telling, then, that Japan, the host of the Kyoto conference and no stranger to the destructive potential of nuclear power, today offers a solid example of a national energy policy that deserves American support.

40. Lynn Martin. THE CHRISTIAN SCIENCE MONITOR. 31 March 1997, On-line. A critical component of Japan's plans for safely and sustainably meeting their energy needs is the expanded use of nuclear energy, providing environmental benefits locally, regionally, and globally. Locally, there is less pollution in Japan's cities and countryside. Regionally, there is less acid rain caused by power production in Japan. Globally, with nuclear power production increasing at the expense of oil and/or coal, greenhouse gas emissions are greatly reduced.

41. Lynn Martin. THE CHRISTIAN SCIENCE MONITOR. 31 March 1997, On-line. A critical component of Japan's plans for safely and sustainably meeting their energy needs is the expanded use of nuclear energy, providing environmental benefits locally, regionally, and globally. Locally, there is less pollution in Japan's cities and countryside. Regionally, there is less acid rain caused by power production in Japan. Globally, with nuclear power production increasing at the expense of oil and/or coal, greenhouse gas emissions are greatly reduced.

42. Lynn Martin. THE CHRISTIAN SCIENCE MONITOR. 31 March 1997, On-line. As the United States prepares for Kyoto, it would do well to look to its host as one source of new thinking in addressing the policy conundrum of sustainable development. This term first appeared in 1987 when the World Commission on Environment and Development (known as the Bruntland Commission) released its report, "Our Common Future." In that report, sustainable development was defined as, "meeting the needs of the present without compromising the ability of future generations to meet their own needs."

43. Lynn Martin. THE CHRISTIAN SCIENCE MONITOR. 31 March 1997, On-line. Japan's state-of-the-art use of nuclear energy will provide US and global planners with an important renewable source of power that will enable the inhabitants of the 21st century to grow and thrive as did their oil-dependent ancestors of the 20th century.

But the US, the originator of nuclear energy and once the industry leader, may have to take a back seat while other nations take the nuclear lead.

44. GLOBAL WARMING NETWORK ONLINE TODAY. 31 October 1996, On-line. Greenhouse gas emissions, a contributing factor to global warming, could decrease with the expanded use of nuclear power, said Hans Blix, director-general of the International Atomic Energy Agency (IAEA) in a speech to the United Nations General Assembly this week.

45. GLOBAL WARMING NETWORK ONLINE TODAY. 31 October 1996, On-line. According to Blix, nuclear power and renewable sources produce the least amount of greenhouse gases compared with coal, gas, and other fossil fuels. "There should now be a general awareness among governments that an expanded use of nuclear power and of renewable sources of energy together with conservation measures could significantly help to restrain greenhouse gas emissions," said Blix.

46. GLOBAL WARMING NETWORK ONLINE TODAY. 31 October 1996, On-line. Japan, South Korea, China and Eastern Europe have increased their use of nuclear power due to environmental concerns.

"However, at present most countries are continuing to expand their use of fossil fuels and are failing to meet targets which they have set for themselves to restrain their emissions of greenhouse gases," Blix added.

47. THE HERALD. 17 July 1996, On-line. "Nuclear power and the renewables are the only energy sources which do not produce carbon dioxide," the forum said in a letter to Mr. Gummer.

Carbon emissions are widely blamed for rising world temperatures.

48. THE HERALD. 17 July 1996, On-line. "If we are serious about combating global warming, we must retain and develop the nuclear option and encourage renewable energy sources as a matter of urgency," said the letter from the forum's director-genera, Roger Hayes.

49. THE ECONOMIST. 6 April 1997 (Reprinted in The Sacramento Bee), On-line. There is something about environmental conferences that brings out the hypocrisy in politicians.

Five years ago, at the Earth summit in Rio de Janeiro, world leaders agonized about a host of planetary problems, from the destruction of forests to the loss of biodiversity. The centerpiece of their efforts was the global-warming treaty, a promise to reduce the output of greenhouse gases from rich countries to 1990 levels by the decade's end. Then they went home, and did virtually nothing about it: Almost all rich countries are expected to overshoot their 2000 target, some by as much as 40 percent.

Meanwhile, the international environmental jamborees continue. Last week a host of non-governmental groups concerned about the future of the planet held a meeting of their own in Rio.

50. THE ECONOMIST. 6 April 1997 (Reprinted in The Sacramento Bee), On-line. Two features of global environmental conferences encourage politicians to make empty promises. The first is that politicians from one country can almost always rely on other countries not to meet their targets either, so the blame is spread thinly. Just as no country is willing to disarm unilaterally, no country will cut greenhouse gases alone (the costs may be high, and the benefit almost zero if others continue to pollute the atmosphere).

51. THE ECONOMIST. 6 April 1997 (Reprinted in The Sacramento Bee), On-line. The second is that most politicians turning up to such conferences these days are environment, rather than treasury or industry, ministers. Their principal concern is therefore to be seen to save the planet. They are content to let other ministries at home devise the policies needed to meet their targets (targets which these ministries in turn politely ignore).

52. THE ECONOMIST. 6 April 1997 (Reprinted in The Sacramento Bee), On-line. Given these uncertainties, it would be stupid to incur huge economic costs to avert global warming.

Nine-tenths of the world's commercial energy comes from fossil fuels, such as oil, coal and gas, which give off greenhouse gases. Alternative sources of energy, such as solar power, are still more expensive.

53. THE ECONOMIST. 6 April 1997 (Reprinted in The Sacramento Bee), On-line. At the same time it would be stupid not to impose measures which will reduce greenhouse-gas emissions at little or no cost; and, in this respect, most governments have failed to take advantage of some huge opportunities. On one estimate, energy subsidies worldwide are worth over $600 billion a year. Many of these subsidies force down the price of fossil fuels, encouraging consumers to burn them wastefully. It is no accident that Britain, which has withdrawn support from its coal industry, is one of the few countries on course to meet its greenhouse-gas targets.

54. THE ECONOMIST. 6 April 1997 (Reprinted in The Sacramento Bee), On-line. A full-scale assault on fossil-fuel subsidies in rich countries would salvage the global-warming negotiations. Only then will governments be able to discuss more costly attempts to cut greenhouse emissions with any credibility.

55. THE ECONOMIST. 6 April 1997 (Reprinted in The Sacramento Bee), On-line. Naturally governments prefer to avoid the political flak which comes with slashing subsidies. But how better to promote an economically sensible reform than as an attempt to save the Earth too?

56. THE ECONOMIST. 6 April 1997 (Reprinted in The Sacramento Bee), On-line. Indeed, they should place more emphasis on the credibility of new targets than on their precise size and timing. For as soon as firms in the $1-trillion-a-year fossil-fuel industry believe governments are serious about cutting greenhouse emissions, they will invest furiously in reducing the costs of alternative sources of energy.

57. INSIDE ENERGY/WITH FEDERAL LANDS. 7 October 1996, On-line. The bill provided $570 million for energy efficiency programs in FY-97, a 6% increase over the previous year, but well below Clinton's request of $735 million. Advocates said the administration deserved credit for insisting that conferees provide at least the funding level approved by the Senate. The House version of the bill would have provided only $523 million for the efficiency programs. House Republican leaders proposed even more drastic cuts in their version of the budget resolution.

"In light of the open hostility these programs have faced from many in the 104th Congress, this is an amazing turnaround," said David Nemtzow, president of the Alliance to Save Energy. "We are glad that many members of Congress recognized the key role played by increasing energy efficiency in environmental protection, national security, economic growth and international competitiveness."

DOE Principal Deputy Assistant Secretary for Energy Efficiency and Renewable Energy Joseph Romm said the department was "absolutely delighted" with the increase, saying it showed strong bipartisan support for the efficiency programs.

58. Joseph Romm and Charles Curtis. THE ATLANTIC MONTHLY. April 1996, On-line. Imagine a world in which the Persian Gulf controlled two thirds of the world's oil for export, with $200 billion a year in oil revenues streaming into that unstable and politically troubled region, and America was importing nearly 60 percent of its oil, resulting in a $100-billion-a-year outflow that undermined efforts to reduce our trade deficit. That's a scenario out of the 1970s which can never happen again, right? No, that's the "reference case" projection for ten years from now from the federal Energy Information Administration.

59. Joseph Romm and Charles Curtis. THE ATLANTIC MONTHLY. April 1996, On-line. Imagine another world in which fossil-fuel use had begun a slow, steady decline; more than a third of the market for new electricity generation was supplied from renewable sources; the renewables industry had annual sales of $150 billion; and the fastest-growing new source of power was solar energy. An environmentalist's fancy, right? No, that's one of the two planning scenarios for three to four decades from now, developed by Royal Dutch/Shell Group, the world's most profitable oil company, which is widely viewed as a bench mark for strategic planning.

60. Joseph Romm and Charles Curtis. THE ATLANTIC MONTHLY. April 1996, On-line. A decade's worth of little-heralded technological advances funded by the Department of Energy have helped to bring such a renewables revolution within our grasp. Yet budget cuts already proposed by Congress would ensure that when renewable energy becomes a source of hundreds of thousands—if not millions—of new high-wage jobs in the next century, America will have lost its leadership in the relevant technologies and will once again be importing products originally developed by U.S. scientists.

61. Joseph Romm and Charles Curtis. THE ATLANTIC MONTHLY. April 1996, On-line. Moreover, Congress's present and planned cuts in advanced transportation and fossil-fuel research and development impede efforts to maximize the nation's conventional-energy resources base.

62. Joseph Romm and Charles Curtis. THE ATLANTIC MONTHLY. April 1996, On-line. Although little can be done to change the first scenario, Congress's actions all but guarantee that if an oil crisis comes, our national response will be reactive, uninformed, and unduly burdensome. Having abandoned the technological means to minimize the crisis, the nation will be left in the next century with little more than its usual responses to energy crises: price controls or other rigid regulations, or unplanned, ineffective attempts to deal with the effects of sharp price or supply fluctuations.

63. Joseph Romm and Charles Curtis. THE ATLANTIC MONTHLY. April 1996, On-line. What's more, cuts in research on clean-energy technologies represent a statement by Congress—conscious on the part of some members, unintentional on the part of others—that global climate change is of little or no concern, and that domestic environmental problems such as urban air quality and industrial waste, require nothing more than existing strategies.

64. Joseph Romm and Charles Curtis. THE ATLANTIC MONTHLY. April 1996, On-line. Yet the nation's "tools" for dealing with pollution are similar to those for dealing with an oil crisis, and new technology usually provides the most cost-effective solution. One example: A relatively small amount of money spent today to develop, test, and deploy highly reflective roofing and road material and plant shade trees could help cool the Los Angeles area by five degrees, reducing annual air-conditioning bills buy more than $150 million. Since smog formation is very temperature-sensitive, such cooling would reduce smog concentrations by 10 percent, which would be comparable to removing three quarters of the cars on the road. The health-related benefits of that smog reduction would be worth $300 million a year. Applied nationally, the energy savings alone could exceed $10 billion a year by 2015.

65. Joseph Romm and Charles Curtis. THE ATLANTIC MONTHLY. April 1996, On-line. Although news coverage of the environment has focused on congressional efforts to roll back environmental regulations, cuts in environmental-technology programs will have as significant an impact on our quality of life in the long run. And by turning a blind eye to the technological solutions to environmental problems, we limit ourselves to far-more-onerous alternatives. The environmental regulations that Congress is rolling back today may become all the more necessary in the not too distant future.

66. Joseph Romm and Charles Curtis. THE ATLANTIC MONTHLY. April 1996, On-line. The programs being cut are not those failures of the past that are often mentioned by critics of federal energy research—for example, the synfuels program of a decade and a half ago. They are instead programs that have been delivering results for years. A report released last June by a blue-ribbon panel of independent energy analysts, led by the energy expert Daniel Yergin, the Pulitzer-prize winning author of *The Prize,* cited dozens of federally funded technological advances that "are generating billions of dol-

67. Joseph Romm and Charles Curtis. THE ATLANTIC MONTHLY. April 1996, On-line.

The Coming Oil Crisis

Given that the most recent war America fought was in the Persian Gulf, let's start by examining the likelihood that an oil crisis will occur in the coming decade. Forecasting is always risky, especially where oil is concerned, but consider what a variety of experienced energy hands from every point on the political spectrum have said in the past year alone. Donald Hodel, who was a Secretary of Energy under Ronald Reagan, has said that we are "sleepwalking into a disaster," and predicts a major oil crisis within a few years. Irwin Stelzer, of the American Enterprise Institute, says that the next oil shock "will make those of the 1970s seem trivial by comparison." Daniel Yergin says, "People seem to have forgotten that oil prices, like those of all commodities, are cyclical and will go up again." James Schlesinger, who was the Secretary of Energy under Jimmy Carter, has said, "By the end of this decade we are likely to see substantial price increases." In March of last year Robert Dole, the Senate majority leader, said in a speech at the Nixon Center for Peace and Freedom, "The second inescapable reality of the post-twentieth-century world is that the security of the world's oil and gas supplies will remain a vital national interest of the United States and of the other industrial powers. The Persian Gulf ... is still a region of many uncertainties. ... In this 'new energy order' many of the most important geopolitical decisions—ones on which a nation's sovereignty can depend—will deal with the location and routes for oil and gas pipelines. In response, our strategy, our diplomacy, and our forward military presence need readjusting." The chairman of the Federal Reserve, Alan Greenspan, not known for being an alarmist, in testimony before Congress last July raised concern that a rising trade deficit in oil "tends to create questions about the security of our oil resources." [ellipses in original]

68. Joseph Romm and Charles Curtis. THE ATLANTIC MONTHLY. April 1996, On-line.

Concerns about a coming oil crisis have surfaced in the financial markets as well. Last October, in an article titled "Your Last Big Play in Oil," *Fortune* magazine listed several billionaires and "big mutual fund managers" who were betting heavily that oil prices would rise significantly. The magazine went on to suggest an investment portfolio of "companies that are best positioned to profit from the coming boom."

69. Joseph Romm and Charles Curtis. THE ATLANTIC MONTHLY. April 1996, On-line.

Fundamental trends in oil demand and supply underlie this emerging consensus. First, the world will probably need another 20 million barrels of oil a day by the year 2010, according to the Energy Information Administration (EIA). The International Energy Agency projects an even greater growth in demand, following the inexorable tide of population growth, urbanization, and industrialization.

70. Joseph Romm and Charles Curtis. THE ATLANTIC MONTHLY. April 1996, On-line.

Second, the world's population is expected to increase by 50 percent by 2020, with more than half those additional people born in Asia and Latin America. And as farm workers move to the city, much more energy and oil will be needed. the fundamentals of urbanization—commuting, transporting raw materials, constructing infrastructure, powering commercial buildings—all consume large amounts of oil and electricity. At the same time, fewer farms will have to feed more people, and so the use of mechanization, transportation, and fertilizer will increase, entailing the consumption of still more energy and oil. An analysis by one of the Department of Energy's national laboratories found that a doubling of the proportion of China's and India's populations that lives in cities could increase per capita energy consumption by 45 percent—even if industrialization and income per capital remain unchanged.

71. Joseph Romm and Charles Curtis. THE ATLANTIC MONTHLY. April 1996, On-line.

Finally, industrialization has an even greater impact on energy use. As countries develop industries, they use more energy per unit of gross national product and per worker. Crucial industries for development are also the most energy-intensive: primary metals; stone, clay, and glass; pulp and paper; petroleum refining; and chemicals. In the United States these industries account for more than 80 percent of manufacturing energy consumption (and more than 80 percent of industrial waste).

72. Joseph Romm and Charles Curtis. THE ATLANTIC MONTHLY. April 1996, On-line.

As *Fortune* has noted, if the per capita energy consumption of China and India rises to that of South Korea, and the Chinese and Indian populations increase at currently projected rates, "these two countries alone will need a total of 119 million barrels of oil a day. That's almost double the world's entire demand today."

73. Joseph Romm and Charles Curtis. THE ATLANTIC MONTHLY. April 1996, On-line.

Barring a major and long-lasting worldwide economic depression, global energy demand will be rising inexorably for the foreseeable future. The Persian Gulf, with two thirds of the world's oil reserves, is expected to supply the vast majority of that increased demand—as much as 80 percent, according to the EIA. Within ten to fifteen years the Persian Gulf's share of the world export market may surpass its highest level to date, 67 percent, which was

attained in 1974. The EIA predicts that in the face of increased demand, oil prices will rise slowly to $24 a barrel (1994 dollars) in 2010. If, instead, they remain low, the Gulf's share of the world export market may rise as high as 75 percent in 2010.

74. Joseph Romm and Charles Curtis. THE ATLANTIC MONTHLY. April 1996, On-line. The growing dependence on imported oil in general and Persian Gulf oil in particular has several potentially serious implications for the nation's economic and national security. First, the United States is expected to be importing nearly 60 percent of its oil by ten years from now, with roughly a third of that oil coming from the Persian Gulf. Our trade deficit in oil is expected to double, to $100 billion a year, and by that time—a large and continued drag on our economic health. To the extent that the Gulf's recapture of the dominant share of the global oil market will make price increases more likely, the U.S. economy is at risk. Although oil imports as a percent of gross domestic product have decreased significantly in the past decade, our economic vulnerability to rapid increases in the price of oil persists. Since 1970 sharp increases in the price of oil have always been followed by economic recessions in the United States.

75. Joseph Romm and Charles Curtis. THE ATLANTIC MONTHLY. April 1996, On-line. Second, the Persian Gulf nations' oil revenues are likely to almost triple, from $90 billion a year today to $250 billion a year in 2010—a huge geopolitical power shift of great concern, especially since some analysts predict increasing internal and regional pressure on Saudi Arabia to alter its pre-Western stance. This represents a $1.5 trillion increase in wealth for Persian Gulf producers over the next decade and a half. That money could buy a tremendous amount of weaponry, influence, and mischief in a chronically unstable region. And the breakup of the Soviet Union, coupled with Russia's difficulty in earning hard currency, means that for the next decade and beyond, pressure will build to make Russia's most advanced military hardware and technical expertise available to well-heeled buyers.

76. Joseph Romm and Charles Curtis. THE ATLANTIC MONTHLY. April 1996, On-line. The final piece in the geopolitical puzzle is that during the oil crisis of the 1970s the countries competing with us for oil were our NATO allies, but during the next oil crisis a new, important complication will arise: the competition for oil will increasingly come from the rapidly growing countries of Asia. Indeed, in the early 1970s East Asia consumed well under half as much oil as the United States, but by the time of the next crisis East Asian nations will probably be consuming more oil than we do.

77. Joseph Romm and Charles Curtis. THE ATLANTIC MONTHLY. April 1996, On-line. What is the appropriate national response to the re-emerging energy-security threat? Abroad the Department of Energy has been working hard to expand sources of oil outside the Persian Gulf region—in the former Soviet Union, for example—and to encourage the privatization of the oil companies in Mexico and other Latin American countries.

78. Joseph Romm and Charles Curtis. THE ATLANTIC MONTHLY. April 1996, On-line. At home the DOE is encouraging greater production by providing royalty relief in the deep waters of the Gulf of Mexico and similar incentives, so that the industry can drill wells that otherwise would not be cost-effective. The DOE is working to reduce the cost for the industry to comply with federal regulations. Finally, the department is spending tens of millions of dollars a year to develop new technologies that will lower the cost of finding and extracting oil-for example, using advanced computing to model oil fields. Still, few expect to reverse the decade-long decline in U.S. oil production. Some would open the Arctic National Wildlife Refuge to drilling, a plan the Clinton Administration has opposed on environmental grounds, but not even that would change our forecasted oil dependency much. This is true even using earlier, more optimistic estimates that the refuge could provide 300,000 barrels of oil a day for thirty years. The EIA projects that within ten to fifteen years the United States will probably be importing thirty times as much—some 10 million barrels of crude oil a day, even if the decline in other domestic production levels off in the next few years.

79. Joseph Romm and Charles Curtis. THE ATLANTIC MONTHLY. April 1996, On-line. Increasing domestic supply, although it may help to slow the rising tide of imports, cannot itself reverse the major trend. And reversing the nation's ever-increasing demand for oil would be difficult. The country is in no mood to enact higher energy taxes in order to bring our energy markets into better balance. To most people, an increase in gasoline taxes of even a few cents a gallon—let alone the amount needed to have a noticeable impact on consumption—is anathema. Similarly, Congress is in no mood for a regulatory approach, such as mandating increased fuel efficiency for cars.

80. Joseph Romm and Charles Curtis. THE ATLANTIC MONTHLY. April 1996, On-line. That leaves one solution for reducing consumption: the technological approach, which draws on America's traditional leadership in research and development. Here tremendous progress has been made. Given the uncertain nature of long-term, high-risk R&D in leapfrog technologies, the prudent approach is to explore a number of possibilities. The DOE has invested in the development of cars and trucks that are highly fuel-efficient, along with cars that run on electricity, on liquid biofuels from crops, crop waste, and municipal solid waste, or on natural gas.

81. Joseph Romm and Charles Curtis. THE ATLANTIC MONTHLY. April 1996, On-line. Technologies are also being developed to make possible a superefficient hybrid vehicle that has both an internal-combustion engine and some kind of energy-storage device, such as a battery or a flywheel. A very advanced hybrid has been described by Amory B. Lovins and L. Hunter Lovins (see "Reinventing the Wheels," January, 1995, *Atlantic*). Supporting technologies include lightweight, superstrong materials and advanced engines, among other things. This research has been undertaken by the Partnership for a New Generation of Vehicles, a collaboration among several federal agencies, the DOE's national laboratories, and the auto industry. The goal of the partnership is to design and construct by 2004 a prototype clean car that has three times the fuel efficiency of existing cars and very low emissions, and also comparable or improved performance, safety, and cost. Such a car would allow domestically produced advanced technologies to replace oil imports.

82. Joseph Romm and Charles Curtis. THE ATLANTIC MONTHLY. April 1996, On-line. Another direction that research is taking is toward advanced batteries for use in electric cars—among them the nickel metal-hydride battery—which promise to double the range achievable with existing lead-acid batteries. In conjunction with advances in clean power generation, described below, these batteries hold the prospect of replacing imported oil with domestically produced electricity.

83. Joseph Romm and Charles Curtis. THE ATLANTIC MONTHLY. April 1996, On-line. The technology that most experts would agree has the best chance over the long term of replacing petroleum use in the transportation sector is fuel cells. These are compact modular devices that generate electricity and heat with high efficiency and virtually no pollution. They run on hydrogen converted from natural gas and other fuels. The National Aeronautics and Space Administration developed early versions of fuel cells for use on space missions. Over the past two decades the DOE has spent tens of millions of dollars on several types of fuel cells that will soon be used to power cars, trucks, utilities, commercial buildings, and industries. The Japanese government has been increasing its fuel-cell budget by an average of 20 percent a year for the past five years, and Japanese companies are less than five years behind U.S. companies in this technologies. The Europeans are considering significantly increasing their fuel-cell funding. Sustained federal support might well give America the lion's share of a multibillion-dollar global market.

84. Joseph Romm and Charles Curtis. THE ATLANTIC MONTHLY. April 1996, On-line. Current DOE programs—unlike those of the late 1970s, which required oil to cost $80 a barrel if they were to be competitive—are aimed at making alternatives competitive even if oil prices decline. The likely outcome of all the programs mentioned above should not be overstated: we will not achieve energy independence in the next fifteen years. What this investment portfolio does offer is a chance in the years thereafter to blunt any foreign threat to raise oil prices dramatically and to limit the economic and geopolitical impact of Persian Gulf oil in particular. At the same time, domestic jobs will be created if money that might have gone overseas to buy foreign oil goes instead to manufacturing cars and trucks or domestic biofuels.

85 Joseph Romm and Charles Curtis. THE ATLANTIC MONTHLY. April 1996, On-line. What's more, the rapid population growth and urbanization of developing nations, coupled with the harsh pollution that characterizes most major urban centers in those nations, ensure a tremendous market for low-emission, superefficient automotive technology.

86. Joseph Romm and Charles Curtis. THE ATLANTIC MONTHLY. April 1996, On-line. Our industrialized competitors have one inherent advantage in the race to develop the supercar: gas prices of $3.00 or $4.00 a gallon. Fuel efficiency matters more in their economies, and vehicles that use alternative fuels will be cost-competitive in their markets sooner. The primary counterbalance to that advantage is U.S. technological leadership in most relevant areas, stemming in part from historically higher levels of R&D spending.

That counterbalance is about to disappear. Congress has cut the proposed fiscal year 1996 allocations for the DOE's advanced-transportation-technology budget by 30 percent. Moreover, the multi-year balanced-budget plan approved by the House and Senate would cut the budget for such technology by 60–80 percent in real terms.

87. Joseph Romm and Charles Curtis. THE ATLANTIC MONTHLY. April 1996, On-line. The fact that the DOE has been collaborating with the auto industry in the Partnership for a New Generation of Vehicles gives some in Congress a thin excuse to label the partnership's programs "corporate welfare." Yet Detroit's car makers agreed to match federal spending while coordinating their corporate research with the DOE's national laboratories in order to address the pressing national problems of oil imports and urban air quality. The last time America ignored the warning signs of growing dependence on imported oil, the Japanese were able to seize a significant share of the U.S. auto market with fuel-efficient cars.

88. Joseph Romm and Charles Curtis. THE ATLANTIC MONTHLY. April 1996, On-line. That the nation's and the world's dependence on Persian Gulf oil will grow over the next decade seems inevitable. This is particularly true since most projections assume con-

tinuing significant technological progress in bringing down the cost of domestic production, in developing alternatives, and in using energy and oil more efficiently. But those projections have not factored in the federal government's plans to withdraw from its role in fostering the development and deployment of those technologies.

89.   Joseph Romm and Charles Curtis. THE ATLANTIC MONTHLY. April 1996, On-line.
The Renewables Revolution
Predicting our energy future beyond 2010 is chancy, but here we have an opportunity to rely on perhaps the most successful predictor in the energy business: Royal Dutch/Shell Group. According to *The Economist*, "The only oil company to anticipate both 1973's oil-price boom and 1986's bust was Royal Dutch/Shell." Anticipating the oil shocks of the 1970s helped Shell to move from being the weakest of the seven largest oil companies in 1970 to being one of the two strongest only ten years later. Anticipating the oil bust was apparently even more lucrative. According to *Fortune*'s ranking of the 500 largest corporations, Royal Dutch/Shell is now not only the most profitable oil company in the world but the most profitable corporation of any kind.

When such a company envisions a fundamental transition in power generation from fossil fuels to renewable energy beginning in two decades, a transition that will have a significant impact on every aspect of our lives, the prediction is worth examining in some detail. Chris Fay, the chairman and CEO of Shell UK Ltd., said in a speech in Scotland last year, "There is clearly a limit to fossil fuel. . . . Shell analysis suggests that resources and supplies are likely to peak around 2030 before declining slowly. . . . But what about the growing gap between demand and fossil fuel supplies? Some oil will obviously be filled by hydro-electric and nuclear power. Far more important will be the contribution of alternative renewable energy supplies." [ellipses in original]

90.   Joseph Romm and Charles Curtis. THE ATLANTIC MONTHLY. April 1996, On-line.
Fay presented a detailed analysis of future trends in energy supply and demand, noting that the fossil-fuel peak in 2030 would occur at a usage level half again as high as today's. Shell's analysis does not rely exclusively on supply limits—after all, for decades people have been worried about such limits, and the supply has continued to expand—but also incorporates a recognition of the tremendous advances that have been made in renewable energy technologies over the past two decades and that are expected to be made over the next two decades.

91.   Joseph Romm and Charles Curtis. THE ATLANTIC MONTHLY. April 1996, On-line.
Although these advances in renewables have received very little media attention, they have persuaded Shell planners that renewables may make up a third of the supply of new electricity within three decades even if electricity from fossil fuels continues to decline in cost. An "Energy in Transition" scenario that they have prepared does not assume price increases in fossil fuels—also, as we have seen, a plausible hypothesis. Nor does Shell assume any attempt by governments to incorporate environmental costs into the price of energy, even though every single independent analysis has found that fossil-fuel generation has much higher environmental costs than non-fossil-fuel generation has. According to Shell's strategic-planning group, "The Energy in Transition future can claim to be a genuine 'Business as Usual' scenario, since its energy demand is a continuation of long historical trend, and the energy is supplied in a way which continues the pattern."

92.   Joseph Romm and Charles Curtis. THE ATLANTIC MONTHLY. April 1996, On-line.
Indeed, in the past fifteen years the Department of Energy, working with the private sector, has reduced the costs of electricity from biomass (such as crops and crop waste) and wind, bringing them into the current range of wholesale costs for coal and other traditional sources of electricity; three to five cents per kilowatt-hour.

93.   Joseph Romm and Charles Curtis. THE ATLANTIC MONTHLY. April 1996, On-line. A quiet revolution has already brought the United States almost eight gigawatts of biomass electrical capacity. Gasifying biomass and using advanced turbines could bring biomass power to 4.5 cents per kilowatt-hour within a decade, according to the DOE's National Renewable Energy Laboratory. Shell projects that by 2010 commercial energy from biomass could provide five percent of the world's power; using Shell's projections, we estimate that the value of that power generation could exceed $20 billion.

94.   Joseph Romm and Charles Curtis. THE ATLANTIC MONTHLY. April 1996, On-line.
Photovoltaic (PV) cells, which convert sunlight into electricity, now cost one tenth of what they did in 1975. The DOE has invested heavily in new thin-film PV panels, which take advantage of U.S. expertise in semiconductor fabrication. Shell expects that PVs, along with fuel cells and small gas-fired power plants, will permit the growth of distributed-power systems. In developing nations distributed sources can obviate the need for huge power lines and other costly elements of an enormous electric-power grid (much as personal computers replace large mainframe computers). PV modules sold worldwide totaled less than four megawatts in 1980 and now exceed 80 megawatts a year; sales continue to grow. The Energy in Transition scenario predicts that photovoltaics and other direct conversions of sunlight will be the most rapidly growing form of commercial energy after 2030. Sales could quickly exceed $100 billion. Shell itself has bought two photovoltaic companies.

95. Joseph Romm and Charles Curtis. THE ATLANTIC MONTHLY. April 1996, On-line. This scenario, a highly credible one given Shell's reputation, is tantalizing, because it holds out the possibility that the world could within a few decades begin to realize the dream of nearly pollution-free energy. Consider also that the United States, which is now the leader in most areas of renewables technology, could simultaneously reduce its dependence on foreign energy supplies, reverse the trend toward an ever-increasing energy trade deficit, and capture a large share of what promises to be perhaps the largest new job-creating sector of the international economy

96. Joseph Romm and Charles Curtis. THE ATLANTIC MONTHLY. April 1996, On-line. This is only a scenario; our actions today can have an impact, either positive or negative. According to Chris Fay, of Shell, "New technologies cannot leap from laboratory to mass market overnight. They must first be tested in niche markets, where some succeed but many fail. Costs fall as they progress down the 'learning curve' with increasing application." The long-term nature of research, and the real potential for failure, are why many options must be pursued at once and why many private-sector companies have been reluctant to invest. Fay observes, "Renewables will have to progress very quickly if they are to supply a major portion of the world's energy in the first half of the next century. . . . They can only emerge through the process of widespread commercial experimentation and competitive optimization." [ellipses in original]

97. Joseph Romm and Charles Curtis. THE ATLANTIC MONTHLY. April 1996, On-line. Federal investments clearly make a difference in technology development and global market share. Consider the case of photovoltaics. In 1955 Bell Laboratories invented the first practical PV cell. Through the 1960s and 1970s investments and purchases by NASA, the Pentagon, and the National Science Foundation helped to sustain the PV industry and gave America leadership in world sales. In 1982 federal support for renewable energy was cut deeply, and within three years Japan became the world leader in PV sales. The Bush Administration began to increase funding for solar energy and, in 1990, collaborated with the American PV industry in efforts to improve manufacturing technology; three years later the United States regained the lead in sales in this rapidly growing industry. The Clinton Administration has accelerated funding for PVs.

Sadly, however, the cuts of the 1980s have taken their toll: in the past decade German and Japanese companies snapped up several major American PV companies, which accounted for 63 percent of the PVs manufactured in the United States. Such purchases represent huge savings for our foreign competitors. They don't have to spend hundreds of millions of dollars to determine which technologies succeed. They need only let the United States do the basic research, then spend a few tens of millions of dollars plucking the winners when the federal government abandons funding or applied research.

98. Joseph Romm and Charles Curtis. THE ATLANTIC MONTHLY. April 1996, On-line. Although many members of Congress argue that the cuts in federal R&D will be made up for by the private sector, historically this hasn't happened. When the government pulls out of an area of technology, it sends a signal to the industrial and financial communities that the are has no long-term promise and that the federal government is not a reliable partner. The situation is especially bad today, because recent studies make clear that private-sector R&D has been fairly flat since 11991, and because U.S. companies have been shifting away from basic and applied research toward incremental product and process improvement—a shift that has been exacerbated by increased international competition and the downsizing of corporate laboratories.

99. Joseph Romm and Charles Curtis. THE ATLANTIC MONTHLY. April 1996, On-line. In addition, whereas the federal government only recently, and temporarily, increased funding for renewable energy, reversing the deep cuts of the 1980s, our foreign competitors have been steadily increasing such funding for a decade and a half. Whereas we once spent several times as much as the rest of the world combined, the rest of the world now significantly outspends us. Moreover, countries such as Germany, Japan, Denmark, and the Netherlands have far greater financial incentives for renewable energy. And their prices for electricity are typically much higher: in 1991 electricity cost Germany's industrial sector 8.8 cents per kilowatt-hour, whereas in the United States it cost 4.9 cents per kilowatt-hour. That means renewable energy will be cost-effective in foreign countries before it is in America.

100. Joseph Romm and Charles Curtis. THE ATLANTIC MONTHLY. April 1996, On-line. The primary competitive advantage the United States has had in renewables is technological leadership by long-term federal spending prior to the early 1980s and then the spending in the early 1990s. Recently Congress cut renewable-energy funding by 30 percent, and its multi-year budget plan calls for overall cuts of 60 percent or more by the year 2002. The cuts will have two effects.

101. Joseph Romm and Charles Curtis. THE ATLANTIC MONTHLY. April 1996, On-line. First, the transition to renewables that Shell envisions will probably be slowed somewhat, since America remains the leader in many relevant renewables technologies and U.S. government funding remains a sizable fraction of R&D funding worldwide. The transition, however, even if slowed, seems inevitable at some point in the middle of the next century.

102. Joseph Romm and Charles Curtis. THE ATLANTIC MONTHLY. April 1996, On-line. Second, when the transition occurs, the United States will miss what may well be the single largest new source of jobs in the next century. Mature areas like automobile manufacturing and aerospace haven't been significant net job producers for the country in two decades. The most highly promoted new area—the information revolution—is unlikely to provide as many jobs as manufacturing can, because making duplicate pieces of information generates many fewer new jobs than manufacturing duplicate pieces of hardware. Yet according to Shell's numbers, annual sales in renewable-energy technologies may hit $50 billion in 2020 and almost $400 billion in 2040. In the later year such an industry would support several million jobs.

103. Joseph Romm and Charles Curtis. THE ATLANTIC MONTHLY. April 1996, On-line. Moreover, as said above, the United States will be importing $100 billion worth of oil annually ten years from now. With prudent federal investment today, that might be the peak, and we might then see a gradual decline as U.S. technology and domestic fuels, including homegrown biomass, replace imported oil. With Congress's cuts, however, we may be only augmenting our debilitating trade deficit in oil with an equally debilitating trade in oil-replacing technologies.

104. Joseph Romm and Charles Curtis. THE ATLANTIC MONTHLY. April 1996, On-line. The renewables revolution, inevitable or not, won't spell the end of the nation's or the world's environmental problems. In Shell's scenario overall fossil-fuel use will increase steadily for decades, peaking in 2030 at a level half again as high as today's, and will not dip below current levels until 2100. If we are to achieve genuine prosperity—higher living standards accompanied by improved environmental quality—we will need to do better.

105. Joseph Romm and Charles Curtis. THE ATLANTIC MONTHLY. April 1996, On-line. This energy-saving, pollution-avoiding approach would be part of a much broader shift in the nation's environmental policy, which is vital if we are to be a prosperous country in the next century. The environmental paradigm that has predominated since the 1960s has been based on the notion that pollution is an inevitable by-product of business and that public- and private-sector efforts should be aimed at cleaning up that pollution after the fact or safely disposing of it in land, water, or the atmosphere.

106. Joseph Romm and Charles Curtis. THE ATLANTIC MONTHLY. April 1996, On-line. This so-called end-of-pipe approach is increasingly being challenged not only on environmental grounds but also on economic ones. Michael Porter, a professor at the Harvard Business School, wrote in the September-October, 1995, issue of the *Harvard Business Review*,

> When scrap, harmful substances, or energy forms are discharged into the environment as pollution, it is a sign that resources have been used incompletely, inefficiently, or ineffectively. Moreover, companies then have to perform additional activities that add cost but create no value for customers: for example, handling, storage, and disposal of discharges.

107. Joseph Romm and Charles Curtis. THE ATLANTIC MONTHLY. April 1996, On-line. Funding for pollution prevention is the best way for the nation to avoid the need for costly environmental regulations. The government has a role in encouraging pollution prevention for several reasons, First, pollution-prevention technologies often benefit each of many companies only a little bit, so no one company has an incentive to spend the necessary money by itself. Second, prevention has many societal benefits: it reduces energy and other resource consumption and improves the environment, among other advantages. Third, and most important, pollution prevention and resource efficiency help companies to shift money from consuming energy and resources to investing in technology and capital equipment, thus creating jobs and economic growth. Indeed, a shift from consumption to investment may be the single most important transformation to the U.S. economy must undergo if we are to remain prosperous in the next century.

108. Joseph Romm and Charles Curtis. THE ATLANTIC MONTHLY. April 1996, On-line. A 1993 analysis for the DOE attempted to quantify the macroeconomic benefits of pollution prevention. The study found that a 10–20 percent reduction in waste by American industry would generate a cumulative increase of $1.94 trillion in the gross domestic product from 1996 to 2010. By 2010 the improvements would be generating two million new jobs, or roughly 1.5 percent of employment in that year. According to the study, this is "a relatively large impact considering that the investments driving it were assumed to be made for purposes other than increasing employment."

109. Joseph Romm and Charles Curtis. THE ATLANTIC MONTHLY. April 1996, On-line. Moreover, this analysis does not include the jobs to be gained from capturing the large and growing export market for clean technologies and processes. Resource inefficiency and environmental degradation are very real limitations on the attempts of developing nations to raise the living standards of their people, especially since most of those nations do not have the abundance of resources with which America is endowed. The World Bank estimates that by 2000 the countries of Asia alone will need to spend

about $40 billion a year on clean technologies. By then the global market for environmental services and technologies is expected to exceed $400 billion. The resource, environmental, and capital constraints on the developing world guarantee a rich export market for the nation that leads the world in developing clean technologies.

110. Joseph Romm and Charles Curtis. THE ATLANTIC MONTHLY. April 1996, On-line.
As Michael Porter wrote in the *Harvard Business Review*,

> We are now in a transitional phase of industrial history in which companies are still inexperienced in handling environmental issues creatively.... The early movers—the companies that can see the opportunity first and embrace the innovation-based solutions—will reap major competitive benefits, just as the German and Japanese car makers did with fuel-efficient cars in the early 1970s. [ellipses in original]

111. Joseph Romm and Charles Curtis. THE ATLANTIC MONTHLY. April 1996, On-line.
That's why foreign governments are forming partnerships with their nations' companies to develop clean technologies: to overcome inexperience and ensure that they reap the benefits of early strength in the field.

The Japanese government is betting heavily on clean technologies and renewable energy. It is vigorously pursuing the Asian environmental market through the Green Aid Plan, which is designed to help Asian countries prevent water and air pollution, recycle waste, conserve energy, and develop alternative energy sources. In 1993 Japan quadrupled funding for the Green Aid Plan, to $120 million.

112. Joseph Romm and Charles Curtis. THE ATLANTIC MONTHLY. April 1996, On-line.
Germany, too, is moving in this direction, with regulations that increasingly push industry toward prevention, recycling, and life-cycle analysis. Proposed or pending regulations throughout Western Europe have implications for U.S. companies, as noted in a 1993 report prepared for the Saturn Corporation by the University of Tennessee Center for Clean Products and Clean Technologies: "European auto manufacturers are the current world leaders in car recycling and the use of life-cycle assessment to design environmentally superior cars."

113. Joseph Romm and Charles Curtis. THE ATLANTIC MONTHLY. April 1996, On-line.
Congress, in contrast, has cut by a third the Department of Energy's proposed budget for the development and deployment of energy-efficient and pollution-prevention technologies—a step that threatens U.S. leadership in this crucial area. Congress has proposed still deeper cuts in its multi-year budget plans—cuts that would deny U.S. companies a great many opportunities to compete and the nation as a whole the opportunity to capture a big piece of a market whose potential is equal to that of renewable energy: several hundred billion dollars a year.

114. Joseph Romm and Charles Curtis. THE ATLANTIC MONTHLY. April 1996, On-line.
Another criticism often leveled at the DOE is that it has had big, expensive failures, such as the synthetic-fuels program, but few successes. The department has learned from experience, however, and its R&D portfolio is diverse, emphasizing small-scale technologies that have in fact been remarkably successful in the past. The recently concluded independent review of the department's energy-research portfolio cited dozens of examples of such technologies, among them a $3 million investment in energy-efficient windows made in the late 1970s, which has already saved U.S. taxpayers more than $1 billion in lower energy bills; a polycrystalline diamond drill that has reduced the cost of drilling for oil by $1 million per well; and many of the advances described above, including photovoltaics.

115. Joseph Romm and Charles Curtis. THE ATLANTIC MONTHLY. April 1996, On-line.
Diversity is the key element of DOE policy today: diversify the world's oil supply, and diversify America's domestic supply and end-use options. Because no one can predict the future with certainty, or know the outcome of R&D in advance, the DOE must invest in many options. The sharp cuts that Congress is pursuing narrow the country's options and leave us far less flexibility to respond to future crises and opportunities.

116. Joseph Romm and Charles Curtis. THE ATLANTIC MONTHLY. April 1996, On-line.
Finally, some argue that government investments are "corporate welfare," a term implying a giveaway with no societal benefits. But the DOE has formed partnerships with the private sector to develop leapfrog technologies—such as the fuel cell, solar energy, and clean industrial, building, and transportation technologies—that will benefit many segments of our society.

117. Joseph Romm and Charles Curtis. THE ATLANTIC MONTHLY. April 1996, On-line.
Americans today have a duty to eliminate the deficit, rooted in the obligation to future generations, but the country also needs to acknowledge that public investment in R&D, far from being corporate welfare, is an investment in America's own future. As the Yergin task force wrote, Americans have an obligation to "assure for future generations that our Nation's capacity to shape the future through scientific research and technological innovation is continually being renewed."

The cuts planned for the energy-efficiency -and-renewable-energy program—30 percent this year and 60–80 percent over the next several years—far exceed the

cuts planned in overall domestic discretionary funding to balance the budget. The impact of such cuts will be enormous.

Perhaps the only way to begin to realize the loss to the future is to look at the past. Federal investment in research and development for national needs has been one of the great success stories in twentieth-century America. Why does the United States retain leadership and strong exports in vital industries like aerospace, computers, and biomedicine? American ingenuity and the private sector have certainly been instrumental in each of these industries. yet these industries have also enjoyed government support for decades. Who can doubt that a sustained high level of federal funding—eight times as much money as America's leading competitor provides—is responsible for U.S. leadership in biomedical and biotechnological research?

118. Joseph Romm and Charles Curtis. THE ATLANTIC MONTHLY. April 1996, On-line. As for computers and software, the Pentagon's Advanced Research Projects Agency "virtually single-handedly created the United States' position of world leadership in computer sciences," according to a Harvard Business School case study on ARPA. And of all R&D dollars spent in the aircraft industry from 1945 to 1984, some 85 percent came from the federal government. In an unexpected benefit of the kind that is common in federal R&D, much of the turbine technology that is today generating electricity and helping to keep down utility rates had its roots in government-funded work on jet engines.

John Preston, formerly the director of technology development for the Massachusetts Institute of Technology, told Congress in 1993, "It seems clear that when the government teams up with academia and industry, and participates throughout the spectrum of technology, the United States becomes dominant in that industry." America's technological lead in most kinds of fuel cells and photovoltaics stems from almost two decades of NASA, National Science Foundation, and Pentagon support, followed by almost two decades of DOE support.

119. Joseph Romm and Charles Curtis. THE ATLANTIC MONTHLY. April 1996, On-line. Some of the most pressing national needs in the coming decades are to reduce the country's huge and growing trade deficit in oil, to minimize any economic or political threat that might arise from the growing dependence on Persian Gulf oil, to prevent pollution, to avoid irreversibly changing the global climate, and to capture a large share of the enormous potential market for energy and environmental technologies. Remarkably, a great many of the same R&D investments can simultaneously achieve all these ends while cost-effectively reducing the energy bills of businesses and consumers. Equally remarkably, Congress demonstrates an overwhelming desire to gut the funding for investments by the energy-efficiency-and-renewable-energy program, although it costs Americans only $4.00 per person a year.

120. Joseph Romm and Charles Curtis. THE ATLANTIC MONTHLY. April 1996, On-line. Nothing is clearer to those who study the matter than that the world is on the verge of a revolution in energy and environmental technologies—a revolution made possible by more than two decades of U.S. government investment. This revolution can be expected to create a number of industries that collectively will provide one of the largest international markets and one of the largest sources of new high-wage jobs in the next century, with annual sales in excess of $800 billion.

Yet just as our foreign competitors are starting to catch on to the major trends in this American-led revolution, Congress wants to pull the federal government out of every relevant technology, leaving America on the sidelines, perhaps for good. Only a misbegotten ideology could conceive a blunder of such potentially historic proportions.

121. Joel L. Silverman. ELECTRICITY JOURNAL. March 1995, On-line. During the past decade or so, economic, political and technological forces have combined to alter the framework of the electric power industry. As a result of this change, a small but growing market for non-utility supplied energy has emerged. This emerging independent electricity market has in turn spawned a variety of nascent electricity-generating industries based on renewable energy technology (RET). These industries include biomass, geothermal, photovoltaics, high-temperature solar thermal, wind and non-utility hydro.

122. Joel L. Silverman. ELECTRICITY JOURNAL. March 1995, On-line. Although the independent power market is relatively robust, the viability of the renewable energy industries competing in this market seems uncertain. While there was a significant growth in renewables during the 1980s, fueled in large part by the political desire to reduce oil dependence and the social heat to reduce pollution, the growth is relative. Renewables have made few inroads in the overall energy mix of the U.S. In spite of the fact that lawmakers and activists have called for increasing the use of renewables nationwide, grid-connected renewable electric capacity (if large scale hydro is factored out) is less than 3 percent of total U.S. capacity.

123. Joel L. Silverman. ELECTRICITY JOURNAL. March 1995, On-line. To some extent, technological and economic factors account for the slow penetration of renewables. But competitive inequities are also to blame. The nature and technology of renewables are markedly different from fossil fuels, and their associated industries are similarly diverse in structure and strategy. Yet companies within these industries must compete almost solely within the given structure of the electric

power industry—an industry dominated by the buying and lobbying power of the utility companies and large independent developers. In this broader industry, the standards for competition are based on metrics geared for fossil fuels, which often do not account for the economic and environmental benefits of renewable energy.

124. Joel L. Silverman. ELECTRICITY JOURNAL. March 1995, On-line. The slow growth of renewables may eventually spell disaster for these industries. Some may assume that renewable energy technologies will automatically be around when called into service. But renewable electricity generating resources are only as viable as the industries that sustain them. A renewables technology is more likely to achieve its market potential if there is a robust, competitive group of firms comprising the industry. This industry would be characterized by: (1) greater investment and more creative approaches to product and manufacturing process research and development (leading to higher quality and lower costs); (2) sound market development strategies, which increase revenues and establish secure, growing markets; (3) the development of long-term business strategies which account for the issues and risks inherent in the industry environment; (4) enhanced global competitiveness; and (5) a flow of investment capital from the private sector. A snowballing occurs; as the industry firms and investors see increases in size and profitability within a reasonable time, the industry becomes more attractive, drawing in new players and investment. Conversely, if there is no market in which to compete, no driver for technological and market innovation, then there is no basis for sustaining individual companies—and essentially no industry.

125. Joel L. Silverman. ELECTRICITY JOURNAL. March 1995, On-line. Which is the reality for renewables? The Union of Concerned Scientists along with the Coalition for Energy Efficiency and Renewable Technologies have argued for "sustained orderly development of renewable technologies," pointing out that while emerging wind and solar technologies are projected to be cost competitive for large-scale development by the year 2000, those industries must be able to sustain themselves and mature in the interim in order to meet that demand. The implication is that RETs may be in trouble.

126. Joel L. Silverman. ELECTRICITY JOURNAL. March 1995, On-line. Events in the once-promising, high-temperature solar thermal industry epitomize the concern. A recent DOE report pointed out that "the boom and bust cycles of the past have caused industry to be reticent about embarking on a major investment program without stronger signals from the marketplace that there will be a stable expansion of the demand for solar electric technologies." Luz, the last remaining competitor of real consequence, filed for bankruptcy in 1992, undone mostly by inconsistent government policies and the lack of strong signals of support. According to Michael Lotker, Luz's former vice president of business development, the failure demonstrated "that a company and a technology must have room to grow if it is to survive, including a stable regulatory environment, a marketplace that values solar technologies' mitigation of fuel price risk, and an overall tax structure that is both stable and equitable for all solar technologies."

127. Joel L. Silverman. ELECTRICITY JOURNAL. March 1995, On-line. Table 1 charts the primary renewable energy sources and the methods used to convert them for electricity generation. (Table 1 omitted) Much of their attraction lies in the fact that renewables are essentially limitless in supply (or, as in the case of geothermal and biomass, so abundant as to be considered nearly inexhaustible); they are largely nonpolluting and otherwise inoffensive to the environment (biomass and hydropower are the exceptions); they are versatile in application size, location and customer; require no importation and (excepting biomass) have no fuel costs. Although renewable-based power plants have high capital costs, they have short construction lead times and generally low variable costs, unlike their fossil-fuel counterparts. Furthermore, technological advancements are bringing the price of most renewable energies almost in line with that for fossil fuels.

128. Joel L. Silverman. ELECTRICITY JOURNAL. March 1995, On-line. The energy potential of renewables alone makes them worthy competitors to traditional energy sources. For instance, the annual wind energy potential of just three states—North Dakota, Montana and Wyoming—could have met the total 1990 U.S. demand for electricity. Sunlight in this country delivers almost 500 times as much energy as we consume every year; and geothermal resources are estimated to be thousands of times greater than U.S. coal reserves. In all probability, only a fraction of these resources is truly exploitable given land constraints, technological limitations, energy conversion efficiencies, environmental impact considerations, and the like. Yet, even that fraction is regarded as more than sufficient to meet forecasted U.S. energy demands.

129. Joel L. Silverman. ELECTRICITY JOURNAL. March 1995, On-line. The current renewable energy industry has its roots in the 1970s when oil embargoes, rising energy prices and increased pollution raised concerns about continued dependence solely on fossil fuels. When world oil prices increased 300 percent in 1974, industrialized nations hastened to develop alternative energy sources, authorizing huge investments in research and project development. In the U.S., federal support took the form of investment tax credits (for both individuals and businesses) and millions in capital research funds, especially for wind and solar power. Federal funding for renewable energy R&D rose from $75.1 million in fiscal 1975 to $718.5 million in fiscal 1980. Private capital

flowed (relatively) freely, encouraged by the strong show of government support and consumer interest. Universities, utilities, and private companies all took advantage of the new research largesse, pioneering technologies and applications, especially in the area of wind and PVs.

130. Joel L. Silverman. ELECTRICITY JOURNAL. March 1995, On-line. While economic forces laid the foundation, regulation built the industry structure. The single most important spur to creation of a commercial renewable power market—in fact, to the development of the whole independent power industry—was undoubtedly the passage in 1978 of the Public Utility Regulatory Policies Act. Under this law, designed to encourage small-scale electric power production, cogeneration and energy conservation, electric utilities are required to buy power from independent producers—defined by PURPA as "qualifying facilities" (QFs)—at a price determined by the "avoided cost" of generation (i.e., the utility's marginal electricity cost). The utilities must also provide backup power as necessary.

131. Joel L. Silverman. ELECTRICITY JOURNAL. March 1995, On-line. Ironically, the environment that fostered the renewables industry has largely eroded just when experience and technology have matured sufficiently to make most renewables a truly cost-competitive, reliable energy alternative. Emphasis on decreased energy consumption, coupled with an abundance of inexpensive natural gas, have overshadowed the perceived need for alternatives to limited fossil fuels. Federal support for renewables declined steadily throughout the 1980s, reaching a low of $140.7 million in fiscal 1989. By 1993, appropriations had crept back up, reaching $327 million, still only a fraction of renewables budgets in the late 1970s. Private investment has languished, scared off by early project failures, overoptimistic cost estimates and diminished government support. This lack of capital, either public or private, has severely hampered technological R&D, industry technology transfer for commercial project testing, and market development activities. In addition, despite the great technological advances made by U.S. companies, foreign firms and governments have in some cases equaled or surpassed the U.S., raising questions as to which countries would dominate the global market as technologies mature.

132. Joel L. Silverman. ELECTRICITY JOURNAL. March 1995, On-line. Changes in power regulation and purchase contracts have also upped the competitive ante. A recent survey found that nearly all states with identified adopted or are in the process of developing competitive bidding schemes for supply sources. Independent power producers as a whole—renewables included—will compete for the same share of the pie. In addition, the 1992 Energy Policy Act encourages development of freer market entry for IPPs, with no restrictions on size or fuel source, such as those imposed by PURPA. Both changes are likely to put renewables at a competitive advantage.

133. Joel L. Silverman. ELECTRICITY JOURNAL. March 1995, On-line. Finally, deregulation in the electric utility industry poses major problems for renewables. New industry structures are being proposed that would provide most customers, especially large buyers, a primary purchase criterion will be price. While renewables are becoming more price competitive, many of their benefits are societal in nature, measured in terms of environmental considerations, diversification of our fuel source mix, etc. Individual buyers will have little interest in the benefits provided by electricity from renewables. Thus, renewables could suffer a major setback under such deregulation scenarios.

134. Joel L. Silverman. ELECTRICITY JOURNAL. March 1995, On-line. The wind-power industry is global, with a small number of players competing for a limited market. During the late 1980s the domestic industry was marked by major shakeouts in the wake of drops in federal funding, loss of tax credits and the suspension of California's Standard Offer 4 (SO4) contracts, which were long-term power purchase contracts from utilities at a fixed price per unit of output. The industry's reputation was damaged by inexperienced and overzealous operators who often oversold their products and couldn't deliver the capacity promised. Both the public and investors looked upon the wind industry with skepticism. Where once dozens of companies were competing, the number of U.S. developers and manufacturers has dropped to a handful. More recently, the Energy Policy Act of 1992 (EPAct) provide a 1.5c/kWh production credit for electricity produced from wind—a substantial bonus for the industry.

135. Joel L. Silverman. ELECTRICITY JOURNAL. March 1995, On-line. As table 2 indicates, industry growth has been relatively robust. (Table 2 omitted) Installed worldwide capacity exceeds 4000 MW. The U.S. market has historically been the largest and fastest growing. California projects alone represented close to 70 percent of 1991 production. Technological improvements in turbines, led by U.S. firms, have brought wind prices down to about 5c/kWh, making wind virtually cost competitive with conventional fuels, even without the EPAct incentive. According to a widely cited Pacific Northwest Laboratories study done in 1990, there is enough wind potential in the U.S. to produce more than the total amount of electricity the nation currently uses.

136. Joel L. Silverman. ELECTRICITY JOURNAL. March 1995, On-line. Wind is an intermittent resource, varying greatly in volume and quality depending upon location and time. Even in prime areas, wind speeds change seasonally and throughout the day, meaning that capacity is not constant. The trick in project development

is to find a wind resource that tends to match a utility's peaking requirements. Most utilities do not have sufficiently detailed information on local wind resources to accurately predict load-carrying capability, and lack of site-specific resource data is a serious impediment to wind's wider acceptance.

137. Joel L. Silverman. ELECTRICITY JOURNAL. March 1995, On-line. Until recently, the only areas economically viable for windpower development were those that had class 5–7 winds—16+ mph. Most of these are concentrated on the Pacific Coast and the Great Plains, but only in California has there been any real commercial development. These sites are often remote, making transmission access a key concern. Moderate wind areas—averaging speeds of 14 mph—are much more widely distributed, but turbine technology could not make efficient use of them. A new generation of turbines, most notably Kenetech Windpower's 33 M-VS, is likely to change that. The turbine uses a variable-speed generator to produce electricity more efficiently over a wider range of speeds that with previous turbines.

138. Joel L. Silverman. ELECTRICITY JOURNAL. March 1995, On-line. With these recent advances in turbine efficiency and reliability has come increased utility interest in wind power. Kenetech Windpower has executed contracts or is in negotiation with utilities in the Pacific Northwest, the Midwest and California, as well as Canada and Western Europe. These contracts total in the hundreds of megawatts.

139. Joel L. Silverman. ELECTRICITY JOURNAL. March 1995, On-line. But, ironically, early advances by U.S. players fostered worldwide technology development efforts that now far surpass domestic investments—and pose considerable competitive threat to U.S. companies. Industry analysts say that European countries will spend $150 million annually through the year 2000 for wind-power R&D and market incentives, creating a market totaling $3.2 billion. European development will likely eclipse that of the U.S. throughout the 1990s, a by-product of EC-member nation policies encouraging renewable energy. By contrast, DOE funding in FY 1992 was $21.5 million—a 91 percent increase of 1991, but a far cry from the high of $60 million in 1980. At stake for U.S. companies is not only the world wind market, but a substantial foreign market as well. To be sure, expansion of European and other foreign markets does not necessarily mean that U.S. companies will lose market share, but they may face greater challenges when competing against foreign companies on those firms' own turf. European firms are also pursuing the U.S. market. A number of these European firms claim to be on par technologically with Kenetech Windpower. U.S. dominance of the wind market could be seriously compromised in the face of these developments.

140. Joel L. Silverman. ELECTRICITY JOURNAL. March 1995, On-line. Outlook and Viability: A few years ago, the outlook for the wind power industry in the United States was extremely tenuous. SO4 contracts were winding down and wind power contracts for new generating capacity were few and intermittent. Currently, the industry's long- and short-term outlooks, while vulnerable, seem extremely favorable.

141. Joel L. Silverman. ELECTRICITY JOURNAL. March 1995, On-line. Wind technology for grid application had already proven itself through established wind farms. Significant groundwork had already been laid in building relationships with utilities through consortiums and joint ventures. But the primary reason for the improved industry outlook is Kenetech Windpower's introduction of the new-generation 33MVS variable speed turbine, which allows wind to effectively compete for new power contracts.

Kenetech has recently signed numerous contracts for wind plants using its new variable speed turbine. FloWind has also been awarded new contracts. These contracts represent the leading edge of a strong growth curve as the domestic wind industry moves out of its emergent stage.

142. Joel L. Silverman. ELECTRICITY JOURNAL. March 1995, On-line. However, there is still vulnerability in the industry, as its future is dependent on the fortunes of Kenetech Windpower. Although no warning signs are apparent, a worst-case scenario would involve a sudden shutdown in market demand or problems with the new-generation turbines that could cause sever problems at Kenetech. This seems unlikely, s Kenetech has already proven itself adept in its financial skills and in preparing for such contingencies. Nevertheless, the domestic industry's dependence on one or two companies introduces a measure of risk.

143. Joel L. Silverman. ELECTRICITY JOURNAL. March 1995, On-line. One word describes the U.S. geothermal industry: concentration. Of the more than 2500 MW of geothermal generating capacity currently in operation in the U.S., all but about 50 MW comes from projects in California and Nevada. Worth nearly $4 billion, these projects rest mostly in the hands of five large companies: Unocal, Calpine, OESI, California Energy (CEC), and Oxbow Geothermal. Entry barriers are clearly high, due in part to the large capital requirements of geothermal projects and the limited supply of geothermal leaseholds. CEC alone controls over 550,000 acres of leaseholds in the West.

144. Joel L. Silverman. ELECTRICITY JOURNAL. March 1995, On-line. Until the mid-1980s, the industry was one of the fastest-growing alternative sources of power generation, and claimed many of the lucrative, high fixed-payment SO4 contracts awarded by California

utilities. Federal tax credits of 15 percent for renewables business investment also helped to fuel the growth. Geothermal project profitability has been generally high (due in no small part to the aforementioned contracts) with typical net cash flows of approximately 44 percent.

145.   Joel L. Silverman. ELECTRICITY JOURNAL. March 1995, On-line. Growth (currently estimated at less than 10 percent annually, from a high of nearly 18 percent) has slowed considerably in the past five years, a function of both falling fuel prices and excess capacity in the industry's major geographic market, the West. Although capacity needs may increase in the future, they will be dampened by utility conservation efforts. Near-term geothermal growth rests heavily with reserved contract set-asides in California, Nevada and the Northwest, totaling less than 500 MW. Two small opportunities totaling 75 MW exist in Hawaii and the Aleutians. Longterm growth will likely be led by Nevada utilities, with expectations of a minimum 25–35 MW awarded to geothermal companies annually. Hawaii also provides some hope, as load growth there is expected to increase 5–10 MW yearly, and geothermal resources are an abundant, economic alternative in the state. At the same time, Hawaii's geothermal industry faces enormous public opposition.

146.   Joel L. Silverman. ELECTRICITY JOURNAL. March 1995, On-line. The geothermal resource is an energy heavy-hitter, with the ability to generate reliable, high-capacity baseload power for between 5c/kWh and 7c/kWh—very near conventional fuel costs. It is one of the most commercialized of all renewable technologies. Utilities are virtually the industry's sole buyer. Supply is vast; hydrothermal reservoirs—the only commercialized resource-represent but a small fraction of geothermal resource potential.

147.   Joel L. Silverman. ELECTRICITY JOURNAL. March 1995, On-line. Geothermal is also one of the least modular and least differentiated of resources. As a high-capacity, baseload generating resource, a hydrothermal reservoir lends itself almost exclusively to large-scale, utility-based applications. Geography also constrains geothermal modularity. Hydrothermal reservoirs are generally in remote locations: In the U.S., they are essentially limited to California, Nevada, Hawaii and Utah (in resource order).

148.   Joel L. Silverman. ELECTRICITY JOURNAL. March 1995, On-line. For the long term, solutions seem to lie with improved technology and the commercialization of other geothermal resources such as geopressured brine and hot dry rock. New, more powerful turbines are available which can add 30 percent more generating power to hydrothermal projects. Two California companies recently signed a joint development agreement specifically for wet steam and brine projects. Even at that, some say commercialization is at minimum 10 years away: Brine extraction technology has been proven and is still being refined. But the commercial barrier is being able to predict a reservoir's productive life with higher confidence. Unfortunately, confirming the quantity and quality of any geothermal resource is difficult, expensive and risky—more so than that for most other resources, according to industry experts.

149.   Joel L. Silverman. ELECTRICITY JOURNAL. March 1995, On-line. Hot dry rock (HDR), the only geothermal resource that is widely dispersed, may hold greater near-term commercial potential—but U.S. development has been hampered by cuts in government spending. That in turn may spell trouble for U.S. competitiveness. The U.S. invented the extraction technology for HDR and has held dominance in the field, but Japan, Russia and several European countries are in active research and development.

150.   Joel L. Silverman. ELECTRICITY JOURNAL. March 1995, On-line. Lack of flexibility. In the leaner, meaner power market, flexibility vis-à-vis application, generating capacity and customer—is nearly as important as capitalization. Geothermal is at a decided disadvantage compared to its renewable power colleagues. Given what's currently commercialized, it cannot offer the variety of applications and locations of sun, wind or biomass power. Hence, it does not have much maneuverability in seeking out new market segments.

151.   Joel L. Silverman. ELECTRICITY JOURNAL. March 1995, On-line. Outlook and Viability: The geothermal industry can sustain itself through an anticipated moderate growth in demand over the remainder of the '90s. However, its long-term future is clouded by significant uncertainty. In the short term, existing SO4 contracts, a number of smaller contracts that seem likely to be signed in Nevada, California, Hawaii and the Northwest, and continued utility involvement in the industry all provide stabilizing influences. Growth through 1999 is expected to average about five percent annually in the U.S.

152.   Joel L. Silverman. ELECTRICITY JOURNAL. March 1995, On-line. In the longer term, the cutoff of R&D funding for emerging geothermal technologies seriously diminishes the prospect for geothermal to become a significant contributor to the nation's renewables mix. Without access to HDR and other geothermal resources, the industry is constrained to the remaining liquid-dominated hydro-thermal reservoirs, which are limited both in size and geographic scope. Even if the industry can remain viable by developing untapped hydrothermal fields, R&D is needed to create a bridge to significant new capacity. Furthermore, a future based on only exploiting hydrothermal resources has its pitfalls. The industry is

down to less than six players from 10 in 1984. One positive consequence of the consolidation is that the remaining firms have a larger capital base, a boon for competing in foreign markets. A worst-case scenario would see little or intermittent growth due to geothermal's inability to compete on a cost basis with other energy sources or due to its failure to attract financing for new geothermal fields.

153. Joel L. Silverman. ELECTRICITY JOURNAL. March 1995, On-line. C. The Photovoltaic Industry. PVs are a $500 million global industry dominated by the U.S., Japan and a few European countries. Large corporate giants like Texas Instruments, Siemens, Canon, Mobil, Amoco and many others are jockeying for position, not only with each other, but with a host of small, well-funded entrepreneurial firms with strong R&D bases. Five manufacturers—Siemens Solar (a U.S. subsidiary of German giant Siemens), Solarex (a privately held U.S. company), Sanyo (Japan), Kyocera (Japan), and Kaneka (Japan)—accounted for 55.3 MW in 1991, more than half of the world's shipments. In the U.S., Siemens Solar and Solarex accounted for 85 percent of U.S. shipments. Although output has doubled since 1985, and the annual growth rate is around 20 percent, all but a few firms are sustaining substantial losses. The reason for these losses is the enormous investment in R&D being made by firms in the industry coupled with intense price competition.

154. Joel L. Silverman. ELECTRICITY JOURNAL. March 1995, On-line. Technology and innovation is the most dynamic aspect of this industry. Different technology tracks compete for dominance of the industry, as new records for efficiency and manufacturing costs are frequently set. Both product and process R&D are crucial to industry development. The two primary technology paths are thin-film silicons and the more well-established crystalline and polycrystalline silicons. The technology demands such intensive levels of R&D that players must commit to one, or at the most two, technologies. Despite this, over $2 billion has been invested by the private sector since the 1970s, a clear indicator that high returns are expected eventually to pay back costs. In any case, companies must decide how to allocate resources between basic R&D and process technology development. The continuing challenge is how to finance both the expensive R&D and the capital-intensive plant and equipment needed to stay competitive.

155. Joel L. Silverman. ELECTRICITY JOURNAL. March 1995, On-line. Modules are bought and sold based on their rated generating capacity, no matter what technology is used. With little quality differentiation, new market development and competition among manufacturers is based primarily on price. This intense price competition, along with the large investments in R&D, has resulted in frequent consolidation, exits, entrances and divestment as companies compete for niches, market share and technological superiority.

156. Joel L. Silverman. ELECTRICITY JOURNAL. March 1995, On-line. In Table 3 once can see that PVs enjoy wider market applications than other RETs. (Table 3 omitted) The three major segments are consumer products, remote power and bulk power generation. Consumer applications, growing at about five MW annually, covers everything from calculators and watches to larger systems, such as batteries and lighting. Remote power currently stands as the largest commercial segment. This market includes stand-alone applications for telecommunications, water pumping, security systems, navigation aids, highway lighting and call boxes, and village power for vital refrigeration, communication and lighting in developing nations. Although fragmented, it holds enormous potential, with an estimated five million needy villages worldwide.

157. Joel L. Silverman. ELECTRICITY JOURNAL. March 1995, On-line. If the geothermal resource is the "hulk" of the renewable industry, then surely photovoltaics are the "lean machines." Modular, lightweight, and portable, PVs lend themselves easily to both large- and small-scale (even minute) applications. Unlike some RETs, they require no other fuel or water, and can eliminate the need for transmission lines— decided competitive advantage. They have proven to be a reliable power source for numerous consumer, space and military uses. The downside is that PVs are still very expensive: about $0.25-0.40/kWh.

158. Joel L. Silverman. ELECTRICITY JOURNAL. March 1995, On-line. The U.S. was once the undisputed R&D and manufacturing leader of the industry, with enormous investments in crystalline silicon technology. However, it has seen its market share slip from 65 percent to 33 percent in the past 10 years. Much of this slippage has been due to the lack of domestic governmental support given to the industry. Over the past 25 years, the U.S. government has invested close to $1 billion in PV R&D and demonstration projects. Unfortunately, support for PVs has dropped by 75 percent since 1980, down to an annual average of $40 million. In notable contrast, both the Japanese and German governments have been aggressive in their encouragement of domestic PV markets. In the late 1980s, Germany's PV budget was 65 percent greater than that of the U.S., while Japan's was 40 percent greater. Currently, the U.S. shares the worldwide market almost equally with Japan and a few European firms. There is some fear in the domestic industry that government cutbacks in R&D funding, the foreign acquisition of major U.S. players and a traditional emphasis on short-term profitability could stem private investment, leaving the market wide open to foreign competitors.

159. Joel L. Silverman. ELECTRICITY JOURNAL. March 1995, On-line. Outlook and Viability: The PV industry is probably the most secure and robust of the renewable energy industries. Opportunities and growth prospects are assured both in the long and short term. Current purchases of PV products are for cost-effective applications, even though the cost of generating electricity from PVs is higher than all the other renewables. The intensive investment in R&D will continue to increase cell efficiencies, opening markets for new cost-effective applications. In addition, manufacturing process innovation and scale economies will continue to drive down prices.

The diversity in types and numbers of industry players, technologies and markets also reduces greatly the risk that the PV industry will have a major setback. And the large number of industry players stimulates innovation in both technology and marketing.

160. Joel L. Silverman. ELECTRICITY JOURNAL. March 1995, On-line. Three broad technologies constitute the major segments of this industry: waste-to-energy, wood and agricultural waste burning and landfill gas production. These methods are used in cogeneration projects, baseload grid sales to utilities and self generation. Biomass more closely resembles traditional power plants, as the technologies are almost identical to conventional coal-powered technologies. It has therefore been perceived as less risky than other RETs, especially by utility buyers. The advantages of biomass fuels relative to coal are the low levels of certain pollutants found in fossil fuels and that there are no net increases in greenhouse gases due to the absorption of CO_2 while the biomass is growing. However, NOx emissions still constitute a significant environmental concern for biomass. Besides geothermal, it is the only RET that can supply predictable energy capacity, and it can do so in reasonably small increments. There is about 9000 MW of biomass capacity in operation in the U.S., mostly from wood-fired projects.

161. Joel L. Silverman. ELECTRICITY JOURNAL. March 1995, On-line. Waste-to-energy projects generate electricity through the incineration of municipal solid waste (MSW) and refuse-derived fuel (RDF). A maturing industry, biomass was one of the fastest growing of the RETs, proving itself a practical option in the face of diminishing landfill capacity. Most players operate as developer/vendors. Competition in this segment is very high. A relatively small pool of municipal contracts is the chief source of business, garnered through stiff bidding and negotiation. However, environmental concerns have put pressure on this segment. Compliance with more stringent air and water quality standards has increased project costs. Citizen opposition to incineration plants is also high; the not-in-my-backyard syndrome has been effective in canceling or delaying projects, mostly on environmental grounds.

162. Joel L. Silverman. ELECTRICITY JOURNAL. March 1995, On-line. While costs may be in a state of flux, supply is uncertain as well. The current emphasis on resource recycling and source reduction—with goals as high as 50 percent of the waste stream in some states—could reduce fuel for waste-to-energy plants, or lessen the quality and hence the burning efficiency of the fuel. Most industry observers believe that even with recycling there will be more waste than can be handled by landfills.

163. Joel L. Silverman. ELECTRICITY JOURNAL. March 1995, On-line. Despite the seeming obstacles, growth will continue in the waste-to-energy segment, driven by the mounting garbage crisis in the U.S. While this industry segment contributes to the renewables mix, its size will always be limited by the pool of contracts and the size of the solid waste stream. The segment's competitive advantage may in fact be that it can be both an electricity-generating alternative and a waste-management alternative—a benefit no other RET can claim. Existing players, especially large companies, are likely to do well. The odds of survival will be increased if developers can integrate resource recycling into their projects, which may lead to further consolidations and mergers or joint ventures.

164. Joel L. Silverman. ELECTRICITY JOURNAL. March 1995, On-line. Outlook and Viability: The biomass industry is very likely to grow significantly. It is already the largest of the renewable industries, and as a baseload supplier of electricity with a conventional, utility-accepted generation technology, it is extremely attractive to utilities. Growth will be fueled by (1) biomass' ability to compete effectively in bidding situations, especially against other renewables; (2) expected significant increases in efficiency due to new generation technologies and particularly (3) development of dedicated feedstock supply systems. At the same time, environmental concerns may somewhat dampen growth prospects. A significant number of firms (IPPs, utilities and E&C firms) with experience in biomass are prepared to facilitate and take advantage of these opportunities. Ultimately, high rates of growth are dependent on the development of DFSS.

165. Joel L. Silverman. ELECTRICITY JOURNAL. March 1995, On-line. As of this writing, over 380 MW of capacity is in operation worldwide, almost all of it in California. The company responsible for developing most of that capacity—Luz International—has gone bankrupt. There is only one company still actively involved as a commercial developer: Industrial Solar Technologies. Unlike Luz, this small, privately held company is focused on industrial heat applications rather than utility grid applications Most other players have either exited totally or are involved with other projects, waiting to see if a viable market ever evolves.

166. Joel L. Silverman. ELECTRICITY JOURNAL. March 1995, On-line. That such a seeming successful company as LUZ should go bankrupt, when it has a virtual monopoly on the market, illustrates one of the chief barriers for solar energy developers: a lack of tax equalization. Under most state tax codes, solar plants face much heavier tax burdens than conventional plants because their fuel supply and sourcing are one and the same. Under most state tax codes, solar collectors are considered capital equipment with the solar field representing real property. The plant thus incurs both a recurring property-tax liability, and sales taxes on the purchase of equipment for plant construction. In contrast, as conventional fuel plants buy their fuel directly, and own no equipment to create the plant's fuel, there are neither property taxes nor sales taxes at the time a plant is built. The effect of this inequity can be staggering: "Solar plants can pay out more than four times the level of taxes paid by a comparably size gas-fired plant," according to Dave Menicucci of Sandia National Labs.

167. Joel L. Silverman. ELECTRICITY JOURNAL. March 1995, On-line. There is however, significant R&D activity in both dish and central receiver technologies. Dishes in conjunction with Stirling engines have high efficiencies, excellent modularity and siting flexibility. These dish/Stirling systems have potential to penetrate off-grid niche markets domestically and especially in international markets. Cummins Power Generation is cost sharing R&D associated with 5 kW dish/Stirling systems with the U.S. Office of Solar Energy Conversion (OSEC). This is a long-term effort that could lead to commercialization of dish/Stirling systems.

168. Joel L. Silverman. ELECTRICITY JOURNAL. March 1995, On-line. Outlook and Viability: The near-term forecast for this industry is bleak. Entrance barriers are excessive; high capital costs and the absence of markets provide little incentive for private investment or entry. What little commercial opportunity exists for medium-high temperature solar thermal systems is almost exclusively in industrial heat applications, not solely electricity generation. Even in industrial applications, low natural gas prices act as disincentive for private companies to invest in solar thermal systems. (Public institutions such as military bases, hospitals, and prisons are likely to be the major niche market.) Although the National Renewable Energy Laboratory feels that worldwide interest in hybrid trough systems (like those of Luz) has increased, most of the activity is in developing nations. International competitors seem to be ahead of the U.S. in this regard. Spain, Germany and Israel are cited by NREL as being particularly aggressive in capitalizing on these emerging markets. Under the circumstances, the U.S. solar thermal industry may have little chance of survival. While the technology itself offers great potential, commercialization of central receivers and parabolic dishes is several years off and faces many hurdles.

169. Joel L. Silverman. ELECTRICITY JOURNAL. March 1995, On-line. Meanwhile, the increasing cost efficiencies and commercialization of competing renewable technologies such as wind and PVs are already outpacing solar thermal systems. Utility emphasis on demand-side management over capacity additions puts any nonmodular (or limited-scale) technology at a competitive disadvantage. As the electric power industry continues to restructure, as needs shift, and as other RETs develop in sophistication and market penetration, there may simply be no need or room for HT solar thermal applications.

170. Joel L. Silverman. ELECTRICITY JOURNAL. March 1995, On-line. One final caveat, however, has to do with distinguishing between policy initiatives characterized as "technology push" as opposed to "market pull." Technology push initiatives are intended to speed up the research, development and demonstration of technologies as their cost and performance effectiveness increases. Market pull initiatives are designed to create market incentives (reducing buyer's costs and barriers) to induce customers and energy suppliers to select renewables. During the 1970s and 1980s, the government emphasized technology push. In the 1990s we are witnessing an increase in emphasis on market pull. This new direction in policy emphasis is consistent with the conclusions reached below. While there are still significant gains to be made through technology development, the renewables industries are also in need of initiatives outside the technology arena, that will sustain and enhance their viability.

171. Joel L. Silverman. ELECTRICITY JOURNAL. March 1995, On-line. Federal and state programs and regulations that create a level playing field relative to fossil fuels will benefit all renewables industries.

The pricing of fossil fuels does not currently recognize the environmental costs absorbed by society in the form of air pollution, greenhouse gases, acid rain, etc. Failure to explicitly acknowledge these external costs deprives renewables of any advantage attributable to their relatively benign environmental impacts. Hohmeyer has estimated that explicit recognition of externalities would add 4–8c/kWh to the cost of electricity generated by fossil fuels. If his estimates are valid, many renewables would not only be competitive with fossil fuels, they would be the low-cost generation option.

172. Joel L. Silverman. ELECTRICITY JOURNAL. March 1995, On-line. The federal government can support and catalyze the exporting of renewables and the development of international markets.

U.S. international competitiveness is at stake in many renewables industries. As can be seen in Table 3, all of the renewable industries have international market opportuni-

ties. International competitiveness provides additional sources of revenues and profits for renewables firms, moving firms up the experience curve while potentially enhancing scale economies. International markets also diversify market risk. These factors contribute to increasing industry viability, as well as creating jobs domestically and favorably impacting the balance of payments.

173.   Joel L. Silverman. ELECTRICITY JOURNAL. March 1995, On-line. Renewables industries are increasingly facing global competitors. Photovoltaics and wind are already global industries. Geothermal and biomass are becoming international in scope. The global players in these industries are looking to U.S. markets for opportunities. U.S. firms may have to develop global strategies to defend themselves and remain viable. In some cases, such as PV and parabolic dishes, niche markets need to be developed. In others, such as wind, biomass and geothermal, large contracts for grid-connected power are at stake. Each industry has specific export strategy needs that must be taken into account.

174.   Joel L. Silverman. ELECTRICITY JOURNAL. March 1995, On-line. DOE has recognized the importance of international markets for renewables. The Export Council for Renewable Energy, jointly funded by DOE and the Agency for International Development, represents all of the domestic renewables industries in attempting to open up international markets in developing countries. Also, DOE has been supporting Project FINESSE (Financing Energy Services for Small Scale Uses), which has catalyzed activities in support of accelerating adoption of renewables and energy efficiency in the developing world. However, the proportion of DOE funds going to international projects is very small. These types of programs would benefit from significant increases in support.

175.   Joel L. Silverman. ELECTRICITY JOURNAL. March 1995, On-line. Wind can now compete head to head with natural gas in many bidding situations. Many new contracts for wind have already been signed and others are in the works. Because of technology gains in wind, along with Energy Policy Act production credits for wind, the wind industry is well situated. Long-term attractiveness can be accelerated by increasing the extent to which states explicitly acknowledge the environmental benefits of wind and by fostering set-asides that ensure the flow of contracts remains steady over time. In states where set-asides are currently unlikely, initiatives that foster demonstration projects would be a precursor to utility adoption. The rate of adoption and the growth of wind power can be expected to accelerate as a wider base of users gains experience with the technology.

Also, in order to reduce the risk associated with the industry being dependent on one major firm, the federal government should continue to fund and perhaps increase its R&D support for other wind firms.

176.   Joel L. Silverman. ELECTRICITY JOURNAL. March 1995, On-line. C. Photovoltaic. The primary role the government can play in the PV industry is to stimulate its markets. On the technology side, the federal government has been extremely supportive of basic R&D. It has also recognized the importance of funding programs to improve manufacturing process efficiency, through its PVMAT (Photovoltaics Manufacturing Technology Initiative) program. The major log-jam for the industry, however, has been on the market development side. There are an enormous number of cost effective applications for PVs in which buyers are not fully aware of the PV option or are reluctant adopters. These opportunities are both domestic and international. The federal government has begun, over the past few years, to address this issue through more intensive market conditioning programs. The role of DOE's Office of Solar Energy Conversion's (OSEC) is "to serve as a facilitator in breaking down barriers and bringing together producers and users in order to accelerate market penetration of solar electric technologies" (PV, biomass and solar thermal). OSEC explicitly recognizes that the success of the U.S. solar industry depends on the private sector's ability to compete with global competition. Also, through its support of PVUSA (Photovoltaics for Utility Scale Application), the DOE is fostering PV penetration of the utility market. At the state level, utility regulators can encourage and provide incentives to their respective utilities to begin to use small-scale, cost-effective PV applications within their systems.

177.   Joel L. Silverman. ELECTRICITY JOURNAL. March 1995, On-line. Because the DOE has a strategy for PVs that seems to address the industry's needs, the only recommendation comes in terms of the level of support being provided to both R&D and market stimulation. The U.S. is falling behind Germany's and Japan's funding levels at a time when firms are establishing global strategic positions and dominance. If the U.S. industry is to remain internationally competitive, and if PVs are to reach the point where they can play a significant role in the bulk power market in a reasonable period of time, increased government funding seems critical to keep pace with foreign competitors.

178.   Joel L. Silverman. ELECTRICITY JOURNAL. March 1995, On-line. By rights, renewable energy industries in this country should be on their last legs. After 12 years of having renewables program budgets cut by the administrations of Presidents Reagan and Bush, with low oil and gas prices and a general lack of interest by most utilities in the benefits of renewables, how could these industries survive? Well, except for one short-term casualty, these industries have developed relatively resilient and adaptive industry structures that are testimony to the entrepreneurship and vibrancy of our competitive system. However, their adaptability seems only to maintain them in a marginal survival mode.

Industry players and potential investors will not aggressively innovate and invest unless they perceive the industry to be attractive over a reasonable period of time. Without this entrepreneurial thrust, in the absence of a level playing field and with the continued institutional barriers, the driving force behind the industry will stagnate or be lost. We cannot afford to let this happen. Jobs, energy security, our natural environment, and our international competitiveness are all at stake. Appropriate interventions must be made to enhance the long-term attractiveness of each of these renewable energy industries.

179. ENERGY CONSERVATION NEWS. 1 February 1996, On-line. Generally, the results indicate that voters strongly favor funding federal research and development programs for renewable energy sources and energy-efficiency measures over programs for nuclear power and fossil fuels. For example, 34% of those polled felt that renewable energy should receive the highest priority for DOE R&D funding. Twenty-one percent felt that highest priority should go to conservation programs. Twenty-six percent of respondents said highest priority should go to fossil or nuclear research. The remaining respondents either had no opinion or said "none of the above".

More than 70% of respondents recognize global warming or climate change as a threat, while better than 75% want to do something about U.S. dependency on foreign oil. Improving the fuel efficiency of cars and light trucks received nearly unanimous (94%) support among the subgroup that said oil imports were a problem, as did developing renewable energy alternatives, with 90% support. Half of the respondents favored retaining the EPA's Energy Star and Green Lights program, a fairly strong level of support in a budget-cutting climate, while 41% favored cutting it.

180. THE CHICAGO TRIBUNE. 6 March 1996, On-line. Ethanol enjoys broad bipartisan support among presidential candidates because this industry is responsible for more than 40,000 jobs across the country. The demand for grain stimulates rural economies. The ethanol industry generates more than $6 billion for the economy every year. The ethanol tax incentive saves the federal government $600 million annually and generates additional revenue in federal, state and local taxes. Name another federal program that can match that record.

181. NATIONAL CENTER FOR POLICY ANALYSIS, EXECUTIVE ALERT. 1996, (http://www.public-policy.org/"nysa/ea) Electric cars will require a number of subsidies. Recharging stations may be financed by taxpayers, and utility customers may be charged more to subsidize reduced electric rates for car owners. Auto companies are testing prototype vehicles that will sell for more than $20,000, but they may initially cost two to three times that much to make. The manufacturers may pass on that extra cost to buyers of other cars.

- DRI/McGraw-Hill estimates that the price of new gasoline-powered cars could rise by $400 to $4,400, while the California Air Resources Board claims the increased cost to new car buyers will be just $200.

- The California-based Reason Foundation estimates that a family of four in California will pay between $160 and $1,030 a year in increased car prices, energy rates and taxes to fund government subsidies.

182. BUSINESS WIRE. 8 November 1995, On-line. "A minute ago, I mentioned the potential cost of home charging. But all the other items cost money, too-lots of money. In fact, Sacramento-based Sierra Research estimates that including infrastructure, the EV mandate will cost nearly $20 billion over the next 15 years.

"Who is going to foot the bill? Ratepayers for one, if the California Public Utility Commission approves the $35 million rate increases in electric and natural gas rates to help pay for the mandate. Californians already pay electric rates 50 percent higher than energy users in the rest of the country. Now they would be asked to pay even more.

"Buyers of conventional vehicles would pay too. Manufacturing mandated electric cars could result in huge surcharges on conventionally fueled cars."

183. THE CHICAGO TRIBUNE. 7 February 1993, On-line. Above all, analysts say, there can be no successful electric-vehicle industry unless there is an infrastructure to support it.

Power stations in the U.S. could handle the additional demands of a modest electric fleet—provided most of the vehicles were recharged at night, in off-peak hours, industry experts say.

But other problems loom building charging stations, standardizing the size of cords, plugs and voltage outlets, and so on, at a potential cost of billions of dollars in public and private funds. Beyond that, if future battery technology still requires four to six hours for a complete recharge, the possibility of charging stations as efficient s today's gas stations is remote.

184. "Air/Water Pollution Reports," ENVIRONMENT WEEK. 3 November 1995, On-line. The American Automobile Manufacturers Association (AAMA) says that "while the world's automakers have made major investments in electric vehicle research and development and believe electric vehicles will fill an important niche in the future, they oppose electric vehicle mandates." The U.S. automakers, far more openly aggressive in the battle than their foreign competitors, argue that battery technology needed to make EVs viable alternatives to gas-powered engines does not exist; that EVs are likely to be much more expensive than conventional cars; and that EV mandates will cost taxpayers because governments will have to subsidize EV sales.

185. DETROIT NEWS. 15 May 1995, On-line. A host of other problems plague the electric vehicle mandate. While the cars themselves may be emission-free, the power plants generating all the electricity needed to propel them certainly are not. And few consumers actually want to own one. Electric cars are underpowered, and are twice as expensive to maintain as gasoline-fueled vehicles.

Nor are they competitive on sticker price, running an estimated $12,000 to $21,000 more than conventional cars. Complying with the mandate will thus force automakers to subsidize this cost by raising prices across the fleet. New car sales would slow, increasing the number of older, more polluting vehicles on the nation's roads.

186. Eric Peters. CONSUMERS' RESEARCH MAGAZINE. August 1995, On-line. Obviously, the effort to build a viable electric car must be successful—or it will be a disaster both for the industry and consumers. Millions of dollars have already been spent in research and development—to say nothing of eventual tooling and other related production costs—by each of the automobile manufacturers. When all is said and done, the total industry commitment to electric vehicles is likely to exceed several billion dollars. To put this in perspective, $1 billion is about what Chrysler spent on bringing the Neon, the company's hugely successful new economy car, from the designer's sketchpad to dealer showrooms. In other words, for the money being spent on EVs, the manufacturers could have designed, built, and sold several entirely new conventional car models.

If electric cars don't sell—or more importantly, if they can't be sold at prices which reflect their true cost to manufacture—the automobile industry will be faced with two equally unpalatable choices: The car companies will have to subsidize the "sale" of electric vehicles—selling the cars below cost and raising prices of conventional cars—to make them more attractive to consumers, or, manufacturers may simply unload fleets of the unsaleable electrics to commercial users (mainly utilities) at tremendous discounts. Either way, the money lost will have to be made up from purchasers of conventional, gasoline-burning cars and trucks.

In other words, you and me.

187. BOSTON HERALD. 1 October 1995, On-line. A study by the libertarian Reason Foundation of the costs per ton of controlling ozone-forming compounds like gasoline vapor concluded that electric vehicles are the most expensive way to clean the air.

Few electric cars will actually hit the road in Massachusetts in the near future. The legislation requires automakers to offer the cars; it doesn't require anyone to buy them. Chances are, most will just sit on dealers' lots or be sold at a loss, because of their poor performance and high cost. Either way, dealers will have to make up the difference by charging more for gasoline-powered vehicles.

Technology simply isn't advanced enough to produce a reasonably priced electric car that's comparable to a gasoline-powered car. Experts say they're years away from a battery breakthrough and there's no reason not to believe them. As Kenneth Green pointed out in the *Los Angeles Times*, "If Detroit's Big Three could make a profitable electric vehicle that consumers wanted to buy, they'd be making it at the behest of their own stockholders." David Power, president of J. D. Power & Associates, a noted auto market research firm, summed it up best: "Mass marketing of electric vehicles in 1998 is doomed to failure."

188. WASHINGTON POST. 18 April 1995, On-line. U.S. automakers, who have long questioned whether electric cars are road-ready, remain adamantly opposed to national marketing of the vehicles.

"If they are introduced on a wide scale before the technology is ready, we could have a big problem on our hands," said Andrew Card, president of the American Automobile Manufacturers Association.

Although the industry's long-term future in California is promising, at present it is still in flux. Last month, for instance, executives at U.S. Electricar, the nation's largest EV manufacturer, announced that it was closing two plants and cutting its work force by 35 percent. Company executives acknowledged they had spread their investments too thin and may have to cut back even further unless Japanese investors give them a fresh infusion of cash.

189. TIMES-UNION. 1 August 1995, On-line. Not surprisingly, automakers are resisting the idea, on the grounds that they can produce a cleaner internal combustion engine that will improve air quality without the need for a drastic overhaul that would produce cars of limited range, speed and appeal to customers.

They have a point. Not only are electric cars unlikely to attract many buyers in the cold Northeast, for example, they are likely to be considered unreliable but they won't be capable of replacing the long-haul gasoline vehicles that clog the superhighways. What's more, the lead batteries that would be required to power these vehicles also will produce pollution of their own in the form of landfill hazards they will pose when disposed of.

Moreover, the need to recharge these batteries will increase the demand for electricity, thereby increasing the strain on capacity. At the moment, utilities aren't worried about the added demand because there's a surplus available. Yet a mass transition to electric vehicles could change that situation dramatically.

190. CNN NEWS TRANSCRIPT. 30 September 1995, On-line. Ed Garsten, Correspondent: They can't go very far or very fast, but electric vehicles are very clean—they product zero pollution, and that's why California will soon mandate sales quotas on electric vehicles. But the automakers have long argued consumers elsewhere in the country just don't want them. Now, the federal government has come up with a plan for the other 49 states that would not mandate electric vehicles to cut pollution, but

the sale of cleaner-burning gasoline-powered cars and trucks.

191. Michael Tucker. BUSINESS AND SOCIETY REVIEW. 22 March 1995, On-line. Calling for a minimum EV sales target is decidedly different from demanding buyers to purchase EVs. CRB's regulation will create EV supply, not demand. Carmakers are concerned about the possibility of being left with parking lots filled with unsold electric vehicles. It is likely, however, that state agencies and electric utilities will provide at least some of the sales needed to reach the 2 percent goal. Affluent two-car families looking to make environmental statements may also find EVs acceptable as a third vehicle. Under these circumstances EVs will be a market niche that may not grow as rapidly as CARB's mandated supply expansion.

192. GAO REPORTS. 4 January 1995, On-line. The ultimate viability of electric vehicles for widespread transportation cannot now be predicted or ensured. Five major barriers to the immediate introduction of electric vehicles are limitations of current battery technology, gaps in required infrastructure, uncertain safety, uncertain market potential, and high initial purchase price. Extensive efforts to eliminate these barriers are inherently risky and will require substantial money, time, and attention.

The U.S. policy toward electric vehicles is fragmented in two ways. First, already limited funds are divided into several small programs across three different federal departments. Second and more importantly, the lack of emphasis on the barriers that can be addressed before a battery breakthrough and that ultimately must be resolved to market a viable vehicle—namely, issues of infrastructure support, market development, and production—leave a gap between state policies mandating electric vehicle markets and federal policies supporting battery technology initiative.

The fragmented U.S. approach, when coupled with other nations' more comprehensive focus on infrastructure, marketing, and production, raises the specter of past U.S. technological successes better commercialized by foreign competitors. The United States may fund the successful development of an advanced battery that other countries could quickly incorporate into marketable, low-cost, performance-tested vehicles. The case of electric vehicles, moreover, could pose a unique risk because of the artificial U.S. market created by state mandates.

193. GAO REPORTS. 4 January 1995, On-line. Electric vehicles—or any single technology—will not solve the world's assorted transportation-related problems. This nation's fuel-neutral energy policy divides funding among many fuel types, in part to ensure that viable alternative fuels will be developed and commercialized. Electric vehicles receive disproportionately less funding compared to other alternatives. They are not fully developed on any dimension and will likely remain so without a balanced national policy that supports all aspects of EV development and infrastructure. While inherent risks are associated with sizable investments in a nascent technology, a more tentative U.S. approach carries another risk: investing the millions of dollars in battery research and then losing early market share in mandated state markets.

194. GAO REPORTS. 4 January 1995, On-line. While currently available electric vehicles use 20 to 35 percent more primary energy than gasoline vehicles, advanced technology electric vehicles are anticipated to reduce U.S. primary energy consumption by 30 to 35 percent in 2010. The United States would save more annually ($2.5 billion) by replacing 10 percent of its vehicle numbers with electric vehicles than any other nation GAO reviewed, while Italy would save the least (approximately $300 million).

195. OCTANE WEEK. 12 June 1995, On-line. While overall energy consumption may rise by a switch from conventional cars to EVs, there is the long-range opportunity of actually extending U.S. energy independence. This may sound contradictory, but the U.S. has an almost unlimited resource in coal and is better positioned in natural gas than oil.

196. THE WASHINGTON TIMES. 1 September 1996, On-line. Mr. Brier makes the bald statement that volatile organic compounds (VOCs) "are produced almost exclusively by internal combustion engines." Actually, more than 50 percent of the VOC burden all across the country comes from natural sources rather than from any sort of combustion process. President Ronald Reagan was not so totally wrong as the Green lobby claimed.

Where did the Great Smoky Mountains get their "smoke"? The Spanish found smog in the San Diego basin they first debarked from their sailing ships. More recently, Atlanta found that rather strenuous control of emissions from automobiles made no impression on local ozone levels. VOCs from trees drove the equation. EEI, at least the scientific side, is well aware of the 10-year-old Georgia Tech study documenting the Atlanta problem.

197. Peter Passell. THE NEW YORK TIMES. 28 November 1996, On-line. The good news here is that the E.P.A. is updating particulate standards to mesh with scientific findings that very small particles (2.5 microns, or millionths of a meter, in diameter) are as damaging to health as larger ones. It is still not clear, however, that all new particles being regulated are equally dangerous. Exposure to nitrogen compounds (from car exhausts and power plants) may be less risky than contact with sulfur compounds (mostly from burning coal).

Even less clear is the logic in linking the regulation of soot and ozone. Killer smog, still a phenomenon in the third world, is typically loaded with both particulates and

ozone. And reducing emissions of one often reduces emissions of the other. But those rationalizations mask big differences.

The epidemiological case against ozone is weak. For example, a study of the horrific air pollution in Mexico City by the widely respected Health Effects Institute found that particulates rather than ozone did most of the medical damage. And while researchers have been looking for the smoking gun on ozone for decades, Paul Portney, president of the economically aware environmental group Resources for the Future, says that "all they've been able to prove is that exposure causes short-term reversible effects," like bronchial irritation and watery eyes.

198. DETROIT NEWS. 24 July 1995, On-line. One would assume, given the nature of these sanctions, that the nation has an ozone problem that demands attention. But it doesn't. Most emissions in this nation come from older cars. As those vehicles move out of the fleet, the air gets cleaner. Ozone and VOC emissions have fallen dramatically during the past 15 years. In a study just released by the Cato Institute, Jonathan Adler and K. H. Jones note that concentrations of the two pollutants have dropped more than 50 percent in the New York-New Jersey-Connecticut region since 1980.

Furthermore, the number of high-ozone days has tumbled since the unusual summer of 1988.

199. Eric Peters. CONSUMERS' RESEARCH MAGAZINE. August 1995, On-line. The clincher is that electric cars probably aren't necessary to improve the quality of the air we breathe anyway. Advances in emissions control technology for gasoline-fueled cars have come within a hair of making them free of harmful pollutants. "Zero emissions" electrics, in this respect, make only small improvements over current "low emission" vehicles. And while electrics totally eliminate certain smog-causing emissions from tailpipes, most of these emissions will continue to come from the stationary sources—factories and such.

"The proportional contribution of stationary sources emissions has been getting larger and larger over the past three decades while the contribution of mobile source cars and light trucks has been getting smaller and smaller, despite a substantial increase in the total number of miles driven each year," says AAA's William Berman.

200. BUSINESS WIRE. 8 November 1995, On-line. "I've talked a lot about obstacles that must be overcome in order to support the widespread introduction of electric vehicles in the state. But the central focus of the EV mandate is—and must be—its effect on California consumers. What do Californians want—and expect—out of electric vehicles? A lot. According to a survey by David S. Bunch and his colleagues, the top four criteria were: (1) a range between recharging of at least 150–200 miles; (2) availability of recharging stations; (3) price of the car; and (4) operating cost.

"Let's take each one of these criteria in turn.

"First, today's lead-acid batteries have a range of 60 or 70 miles on a good day—and considerably less on a cold day or on a hot day with the air conditioner on.

"Second, as I mentioned before, the infrastructure needed to support widespread, convenient recharging is nonexistent—certainly when compared to the approximately 8,000 gasoline stations in the state. And recharging a battery may never be comparable to refueling a gas tank. Today, a motorist can fill a gas tank with around 300 miles of energy in less than five minutes. So far, nobody has been able to get energy from a power grid to a storage facility in the car in anywhere near five minutes. In fact, filling today's lead-acid batteries with enough energy to drive 100 miles is like filling a gas tank using a straw.

"Third, today's electric vehicles are far more costly to produce than conventional automobiles—in most cases twice as much. Until battery technology improves substantially, electric vehicles are expected to continue to be excessively costly.

"Finally, with respect to operating cost, while some studies have shown that electricity for recharging is cheaper than gasoline, the higher production cost combined with periodically replacing the batteries—at anywhere between $2,000 and $6,000 a piece every two to three years for lead-acid batteries—makes electric vehicles more expensive to operate than conventional cars."

201. Karp and Gaulding. HUMAN RELATIONS. May 1995, On-line. But the car has its limitations: it doesn't drive as far or as fast; it has no past record of performance; it's smaller than most other vehicles on the road making it less safe in a collision, etc. Particularly worrisome is the fact that if enough of them are not sold, recharging stations may not materialize, and the model will be discontinued, leaving you with a car nobody can fix or find parts for. Not only is this outcome bad for you, but also it does little to reduce air pollution because too few electric cars are purchased to make a difference. The choice of contributing to the collective good when others fail to do so is appropriately known as the "sucker's payoff" (Rapoport, 1967; Bruins et al. 1989). Therefore, apparently, it is not in your best interest to buy this car. The eventual costs to you personally outweigh the benefits, which can be defined simply as your own minuscule contribution to enhanced air quality.

The environment would be better off, however, if people began buying electric cars. Collectively, we pay the price for our failure to make these environmentally sound purchases. Therefore, you might be content buying an electric car as long as millions of others do so, and the air quality was measurably improved. Better still, you would rather have thousands of other people buy electric cars, taking the risk on this new technology, while you buy a conventional vehicle. This attractive option is called "free-

riding" (Olson, 1965), and is indicative of one's willingness to enjoy the benefits of others' cooperation without paying the cost. Finally, as additional consideration is the uncertainty of the future environmental consequences compared with the certainty of cost-savings by avoiding this expensive purchase.

From the perspective of the anonymous and autonomous decision maker, it does not make sense to purchase the electric car. However, the aggregated consequence is collectively damaging. Because of an inherent conflict of motives between what is best for the individual and what is best for society (the social trap), and between what is best in the present and what is best in the future (the temporal trap), the social dilemma has no straightforward solution. Figure 1 illustrates this conflict by presenting a two-by-two payoff matrix in which each quadrant is split (diagonally) between decision outcomes for self and for others. As self and others attempt to free ride, the worst of the four collective outcomes (self + others) results.

202. THE SACRAMENTO BEE. 28 March 1996, On-line. Air board officials plan to open a two-day public hearing on the repeal today with the explanation that the currently mandated 1998 launch of 22,000 electric vehicles in California will likely "poison the well" for future sales.

Officials fear the quota is too high for consumer appetites, given the relatively high price and low driving range of these lead-acid-battery-powered vehicles. Unsold inventory in the initial years could kill the program in its infancy, they say.

203. BUSINESS WIRE. 8 November 1995, On-line. "In order to sell electric vehicles, some have suggested that manufacturers subsidize them in some manner—for example, raising the price of gasoline-powered vehicles to new car buyers. Increasing the cost of motor vehicles would slow vehicle turnover. That, in turn, would keep more high-polluting vehicles on the road longer. Ironically, forcing electric vehicles on the market before they are ready would hurt air quality. According to Sierra Research, this effect will last more than 15 years. Thus, Californians are going through a great deal of trouble and will spend a great deal of money—and the most likely outcome will be dirtier, not cleaner air.

204. Lester B. Lave, Christopher T. Hendrickson, and Francis Clay McMichael. SCIENCE. 19 May 1995, On-line. We focus on the environmental consequences of producing and reprocessing large quantities of batteries to power electric cars. For vehicles that are to be mass produced in late 1997, lead-acid batteries are likely to be the only practical technology. Smelting and recycling the lead for these batteries will result in substantial releases of lead to the environment. Lead is a neurotoxin, causing reduced cognitive function and behavioral problems, even at low levels in the blood. Environmental discharges of lead are a major concern. For example, eliminating tetraethyl lead (TEL) from U.S. gasoline greatly reduced blood-lead levels in children.

Alternative battery technologies that are currently available include nickel-cadmium and nickel metal hydride batteries, which are much more expensive than lead-acid batteries. In addition, nickel and cadmium are highly toxic in humans and the environment. Technologies such as sodium-sulfur and lithium-polymer batteries are unlikely to be commercially available for years.

205. Peter Passell. THE NEW YORK TIMES. 29 AUGUST 1996, On-line. More sobering still, all this illustrates how attacking one pollution problem can worsen another. The Environmental Protection Agency views the elimination of lead additives from gasoline as a great victory, with estimated benefits of roughly $100 billion annually in reduced health costs and lower rates of disease. That could be at risk unless the lead and battery industries can devise ways to contain virtually all the emissions from manufacturing and recycling the heavy metal.

But Mr. Lave wonders why anyone, other than the wannabe Bill Gateses of the battery world, would willingly take the chance with toxic lead. "The fact is," he said, "we'd not get any benefit from electric cars."

206. Eric Peters. CONSUMERS' RESEARCH MAGAZINE. August 1995, On-line. In point of fact, studies suggest that the added demand for electricity from coal-fired utility plants will result in a manifold increase in so-called "stationary source" pollution. This is especially true in the Northeast, where the coal burned is of the high-sulfur type that is by nature the most polluting.

Estimates of the increase in sulfur dioxide emissions— which cause acid rain—vary from 17% to 2,100%, according to figures compiled by the National Conference of State Legislatures. Yet big electric utilities—which hope to find a captive audience for their excess generating capacity—continue to advocate EVs and EV mandates. And for the most part, "environmentalist" groups have been strangely silent on the question of electric vehicles' potential for contributing to worsening air quality.

207. Eric Peters. WASHINGTON TIMES. 21 August 1995, On-line. A sad thing about all the commotion surrounding pie-in-the-sky "alternative" fuel vehicles—especially electric ones—is that all the racket has squelched some genuinely promising pollution control technologies that would allow Americans to continue their love affair with the automobile—without despoiling the planet.

Perhaps the most interesting of these nascent developments is the so-called "smog-eating radiator" that Ford Motor Co. recently announced it will begin installing in a fleet of test cars for real-world evaluation.

208. Eric Peters. WASHINGTON TIMES. 21 August 1995, On-line. Instead of the usual 10 or more

years it takes for something like engine computers or fuel injection to become commonplace—and therefore have a meaningful impact on air quality–the smog eating radiator could be put into widespread service in a fraction of that time.

Now here's the kicker. Studies performed by Sierra Research suggest that "if the catalyst systems [e.g., smog-eating radiators] are employed on all vehicles, they could reduce more ozone and carbon monoxide than other clean air programs such as the electric car . . . reformulated gas . . . and employee commute options [mandatory carpools]."

For example, the Sierra study estimates that if 9 million vehicles travelling 266 million miles per day in Los Angeles were equipped with smog-eating radiators, the air flowing through these radiators—and being purified of harmful ozone and carbon monoxide along the way—would be equivalent to all the air across the entire Los Angeles area up to a height of about 15 feet.

In hard numbers, the smog-eating radiator has the potential to reduce peak levels of ozone by 4.5 parts per billion in Los Angeles, and carbon monoxide levels by more than 12 percent, according to Sierra's findings.

Not only is this a tremendous potential reduction in dirty air—it would come at an affordable cost: just $500–$1,000 per vehicle (new or old). That's considerably cheaper than electric cars (expected retail price: $20,000–30,000), and much less inconvenient than forced carpooling. [ellipses in original]

209. Tom Lankard. AUTOMOTIVE INDUSTRIES. August 1995, On-line. However, a rewrite of California's standards could actually mean cleaner air for the OTC states, and the rest of the country. A compromise proposal by automakers to sell cars outside of California that meet that state's LEV standard would be the best for all concerned, according to an EPA report. Automobile emissions everywhere would be less than EPA requirements. And, in the Northeast, not only would this mean fewer emissions from fossil-fuel burning electrical generation plants, but that cars driven into the state would be cleaner sooner than under the OTC LEV plan.

210. "Air/Water Pollution Reports," ENVIRONMENT WEEK. 29 September 1995, On-line. U.S. automakers, trying to kill ZEV mandates, have offered to build a "49-state" car having emissions between California standards and those required in the rest of the nation. EPA has been backing the effort and on Sept. 28 formally proposed a rule sanctioning the 49-state vehicle. EPA Administrator Carol Browner, drawing disagreement from health groups and some Northeast air regulators, said the vehicle would be better for the region than the California approach and would help air quality nationwide.

211. OCTANE WEEK. 2 October 1995, On-line. Because the Clean Air Act precludes EPA from requiring new exhaust standards before model year 2004 and the proposed plan would begin in 1997, it is voluntary. However, once the manufacturers commit to the program, the standards would be enforceable just as the other vehicle emission programs are now, the agency said. The program would begin in 13 OTC states with the 1997 model year for gasoline and diesel, and nationwide with model year 2001, except for California, which already requires sale of a cleaner car.

212. Katherine Culbertson. OIL DAILY. 29 September 1995, On-line. Though the auto industry appears not to be budging, one EPA official on Thursday said she believes it is closer to accepting the 49-state car plan as is.

Under the 49-state plan, EPA said nationwide nitrous oxide emissions should be reduced by 400 tons per day in 2005 and 1,200 tons per day in 2015. Organic gas emissions would be cut by 279 tons per day in 2005 and 778 tons per day 10 years later.

213. "Air/Water Pollution Reports," ENVIRONMENT WEEK. 3 November 1995, On-line. The three states' ZEV mandates are running into stiff opposition from the auto and petroleum industries. The battle takes a variety of forms, from lobbying campaigns and research papers to trying to negotiate the automakers' 49-state-car plan with New York and Massachusetts, with EPA acting as a broker. Under this approach, the two states—and the rest of the Northeast—would not have EV mandates. In return, Detroit would begin marketing—everywhere but California—a vehicle with emissions levels exceeding CAA requirements. On the western front, AAMA has waged a vigorous grass roots campaign in an effort to get the California mandate repealed. Late last month, the automakers won a battle in Canada when British Columbia agreed to drop its flirtation with a provincial EV mandated.

214. Julie Smyth. TIMES-UNION. 20 July 1995, On-line. Automakers are lobbying Pataki to abandon the mandate, promising to create a lower-emission, gasoline-burning car in return. That vehicle, the so-called "49-state car," would be made available in all states but California as a way to reduce auto emissions at higher levels than federal law requires without jeopardizing the auto and petroleum industries.

Esper said release of that car would reduce auto pollution by necessary levels in the Northeast without forcing automakers to produce cars no one wants.

"Manufacturers will do whatever the law tells them to do, even if that means making a car that no one will buy and losing lots of money," he said. "We think the 49-state car is more effective, and we can't pursue both programs at the same time."

215. THE SAN FRANCISCO CHRONICLE. 15 February 1996, On-line. Dunlap said the agreement would not sacrifice any reductions in pollutants. As part of the pact, automakers would agree to introduce cleaner gasoline-powered cars in the rest of the country starting in

2001—three years earlier than would be required under federal clean air laws.

Low-emission vehicles, which cut tailpipe pollutants by 70 percent from current federal standards, are required in California by 2000. California would benefit from nationwide adoption of those standards because nearly 20 percent of cars registered here come in from out of state.

216. 21ST CENTURY FUELS. 1 October 1995, On-line. While the proposed rulemaking does not propose any new fuel requirements for states, such as nationwide RFG, it does suggest a multi-party process to resolve fuel issues in future, if deemed necessary.

Under the proposal, manufacturers may use California Phase II gasoline to show compliance with emission standards, the same option allowed to California in implementing its regulations.

Using this ultra-clean, low-sulfur RFG would reduce manufacturers' cost of demonstrating compliance. However, the agency is not proposing any in-use changes for vehicle fuel, it said.

The agency believes the national program would result in equivalent or better emission reductions than the original plan, while also providing benefits for other states.

Automakers negotiated for the nationwide plan because it would "harmonize" federal with California regulations so they could design and test just one set of vehicle standards nationwide.

217. THE SAN FRANCISCO CHRONICLE. 15 February 1996, On-line. In addition, the new gas will reduce cancer risk related to gasoline exposure by 30 percent, mostly by halving the amount of benzene, a known human carcinogen.

Air quality officials say the fuel will be 20 times more effective at reducing smog than the state's controversial electric car mandates, which requires that electric cars account for 10 percent of all vehicles sold in the state by 2003.

And unlike the battle over electric cars, the switch has the full cooperation of the oil industry. "Everyone wants this to work—the environmental community, the oil industry and the state," Lee said. The refinery overhaul created 20,000 temporary construction jobs and thousands of permanent jobs to run expanded facilities.

218. THE SAN FRANCISCO CHRONICLE. 15 February 1996, On-line. In its most ambitious step toward cleaner air in more than 20 years, California is switching to a cleaner burning gasoline.

Over the past month, many of the state's 12,000 service stations have begun replacing current blends with a newly formulated fuel expected to cut smog 15 percent. All must switch by June 1.

The new gas is the most significant smog-reducing attempt since California began requiring catalytic converters in 1975. Unlike the catalytic converter, which was phased in over 10 years, its effect will be almost instantaneous.

Adopting the new gas is equivalent to taking 3.5 million of the state's 24 million cars off the road, according to the state Air Resources Board. "In terms of immediate results, we've never had one single program that gets emission reductions of this magnitude," said air board spokesman Allan Hirsch.

219. LOS ANGELES TIMES. 14 January 1996, On-line. In California's epic quest for cleaner air, 1996 will stand out: Almost instantaneously, all 24 million cars—from old clunkers to brand-new luxury cars—will pollute dramatically less as the entire state switches to a new formula for gasoline.

Over the next couple of months, all service stations in California will replace their old blends with the world's cleanest-burning gasoline to comply with a state air-quality regulation.

With automobiles blamed for more than half of the smog blanketing the Los Angeles Basin, the new recipes for gasoline are the most sweeping and effective effort to combat smog since cars were equipped with catalytic converters 20 years ago. Cleaner-burning gasoline is equivalent to removing 3-1/2 million cars from California roads—nearly half of them in smoggy Los Angeles, Orange, Riverside and San Bernardino counties.

"The air quality benefits of cleaner burning gasoline are real and substantial, and the effects will be immediate and statewide," said John Dunlap, chairman of the Air Resources board, which set the new fuel rule. "This is the last big (smog-cutting) measure for California. We won't have another one that will have as broad an immediate impact."

220. THE SAN FRANCISCO CHRONICLE. 15 February 1996, On-line. Air quality officials have gone to great lengths to ensure that this transition to the new fuel is smooth and avoids the problems that plagued the earlier introduction of a new diesel fuel.

The state Air Resources Board mandated the cleaner-burning gas in 1991 as a key component of its plan to bring Los Angeles, Sacramento and other chronically smog-choked areas into compliance with new, stricter federal clean air standards by 2010.

The new fuel is expected to make up about 25 percent of the required reductions in ozone, the main constituent of smog.

In the Bay Area, which last April became the largest metropolitan area to attain the new standards, the new gas will help the nine-county region stay in compliance while still allowing for future growth, air quality officials say.

"It's a huge, huge air pollution benefit for the Bay Area, both in terms of ozone and other air pollutants," said Terry Lee, a spokeswoman for the Bay Area Quality Management District.

CHAPTER VII

WHO'S WHO

AAMODT, JASON
    Law Student

AARTS, PAUL
    Professor of International Relations, University of Amsterdam

ABRAHAMSON, DEAN
    Professor, University of Minnesota's Humphrey Institute of Public Affairs.

ADLER, JONATHAN
    Competitive Enterprise Institute

ADLER, STEVEN
    Environmental Studies Enterprise Institute

AKARCA, PAUL

AKINS, JAMES E.
    Former U.S. Ambassador to Saudi Arabia

AL-CHALABI, FADHIL
    Former Deputy General, Secretariat of OPEC

AL-HASSAN, OMAR
    Chairman, Gulf Center of Strategic Studies, London

ALBRITTON, D.
    NOAA

ANDERSON, FREDERICK
    Dean of the American University Law School

ANDERSON, KYM
    Member and Director of Economic Research Division of GATT Secretariat

ANDREWS, RICHARD N. L.
    Professor, University of North Carolina School of Public Health

ARBATOV, GEORGI
    Director, Russian Academy of Sciences

ARMITAGE, ST. JOHN Diplomat Specializing in Middle East Affairs

ARNOLD, MATTHEW
    Professor of Government, Indiana University

ARRANDALE, TOM
    Staff Writer, *Governing*

ASSENZA, MARIT P.
    Management Consultant, University of Oslo.

ATCHESON, JOHN
    Director of Policy Analysis for Energy Efficiency Department, Department of Energy

BACEVICH, A. J.
    SAIS, Johns Hopkins University

BAILEY, RONALD
    Science writer

BAKER, KENNETH
    Director, Office of Nonproliferation and National Security

BALLING, ROBERT C.
    Professor of Geography, Arizona State University; Director, Laboratory of Climatology, Arizona State University

BAR-SIMAN-TOV
    Chairman, Department of International Relations, Hebrew University

BARKET, DAVID
    Professor of Economics, Massachusetts Institute of Technology

BARKSDALE, YVETTE M.
    Professor, John Marshall Law School

BARRETT, SCOTT
    Professor of Economics, London School of Business

BARRY, JOHN
    Research Assistant, Heritage Foundation

BARTON, PAUL
    Staff Writer, Gannett News Service

BATES, N. H.
    Researcher, International Soil Reference and Information Center

BAYNE, PHILLIP
    President of Nuclear Energy Institute

BELGRAVE, ROBERT
    Royal Institute of International Affairs, Centre of Strategic Studies

BERGSTEN, C. FRED
    Director of Institute for International Economics; Chairman, Competitiveness Policy Council

BERTEL, EVELYNE
    Staff Member, OECD Nuclear Energy Agency

BHAGWATI, JAGDISH
    Professor of Economics and Political Science, Columbia University; Former Economic Policy Advisor to Director-General of GATT

BHATTACHARYA, N. C.
>Plant Physiologist, Department of Agriculture

BIERBAUM, ROSINA
>Director, Oceans and Environment Program, OTA

BIOLSI, ROBERT
>Manager of Options, New York Mercantile Exchange

BLACK, BERNARD S.
>Professor of Law, Columbia University

BLACKHURST, RICHARD
>Member and Director of Economic Research Division of GATT Secretariat

BLACKWELL, JAMES
>Senior Fellow, CSIS

BLIX, HANS
>General Director, IAEA

BODE, DENISE
>Independent Petroleum Association of America

BOHI, DOUGLAS
>Head of Energy and Natural Resources

BOLTHO, ANDREA
>Fellow and Tutor in Economics, Magdalen College, Oxford University

BOND, JON
>Professor of Political Science, Texas A&M University

BONIFANT, BEN
>Professor of Government, Indiana University

BRIDGES, E. M.
>Researcher, International Soil Reference and Information Center.

BROSCIOUS, CHUCK
>Director of Environmental Defense Institute

BROWN, LESTER
>President of Worldwatch Institute

BROWN, PAUL
>Staff Writer, *The Guardian*

BROWNER, CAROL M.
>Administrator, United States Environmental Protection Agengy

BRYAN, RICHARD
>United States Senator, Nevada

BULL-BERG, HANS JACOB
>Royal Norwegian Ministry of Petroleum and Energy

CABLE, VINCENT
>Chief Economist, Shell International

CAGIN, SETH
>Ozone Historian

CALLEO, DAVID
>Director of European Studies, Johns Hopkins University

CANTLON, JOHN
>Chairman of the Nuclear Waste Technical Review Board

CASE, J. CALE
>Consultant, Palmer Bellevue

CHAYES, ABRAM
>Harvard Law School

CHUBIN, SHAHRAM
>Specialist on Strategic Studies

CHUPKA, MARC W.
>Assistant Secretary for Policy and International Affairs, Department of Energy

CLARK, SUSAN
>Chairperson of the Florida Public Service Commission

COHN, PAMELA
>B.A. in Justice, The American University; J.D., Emory University Law School

COPAKEN, ROBERT
>Senior Political Economist, Office of Energy Assessments

COPLAND, DALE C.
>Assistaant Professor of Government and Foreign Affairs, University of Virginia

CORDESMAN, ANTHONY
>Professor of National Security, Georgetown University

COULTER, ARTHUR
>Staff Writer, *News and Record*

COWEN, ROBERT C.
>*The Christian Science Monitor*

CRAIG, LARRY
>United States Senator; Chairman of the Senate Repulican Policy Committee

CURTIS, CHARLES
>Policy Analyst, Department of Energy

CUTLER, WALTER
>Former U.S. Ambassador to Saudi Arabia

DANIELS, ALEX
>Staff Writer, *Governing*

DARMSTADTER, JOEL
>Resources for the Future

DAVIS, ROBERT E.
: Professor of Environmental Science, University of Virginia

DECICCO, JOHN
: American Council for an Energy Efficient Economy

DEELSTRA, TJEERD
: International Institute for the Urban Environment

DELORS, JACQUES
: President, Commission of the European Communities

DOLLAR, DAVID
: Professor of Economics and International Relations

DOLLEY, STEVEN
: Research Director, Nuclear Control Institute

DOWLING, EDWARD
: Professor of Economics

DOWNS, GEORGE W.
: Professor of World Politics, Princeton University

DRAY, PHILIP.
: Ozone Historian

DREYFUS, DANIEL
: Director of the Office of Civilian Radioactive Waste Management

DRYSDALE, ALASTAIR
: Professor of Geography, University of New Hampshire

DUNLAP, RILEY
: Environmental Sociologist, Washington State University

DUNLAP, SHELDON
: Environmental Sociologist, Washington State University

DUNN, LEWIS
: Vice-President, SAIC

EASTON, ROBERT
: Chairman, Managed Futures Association

EBINGER, CHARLES
: Royal Institute of International Affairs, Centre of Strategic Studies

EHRLICH, ANNE
: Center for Conservation Biology; Biological Scientist, Stanford University

EHRLICH, PAUL
: Center for Conservation Biology; Biological Scientist, Stanford University

ETESHAMI
: Senior Lecturer, Middle East Politics, University of Durham

EILTS, HERMANN FREDERICK
: Former Ambassador to Saudi Arabia

EISMA, DOEKE.
: Professor of Marine Sedimentology, University of Utrecht

ELIAS, CARLOS
: Professor of Economics, Manhattan College

ENSIGN, REPRESENTATIVE
: United States Representative

ESTY, DANIEL C.
: Senior Fellow and Former Deputy Assistant Administrator for Policy, Planning and Evalution, The Enviornmental Protection Agency

FAIRLEY, PETER
: Staff Writer, *Chemical Week*

FARER
: Professor, Law and International Relations, The American University

FEAGLE, ROBERT G.
: Professor of Atmospheric Sciences, University of Washington

FELD, LOWELL
: Analyst, EIA

FERRONE, BOB
: Design and Manufacturing Engineer

FEUERWERGER
: Washington Institue for Near East Policy

FIKSEL, JOSEPH
: Senior Director of Environmental Management, Battelle Memorial Institute

FINKEL, ADAM M.
: Director, Rational Risk Reduction Program

FISHER, LOUIS
: SOP Specialist, Congressional Research Service

FLAVIN, CHRISTOPHER
: Worldwatch Institute

FRANKEL, CARL
: Editor of U.S. Affairs, *Tomorrow*

FRENCH, HILARY
: Senior Researcher at Worldwatch Institute

FRIED, EDWARD
: Brookings Institute

FRIEDLANDER, SHELDON
: Professor of Chemical Engineering, University of California at Los Angeles

FUTRELL, J. WILLIAM
: President, Environmental Law Institute; Former Professor, University of Georgia Law School

GARDINER, DAVID
: Assistant Administrator for Policy, Planning, and Evaluation, EPA

GATES, GARY
: Vice-President, Omaha Public Power District

GEIPEL, GARY
: Hudson Institute

GELBSPAN, ROSS
: Editor and Reporter on Environmental Affairs, *Harper's* Magazine

GELLER, HOWARD
: American Council for an Energy Efficient Economy

GEORGE, EMMIT
: Commissioner of the Iowa State Utilities Board

GEORGIOU, GEORGE
: Professor of Economics, Towson State University

GEYER, RICHARD A.
: Editor, A Really Big Climate Forum

GLASER, CHARLES
: Professor of Public Policy Studies

GLAZYEV, SERGEI
: Head of Economic Security, Russia

GOETZ, ROLAND
: Federal Institute for East European and International Studies

GORNEY, CAROLE
: Professor of Public Relations, Lehigh University

GOTBAUM, JOSHUA
: Secretary for Economic Security, Department of Defense

GOTTLIEB, DANIEL W.
: Staff Writer, *Purchasing*

GREEN, ERIC MARSHALL
: Economic Analyst, Eisenhower Institute of World Affairs

GREEN, MELVIN
: Executive Director, ASME

HAHN, ROBERT
: Amercan Enterprise Institute

HAHN, WALTER
: Center for Defense Journalism, Boston University

HANSEN, KENT
: Professor of Nuclear Energy, Massachusetts Institute of Technology

HANSON, JOHN
: Analyst for Beveridge & Diamond P.C.

HARMON, WILLIS
: President, Institute of Noetic Sciences

HART, DAVID
: Department of Government, Harvard University.

HART, KATHLEEN
: Staff Writer, *Nucleonic Week*

HARTSHORN, J. E.
: OPEC Industry Analyst

HELMS, JESSE
: United States Senator (R.-N.C.)

HEYDLAUFF, DALE E.
: Vice President, Environmental Affairs, American Electric

HILTON, FRANCIS
: Professor of Economics

HINNEBUSCH, RAYMOND
: Chairman, Department of Political Science and International Relations, College of St. Catherine, St. Paul, Minn.

HIRSCH, ROBERT
: Staff Writer, *Public Utilities Fortnightly*

HODEL, DONALD
: Director, Summit Group International; Former U.S. Secretary of Energy

HODGE, IAN
: Professor of Economics, University of Wisconsin at Madison

HOEKMAN, BERNARD
: Member of Economic Research Division of GATT Secretariat

HOMER-DIXON, THOMAS
: Assistant Professor of Political Science, University of Toronto

HOOD, JOHN
: Senior Fellow, The Heritage Foundation

HOPE, CHRIS
: Judge Institute of Management Studies, Cambridge University

HORNER, DANIEL
: Nuclear Control Institute

HOWSE, ROBERT
: University of Toronto

HUMPHREY, CRAIG
: Professor of Political Science, Pennsylvania State University

HUMPHRIES, MICHAEL E.
    Petroleum Finance Company

IDINOPULOS, TOM
    Professor of Religious Studies, Miami University (Ohio)

IDSO, KEITH E.
    Arizona State University

IDSO, SHERWOOD
    Research Physicist, USDA Agricultural Research Service.

INBAR, DAVID
    Associate Professor of Political Science, Bar-Ilan University

JACKSON, JOHN H.
    Professor of Law, University of Michigan; Advisory Committee, Center for International Environmental Law's Trade and Environment Program

JACKSON, SHIRLEY
    Chairman, National Regulatory Commission

JAFFE, MARK
    Staff Writer, *Atlanta Journal & Constitution*

JOHNSON, CATHY MARIE
    Professor, Williams College

JOHNSON, J. BENNETT
    United States Senator

JONES, CHARLES O.
    Brookings Institution

JONES, RUSSELL O.
    Senior Economist, American Petroleum Institute

JORGENSON, DALE W.
    Professor of Government, Harvard University

JUHN, POONG ELI
    Director of the IAEA Division of Nuclear Power

KAM
    Research Associate, JCSS

KARP, AARON
    Adjunct Professor, Old Dominion University

KEITHLY, DAVID
    Defense Intelligence College

KEMP, GEOFFREY
    Carnegie Endowment for International Peace

KHADDURI
    Editor, *Middle East Economic Survey*

KHALIL, M. A. K.
    Professor of Environmental Science and Engineering, University of Oregon

KHALILZAD, ZALMAY
    Program Director, RAND

KHATOR, RENU
    Professor, University of South Florida

KHRIPUNOV, IGOR
    Director of NIS programs; Professor, University of Georgia (Republic of Georgia)

KIDNEY, STEVE
    Energy Daily

KISH, RICHARD
    Professor of Finance and Economics, Lehigh University

KOBER, STANLEY
    Adjunct Scholar, Cato Institute

KRAPELS, EDWARD
    President, Energy Security Analysis

KRENT, HAROLD
    Professor, Chicago-Kent Law School

KRIZ, MARGARET
    Staff Writer, *National Journal*

KRUEGER, ANNE O.
    Professor of Economics, Duke University

KUPCHAN
    Assistant Professor of Politics, Princeton University

LAITNER, SKIP
    American Council for an Energy Efficient Economy

LANCASTER, JUSTIN
    Research Fellow, Department of Physics, Harvard University.

LANDY, MARC K.
    Professor, Boston College

LAROCCA, LARRY
    Former New York State Commissioner of Energy Aid

LASHOF, DANIEL
    Senior Scientist, NRDC.

LAZARUS, RICHARD J.
    Professor, Washington University Law School, St. Louis

LEAVER, RICHARD
    Professor, Australian National University

LEDONNE, PAT
    Program Analyst, Global Environment and Trade Staff

LEE, HENRY
    Professor of Government, Harvard University

LEIDY, MICHAEL
:   Member of Economic Research Division of GATT Secretariat

LEVENTHAL, PAUL
:   President, Nuclear Control Institute

LEVEQUE, FRANÇOIS
:   Director of the Center for National Resource Economics, Paris, France

LEVINSON, MARC
:   Economic Analyst, Newsweek

LICHTBLAU, JOHN
:   Chairman, Petroleum Industry Research Foundation

LIEBER, ROBERT
:   Professor of Government, Georgetown University

LINDZEN, RICHARD
:   Professor of Meteorology, Massachusetts Institute of Technology

LIPSON
:   Associate Professor of Political Science, University of Chicago

LLOYD, PETER J.
:   Professor, Dean of Economics and Commerce, University of Melbourne, Australia

LONG, DAVID
:   United States Foreign Service Officer, ret.

LONG, FREDERICK
:   Professor of Government, Indiana University

LOUVISH, MISHA
:   Political Analyst

MAHLMAN, J. D.
:   NOAA

MAKHIJANI, ARJUN
:   PhD in Nuclear Physics; Director of IEER

MAKINEN, GAIL
:   Specialist in Economic Policy

MALIK, KENAN
:   *Australia Economic Review*

MALKIN, MICHELLE
:   Editorial Columnist, *Seattle Times*

MARCHANT, GARY
:   Associate, Kirkland and Ellis law firm, Chicago

MARTIN, LYNN
:   Former U.S. Secretary of Energy

MATHERS, JENNIFER
:   Department of International Politics, University of Wales

MATTHEWS, KEN
:   London School of Economics

MAUL, PHILIP
:   Intera Information Technologies

MAZARR, MICHAEL

McALPINE, JAN C.
:   Team Leader, Global Environment and Trade Staff

McCAIN, JOHN
:   United States Senator (R-Ariz.)

McCORMICK, WILLIAM
:   CEO of CMS Energy Corporation

McFARLAND, MACK.
:   Principal Consultant for Environmenta Programs, Dupont

McGOVERN, GEORGE
:   President, Middle East Council

MEDHURST, JAMES
:   Director of ECOTEC Research and Consulting Ltd.

MICHAELS, PATRICK. J.
:   Associate Professor of Environmental Sciences, University of Virginia

MIKULSKI, BARBARA
:   United States Senator, Maryland

MILLER, ALAN
:   Executive Director, Center for Global Change, University of Maryland

MILLER, KEN
:   Gannett News Service

MILLERD, FRANK
:   Professor of Economics, Wilfrid Laurier University

MINTZ, JOEL A.
:   Professor, Nova-Southeastern Law School

MISKEL, JAMES
:   Professor of National Security, Naval War College

MITCHELL, RONALD
:   Professor of Political Science, University of Oregon

MOHAMMADIOUN, MINA
:   Senior Economist and Headof the Natural Resources Burea of Business Research

MOORE, THOMAS GALE
:   Senior Fellow, Hoover Institution

MOORHEAD, CARLOS J.
: United States Representative

MOOSA, IMAD A.
: Professor of Management, Seffield University

MORAN, THEODORE
: Professor of International Relations

MORICI, PETER
: Professor of International Business, University of Marylaand; Former Director of U.S. ITC

MORRIS, ROBERT J.
: United States Business Observer to OECD on Trade and Environmental Linkages

MORRISETTE, PETER M.
: Research Consultant, Human Dimension of Global Change

MUNGER, FRANK
: Staff Writer, *Knoxville* (Tenn.) *News-Sentinel*

MYERS, NORMAN
: Author on Environment

NAGEL, ROGER
: Professor of Manufacturing Systems Engineering, Lehigh University

NANDA, NED
: Director of Intenational Legal Studies, University of Denver

NISHIHARA, MASASHI
: Professor at National Institute for Defense Studies

NIVOLA, PIETRO
: Senior Fellow, Brookings Institute

NOEST, VERA
: Department of Ecological Botany, Uppsala University

NORBERG-BOHM, VICKI
: Department of Urban Planning, Massachusetts Institute of Technology

NORDHAUS, WILLIAM D.
: Yale University

O'KEEFE, WILLIAM F.
: Vice-president, American Petroleum Institute

O'LEARY, HAZEL R.
: United States Secretary of Energy

OKINO, HIDEAKI
: Royal Institute of International Affairs, Centre of Strategic Studies

OLSON, LAURA KATZ
: Professor of Government, Lehigh University

ONAL, HAYRI
: Assistant Professor of Agricultural Economics, University of Illinois

OSTROM, CHARLES
: Professor of Political Science, Michigan State University

OWEISS, IBRAHIM
: Professor of Economics, Georgetown University

PARSON, EDWARD
: Professor of Government, Harvard University

PEAY, J. H. BEDFORD
: Commander in Chief, U.S. Central Command

PENA, FEDERICO
: U.S. Secretary of Energy

PERCIVAL, ROBERT V.
: Professor, University of Maryland Law School

PFAHL, ROBERT
: Director of Advanced Manufacturing Technology, Motorola

PIERCE, RICHARD J., JR.
: Professor of Law, Columbia University

PIGGOT, JOHN
: Professor of Economics, University of New South Wales, Australia

POPOFF, FRANK
: Chairman of Dow Chemical

PORTNEY, KENT
: Professor of Political Science, Tufts University

PORTNEY, PAUL R.
: Vice-President Resources for the Future

PROBST, KATHERINE N.
: Center for Risk Management

RAMANTHAN, V. (RAM)
: Scripps Institute of Oceanography

RAY, DIXIE LEE
: Professor of Zoology, Washington State University

RAZUVAYEV, VLADAMIR
: Head of Center for Eurasian Studies, Russia

REES, WILLIAM E.
: School of Community and Regional Planning, University of British Columbia, Vancouver

REICH, ROBERT B.
: Professor, John F. Kennedy School of Government, Harvard University

REILLY, KATHLEEN
: Law Student, Indiana University

REINSCH, WILLIAM
    Undersecretary of Export Administration, Department of Energy

REITZE, ARNOLD W.
    Professor of Law, George Washington University

REPETTO, ROBERT
    Vice-President and Director, Program in Economics and Population, World Resources Institute

RHODES, JAMES
    CEO, Virginia Power

RICHARDS, DEANNA
    Senior Program Officer, The National Academy of Engineering

ROBERTS, MARC J.
    Professor, Harvard School of Public Health

ROBERTS, THOMAS C.
    Director, Legislative Analysis Division, EPA

ROBINSON, GLEN O.
    Professor, University of Virginia Law School

ROMM, JOSEPH
    Policy Analyst, Department of Energy

ROSENBAUM, WALTER A.
    Professor, University of Florida

ROSS, JOHN
    Staff Writer, *Smithsonian*

ROSSI, JAMES
    Professor, Chicago-Kent Law School

RUCKELSHAUS, WILLIAM D.
    Former Administrator, United States Environmental Protection Agency

RUGGIERO, RENATO
    Director-General, World Trade Organization

SACHS, NOAH
    Policy Analyst and Project Scientist, IEER

SAMIL, MASSOOD
    Professor of Business, New Hampshire College

SANDA, KRISTA
    Commissioner of the Minnesota Department of Public Service

SANDS, KRIS
    Founding Member of the Nuclear Waste Strategy Coalition

SCHLESINGER, JAMES
    Center for Strategic and International Studies

SCHULTZ, JENNIFER
    Lecturer in Law, Monash University, Victoria, Australia

SCHUURMANS, C. J. E.
    Royal Netherlands Meteorological Institute

SEBENIUS, JAMES K.
    Professor of Business, Harvard University.

SEGAL, SCOTT
    Professor of Environmental Management, University of Maryland

SEIDENFELD, MARK
    Professor, Florida State Law School

SEITZ, ROGER
    Staff Member, IAEA Division of Radiation and Waste Safety

SELIN, IVAN
    Chairman of the US NRC

SERAFY, SALAH EL
    Economist, World Bank

SHANAHAN, JOHN
    Policy Analyst, The Heritage Foundation

SHANLEY, ROBERT A.
    Professor, University of Massachusetts

SHAPIRO, ROBERT
    Professor of Political Science, Columbia University

SHERMAN, WILLIAM
    Northeast High Level Radioactive Waste Transportation Task Force; Vermont Department of Public Service

SHIMBERG, STEVEN
    Chief Counsel, Senate Committee on Environment and Public Works

SILVERMAN, JOEL L.
    Energy Industry Analyst

SINGER, S. FRED
    Professor of Environmental Science, University of Virginia

SINISI, J. SEBASTIAN
    Denver Post

SKINNER, SAMUEL
    President of Commonwealth Edison

SMITH, JOHN KENLY
    Professor of History of American Chemical Technology, Lehigh University

SMITH, MICHAEL
    Former U.S. Ambassador to GATT; President, International Trade and Consulting Firm SIS Advanced Strategies

SMITH, ZACHARY A.
  Professor, Northern Arizona University

SNAPE, RICHARD H.
  Professor of Economics, Monash University, Melbourne, Australia

SPENCER, ROY W.
  Atmospheric Scientist, NASA

SPETH, JAMES GUSTAVE
  President, World Resources Institute

STARHEIM, FRED
  Environmental Scientist

STAVINS, ROBERT
  Professor of Government, Harvard University

STENZEL, PAULETTE L.
  Professor of Business, Law and Public Policy, University of Michigan

STEPHENS, CANDICE
  Principal Administrator, OECD Economics Division

STEVENS, PAUL
  Centre for Petroleum and Mineral Law, University of Dundee, Scotland

STEWART, RICHARD B. C.
  Assistant Attorney General, Environment and Natural Resources Division, Department of Justice

STOBE, TALCOTT
  United States Deputy Secretary of State

STRAM, BRUCE
  Enron Corporation

SUSSMAN, ROBERT
  Deputy Administrator, EPA

SWITZER, JACQUELINE VAUGHN
  Professor, Southern Oregon State

TAYLOR, GREGG
  Staff Writer, *Nuclear News*

TAYLOR, WILLIAM III
  Assistant Secretary, Congressional, Intergovernmental and International Affairs, DOE

TEITELBAUM, JOSHUA
  Fellow at Moshe Dayan Center for Middle Eastern and African Studies, Tel Aviv University

TEMPEST, PAUL
  Director, World Petroleum Congress

THOMAS, CHRISTOPHER
  Director of International Trade Policy Consultants Inc., Ottawa

THOMAS, STEPHEN R.
  Professor, Fordham University

THORNTON, ROBERT
  Professor of Economics, Lehigh University

THUROW, LESTER
  Professor of Economics

TIERNEY, SUSAN F.
  Assistant Secretary for Policy, Department of Energy

TIMMONS, HEATHER
  Staff Writer, *Chemical Marketing Reporter*

TIMMS, CHARLES, JR.
  Private Counsel

TIPPEE, ROBERT
  Managing Editor, *Oil & Gas Journal*

TOBEY, JAMES
  Administrator of the Environmental Directorate of the OECD

TONELSON, ALAN
  Economic Strategy Institute

TREBILCOCK, MICHAEL J.
  University of Toronto

TSAI, HUI-LIANG
  Research Associate, Center for Yugo-American Studies

TUCKER, ROBERT W.
  Council on Foreign Relations

TUNALI, ODIL
  Worldwatch Institute

UPTON, JOHN
  United States Representative

van der LINDE, CLAAS
  Professor of Economics, St. Gallen University, Switzerland

van der MAAREL, EDDY
  Department of Ecological Botany, Uppsala University.

van der MEULEN, FRANK
  Department of Physical Geography and Soil Science, University of Amsterdam

van LEEUWEN, MARTIN
  Senior Researcher, Netherlands Institute of International Relations

VASCONCELLOS, GERALDO
  Professor of Finance and Economics, Lehigh University

VIG, NORMAN
: Professor of Political Science, Carleton College

VIJUK, ROBERT
: Project Manager, Westinghouse Energy System Business Unit

von WEIZSACKER, ERNST ULRICH
: Director of the United Nations Center for Science and Technology Development

WALD, PATRICIA M.
: District of Columbia Circuit Court Judge

WALKER, SANDRA L.
: Master of Laws and Policy Advisor, Trade Policy Branch, Government of Ontario

WATSON, JAMES
: Graduate Student, Princeton University

WATSON, ROBERT T.
: Associate Director for Environment, OTA

WEIDENBAUM, MURRAY
: Chair of the Center for the Study of American Business, Washington University, St. Louis

WEIMER, DAVID L.
: Professor of Policy Analysis, University of Rochester

WHALLEY, JOHN
: Professor of Ecoomics, University of Western Ontario, Canada

WIGLE, RANDALL
: Professor of Economics, Wilfred Laurier University

WILCOXEN, PETER J.
: Professor of Economics, University of Texas at Austin

WINTERS, L. ALAN
: Head of Department of Economics, University of Birmingham, England; Co-Director of International Trade Program at Centre for Economic Policy Research, London

WIRTH, DAVID A. LAW
: Professor, Washington and Lee University

WITTWER, SYLVAN.
: Professor of Horticulture, Michigan State University

WOLFE, BERTRAM
: Energy Consultant; Retired Vice President and General Manager of GE Nuclear Energy

WOODWARD, JENNIFER
: American University Law Review.

YAIR, EVRON
: Professor of Political Science, Tel Aviv University

YERGIN, DANIEL
: President, Cambridge Energy Research Assoc.

ZAPLER, MIKE
: Staff Writer, States News Service

ZECKHAUSER, RICHARD
: Professor of Government, Harvard University

ZUBOK, VLADISLAV
: Russian historian

ZUNES, STEPHEN
: Director, Institute of New Middle East Policy

CHAPTER VIII

BIBLIOGRAPHY

BOOKS

Bailey, Ronald, ed. *The True State of the Planet.* The Free Press, 1995.

Brown, Lester R., Christopher Flavin, and Sandra Postel. *Saving the Planet: How to Shape an Environmentally Sustainable Global Economy.* W. W. Norton & Company, 1991.

Conca, Ken, Michael Alberty and Geoffrey D. Dabelko, eds. *Green Planet Blues: Environmental Politics from Stockholm to Rio.* Westview Press, Inc., 1995.

Energy & Pollution Control Opportunities to the Year 2000. Proceedings of the 16th World Energy Engineering Congress 3rd Environmental Technology Expo, 1993.

Feldman, David L., ed. *Global Climate Change and Public Policy.* Nelson-Hall Publishers, 1994.

Forster, Bruce A. *The Acid Rain Debate: Science and Special Interests in Policy Formation.* Iowa State University Press, 1993.

Graham, John D., and Jonathan Baert Wiener, eds. *Risk versus Risk: Tradeoffs in Protecting Health and the Environment.* Harvard University Press, 1995.

Greve, Michael S. *The Demise of Environmentalism in American Law.* The AEI Press, 1996.

Grubb, Michael. *Energy Policies and the Greenhouse Effect: Volume One: Policy Appraisal.* The Royal Institute of International Affairs, 1990.

Grubb, Michael, et al. *Energy Policies and the Greenhouse Effect: Volume Two: Country Studies and Technical Options.* The Royal Institute of International Affairs, 1991.

Hickey, James E., Jr., and Linda A. Longmire, eds. *The Environment: Global Problems, Local Solutions.* Greenwood Press, 1994.

Holland, Kenneth M., F. L. Morton, and Brian Galligan, eds. *Federalism and the Environment: Environmental Policymaking in Australia, Canada, and the United States.* Greenwood Press, 1996.

Hu, Howard, Arjun Makhijani, and Katherine Yih, eds. *Plutonium: Deadly Gold of the Nuclear Age.* International Physicians Press, 1992.

Idso, Sherwood B. *CO_2 and the Biosphere: The Incredible Legacy of the Industrial Revolution.* Special Publication, Department of Soil, Water and Climate, University of Minnesota, 1995.

Improving the Environment: An Evaluation of DOE's Environmental Management Program. National Academy Press, 1995.

Jenkins, Glenn, and Ranjit Lamech. *Green Taxes and Incentive Policies: An International Perspective.* ICS Press, 1994.

Johnson, Huey D. *Green Plans: Greenprint for Sustainability.* University of Nebraska Press, 1995.

Kempton, Willett, James S. Boster, and Jennifer A. Hartley. *Environmental Values in American Culture.* The MIT Press, 1995.

Lee, Henry, ed. *Shaping National Responses to Climate Change: A Post-Rio Guide.* Island Press, 1995.

Mather, John R., and Galina V. Sdasyuk, eds. *Global Change: Geographical Approaches.* University of Arizona Press, 1991.

Mintz, Joel A. *Enforcement at the EPA: High Stakes and Hard Choices.* University of Texas Press, 1995.

Myers, Norman, and Julian L. Simon. *Scarcity or Abundance? A Debate on the Environment.* W. W. Norton and Company, 1994.

Policy Implications of Greenhouse Warming: Mitigation, Adaptation, and the Science Base. National Academy Press, 1992.

Portney, Kent E. *Controversial Issues in Environmental Policy: Science vs. Economics vs. Politics.* Sage Publications, Inc., 1992.

Rosen, Louis, and Robert Glasser, eds. *Climate Change and Energy Policy.* The American Institute of Physics, 1991.

Solbrig, O.T., H. M. van Emden, and P. G. W. J. van Oordt, eds. *Biodiversity and Global Change.* CAB International, 1994.

Vig, Norman J., and Michael E. Kraft, eds. *Environmental Policy in the 1990s: Towards a New Agenda.* 2nd Ed. CQ Press, 1994.

Weale, Albert. *The New Politics of Pollution.* Manchester University Press, 1992.

Woodwell, George M., and Fred T. Mackenzie, eds. *Biotic Feedbacks in the Global Climatic System: Will the Warming Feed the Warming?* Oxford University Press, 1995.

NEWSPAPERS

Benson, Robert W. "Lead Pollution from Electric Cars? Look a Little Closer at the Facts." *The Christian Science Monitor*. June 9, 1995, p. 19.

Jordan, J. Phillip, and Joel deJesus. "Recovering the Costs of Investment Stranded on Old Shores." *Legal Times*. February 17, 1997, p. S34.

Matthews, Jessica. "Ignoring Global Warming Will Be a Costly Game." *Sun-Sentinel* (Fort Lauderdale). January 31, 1996.

Matthews, Jessica. "In Denial About Global Warming." *The Washington Post*. January 29, 1996.

Merrill, Stephen. "What's the Best Way to Deregulate Providers of Electric Power? Let the Individual States Deregulate Electricity According to Their Regional Needs." *The Washington Times*. May 12, 1997, p. 24.

O'Driscoll, Mary. "No Need to Reopen the Clean Air Act, Green Groups Contend." *The Energy Daily*. February 14, 1997.

Wamsted, Dennis. "Climate Models Still Have Major Shortcomings." *The Energy Daily*. September 6, 1996.

Yarnall, Louise. "A Global Vision; In the Latest Phase of His Geosphere Project, Tom Van Sant Envisions a Computer Accessible Library Where a Fragile Planet's Workings Come to Life." *Los Angeles Times*. February 4, 1996, p. E1.

PERIODICALS

Andrabi, Tahir. "Seigniorage, Taxation, and Weak Government." *Journal of Money, Credit & Banking*. February 1997, Vol. 29, No. 1, p. 105.

Antle, John. M. "Climate Change and Agriculture in Developing Countries." *American Journal of Agricultural Economics*. August 1995, Vol. 77, No. 3, p. 741.

"Balancing America's Budget: Ending the Era of Big Government." *Heritage Foundation Reports*. April 1997, p. 1.

Baskin, Yvonne. "Under the Influence of Clouds." *Discover*. September 1995, Vol. 16, No. 9, p. 62.

"British, U.S. Scientists Concur Temperatures on the Rise." *World Environment Report*. January 17, 1996, p. 35.

Brower, Michael C., Stephen D. Thomas, and Catherine M. Mitchell. "Lessons from the British Restructuring Experience." *Electricity Journal*. April 1997, Vol. 10, No. 3, pp. 40–51.

"California PV Industry Plan Disrupts Talks on Allocation of Renewables Subsidies." *Utility Environment Report*. December 20, 1996, p. 14.

Cane, Mark A., et al. "Twentieth-century Sea Surface Temperature Trends. *Science*. February 14, 1997, Vol. 275, p. 957.

"CEC Proposal for Renewables Subsidies Seen as Favoring Emerging Technologies." *Utility Environment Report*. January 31, 1997, p. 11.

"Cellulose-to-Ethanol Plants Offer Efficiency, Lower Emissions." *Oxy-Fuel News*. January 13, 1997, Vol. 9, No. 1.

"Charging Ahead: The Business of Renewable Energy and What It Means for America." *Publisher's Weekly*. April 7, 1997, Vol. 244, No. 14, p. 84.

"Citizens Group Wants to Stop Tax Breaks for U.S. Oil Industry." *Oxy-Fuel News*. October 14, 1996, Vol. 8, No. 40.

"Coal, the 'New' Energy Flavor." *International Coal Report*. March 10, 1997.

"Community Policy on Renewable Sources of Energy and the World Solar Summit Process." Speech by H. E. Christos Papoutsis at World Solar Summit, Harare, Zimbabwe. *Rapid*. September 16, 1996.

"Consortium Begin $14M Project to Grow Hybrid Willows for Power." *Air/Water Pollution Report's Environment Week*. November 11, 1996, Vol. 34, No. 44.

"Cost Cutting to a 'New Market'." *International Gas Report*. February 7, 1997.

Cowan, Stuart. "Bioengineering." *Whole Earth Review*. March 22, 1996, No. 89, p. 48.

"DOE Core Budget Flat; Funding for Building, Privatization Added." *New Technology Week*. February 10, 1997, Vol. 11, No. 6.

Dowie, Mark. "A Sky Full of Holes: Why the Ozone Layer Is Torn Worse Than Ever." *The Nation*. July 8, 1996, Vol. 263, No. 2, p. 11.

Downing, Tom. "Damn This Heat; Global Warming." *New Statesman & Society*. August 11, 1995, Vol. 8, No. 365, p. 18.

Ehrlich, Paul R., and Anne H. Ehrlich. "Ehrlichs' Fables; Environmental Management and Government Regulations. *Technology Review*. January 1997, Vol. 100, No. 1, p. 38.

"EIA Projects Nuclear Growth Through 2015." *Power Engineering*. January 1997, Vol. 101, No. 1, p. 12.

"Electricity Deregulation: Separating Fact from Fiction in the Debate over Stranded Cost Recovery." *Heritage Foundation Reports*. March 11, 1997, No. 20.

"Energy." *Power Europe*. March 28, 1977.

"Energy: Industry Divided over Renewables Benefits." *European Report.*. March 26, 1997.

"Energy Supply Research & Development." *The Energy Report*. February 10, 1997, Vol 25, No. 6

"EU/East Europe: Balancing the Needs for Growth Against Climate Change." *Euro-East*. March 18,1997, No. 53.

"Feedback." *New Scientist*. April 26, 1984, p. 84.

Ferguson, Kelly, et al. "Emerging Commodity Players Shape North America's Pulp, Paper Future." *Pulp & Paper*. September 1996, Vol. 70, No. 9, p. 63.

Gladwin, Thomas N., et al. "Shifting Paradigms for Sustainable Development: Implications for Management Theory and Research. *Academy of Management Review*. October 1995, Vol. 20, No. 4, p. 874.

Grams, Senator Rod, and Kris Sanda. "DOE Inaction on Nuclear Waste Disposal Threatens Budget Process." *Legal Backgrounder*. May 2, 1997, Vol. 12, No. 15.

Gulen, S. Gurcan. "Regionalization in the World Crude Oil Market." *The Energy Journal*. April 1997, Vol. 18, No. 2, p. 109.

Handyside, Gillian. "Greenpeace Attacks European Energy Subsidies." *The Reuter European Community Report*. May 20, 1997, BC Cycle.

Harrison, Tom. "Nuclear's Advocates Face Big Job Convincing Skeptics It Can Compete." *Nucleonics Week*. March 6, 1997, Vol. 38, No. 10, p. 16.

Hecht, Jeff. "Triumph of Dogma over Reason: Why Are Politicians Forcing Scientists to Give the Answers They Want to Hear?" *New Scientist*. January 27, 1996, p. 49.

Hiruo, Elaine. "Murkowski Charges President Lacks 'Political Will' to Pick a Storage Site." *Nuclear Fuel*. February 10, 1997. Vol. 22, No. 3, p. 10.

Innes, Robert, and Sally Kane. "Agricultural impacts of global warming." *American Journal of Agricultural Economics*. August 1995, Vol. 77, No. 3, p. 747.

Isaka, Satoshi. "Energy Experts Promote Asian Nuclear Network: Region's Patchwork of Priorities Keeps Talks Tentative." *The Nikkei Weekly*. April 28, 1997, p. 1.

Jenkins, John H. "State of European Refining Industry Is Less Bleak Than Reported." *Oil & Gas Journal*. April 28, 1997, p. 49.

Jordan, J. Phillip and Joel deJesus. "Recovering the Costs of Investment Stranded on Old Shores." *New Jersey Law Journal*. March 17, 1997, p. 33.

Kanner, Marty. "Federal Restructuring Legislation: A Recipe for Successful Action." *Electricity Journal*. March 1997, Vol. 10, No. 2, pp. 20–27.

Karl, Thomas R., et al. "Testing for Bias in the Climate Record. *Science*. March 29, 1996, Vol 271, No. 5257, p. 1879.

Kauffman, Richard. "Sun Spot." *Cost Engineering*. October 1996, Vol. 38, No. 10, pp. 11–12.

Knight, Peter. "Better Than It Looks on Paper." *New Scientist*. September 28, 1996, p. 16.

"Last Minute Electric Industry Restructuring Bills Set Stage for Next Year's Congressional Restructuring Battles on the Hill." *Foster Electric Report*. October 16, 1996, Report No. 97, p. 3.

Lee, Henry, and Negeen Darani. "Electricity Restructuring and the Environment." *Electricity Journal*. December 1996, Vol. 9, pp. 10–15.

Lindzen, Richard. "How Political Advocates Misuse Science." *Consumers' Research Magazine*. September 1996, Vol. 79, No. 9, p. 19.

MacDonald, Neil. "NRC Draws Link Between S&T and Environmental Concerns." *Federal Technology Report*. October 10, 1996, p. 6.

Maier, Leo. "Letting the Land Rest." *OECD Observer*. December 10, 1996, No. 203, p. 12.

Manning, Robert A. "PACATOM: Nuclear Cooperation in Asia." *The Washington Quarterly*. Spring 1997, Vol. 20, No. 2, p. 215.

Mirsky, Steve. "A Swelling of Zoologists." *Sea Frontiers*. March 22, 1996, Vol. 42, No. 1, p. 10.

Motavalli, Jim. "Some Like It Hot: Global Warming Is Not Just a Scientific Prediction—It's Also a Hot Political Football." *E*. January 1996, Vol. 7, No. 1, p. 28.

"Murkowski Pledges to Push Ahead with Stand-alone PUHCA Reform Legislation as the Prospects for Passing a Comprehensive Restructuring Bill This Congress Are Increasingly Uncertain." *Foster Electric Report*. April 30, 1997, Report No. 111, p. 1.

Myers, Edward M., et al. "Using Market Transformation to Achieve Energy Efficiency: The Next Steps." *Electricity Journal*. May 1997, Vol. 10, No. 4, pp. 34–41.

Neely, Christopher J., and Christopher J. Waller. "A Benefit-cost Analysis of Disinflation." *Contemporary Economic Policy*. January 1997, Vol. 15, No. 1, pp. 50–64.

"Nuclear Industry Casts Critical Eye on 1997 As a 'Watershed' Year." *Energy Report*. January 6, 1997, Vol. 25, No. 1.

"Nuclear Power: PBS Program Questions 'Irrational' Fears." *Greenwire*. April 22, 1997.

"Outlook: Record Oil Imports Projected for 1997." *Energy Conservation News*. January 1, 1997, Vol. 19, No. 6.

Pinchot, Gifford, and Elizabeth Pinchot. "Organizations That Encourage Integrity." *Journal for Quality and Participation*. March 1997, Vol. 20, No. 2, pp. 10–19.

Ponnuru, Ramesh. "Charge of the Light Brigade." *National Review*. June 2, 1997, Vol. 49, No. 10, p. 27.

"Position of the American Dietetic Association: Natural Resource Conservation and Waste Management." *Journal of the American Dietetic Association*. April 1997, Vol. 97, No. 4, p. 425.

"Pundits Say Political Dynamics Won't Be Same in 105th Congress." *Nucleonics Week*. November 14, 1996.

Rauber, Paul. "Does He Deserve Your Vote? Bill Clinton's Environmental Record." *Sierra*. September 1996, Vol. 81, No. 5, p. 38.

"Renewables Debate in California May Turn on Rebates for Industrial Customers." *Industrial Energy Bulletin*. January 24, 1997, p. 1.

"Renewables Jockeying for Position in California Direct-Access Market." *Power Markets Week*. December 30, 1996, p. 3.

"Rep. Delay Introduces Bill to Create Retail Electric Choice, Repeal PUHCA and PURPA, and Prohibit Stranded Cost Recovery; Rep. Pallone's Bill Would Tie Restructuring to Clean Air Act Amendments, November 1, 1996." *Foster Natural Gas Report*. October 17, 1996, Report No. 2102, p. 21.

"Research Review." *Journal of the Society of Insurance Research*. Spring 1997.

"RJB Mining Sets Sights on Developing Clean Coal Units." *Coal Week International*. March 4, 1997, Vol. 18, No. 9, p. 10.

Root, Terry L., et al. "Ecology and Climate: Research Strategies and Implications." *Science*. July 21, 1995, Vol. 269, No. 5222, p. 334.

Scarlett, Lynn. "Evolutionary Ecology: A New Environmental Vision." *Reason*. May 1996, Vol. 28, No. 1, p. 20.

Simon, Julian L., et al. "The Global Envirnment: Megaproblem or Not?" *The Futurist*. March 13, 1997, Vol. 31, No. 2, p. 17.

Steer, Andrew. "Accord Likely on Danish Budget." *Finance & Development*. December 1996, Vol. 33, No. 4, p. 4.

Swallow, Stephen K. "Resource Capital Theory and Ecosystem Economics: Developing Nonrenewable Habitats with Heterogeneous Quality." *Southern Economic Journal*. July 1996, Vol. 63, No. 1; p. 106.

"Sweden's Alternative Measures." *Power Europe*. March 228, 1997.

"Technology." *Coal Week International*. March 4, 1997, Vol. 18, No. 9, p. 10.

Thurow, Lester C. "One World, Ready or Not: The Manic Logic of Global Capitalism." *The Atlantic Monthly*. March 1997, Vol. 279, No. 3, p. 97.

Tickell, Oliver. "Healing the Rift." *The Independent*. August 4, 1996, p. 40.

Tickell, Oliver. "Little Life Savers." *The Independent*. October 27, 1996, p. 48.

Tilson, Donn James. "Promoting a 'Greener' Image of Nuclear Power in the U.S. and Britain." *Public Relations Review*. March 22, 1996, Vol. 22, No. 1, p. 63.

"Utility Restructuring: State Subsidy Plan Rewards Wind; Fails to Satisfy All Renewables." *International Solar Energy Report*. April 11, 1997, Vol. 23, No. 7.

Whitman, David, et al. "A Bad Case of the Blues." *U.S. News & World Report*. March 4, 1996, Vol. 120, No. 9, p. 54.

Wolfe, Bertram. "An Apology to Dan Rather—But Are Mr. Rather and His Colleagues Still Dangerous?" *Nuclear News*. April 1997, p. 38.

Wood, Jon. "Frozen Arctic Projects Are Heating UP." *Oil & Gas Investor*. February 1997, Vol. 17, No. 2, pp. 52–55.

PRESS RELEASES

"Environment." Council of Energy Ministers 1990th Council Meeting, Brussels. *Rapid*. March 3, 1997.

RADIO BROADCASTS

Coyle, Kevin, et al. "NPR, Talk of the Nation." Transcript No. 97050601–211, News Feature. Tuesday, May 6, 1997, 2:00 p.m. EDT.

Ehrlich, Paul, et al. "Environmental Debate, Is All Bad?" NPR Talk of the Nation. Transcript No. 96122702–111, Friday, December 27, 1996, 2:00 p.m. EST.

Makhijani, Arjun, et al. "Nuclear Waste Disposal." NPR Talk of the Nation. Transcript No. 97042201–211, April 22, 1997, 2:00 p.m. EDT.

NEWS SERVICES

Baker, James D. Prepared testimony before the House Committee on National Security, Subcommittee on Research and Development; House Resources Committee, Subcommittee on Military Research and Development; House Resources Committee, Subcommittee on Fisheries, Wildlife and Oceans; House Committee on Science, Subcommittee on Energy and Environment. *Federal News Service*. January 25, 1996.

Baliunas, Sallie. "Global Climate Change." Testimony before Senate Energy Committee. September 17, 1996.

"Critical Mass: Environmental and Consumer Groups Release Agenda for Electric-industry Restructuring. *M2 Presswire*. February 18, 1997.

Davis, Robert E. Prepared statement before House Committee on Science. *Federal News Service*. March 6, 1996.

Frank, Steve. "Workshop on Competitive Change in the Electronic Power Industry." Prepared statement before Senate Committee on Energy and Natural Resources. *Federal News Service*. March 6, 1997.

Franklin, Allen. "What Are the Issues Involved in Competition?" Prepared statement before Senate Committee on Energy and Natural Resources. *Federal News Service*. March 6, 1997.

Gibbons, John H. Prepared statement before House Committee on Appropriations, Subcommittee on Veterans' Affairs, HUD, and Independent Agencies. *Federal News Service*. April 30, 1996.

"FY 97 USAID Appropriations." Hearing of the Foreign Operations, Export Finaning and Related Programs Subcommittee of the Hous Appropriations Committee. *Federal News Service*. April 24, 1996.

Krebs, Martha A. "FY 1998 Appropriations Hearings. Prepared statement before House Appropriations Committee, Energy and Water Development Subcommittee. *Federal News Service*. March 13, 1997.

McCracken, Michael C. "U.S. Global Change Research Programs Data Collection and Scientific Priorities." Prepared statement before House Committee on Science. *Federal News Service*. March 6, 1996.

Ramanthan, V. (Ram). "Anthropogenic Climate Change: Areas of Uncertainty." Prepared statement before Senate Energy and Natural Resources Committee. Testimony at the United States Senate Hearing on "U.S. Climate Change Policy. *Federal News Service*. September 17, 1996.

GOVERNMENT DOCUMENTS

Baker, James D. "National Military Research and Development Leveraging National Oceanographic Capabilities." FDCH Congressional Testimony. January 25, 1996.

Baliunas, Sallie. "Energy Global Climate Change." FDCH Congressional Testimony. September 17, 1996.

Curtis, Charles B. "Fiscal Year 98 Budget: Energy Department." FDCH Capitol Hill Hearing Testimony. February 11, 1997.

Fulton, Kathleen. "Technology in the Classroom." FDCH Congressional Testimony. May 6, 1997.

Gibbons, John H. "FY 97 VA/HUD/Independent Agencies Appropriations." FDCH Congressional Testimony. April 30, 1996.

Guerrero, Peter. "Science, Energy and Environment Potential Impacts of Global Climate Changes." FDCH Congressonal Testimony. November 16, 1995.

Hinchman, James F. "Addressing the Deficit—Budgetary Implications of Selected GAO Work for Fiscal Year 1998." Report to the Congress. *GAO Reports*. March 14, 1997. GAO/OCG-97-2, March 14, 1997, GAO Reports.

Krebs, Martha A. "FY 98 Budget for Energy Research Programs." FDCH Congressional Testimony. March 13, 1997.

Romm, Joseph J. "Energy Efficiency and Renewable Energy, U.S. Department of Energy Senate Agricultulture, Nutrition and Forestry Renewable Fuels and U.S. Energy Policy." FDCH Congressional Testimony, October 2, 1996.

White, James R. "Tax Policy—Effects of the Alcohol Fuel Tax Incentives." Report to the Chairman, House Committee on Ways and Means. *GAO Reports*. March 13, 1997, GAO/GGD-97-41, March 6, 1997 GAO Report.